AF323437

COMPOSITES AND THEIR APPLICATIONS

Edited by **Ning Hu**

Composites and Their Applications
http://dx.doi.org/10.5772/3353
Edited by Ning Hu

Contributors

K.D. Mohd Aris, F. Mustapha, S.M. Sapuan, D.L. Majid, Viktor Mykhas'kiv, Jian Cai, Lei Qiu, Shenfang Yuan, Lihua Shi, PeiPei Liu, Dong Liang, Hyun-Sup Jee, Jong-O Lee, Yaolu Liu, Alamusi, Jinhua Li, Huiming Ning, Liangke Wu, Weifeng Yuan, Bin Gu, Ning Hu, Khaled R. Mohamed, Haiyan Li, Jianying Li, Xiaozhou Liu, Alex Fok, Adil Sbiai, Abderrahim Maazouz, Etienne Fleury, Henry Sautereau, Hamid Kaddami, H.P.S. Abdul Khalil, M. Jawaid, A. Hassan, M.T. Paridah, A. Zaidon, N.M. Chikhradze, L.A. Japaridze, G.S. Abashidze, Zulkhair Mansurov, Ilya Digel, Makhmut Biisenbaev, Irina Savitskaya, Aida Kistaubaeva, Nuraly Akimbekov, Azhar Zhubanova, Marjan S. Ranđelović, Aleksandra R. Zarubica, Milovan M. Purenović, Yun Lu, Liang Hao, Hiroyuki Yoshida, Alexander Horoschenkoff, Christian Christner, Maria Mingallon, Sakthivel Ramaswamy, Antonio C. de Oliveira, Ligia S. de Oliveira

CBS, Edition **2017**

Published by InTech

Janeza Trdine 9, 51000 Rijeka, Croatia

All chapters are distributed under the Creative Commons Attribution 3.0 license, which allows users to download, copy and build upon published articles even for commercial purposes, as long as the author and publisher are properly credited, which ensures maximum dissemination and a wider impact of our publications. After this work has been published by InTech, authors have the right to republish it, in whole or part, in any publication of which they are the author, and to make other personal use of the work. Any republication, referencing or personal use of the work must explicitly identify the original source.

Copyright © 2012 InTech

Notice

Statements and opinions expressed in the chapters are these of the individual contributors and not necessarily those of the editors or publisher. No responsibility is accepted for the accuracy of information contained in the published chapters. The publisher assumes no responsibility for any damage or injury to persons or property arising out of the use of any materials, instructions, methods or ideas contained in the book.

Publishing Process Manager Marina Jozipovic
Technical Editor InTech DTP team
Cover InTech Design team

Additional hard copies can be obtained from orders@intechopen.com

Composites and Their Applications, Edited by Ning Hu
p. cm.
ISBN 978-953-51-0706-4

Contents

Preface IX

Section 1 **Health Monitoring
for Composite Material Structures** **1**

Chapter 1 **A Structural Health Monitoring of
a Pitch Catch Active Sensing of PZT Sensors
on CFRP Panels: A Preliminary Approach** **3**
K.D. Mohd Aris, F. Mustapha, S.M. Sapuan and D.L. Majid

Chapter 2 **Numerical Simulation of Wave Propagation
in 3D Elastic Composites with Rigid
Disk-Shaped Inclusions of Variable Mass** **17**
Viktor Mykhas'kiv

Chapter 3 **Structural Health Monitoring for Composite Materials** **37**
Jian Cai, Lei Qiu, Shenfang Yuan, Lihua Shi,
PeiPei Liu and Dong Liang

Chapter 4 **Acoustic Emission of Composite Vessel** **61**
Hyun-Sup Jee and Jong-O Lee

Chapter 5 **Locating Delamination in Composite Laminated Beams
Using the Zero-Order Mode of Lamb Waves** **91**
Yaolu Liu, Alamusi, Jinhua Li, Huiming Ning,
Liangke Wu, Weifeng Yuan, Bin Gu and Ning Hu

Section 2 **Bio-Medical Composites and Their Applications** **111**

Chapter 6 **Biocomposite Materials** **113**
Khaled R. Mohamed

Chapter 7 **Non-Destructive Examination of Interfacial
Debonding in Dental Composite Restorations
Using Acoustic Emission** **147**
Haiyan Li, Jianying Li, Xiaozhou Liu and Alex Fok

Section 3   Natural Fiber, Mineral Filler Composite Materials   169

Chapter 8   **TEMPO-Mediated Oxidation of Lignocellulosic Fibers from Date Palm Leaves: Effect of the Oxidation on the Processing by RTM Process and Properties of Epoxy Based Composites   171**
Adil Sbiai, Abderrahim Maazouz, Etienne Fleury, Henry Sautereau and Hamid Kaddami

Chapter 9   **Oil Palm Biomass Fibres and Recent Advancement in Oil Palm Biomass Fibres Based Hybrid Biocomposites   209**
H.P.S. Abdul Khalil, M. Jawaid, A. Hassan, M.T. Paridah and A. Zaidon

Chapter 10   **Properties of Basalt Plastics and of Composites Reinforced by Hybrid Fibers in Operating Conditions   243**
N.M. Chikhradze, L.A. Japaridze and G.S. Abashidze

Section 4   Catalysts and Environmental Pollution Processing Composites   269

Chapter 11   **Heterogeneous Composites on the Basis of Microbial Cells and Nanostructured Carbonized Sorbents   271**
Zulkhair Mansurov, Ilya Digel, Makhmut Biisenbaev, Irina Savitskaya, Aida Kistaubaeva, Nuraly Akimbekov and Azhar Zhubanova

Chapter 12   **New Composite Materials in the Technology for Drinking Water Purification from Ionic and Colloidal Pollutants   295**
Marjan S. Ranđelović, Aleksandra R. Zarubica and Milovan M. Purenović

Chapter 13   **Mechanical Coating Technique for Composite Films and Composite Photocatalyst Films   323**
Yun Lu, Liang Hao and Hiroyuki Yoshida

Section 5   Other Applications of Composites   355

Chapter 14   **Carbon Fibre Sensor: Theory and Application   357**
Alexander Horoschenkoff and Christian Christner

Chapter 15   **Bio-Inspired Self-Actuating Composite Materials   377**
Maria Mingallon and Sakthivel Ramaswamy

Chapter 16   **Composite Material and Optical Fibres   397**
Antonio C. de Oliveira and Ligia S. de Oliveira

Preface

Composites are engineered or naturally occurring materials made from two or more constituent materials with significantly different physical or chemical properties which remain separate and distinct within the finished structure. Basically, they can be categorized into two major types, i.e., structural composites with outstanding mechanical properties and functional composites with various outstanding physical, chemical or electrochemical properties. They have been widely used in a wide variety of products, e.g., advanced spacecraft and aircraft components, boat and scull hulls, sporting goods, sensor/actuator, catalysts and pollution processing materials, bio-medical materials, and batteries, etc.

This book focuses on the properties and applications of various composites and the solutions for some encountered problems in applications, e.g., composites structural health monitoring. The book has been divided into five parts, which deal with: health or integrity monitoring techniques of composites structures, bio-medical composites and their applications as dental or tissue materials, natural fiber or mineral filler reinforced composites and their property characterization, catalysts composites and their applications, and some other potential applications of fibers or composites as sensors, etc., respectively.

A list of chapters is given below along with short descriptions by providing a glimpse on the content of each chapter.

Part 1. Health Monitoring for Composite Material Structures

Chapter 1. A Structural Health Monitoring of a Pitch Catch Active Sensing of PZT Sensors on CFRP Panels: A Preliminary Approach
In this chapter, the Lamb waves based on structural health monitoring techniques by using PZT sensor network are described. The focus is put on the repaired locations and surface structural integrity monitoring in composites structures.

Chapter 2. Numerical Simulation of Wave Propagation in 3D Elastic Composites with Rigid Disk-Shaped Inclusions of Variable Mass
This chapter presents a work on the numerical simulation of wave propagation in 3D elastic composites and the interaction between the waves and embedded inclusions or damages. A novel boundary element method is proposed to carry out this analysis.

For applications of waves based techniques in the structural health monitoring field, some important fundamental information is provided for deep understanding the wave propagation behaviours in composites with inclusions or damages.

Chapter 3. Structural Health Monitoring for Composite Materials
In this chapter, structural health monitoring (SHM) for composite materials is mainly focused on. The common sensors in SHM and some typical SHM methods are reviewed along with some SHM examples realized on composite structures.

Chapter 4. Acoustic Emission of Composite Vessel
In this chapter, an acoustic emission based technique to evaluate damages in a composite fuel tank is presented in detail by carrying out a massive amount of experimental studies.

Part 2. Bio-medical Composites and Their Applications

Chapter 5. Biocomposite Materials
In this chapter, the composites of ceramics with natural degradable polymers are described by using several particle composites based on degradable biopolymers as example. Their physical, chemical and biological properties and applications in bone structures and bone tissue engineering are described.

Chapter 6. Non-destructive Examination of Interfacial Debonding in Dental Composite Restorations Using Acoustic Emission
This chapter is to present a study on the development of a new method to evaluate the interfacial debonding of dental composite restorations. This non-destructive method based on the acoustic emission technique is evaluated for its use to monitor *in-situ* the interfacial debonding of composite restorations during polymerization.

Part 3. Natural Fiber, Mineral Filler Composite Materials

Chapter 7. TEMPO-mediated Oxidation of Lignocellulosic Fibers From Date Palm Leaves: Effect of the Oxidation on the Processing by RTM Process and Properties of Epoxy Based Composites
In this chapter, TEMPO-mediated oxidation technique for processing lignocellulosic fibers is described. Moreover, the effects of this technique on the thermal, mechanical properties of the lignocellulosic fiber based composites and the fabrication process of the composites are explored in detail.

Chapter 8. Oil Palm Biomass Fibres and Recent Advancement in Oil Palm Biomass Fibres based Hybrid Biocomposites
This chapter is to give an overview on some main results of physical, mechanical, electrical, and thermal properties obtained from oil palm fibres based hybrid composites, which are promising in their applications to automotive sector, building industry etc.

Chapter 9. Properties of Basalt Plastics and of Composites Reinforced by Hybrid Fibers in Operating Conditions

In this chapter, some new research results of a new type of composite materials based on basalt, carbon, glass and polymeric resin, are presented. In particular, a long-term resistance property of the material in corrosive media and at atmospheric action has been focused on.

Part 4. Catalysts and Environmental Pollution Processing Composites

Chapter 10. Heterogeneous Composites on the Basis of Microbial Cells and Nanostructured Carbonized Sorbents

In this chapter, heterogeneous composite materials obtained by immobilization of microorganisms on carbonized sorbents with nanostructured surface are described with the provided evidences collected in in-vivo and in-vitro studies, which strongly suggest that the use of the nano-structured carbonized sorbents as delivery vehicles for the oral administration of probiotic microorganisms has a very big potential for improving functionality, safety and stability of probiotic preparations.

Chapter 11. New Composite Materials in the Technology for Drinking Water Purification from Ionic and Colloidal Pollutants

Due to their positive textural properties and high specific surface area, in this chapter, composite materials working as adsorbents or electrochemically active materials in water purification for deposition of some pollutants from water are described. Especially, three new/modified bentonite based composite materials are explored in detail, where bentonite is a natural and colloidal alumosilicate with particle size less than 10 μm, which is effectively used as sorbent for heavy metals and other inorganic and organic pollutants from water.

Chapter 12. Mechanical Coating Technique for Composite Films and Composite Photocatalyst Films

This chapter presents a newly developed mechanical coating technique (MCT). By comparing with the traditional film coating techniques such as PVD and CVD, this technique, i.e., MCT, shows many advantages including inexpensive equipments, simple process, low preparation cost and large specific area, among others. It can not only fabricate metal/alloy films but also non-metal/metal composite films such as TiO_2/Ti composite photocatalyst films.

Part 5. Other Applications of Composites

Chapter 13. Carbon Fiber Sensor: Theory and Application

This chapter presents the piezoresistive carbon fiber sensor (CFS) consisting of a single carbon fiber working as a strain sensor, which is embedded in a sensor carrier (GFRP patch) for electrical isolation. It has been demonstrated that based on the integral strain measurement method, the CFS is an excellent sensor to detect delamination and matrix cracks in multidirectional reinforced laminates.

Chapter 14. Bio-Inspired Self-Actuating Composite Materials
In this chapter, the research to integrate sensing and actuation functions into a fibre composite material system is described. In this system, which displays adaptive 'Integrated Functionality', fiber composites are anisotropic and heterogeneous, offering the possibility for local variations in their material properties. Embedded fiber optics are used to sense multiple parameters and shape memory alloys integrated into composite material are used for actuation.

Chapter 15. Composite Material & Optical Fibres
In this chapter, development of a special composite formed from a mixture of EPO-TEK 301-2 and some refractory material oxide in nano-particle form, cured and submitted to a customized thermal treatment is described. This material is more resistant and harder than EPO-TEK 301-2 and is found to be well suited to the fabrication of optical fiber arrays from the aspects of CTE matching, machining ability, bonding to glass and ease of polishing, etc.

Acknowledgements

I would like to express my sincere appreciation to the authors of the chapters in this book for their excellent contributions and for their efforts involved in the publication process. I do believe that the contents in this book will be helpful to many researchers in this field around the world.

Ning Hu, Ph.D.
Professor, Department of Mechanical Engineering,
Chiba University,
1-33 Yayoi-cho, Inage-ku, Chiba 263-8522,
Japan

Health Monitoring for Composite Material Structures

A Structural Health Monitoring of a Pitch Catch Active Sensing of PZT Sensors on CFRP Panels: A Preliminary Approach

K.D. Mohd Aris, F. Mustapha, S.M. Sapuan and D.L. Majid

Additional information is available at the end of the chapter

http://dx.doi.org/10.5772/48097

1. Introduction

At present, the advanced composite materials have gained it acceptance in the aerospace industries. The content of these materials has increased dramatically from less than 5% in the late eighties to more than 50% at the beginning at this decade. [1] The materials offer high strength to weight ratio, high strength to weight ratio, corrosion resistance, high fatigue resistance etc. These benefits have transformed the aviation world traveling to better fuel consumption, endurance and more passengers. However, the use of these materials has posed new challenges such as impact, delamination, barely visible internal damage (BVID) etc. Before a part or component being used on the actual structure, they are being tested from small scale to the actual scale in a controlled environment either at lab or test cell. However the attributes imposed during the operation sometimes shows different behavior when the actual operations are performed due to environment factors, human factors and support availability. To ensure the safety is at the optimum level, the continuous conditional monitoring need to be carried out in order to ensure the component operate within the safety margin being placed by the aircraft manufacturers. [2] One of the areas under investigation is the structural integrity assessment through the use of non-destructive inspections (NDI). The NDI allows aircraft operator to seek information on the aircraft structure reliability by inspecting the structure without having to remove it. There are many types of inspection methods which are limited to materials, locations and accuracy depends on methodology applied. [3] Few of popular techniques are eddy current, ultrasonic, radiography, dye penetrant which have been existence in quite a time. However due to composite material applications new methods have emerged in order to improve detection to attain converging results such as tap test, laser shearography, phase array etc.. So far, these methods prove its effectiveness and consistency in finding the anomalies.

However these techniques require total grounding of the aircraft and the inspection are manually intensified. The only clue where to inspect the area from the occurrence report, maintenance schedule or mandatory compliance by the authority. New inspection paradigm need to be developed as defects will arise in the non-conventional ways as the composite materials being used in the pressurized area such as in Boeing 787 and Airbus A350 aircrafts. Therefore, the available methods need to be systematically chosen depends on thin laminate, thick laminate or sandwich structure.[4] The active monitoring offers continuous monitoring either by interrogating or listen to the structure behavior. Embedded sensor and on surface sensors offers the advantages and disadvantages that yet not being explored fully and can accommodate the NDI techniques. The structure integrity will behave differently as the structure being modified and repair to ensure continuation of the aircraft operation and prolong its service life. The aircraft structural health monitoring (SHM) is one of the conditioning monitoring that has gained its usefulness. Such health monitoring of a component has been successfully being used in the aircraft avionics systems, engine management systems, rotary blade systems etc. Since the SHM is still at its infant stage, several methodology and detection methods are been explored to suite the monitoring purposes. Acoustic emission, fiber bragg grating, compact vacuum monitoring etc. are being investigated for their potential. [5] Therefore the paper is focusing on issues on the implementation of the SHM at post repair through the use of PZT sensor by using guided waves as a method of monitoring for active and passive structural surface conditions.

2. Theoretical background

The use of advanced composite materials has shifted the paradigm in aircraft structure design, operation and maintenance philosophy. A simple stop drills procedure is used to prevent further propagation of crack or by removing the damage area and replacing the damage area. This procedure are well written in typical aircraft structural repair manual (SRM) under Chapter 50-xx-xx found in the ATA 100 (Air Transport Associations) [6]. The procedure above can only be applied to metallic structure since the behavior is isotropic in which properties such as damage tolerance, fracture mechanics and fatigue can be predicted although the repair has been done on the damaged structure. The composite structures are made up from various constituents that are laid up and bonded together with the assistance of pressure and temperature at predetermine times. During operations, the aircraft structures are subjected to damages due to impact, environmental, residual imperfections, delaminations that reduces the structural integrity of the aircraft [7]. Typically, there are four types of repair applied to the composite structures. There are external bonded patches, flush or scarf bonded repair, bolted patch and bonded patches [8]. This operation requires the strength to be returned back to the original strength [9]. Due to the orientation, number of plies and materials used the level of recovery of the operating strain is much dependent on the stiffness of the laminates. The governing equation for the actual load to be transmitted to the new repaired laminates are given by the equation below [10] & [11]

$$P = e_a E_x t \tag{1}$$

Where, P, e_a, E_x, and t are actual load, ultimate design strain, modulus in the primary loading direction and the laminate thickness respectively. A simple calculation of the strength of materials can be applied to scrutinized the scarf join for the maximum allowable stress [10] & [11]. The equation is given by

$$P_{max} = \sigma_u t \leq \frac{\tau_p t}{\sin\theta\cos\theta} \qquad (2)$$

Where P_{max}, σ_u, t, θ and τ_p is the maximum load, ultimate stress, thickness, shear stress and scarf angle respectively. By solving the value of θ, the scarf angle is found to be at 2^0 or at 1:30 ratio in order to attained minimum ultimate stress for the repair structure strength to be similar with the parent structure.

Studies have shown the use of PZT sensors on experimental aircraft component such as flaps and wings are promising [12] and [13]. For this experiment, an aircraft spoiler was used as the experimental subject by mounting the sensor arbitrarily on the spoiler's surface. The sensor can also be used to detect the surface condition of normal, damaged and repaired structures.

Most of the structural damage diagnoses were predicted by using analytical or finite element modeling [14], [15] and [16]. Although the results were accepted but it requires a powerful computing hardware, labor intensive interaction and modeling errors before a solution can be converged. Another method is to utilize the statistic to evaluate the captured data. However large amount of data are required to achieve higher reliability and probability to converge to the intended solution. The statistical approach utilizes supervised and unsupervised learning in order to process the data. [17] and [18] The supervised learning uses data as its references and the unsupervised learning uses to cluster the data and group them for selective conditions. The approach can be achieved by using the Statistical Pattern Recognition [19]. The principles in SPR are:-

1.   Operational Evaluation,
2.   Data Acquisition & Cleansing,
3.   Feature Extraction & Data Reduction and
4.   Statistical Model Development or Prognosis

Only no 1 and 2 were concerned in this paper.

Outlier Analysis is one of the method applied in SPR. The OA is used as the detection of cluster, which deviates from other normal trend cluster. One of the most common discordance tests is based on the deviation statistic [19] given by

$$z_i = \frac{d_i - \bar{d}}{\sigma} \qquad (3)$$

where z_i is the outlier index for univariate data, d_i is the potential outlier and $\bar{d}$ and σ are the mean sample and standard deviation. The multivariate discordance test was known as Mahalanobis square distance given by

$$Z_i = (\{x_i\} - \{\bar{x}\})^T [S]^{-1} (\{x_i\} - \{\bar{x}\}) \tag{4}$$

where Z_i is the outlier index for multivariate data, x_i is the potential outlier vector and $\bar{x}$ is the sample mean vector and e is the sample co-variance matrix [20] and [21]. The result of the above equation is congregated when the distance of a data vector is higher than a preset threshold level.

3. Experimental setup

There were two experimental procedures were taken place. The first was the study of the wavelet through an aircraft part at normal, damaged and repaired conditions. The second is to observed the guided Lamb wave behavior when subjected to tensile loading for the three conditions stated above.

The APC 850 PZT sensor from APC International Inc. was used for both experiments. The properties of the sensors are shown in Table 1 below. Two sensors were used as an actuator and receiver with a diameter of 10mm and thickness of 0.5mm. The pitch catch active sensing was used to obtain the data at the receiving sensors. The sensors were placed at 100mm apart due to the optimum wave attenuation from the actuator to the receiver. The actuator was connected to a function generator where a selected input variable were set and the receiver were connected to the oscilloscope for data mining and further processing.

Description	Value
Voltage limit AC/DC	8/ 15 V
Output Power	20 watts/ inch
Relative dielectric constant	1750
Dielectric loss	1.4%
Curie Temperature	360°C
Density	7.7 X 10^3 kg/m^3
Young's Modulus	6.3 X 10^{10} N/m^2

Table 1. APC-850 properties [22]

3.1. Aircraft component analysis

An aircraft spoiler was used for this research. The use of the structure is only arbitrary at this stage. It is use to seek the workability of the sensor upon trial on several flat panels. Three conditions were introduced to the panel which is the undamaged/ parent, damaged and repaired area. The undamaged/ parent was the area free from any defects. The undamaged area is the original conditions or controlled area. The damaged area was damage caused by impact that removes the top laminate. It was made by impacting the faced planes with a blunt object and creating damage less than 40mm diameter fracture. The level of impact is not an interest in this particular testing due to the studied conditions is only applicable to small surface damage due to impact. The repaired area was where the damage plies were removed and replaced in accordance with the SRM [23]. The damaged area was repaired by scarfing method.

Figure 1. Locations of the structural conditions and PZT sensor placements

The repair was conducted by using hot bonder from Heatcon Inc. The Hexply® M10/38%/UD300/CHS/460mm CFRP pre-preg system from Hexcel Corp was used for the repair process. Care and take were observed to ensure similar procedures as per SRM recommendation. All plies were cut according to the sizes required and laid up accordingly. The affected area were vacuum bag as per Figure 2 and cured at 120°C at atmospheric pressure for 120 minutes. All vacuum bag materials were removed once the cycle ended.

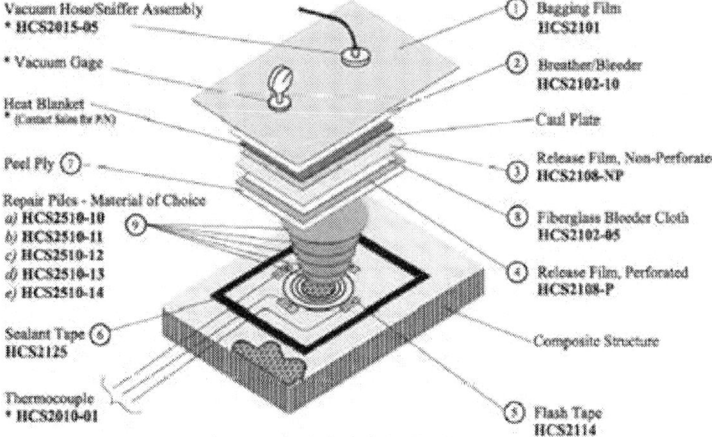

Figure 2. Hot bonder materials sequence for repairing aircraft composite parts. [24]

The PZT sensors were placed at 100mm apart for the three studied conditions. For damaged condition, the sensors were placed in between the damage area and for the repaired area, the sensors were replace across the actual and the repair doubler surface. This is to ensure distance consistency of 100mm between the sensors. One of the sensors acted as an actuator. The actuator controls the surface guided in the form of elastic perturbation through the surface guided wave across the panel. The wave was controlled by a function generator with the setup as per Table 2.

Parameter	Unit	Parameter	Unit
Frequency	250kHz	Symmetrical	50%
Voltage	+10V	Time Generation	3ms at 333.3 Hz
V_{pp}	50Ω	Burst Count	5
Phase	0^0		

Table 2. Actuating setting parameter.

The receiving sensor modulated as the guided wave reached and transmit the energy to electrical signal. The received signals were saved for post processing by using oscilloscope. The arrangement of the equipment is shown in Figure 4.

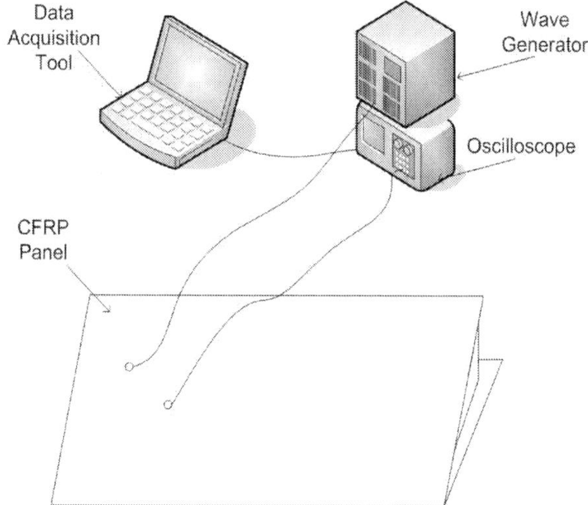

Figure 3. Aircraft spoiler with PZT sensor on specimens set-up.

3.2. Tensile testing

Further investigation was conducted by using 2 sets of tensile testing specimens with condition of normal and repair attached with pair of PZT sensors.Tthree composite plate of 300mm by 300mm were fabricated by using the Hexply® M10/38%/UD300/CHS/460mm from Hexcel Corps. The ply orientation was set to [0/90]s2 orientation to produce a balanced symmetrical flat monolithic structure. The parent specimen was subjected to one time curing. However the repaired specimens undergone for secondary curing once the damage area was removed and new replacement plies were laid up. Both initial and secondary bonding was cured in accordance with Aircraft Structural Repair Manual (SRM). A scarf cutting technique was used to remove the damage and replaced the affected its areas. Curing was achieved by using the Heatcon HCS4000 hot bonder with assisted consolidation from vacuum bag as per Figure 2. The parameters were ramp rate at 3^0C/min, dwell time at 120 minutes, dwell temperature at 121^0C, cooling rate at 3^0C/min and vacuum pressure attained at 22 in mg/ 1 bar.

Once cured, the panels were cut into specimen size according to ASTM D638 standard with five specimens prepared for each conditions by using Shimadzu AGx-50kN Universal Testing Machine as per Figure 5. The specimens were clamped on both ends. The data for mechanical properties were collected by using the Trapezium-X software came with the UTM machine. For the wavelet pitch-catch analysis, two APC 850 PZT smart sensors were affixed at 100mm apart and symmetrical to each other. Similar connection with the spoiler's test was applied to the relevant apparatus for data mining and post processing. The data from the sensor was interrogated and collected at three stages which were at the beginning of the test, within the elastic range, after the detection of the first ply failure and prior to separation.

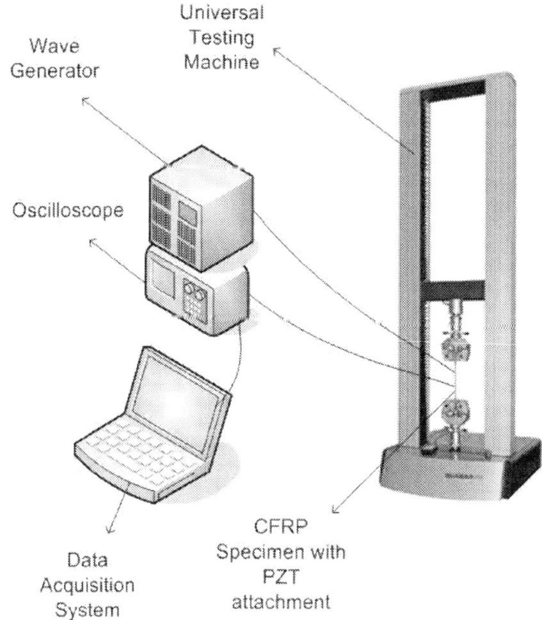

Figure 4. Tensile test with PZT sensor on specimens set-up.

4. Results and discussion

The results for both experiments are being presented into two sections. The initial test was evaluated upon the Vpp from the wavelet analysis. Further overlaying pattern are also being presented. The latter testing involved with the tensile testing and only the results from wavelet analysis are shown accordingly.

4.1. Aircraft component analysis

Statistical pattern recognition was used to analyze the lamb wave generated by the PZT actuator. [15] A total of 100 wave packets were taken for each conditions stated. Each wave packets consisted of 25000 points by default from the oscilloscope. From the 25000 points, it was then grouped to 1000 intervals data set for analysis. There were two significant spike occurred each at point 12000 ~ 13000 and 18000 ~19000 as shown in Figure 4. The reduction of data intervals were applied in order to assist the further analyze the distributions. By judgment, the first group of the spike was concerned and the data packet was zoomed again in 500 data intervals. Figure 6 shows the actuating signals for each of the testing. Consistence settings are required to ensure the wavelet generates similar wave perturbation throughout the experiment.

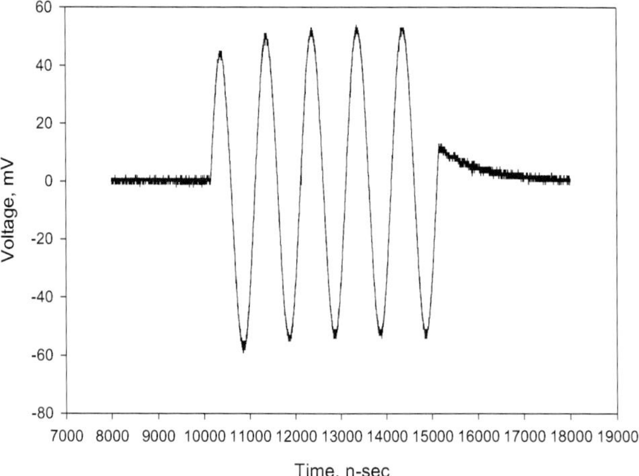

Figure 5. Actuating signals

Figure 7 shows the results of the receiving wave packet upon synthesized by the points for each condition. Different behavior from the voltage (Vpp) and complete time of flight cycle are shown which characterized the evaluated conditions.

Then, each of the receiving structural conditions wavelet data were compared between to ensure the signals was homogeneous to each other on the timeline basis. Since this is the unsupervised learning process the clusters were assigned to separate three conditions as stated in the methodology. The V_{PP} or voltage peak to peak is the attribute to distinguish the conditions. More than 50 V_{PP} values were collected and tabulated. The scattering of the V_{PP} for the three conditions were examined and is shown in Figure 8.

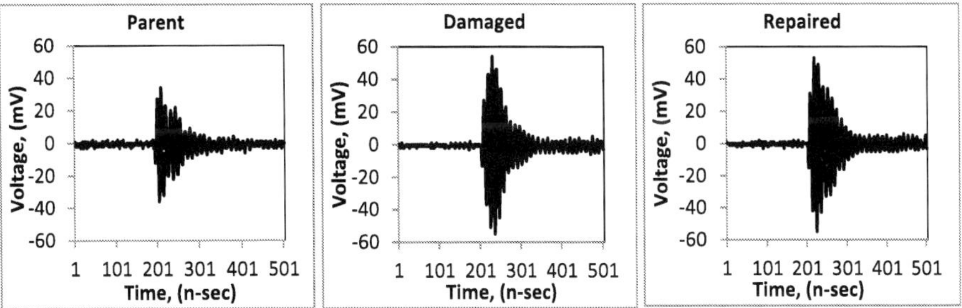

Figure 6. Wave packet samples from a) undamaged, b) damaged and c) repaired structural conditions after synthesized.

The V_{pp} showed similar values for the undamaged and damaged structure condition. It was assumed that the wave travel without discontinuity due to partial damage at that particular area. However several other types of damage need to be examined before any conclusive evidence can be finalized.

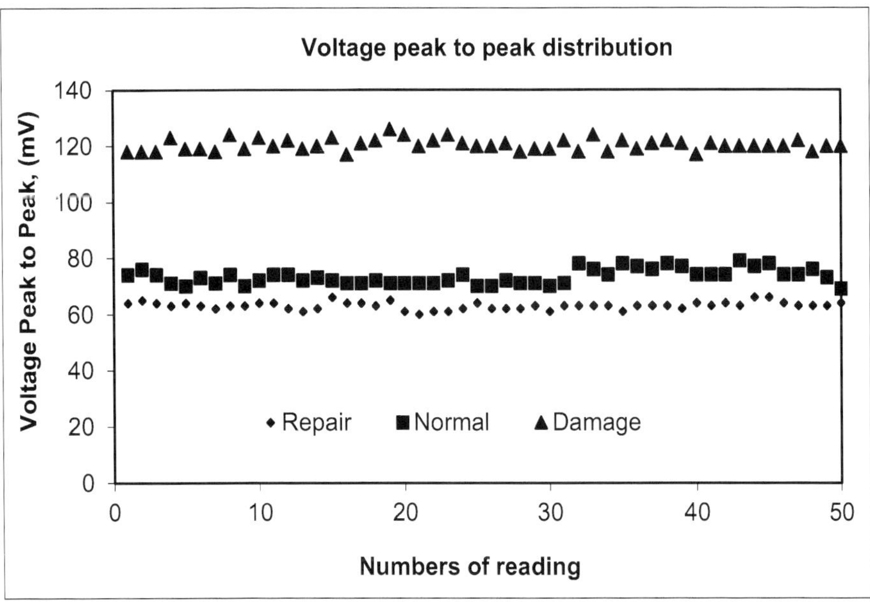

Figure 7. V_{pp} distribution among the structural conditions

A further post processing was carried out by overlapping the three conditions in one graph with similar time-domain comparison. The most common interest point lies within points 12250 ~ 12750. This was the first spike seen in the wave packet. In the earlier V_{pp} comparison, the distribution data between the damage and undamaged were identical. Therefore it was difficult to interpret the data for the latter machine learning process.

However, when all three data was overlapped, a significant different can be seen as shown in Figure 9. The undamaged signals appear at the initial time frame indicated that there was a clean surface wave traveling from the actuator to the receiver. However, once the partial damage was introduced, the spike appear later about 200nsec due to the discontinuity of the spoiler surface. The unaffected wave bifurcated to the receiver with delay. For the repair condition, since the surface integrity has been restored by the flush repair, the continuity of the surface wave was preserved again with delay about 50 nsec .

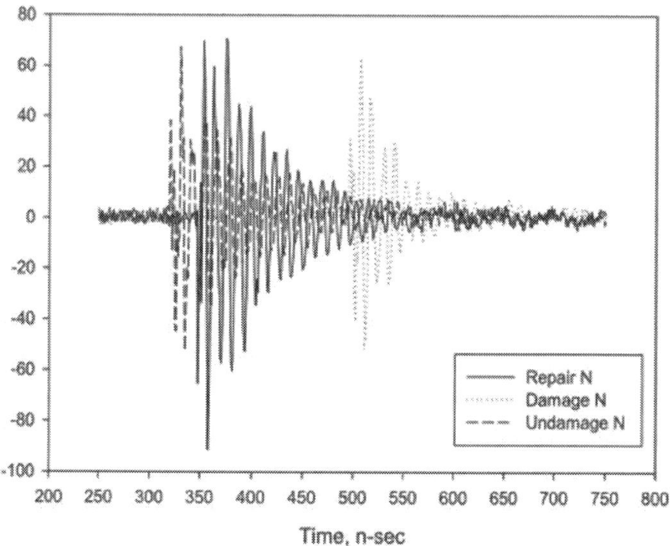

Figure 8. Overlay Outlier Pattern for different structural condition

4.2. Tensile testing

The tensile test results showed a similar behavior towards a brittle stress-strain curve. Below are the results of the tensile test on normal and repaired specimens. All specimens were found to have breakage at center. The fiber breakage for parent specimens occurred at arbitrary layer. However, for the repaired specimens, the breakage occurred at the intermediate bonding layer between the original layers and repair plies. This is due to the fiber discontinuity and the matrix is the only medium to transfer the stress from the parent surface to the repair surface.

Based of the wavelet analysis from the Sigmaplot software, both conditions showed a significant changes at the investigated stages. At the elastic range, the wavelet shows a significant solid wave at the respective time of flight. However a slight change appeared after the first ply failure occurred. The V_{PP} values were found to be higher and there are also a distinct echo developed after the main wave packets. At the end of the testing, the signal lost its signature due to damage upon breakage of the panels. An online monitoring during the course of the test shown a good unique characteristics for anomalies to be identified.

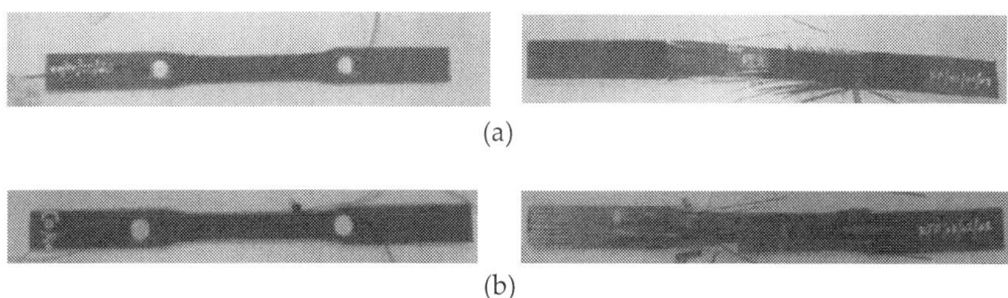

(a)

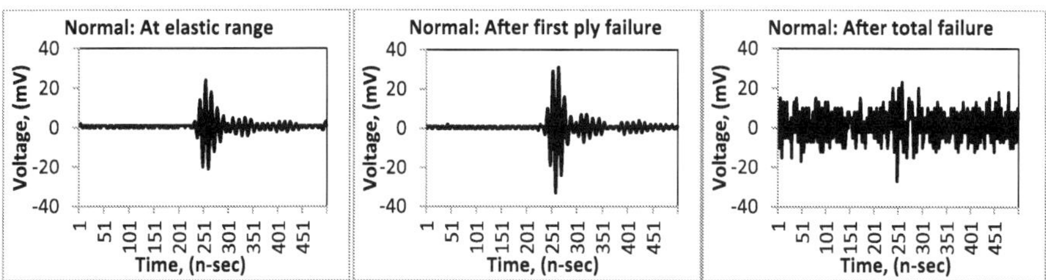

(b)

Figure 9. CFRP result before and after tensile test for parent specimens for a) normal and b) repaired specimens

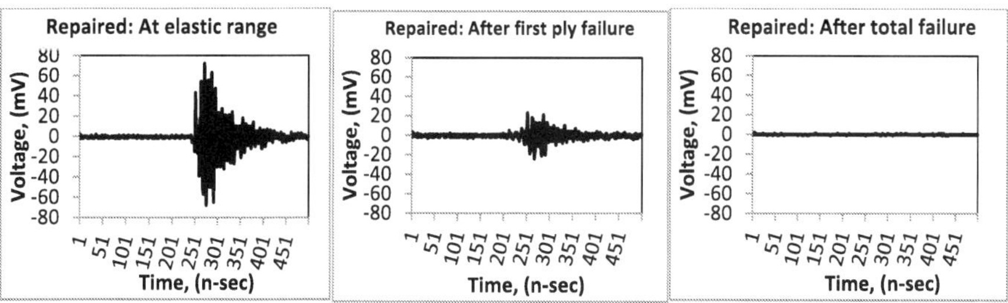

Figure 10. Receiving signals for parent panel with a) at elastic range, b) after first ply failure and c) at failure.

Figure 11 shows the behavior of the full repair panel throughout the testing. The degradation of the signal indicates the lamb wave attenuation has lost due to separation of the repair plies. This can be seen by 50% reduction of the Vpp at the initial of the testing. The significant reduction of the signal strength correlates with the structural integrity lost as the test reached the total fractured. Towards the end of the testing, all the specimen failed at the center and disintegration of the sensor due to the failure of the panels.

Figure 11. Receiving signals for parent panel with a) at elastic range, b) after first ply failure and c) at failure.

5. Discussion

Both results from the experiments shows a promising indicator on the usage of PZT sensors to monitor structure integrity of the aerospace components and controlled testing specimen. Results from the spoiler shows:-

1. The Vpp value for all tested condition showed a significant different due to the signal intensity once it passed the tested conditions. Although the undamaged and damaged signals are almost identical, the repair area shows a higher Vpp values. This might due to the additional plies of the repair and the bouncing of the intensification of the signal to travel the tested structure. The time of flight for the repaired reading was found to be delayed from the two conditions. The additional ply or doubler may contribute to the delay. This can promote detection of hidden repair area. The outlier behavior of the Vpp is an initial indications that PZT sensors can be used to detect interested conditions.

2. The overlay patter analysis show a promising results as it can bifurcate each of the conditions. The outlier analysis can be applied to differentiate the condition of the surface integrity. Due to separation of the fibers, the time for the Lamb wave to travel from the actuator to the receiver has been delayed and takes a longer time with the reduction of the Vpp value when compared to the repaired area.

3. The tensile test indicates that the materials behave in accordance with typical brittle materials. The breakage at the center indicates a good distribution of the load during the testing. The wavelet signals at the elastic area, within the first ply failure and total failure shows good indications of how the Lamb wave behaves before the specimen fails. However, the method of taking the PZT sensor receiver reading need to be improved as the fluctuation of the signals is unbearable.

4. Further post processing techniques need to be used in order to further scrutinize the behavior of the data. A more confident result can be further utilize for more converging result such as by generic algorithm, neural network etc. in order to enhance the prognosis of the structure at later stages.

6. Conclusion

As conclusion, both experiments has proved that PZT sensors can be used to detect anomalies of the CFRP structure either passive or active sensing. In passive sensing, the data received data is very stable and shows a significant consistence reading at any duration. The latter experiment shows that the ability of the sensors to sense structural integrity of a normal and repaired specimens. However, further investigations are required to this robust detection system in order to ensure the results are established. This can be done by comparing the results through various techniques, statistical methods and analytical analysis.

Author details

K.D. Mohd Aris
Universiti Kuala Lumpur, Malaysian Institute of Aviation Technology, Jalan Jenderam Hulu, Selangor, Malaysia

F. Mustapha, S.M. Sapuan, D.L. Majid
Universiti Putra Malaysia, Serdang, Selangor, Malaysia

Acknowledgement

This research is part of Ministry of Science and Technology, Malaysia (MOSTI) SF000064 Escience Fund grant and Spirit Aerosystem (Malaysia) Inc. for donating the aircraft spoiler.

7. References

[1] The World Wide Composite Industry Structure, Trends and Innovation: New 2010 release, JEC Composite, 2010

[2] Soutis C. and Diamanti K., Structural Health Monitoring Techniques for Aircraft Composite Structures, Progress in Aerospace Sciences 46(2010), 342 -353)

[3] Baker A., Dutton S. and Kelly D., Joining of Composite Structure, Composite Matrials for Aircraft Structure, AIAA Inc., 2004. 290 ~295.

[4] Baker A., Dutton S. and Kelly D., Repair Technology, Materials for Aircraft Structure, AIAA Inc., (2004), 290 ~295.

[5] Mujica L.E. et al., Impact Damage Detection in Aircraft Composites Using Knowledge-Based Reasoning, Structural Health Monitoring (2008); 7; 215~230,

[6] Xie Jian, Lu Yao, Study on Airworthiness Requirements of Composite Aircraft Structure for Transport Category Aircraft in FAA, Procedia Engineering, Volume 17 (2011), 270-278

[7] Kim I. and Park C.Y., Prediction of Impact Forces on an Aircraft Composite Wing, J. of Intelligent Materials Systems and Structures, Vol 19, (2008), 319 ~ 324,

[8] S.M. Sapuan, F. Mustapha, D.L. Majid, Z. Leman, A.H.M. Ariff, M.K.A. Ariffin, M.Y.M. Zuhri, M.R. Ishak and J. Sahari, Fiber Reinforced Composite Structure with Bolted Joint – A Review, Key Engineering Materials, 471-472, pg 939-944

[9] G. Goulios, Z. Marioli-Riga, Composite patch repairs for commercial aircraft: COMPRES, Air & Space Europe, Volume 3, Issues 3–4, (2001), 143-147

[10] S.B. Kumar, S. Sivashanker, Asim Bag, I. Sridhar, Failure of aerospace composite scarf-joints subjected to uniaxial compression, Materials Science and Engineering: A, Volume 412, Issues 1–2, (2005), 117-122

[11] Dan He, Toshiyuki Sawa, Takeshi Iwamoto, Yuya Hirayama, Stress analysis and strength evaluation of scarf adhesive joints subjected to static tensile loadings, International Journal of Adhesion and Adhesives, Volume 30, Issue 6, (2010), 387-392

[12] Chang F. K and Ihn J. B., Pitch Catch Active Sensing Methods in Strucural Health Monitoring for Aircraft Structures, Structural Health Monitoring (2008) Vol 7, 5~ 19.

[13] Inman D.J. et al, Damage Prognosis for Aerospace, civil and Mechanical Systems, John Wiley and Sons Ltd., 2005

[14] Ostachowicz W. M., Damage Detection of Structures Using Spectral Finite Element Method, Computers and Structures 86 (2008), 454 ~ 462.

[15] Kesavan A., John S and Herszberg, Structural Health Monitoring of Composite Structures Using Artificial Intelligence Protocol, Journal of Intelligent Material Systems and Structures; 19;63, 63 ~ 72

[16] Worden K, Manson G and Filler N. R. J. , Damage Detection Using Outlier Analysis, Journal of Sound and Vibration 229(3) (2000), 647 ~ 667),

[17] Webb A.R, Statistical Pattern Recognition, John Wiley and Sons Ltd, 2002

[18] F. Mustapha, G. Manson, K. Worden, S.G. Pierce, Damage location in an isotropic plate using a vector of novelty indices, Mechanical Systems and Signal Processing, Volume 21, Issue 4, May 2007, Pages 1885-1906

[19] Charles R Farrar and K. Worden, An introduction to structural health monitoring, Phil. Trans. R. Soc. A 15 February 2007 vol. 365 no. 1851 303-315

[20] Ihn J and Chang F. K., Pitch Catch Active Sensing Methods in Structural Health Monitoring for Aircraft Structures, Structural Health Monitoring 2008 Vol. 7, 1 ~ 19.

[21] Webb A.R, Statistical Pattern Recognition, John Wiley and Sons Ltd, 2002) (Park et al, An Outlier Analysis Framework for Impedance Based Structural Health Monitoring System, Journal of Sound and Vibration 286 (2005), 229 ~ 250

[22] Piezoelectric Ceramics: Principles and Applications, APC International Ltd, (2008)

[23] Boeing 737-300 Structural Repair Manual, The Boeing Company Inc., 1996

[24] Heatcon Composite Systems: Composite Repair Solutions, Product Catalog, 2011

Numerical Simulation of Wave Propagation in 3D Elastic Composites with Rigid Disk-Shaped Inclusions of Variable Mass

Viktor Mykhas'kiv

Additional information is available at the end of the chapter

http://dx.doi.org/10.5772/48113

1. Introduction

In many practical situations elastic composites are subjected to dynamic loadings of different physical nature, which origin the wave propagation in such structures. Then overall dynamic response of composite materials is characterized by the wave attenuation and dispersion due to the multiple wave scattering, in the local sense these materials are exhibited also by the dynamic stress intensification due to the wave interaction with the composite fillers. Essential influences on the mentioned phenomena have the shapes and the space distributions of inclusions, i.e. composite architecture, as well as matrix-inclusion materials characteristics. In this respect the numerical investigation of elastic wave propagation in the composite materials with inclusions of non-classical shape and contrast rigidity in comparison with the matrix material is highly demanded. A deep insight into their dynamic behavior, especially on the microscale, is extremely helpful to the design, optimization and manufacturing of composite materials with desired mechanical qualities, fracture and damage analysis, ultrasonic non-destructive testing of composites, and modeling of seismic processes in complex geological media.

The macroscopic dynamic properties of particulate elastic composites can be described by effective dynamic parameters of the equivalent homogeneous effective medium via a suitable homogenization procedure. Generally speaking, the homogenization procedure to determine the effective dynamic properties of particulate elastic composites is much more complicated than its static counterpart because of the inclusion interactions and multiple wave scattering effects. For small inclusion concentration or dilute inclusion distribution, their mutual interactions and the multiple wave scattering effects can be neglected approximately. In this case, the theory of Foldy [1], the quasi-crystalline approximation of

Lax [2] and their generalizations to the elastic wave propagation [3-5] can be applied to determine the effective wave (phase) velocities and the attenuation coefficients in the composite materials with randomly distributed inclusions. In these models, wave scattering by a single inclusion has to be considered in the first step. Most previous publications on the subject have been focused on 3D elastic wave propagation analysis in composite materials consisting of an elastic matrix and spherical elastic inclusions (for example, see [6,7]). Aligned and randomly oriented ellipsoidal elastic inclusions have been considered in [8-10] under the assumption that the wavelength is sufficiently long compared to the dimensions of the individual inclusions (quasi-static limit). As special cases, the results for a random distribution of cracks and penny-shaped inclusions can be derived from those for ellipsoidal inclusions. In the long wavelength approximation, analytical solutions for a single inclusion as a series of the wave number have been presented in these works. However, this approach is applicable only for low frequencies or small wave numbers. For moderate and high frequencies, numerical methods such as the finite element method or the boundary element method can be applied. By using the boundary integral equation method (BIEM) or the boundary element method (BEM) in conjunction with Foldy's theory the effective wave velocities and the wave attenuations in linear elastic materials with open and fluid-filled penny-shaped cracks as well as soft thin-walled circular inclusions have been calculated in [11,12]. Both aligned and randomly oriented defect configurations have been studied, where a macroscopic anisotropy for aligned cracks and non-spherical inclusions appears. Previous results have shown that distributed crack-like defects may cause a decrease in the phase velocity and an increase in the wave attenuation. The efficiency and the applicability ranges of 2D homogenization analysis of elastic wave propagation through a random array of scatters of different shapes and dilute concentrations based on the BEM and Foldy-type dispersion relations were demonstrated also by many authors, for instance, in the papers [13,14]. In 3D case this approach was applied for the numerical simulation of the average dynamic response of composite material containing rigid disk-shaped inclusions of equal mass only [15]. Dynamic stresses near single inclusion of such type under time-harmonic and impulse elastic waves incidence where also investigated [16-18].

In this Chapter the effective medium concept is extended to the time-harmonic plane elastic wave propagation in an infinite linear elastic matrix with rigid disk-shaped movable inclusions of variable mass. Both time-harmonic plane longitudinal and transverse waves are considered in the analysis. The solution procedure consists of three steps. In the first step, the wave scattering problem is formulated as a system of boundary integral equations (BIEs) for the stress jumps across the inclusion surfaces. A BEM is developed to solve the BIEs numerically, where the kinetics of the inclusion and the "square-root" singularity of the stress jumps at the inclusion edge are taken into account properly. The improved regularization procedure for the obtained BIEs involving the analytical evaluation of regularizing integrals and results of mapping theory is elaborated to ensure the stable and correct numerical solution of the BIEs. The far-field scattering amplitudes of elastic waves induced by a single inclusion are calculated from the numerically computed stress jumps. In the second step, the simple Foldy-type approximation [1] is utilized to calculate the complex effective wave numbers for a dilute concentration of inclusions, where their interactions and multiple wave scattering can be neglected. The averages of the forward scattering amplitudes over 3D inclusion orientations or directions of the wave incidence and over inclusions masses are included into the resulting homogenization formula (dispersion relations). Finally, the effective wave velocity and the attenuation coefficient are obtained by taking the real and the imaginary parts of the effective wave numbers. To investigate the influence of the wave frequency on the effective dynamic parameters, representative numerical examples for longitudinal and transverse elastic waves in infinite elastic composite materials containing rigid disk-shaped inclusions with aligned and random orientation, as well as aligned, normal and uniform mass distribution are presented and discussed. Besides the global dynamic parameters, the mixed-mode dynamic stress intensity factors in the inclusion vicinities are calculated. They can be used for the fracture or cracking analysis of a composite.

2. Boundary integral formulation of 3D wave scattering problem for a single massive inclusion

Let us consider an elastic solid consisting of an infinite, homogeneous, isotropic and linearly elastic matrix specified by the mass density ρ, the shear modulus G and Poisson's ratio ν, and a rigid disk-shaped inclusion with the mass M, which thickness is much smaller than the characteristic size of its middle-surface S. The center of the Cartesian coordinate system $Ox_1x_2x_3$ coincides with the mass center of the inclusion (see Figure 1), within the described geometrical assumptions the limit values $x_3 = \pm 0$ correspond to the opposite interfaces between the matrix and the inclusion, where a welded contact is assumed. The stress-strain state in the solid is induced by harmonic in the time t plane longitudinal L-wave or transverse T-wave with the frequency ω, the constant amplitude U_0, the phase velocities c_L and c_T, and the wave numbers $\chi_L = \omega/c_L$ and $\chi_T = \omega/c_T$, respectively. The displacement vector $\mathbf{u}^{in} = (u_1^{in}, u_2^{in}, u_3^{in})$ of the incident wave is given by the relation

$$\mathbf{u}^{in}(\mathbf{x}) = \mathbf{U}\exp\left[i\chi(\mathbf{n}\cdot\mathbf{x})\right] \tag{1}$$

Here and hereafter the common factor $\exp(-i\omega t)$ is omitted, χ is the wave number of the incident wave, $\mathbf{n} = (\sin\theta_0, 0, \cos\theta_0)$ is the direction of propagation of the incident wave, θ_0 is the angle characterizing the direction of the wave incidence, and $\mathbf{U}$ is the polarization vector with $\chi = \chi_L$ and $\mathbf{U} = U_0\cdot\mathbf{n}$ for the L-wave and $\chi = \chi_T$, $\mathbf{U} = U_0\cdot\mathbf{e}$ and $\mathbf{n}\cdot\mathbf{e} = 0$ for the T-wave.

By using the superposition principle, the total displacement field $\mathbf{u}^{tot}$ in the solid can be written in the form

$$\mathbf{u}^{tot}(\mathbf{x}) = \mathbf{u}^{in}(\mathbf{x}) + \mathbf{u}(\mathbf{x}), \tag{2}$$

where $\mathbf{u} = (u_1, u_2, u_3)$ is the unknown displacement vector of the scattered wave, which satisfies the equations of motion and the radiation conditions at infinity (these well-known governing relations of elastodynamic theory can be found in [19]).

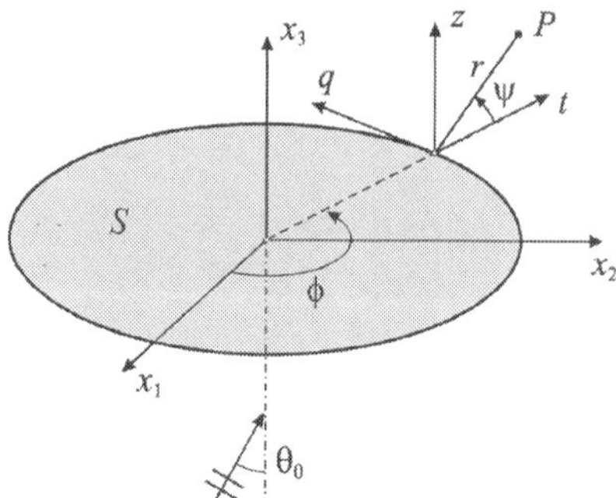

Figure 1. Single disk-shaped inclusion subjected to an incident elastic wave.

The inclusion is regarded as a rigid unit and its motion is described by the translation $\mathbf{u}^0 = (u_1^0, u_2^0, u_3^0)$ and the rotation with respect to the coordinate axes with the angles Ω_1, Ω_2 and Ω_3, respectively. Then the displacement components in the domain S can be represented by

$$u_j(\mathbf{x}) = -u_j^{in}(\mathbf{x}) + u_j^0 + (-1)^j \Omega_3 x_{3-j}, \quad j = 1, 2,$$

$$u_3(\mathbf{x}) = -u_3^{in}(\mathbf{x}) + u_3^0 + \Omega_1 x_2 - \Omega_2 x_1, \quad \mathbf{x}(x_1, x_2, \pm 0) \in S. \tag{3}$$

In order to obtain the integral representations for the displacement components we apply the Betty-Rayleigh reciprocity theorem in conjunction with the properties of the elastodynamic fundamental solutions. As a result, the displacement components of the scattered waves can be written in the form [18]:

$$u_j(\mathbf{x}) = \frac{1}{4\pi G} \iint_S \left\{ \Delta\sigma_j(\xi) \frac{\exp\left(i\chi_T |\mathbf{x} - \xi|\right)}{|\mathbf{x} - \xi|} - \frac{1}{\chi_T^2} \frac{\partial}{\partial x_j} \left[\Delta\sigma_1(\xi) \frac{\partial}{\partial x_1} + \Delta\sigma_2(\xi) \frac{\partial}{\partial x_2} + \right. \right.$$

$$\left. \left. + \Delta\sigma_3(\xi) \frac{\partial}{\partial x_3} \right] \left[\frac{\exp\left(i\chi_L |\mathbf{x} - \xi|\right)}{|\mathbf{x} - \xi|} - \frac{\exp\left(i\chi_T |\mathbf{x} - \xi|\right)}{|\mathbf{x} - \xi|} \right] \right\} dS_\xi, \quad j = 1, 2, 3, \tag{4}$$

where the displacement continuity conditions across the inclusion are used, $|\mathbf{x} - \xi|$ is the distance between the field point $\mathbf{x} = (x_1, x_2, x_3)$ and integration point $\xi = (\xi_1, \xi_2, 0)$, and $\Delta\sigma_j$ $(j = 1, 2, 3)$ are the jumps of the interfacial stresses across the inclusion, which are defined by

$$\Delta\sigma_j(\mathbf{x}) = \sigma_{j3}^-(\mathbf{x}) - \sigma_{j3}^+(\mathbf{x}), \quad j = 1, 2, 3, \quad \mathbf{x} \in S, \quad \sigma_{j3}^\pm(\mathbf{x}) = \lim_{x_3 \to \pm 0} \sigma_{j3}(\mathbf{x}). \tag{5}$$

Eqs. (5) together with the equations of motion of the inclusion as a rigid unit yields the following relations between the translations and the rotations of the inclusion and the stress jumps $\Delta\sigma_j$:

$$u_j^0 = \frac{1}{\omega^2 M}\iint_S \Delta\sigma_j(\xi)\,dS_\xi\,, \quad j=1,2,3\,,$$

$$\Omega_j = \frac{(-1)^{j+1}}{\omega^2 M i_j^2}\iint_S \xi_{3-j}\,\Delta\sigma_3(\xi)\,dS_\xi\,, \quad j=1,2\,, \tag{6}$$

$$\Omega_3 = \frac{1}{\omega^2 M i_3^2}\iint_S \left[\xi_1\,\Delta\sigma_2(\xi) - \xi_2\,\Delta\sigma_1(\xi)\right]dS_\xi\,,$$

where i_j is the radius of inertia of the inclusion with respect to the x_j-axis.

The displacement components in the matrix and the kinematical parameters of the inclusion are related to the stress jumps across the inclusion by the relations (4) and (6). Substitution of Eqs. (4) and (6) into Eqs. (3) results in three boundary integral equations (BIEs) for the stress jumps as

$$\iint_S \Delta\sigma_3(\xi)R_3(x,\xi)\,dS_\xi = -4\pi G\chi_T^2\, u_3^{in}(x),\quad x\in S\,,$$

$$\iint_S \left[\Delta\sigma_j(\xi)R_j(x,\xi) + \Delta\sigma_{3-j}(\xi)R_{j(3-j)}(x,\xi)\right]dS_\xi = -4\pi G\chi_T^2\, u_j^{in}(x),\quad j=1,2,\ \ x\in S \tag{7}$$

In Eq. (7), the kernels R_j, R_{12} and R_{21} have the form

$$R_j(x,\xi) = L_1\big(|x-\xi|\big) - \frac{(x_j-\xi_j)^2}{|x-\xi|^2}L_2\big(|x-\xi|\big) - \frac{4\pi\rho}{M}\left(1 + \frac{\xi_{3-j}x_{3-j}}{i_3^2}\right),\quad j=1,2\,,$$

$$R_3(x,\xi) = L_1\big(|x-\xi|\big) - \frac{4\pi\rho}{M}\left(1 + \frac{\xi_1 x_1}{i_2^2} + \frac{\xi_2 x_2}{i_1^2}\right),$$

$$R_{kj}(x,\xi) = -\frac{(x_1-\xi_1)(x_2-\xi_2)}{|x-\xi|^2}L_2\big(|x-\xi|\big) + \frac{4\pi\rho}{M}\frac{\xi_k x_j}{i_3^2}\,,\quad k,j=1,2,\ \ k\neq j\,, \tag{8}$$

$$L_j(d) = l_{j1}(d)\frac{\exp(i\chi_L d)}{d^3} - l_{j2}(d)\frac{\exp(i\chi_T d)}{d^3}\,,\quad j=1,2\,,$$

$$l_{11}(d) = 1 - i\chi_L d,\quad l_{12}(d) = 1 - i\chi_T d - \chi_T^2 d^2\,,$$

$$l_{21}(d) = 3 - 3i\chi_L d - \chi_L^2 d^2\,,\quad l_{22}(d) = 3 - 3i\chi_T d - \chi_T^2 d^2\,.$$

The problem governed by the BIEs (7) can be divided into an antisymmetric problem and a symmetric problem. The antisymmetric problem corresponding to the transverse motion of the inclusion is described by first equation of the BIEs (7) for the stress jump $\Delta\sigma_3$. After the solution of this equation the displacement u_3^0 and the rotations Ω_1 and Ω_2 can be obtained by using the relations (6). The symmetric problem corresponds to the motion of the inclusion in its own plane, which is governed by the last two equations of the BIEs (7) for the stress jumps $\Delta\sigma_1$ and $\Delta\sigma_2$. After these quantities have been computed by solving these equations, the kinematical parameters u_1^0, u_2^0 and Ω_3 can be obtained by using the relations (6).

The kernels of the BIEs (7) contain weakly singular integrals only. To isolate these singularities explicitly we rewrite the BIEs (7) as

$$A\iint_S \frac{\Delta\sigma_3(\xi)}{|\mathbf{x}-\xi|}dS_\xi + \iint_S \Delta\sigma_3(\xi)\left[\frac{1}{\chi_T^2}R_3(\mathbf{x},\xi) - \frac{A}{|\mathbf{x}-\xi|}\right]dS_\xi = -4\pi G u_3^{in}(\mathbf{x}), \quad \mathbf{x}\in S,$$

$$\iint_S \frac{\Delta\sigma_j(\xi)}{|\mathbf{x}-\xi|}\left[A + B\frac{(x_j-\xi_j)^2}{|\mathbf{x}-\xi|^2}\right]dS_\xi + B\iint_S \Delta\sigma_{3-j}(\xi)\frac{(x_1-\xi_1)(x_2-\xi_2)}{|\mathbf{x}-\xi|^3}dS_\xi +$$

$$+\iint_S \left\{\Delta\sigma_j(\xi)\left[\frac{1}{\chi_T^2}R_j(\mathbf{x},\xi) - \frac{A}{|\mathbf{x}-\xi|} - B\frac{(x_j-\xi_j)^2}{|\mathbf{x}-\xi|^3}\right] + \Delta\sigma_{3-j}(\xi)\left[\frac{1}{\chi_T^2}R_{j(3-j)}(\mathbf{x},\xi) - \right. \right. \tag{9}$$

$$\left. \left. -B\frac{(x_1-\xi_1)(x_2-\xi_2)}{|\mathbf{x}-\xi|^3}\right]\right\}dS_\xi = -4\pi G u_j^{in}(\mathbf{x}), \qquad j=1,2, \quad \mathbf{x}\in S,$$

where

$$A = (3-4\nu)/\left[4(1-\nu)\right], \quad B = 1/\left[4(1-\nu)\right]$$

In Eq. (9), the last integrals on the left-hand sides exist in the ordinary sense. This fact follows from an analysis of the integrand in the limit $\xi\to\mathbf{x}$. Therefore, in the numerical evaluation of these integrals it is sufficient to perform the integration over S_x^0 by excluding a small region (the neighborhood of the $\mathbf{x}$-point) around $\mathbf{x}$ from S.

The singularities of the BIEs (9) are identical to those of the corresponding BIEs for the static inclusion problems, which have been investigated in [20] both for the antisymmetric and symmetric cases. The local behavior of the stress jumps at the front of the inclusion is also the same as in the static case. For a circular disk-shaped inclusion, the stress jumps have a "square-root" singularity, which can be expressed as

$$\Delta\sigma_j(\mathbf{x}) = \frac{f_j(\mathbf{x})}{\sqrt{a^2 - x_1^2 - x_2^2}}, \quad j=1,2,3, \quad \mathbf{x}\in S, \tag{10}$$

where $f_j(\mathbf{x})$ are unknown smooth functions, and a is the radius of the inclusion.

Substitution of Eq. (10) into Eq. (9) results in a system of BIEs for the functions $f_j(\mathbf{x})$. These BIEs have a weak singularity $1/|\mathbf{x}-\xi|$ at the source point $\xi=\mathbf{x}$ and a "square-root" singularity at the edge of the inclusion. To regularize the singular BIEs, the following integral relations for the elastostatic kernels are utilized when $\mathbf{x}\in S$:

$$\iint_S \frac{f(\xi)}{\sqrt{a^2-\xi_1^2-\xi_2^2}\cdot|\mathbf{x}-\xi|}dS_\xi = \pi^2 f(\mathbf{x}) + \iint_{S_x^0}\frac{f(\xi)-f(\mathbf{x})}{\sqrt{a^2-\xi_1^2-\xi_2^2}\cdot|\mathbf{x}-\xi|}dS_\xi,$$

$$\iint_S \frac{f(\xi)}{\sqrt{a^2-\xi_1^2-\xi_2^2}}\cdot\frac{(x_j-\xi_j)^2}{|\mathbf{x}-\xi|^3}dS_\xi = \frac{\pi^2}{2}f(\mathbf{x}) + \iint_{S_x^0}\frac{\left[f(\xi)-f(\mathbf{x})\right]}{\sqrt{a^2-\xi_1^2-\xi_2^2}}\cdot\frac{(x_j-\xi_j)^2}{|\mathbf{x}-\xi|^3}dS_\xi, \qquad (11)$$

$$\iint_S \frac{f(\xi)}{\sqrt{a^2-\xi_1^2-\xi_2^2}}\cdot\frac{(x_1-\xi_1)(x_2-\xi_2)}{|\mathbf{x}-\xi|^3}dS_\xi = \iint_{S_x^0}\frac{\left[f(\xi)-f(\mathbf{x})\right]}{\sqrt{a^2-\xi_1^2-\xi_2^2}}\cdot\frac{(x_1-\xi_1)(x_2-\xi_2)}{|\mathbf{x}-\xi|^3}dS_\xi.$$

Here the special integral identities, taken from [20], are used, namely:

$$\iint_S \frac{dS_\xi}{\sqrt{a^2-\xi_1^2-\xi_2^2}\,|\mathbf{x}-\xi|} = \pi^2,$$

$$\iint_S \frac{(x_1-\xi_1)^k(x_2-\xi_2)^j}{\sqrt{a^2-\xi_1^2-\xi_2^2}\,|\mathbf{x}-\xi|^3}dS_\xi = \begin{cases} \pi^2/2, & \text{when } k=2,\ j=0 \text{ or } k=0,\ j=2, \\ 0, & \text{when } k=1,\ j=1. \end{cases} \qquad (12)$$

Next we perform the following transformation of the variables:

$$\begin{cases} x_1 = a\sin y_1\cos y_2, \\ x_2 = a\sin y_1\sin y_2, \end{cases} \quad \begin{cases} \xi_1 = a\sin\eta_1\cos\eta_2, \\ \xi_2 = a\sin\eta_1\sin\eta_2, \end{cases} \qquad (13)$$

where $\mathbf{y}=(y_1,y_2)$ and $\eta=(\eta_1,\eta_2)$ are new variables in the rectangular domain $\tilde{S}:\{0\le y_1,\eta_1\le\pi/2;0\le y_2,\eta_2\le 2\pi\}$. Equation (13) transforms the circular integration domain to a rectangular integration domain and eliminates the "square-root" singularity at the front of the inclusion corresponding to $\eta_1=\pi/2$.

By applying Eqs. (11)-(13) to the BIEs (9) we obtain their regularized version as

$$A\tilde{f}_3(\mathbf{y})\left[\pi^2-\iint_{\tilde{S}_y^0}\frac{\sin\eta_1}{R(\mathbf{y},\eta)}dS_\eta\right]+\frac{1}{\vartheta^2}\iint_{\tilde{S}_y^0}\tilde{f}_3(\eta)\tilde{R}_3(\mathbf{y},\eta)\sin\eta_1 dS_\eta = -4\pi G\tilde{u}_3^{in}(\mathbf{y}),\ \mathbf{y}(y_1,y_2)\in\tilde{S},$$

$$\tilde{f}_j(\mathbf{y})\left\{A\left[\pi^2-\iint_{\tilde{S}_y^0}\frac{\sin\eta_1}{R(\mathbf{y},\eta)}dS_\eta\right]+B\left[\frac{\pi^2}{2}-\iint_{\tilde{S}_y^0}\Phi_j(\mathbf{y},\eta)\sin\eta_1 dS_\eta\right]\right\}- \qquad (14)$$

$$-B\tilde{f}_{3-j}(\mathbf{y})\iint_{\tilde{S}_y^0}\Psi(\mathbf{y},\eta)\sin\eta_1 dS_\eta + \frac{1}{\vartheta^2}\iint_{\tilde{S}_y^0}\left[\tilde{f}_j(\eta)\tilde{R}_j(\mathbf{y},\eta)+\right.$$

$$+\tilde{f}_{3-j}(\eta)\tilde{R}_{j(3-j)}(\mathbf{y},\eta)\Big]\sin\eta_1 dS_\eta = -4\pi G\,\tilde{u}_j^{in}(\mathbf{y}),\quad j=1,2,\quad \mathbf{y}(y_1,y_2)\in\tilde{S}\,,$$

where

$$\tilde{f}_j(\mathbf{y})=f_j(\mathbf{x}),\;\; \tilde{u}_j^{in}(\mathbf{y})=u_j^{in}(\mathbf{x}),\;\; \tilde{R}_j(\mathbf{y},\eta)=a^3 R_j(\mathbf{x},\xi),\quad j=1,2,3,$$

$$\tilde{R}_{kj}(\mathbf{y},\eta)=a^3 R_{kj}(\mathbf{x},\xi),\;\; k,j=1,2,\;\; k\neq j. \tag{15}$$

In Eq. (15), $\mathbf{x}$ and ξ are defined by Eq. (13), $\vartheta=\chi_T a$ is the normalized wave number of the T-wave, $\tilde{S}_y^0$ is the mapping of the domain S_x^0 due to the transformation (13) (in the domain $\tilde{S}_y^0$ the points $\mathbf{y}$ and η do not coincide), and

$$R(\mathbf{y},\eta)=\Big[\sin^2 y_1+\sin^2\eta_1-2\sin y_1\sin\eta_1\cos(y_2-\eta_2)\Big]^{1/2},$$

$$\Phi_1(\mathbf{y},\eta)=\big(\sin y_1\cos y_2-\sin\eta_1\cos\eta_2\big)^2\big[R(\mathbf{y},\eta)\big]^{-3}, \tag{16}$$

$$\Phi_2(\mathbf{y},\eta)=\big(\sin y_1\sin y_2-\sin\eta_1\sin\eta_2\big)^2\big[R(\mathbf{y},\eta)\big]^{-3},$$

$$\Psi(\mathbf{y},\eta)=\big(\sin y_1\cos y_2-\sin\eta_1\cos\eta_2\big)\big(\sin y_1\sin y_2-\sin\eta_1\sin\eta_2\big)\big[R(\mathbf{y},\eta)\big]^{-3}$$

For the discretization of the domain $\tilde{S}$, a boundary element mesh with equal-sized rectangular elements is used. For simplicity, constant elements are adopted in this analysis. By collocating the BIEs (14) at discrete points coinciding with the centroids of each element, a system of linear algebraic equations for discrete values of $\tilde{f}_j$ is obtained. After solving the system of linear algebraic equations numerically, the stress jumps $\Delta\sigma_j$ across the inclusion can be obtained by the relations (10) and (15).

The far-field quantities of the scattered elastic waves can be computed from the stress jumps $\Delta\sigma_j$. For this purpose we use the asymptotic relations for an observation point far away from the inclusion, namely $|\mathbf{x}-\xi|\approx|\mathbf{x}|-(\mathbf{x}\cdot\xi)/|\mathbf{x}|$ and $|\mathbf{x}-\xi|^{-1}\approx|\mathbf{x}|^{-1}$, when $|\mathbf{x}|\to\infty$. By substituting of these relations into the integral representation formula (4) and introducing the spherical coordinate system with the origin at the center of the inclusion as

$$x_1=R\sin\theta\cos\varphi,\;\; x_2=R\sin\theta\sin\varphi,\;\; x_3=R\cos\theta,\; 0\leq\theta,\varphi\leq 2\pi, \tag{17}$$

the asymptotic expressions for the scattered radial u_R and tangential u_θ,u_φ displacements in the far-field are obtained in the form

$$u_R(R,\theta,\varphi)=\frac{\exp(i\chi_L R)}{4\pi R}F_L(\theta,\varphi,\gamma_0),\; R\to\infty\,,$$

$$u_\theta(R,\theta,\varphi)=\frac{\exp(i\chi_T R)}{4\pi R}F_{TV}(\theta,\varphi,\gamma_0),\; R\to\infty\,, \tag{18}$$

$$u_\varphi\left(R,\theta,\varphi\right) = \frac{\exp\left(i\chi_T R\right)}{4\pi R}F_{TH}\left(\theta,\varphi,\gamma_0\right), \quad R \to \infty$$

Here, F_L, F_{TV}, and F_{TH} are the longitudinal, vertically polarized transverse, and horizontally polarized transverse wave scattering amplitudes, respectively, which are related to the inclusion of normalized mass $\gamma_0 = M/\left(\rho a^3\right)$. They are given by

$$F_L\left(\theta,\varphi,\gamma_0\right) = \frac{1-2\nu}{2(1-\nu)G}\sum_{j=1}^{3}p_j\iint_S \Delta\sigma_j\left(\xi\right)\exp\left[-i\chi_L\left(\mathbf{p}\cdot\xi\right)\right]dS_\xi \ ,$$

$$F_{TV}\left(\theta,\varphi,\gamma_0\right) = \frac{1}{G}\sum_{j=1}^{3}r_j\iint_S \Delta\sigma_j\left(\xi\right)\exp\left[-i\chi_T\left(\mathbf{p}\cdot\xi\right)\right]dS_\xi \ , \qquad (19)$$

$$F_{TH}\left(\theta,\varphi,\gamma_0\right) = \frac{1}{G}\sum_{j=1}^{3}\tau_j\iint_S \Delta\sigma_j\left(\xi\right)\exp\left[-i\chi_T\left(\mathbf{p}\cdot\xi\right)\right]dS_\xi \ ,$$

where p_j, r_j, τ_j $(j=1,2,3)$ are the coordinates of the spherical unit vectors $\mathbf{p} = (\sin\theta\cos\varphi, \sin\theta\sin\varphi, \cos\theta)$, $\mathbf{r} = (\cos\theta\cos\varphi, \cos\theta\sin\varphi, -\sin\theta)$ and $\tau = (-\sin\varphi, \cos\varphi, 0)$.

The forward scattering amplitudes are defined as the values of $F_Z\left(\theta,\varphi,\gamma_0\right)$ $(Z = L, TV, TH)$ in the direction of the wave incidence, i.e., $F_Z\left(\theta_0, 0, \gamma_0\right)$.

Thus, the scattering problem in the far-field is reduced to the numerical solution of the BIEs (14) and the subsequent computation of the scattering amplitudes by using Eq. (19), where the transformation or mapping relations (13) have to be considered.

For the convenient description of the wave parameters in the inclusion vicinity let us introduce the local coordinate system $Otqz$ with the center in the inclusion contour point, so that the value $z = 0$ corresponds to the inclusion plane, the axes Ot and Oq lie in the normal and tangential directions relative to the inclusion contour line, respectively, as depicted in Figure 1. Then the corresponding displacement and stress components at the arbitrary point P near the inclusion in the plane $q = 0$ can be approximated as [21]:

$$u_z(r,\psi,\phi) = \frac{\sqrt{2r}}{2G}\sin\frac{\psi}{2}\left(1-\sin^2\frac{\psi}{2}\right)K_I(\phi) + \frac{\sqrt{2r}}{2G}\cos\frac{\psi}{2}\left(2-4\nu+\cos^2\frac{\psi}{2}\right)K_{II}(\phi) + O(r^{3/2}) \ ,$$

$$u_t(r,\psi,\phi) = -\frac{\sqrt{2r}}{2G}\cos\frac{\psi}{2}\left(4-4\nu-\cos^2\frac{\psi}{2}\right)K_I(\phi) - \frac{\sqrt{2r}}{2G}\sin\frac{\psi}{2}\left(1-\sin^2\frac{\psi}{2}\right)K_{II}(\phi) + O(r^{3/2}) \ ,$$

$$u_q(r,\psi,\phi) = \frac{\sqrt{2r}}{2G}\cos\frac{\psi}{2}K_{III}(\phi) + O(r^{3/2}) \ ,$$

$$\sigma_{zz}(r,\psi,\phi) = -\frac{1}{\sqrt{2r}}\cos\frac{\psi}{2}\left(1-2\nu-\sin\frac{\psi}{2}\sin\frac{3\psi}{2}\right)K_I(\phi) +$$

$$+\frac{1}{\sqrt{2r}}\sin\frac{\psi}{2}\left(2-2v-\cos\frac{\psi}{2}\cos\frac{3\psi}{2}\right)K_{II}(\phi)+O(1),$$

$$\sigma_{tt}(r,\psi,\phi)=\frac{1}{\sqrt{2r}}\cos\frac{\psi}{2}\left(3-2v-\sin\frac{\psi}{2}\sin\frac{3\psi}{2}\right)K_{I}(\phi)+\frac{1}{\sqrt{2r}}\sin\frac{\psi}{2}\left(2v+\cos\frac{\psi}{2}\cos\frac{3\psi}{2}\right)K_{II}(\phi)+O(1),$$

$$\sigma_{qq}(r,\psi,\phi)=\frac{2v}{\sqrt{2r}}\cos\frac{\psi}{2}K_{I}(\phi)+\frac{2v}{\sqrt{2r}}\sin\frac{\psi}{2}K_{II}(\phi)+O(1), \qquad (20)$$

$$\sigma_{zq}(r,\psi,\phi)=\frac{1}{\sqrt{2r}}\sin\frac{\psi}{2}K_{III}(\phi)+O(1),$$

$$\sigma_{tq}(r,\psi,\phi)=-\frac{1}{\sqrt{2r}}\cos\frac{\psi}{2}K_{III}(\phi)+O(1),$$

$$\sigma_{zt}(r,\psi,\phi)=\frac{1}{\sqrt{2r}}\sin\frac{\psi}{2}\left(2-2v+\cos\frac{\psi}{2}\cos\frac{3\psi}{2}\right)K_{I}(\phi)+$$

$$+\frac{1}{\sqrt{2r}}\cos\frac{\psi}{2}\left(1-2v+\sin\frac{\psi}{2}\sin\frac{3\psi}{2}\right)K_{II}(\phi)+O(1)$$

Here r and ψ are the polar coordinates of the point P, ϕ is the angular coordinate of the inclusion contour point (see Figure 1), K_I, K_{II} and K_{III} are the mode-I, II, and III dynamic stress intensity factors in the inclusion vicinity.

By using the Eq. (20) the $K_{I,II,III}$-factors can be defined directly from the stress jumps $\Delta\sigma_j$ or the solutions of BIEs (7) by the following relations:

$$K_{I}(\phi,\gamma_0)=-\frac{1}{4(1-v)\sqrt{a}}\left\{\sqrt{a^2-x_1^2-x_2^2}\left[\Delta\sigma_1(x)\cos\phi+\Delta\sigma_2(x)\sin\phi\right]\right\}\Big|_{\substack{x_1=a\cos\phi;\\x_2=a\sin\phi}},$$

$$K_{II}(\phi,\gamma_0)=-\frac{1}{4(1-v)\sqrt{a}}\left[\sqrt{a^2-x_1^2-x_2^2}\,\Delta\sigma_3(x)\right]\Big|_{\substack{x_1=a\cos\phi;\\x_2=a\sin\phi}}, \qquad (21)$$

$$K_{III}(\phi,\gamma_0)=\frac{1}{4\sqrt{a}}\left\{\sqrt{a^2-x_1^2-x_2^2}\left[\Delta\sigma_1(x)\sin\phi-\Delta\sigma_2(x)\cos\phi\right]\right\}\Big|_{\substack{x_1=a\cos\phi;\\x_2=a\sin\phi}},$$

where the dependence of $K_{I,II,III}$-factors on the inclusion mass also is fixed by the variable γ_0.

3. Dispersion relations for distributed inclusions of variable mass

We consider now a statistical distribution of rigid disk-shaped micro-inclusions in the matrix. The location of the micro-inclusions is assumed to be random, while their orientation is either completely random or aligned, see Figure 2. In the case of aligned inclusions, it is postulated that the inclusions are parallel to the $x_1 - x_2$ -plane. The radius a of the inclusions is assumed to be equal, while their masses can be variable due to the different material properties of inclusions and their geometric aspect ratios.

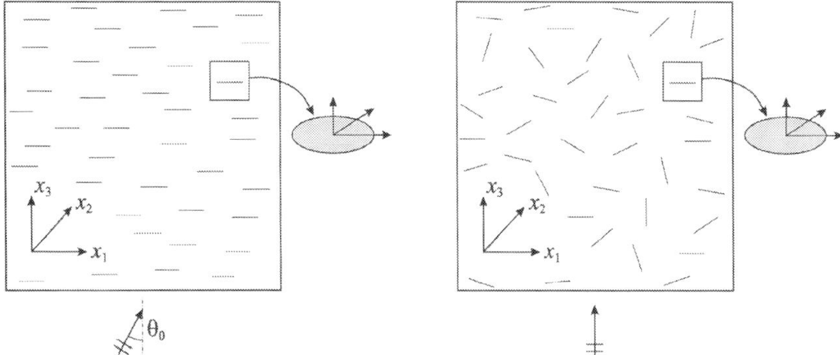

Figure 2. Multiple disk-shaped inclusions in the elastic wave field: parallel inclusions with the angle of wave incidence (left); randomly oriented inclusions (right).

The average response of the composite materials to the wave propagation is characterized by the geometrical dispersion and attenuation of waves due to the wave scattering process. To describe these phenomena within the coherent wave field, the dynamic properties of the composite can be modeled by a complex and frequency-dependent wave number $K(\omega)$ as

$$K(\omega) = \frac{\omega}{c^*(\omega)} + i\alpha(\omega), \tag{22}$$

where $c^*(\omega)$ is the effective phase velocity and $\alpha(\omega)$ is the attenuation coefficient for the wave of corresponding mode. With Eq. (22), the amplitude of a plane time-harmonic elastic wave propagating in the **n**-direction can be expressed as

$$\mathbf{u}(\mathbf{x},\omega) = \mathbf{U}\exp\left[iK(\mathbf{n}\cdot\mathbf{x})\right] = \mathbf{U}\exp\left[-\alpha(\omega)(\mathbf{n}\cdot\mathbf{x})\right]\exp\left[i\omega(\mathbf{n}\cdot\mathbf{x})/c^*(\omega)\right] \tag{23}$$

For low concentration of inclusions or small number density, the interaction or multiple scattering effects among the inclusions can be neglected. Under these assumptions the complex effective wave numbers K_Z $(Z = L, T)$ of plane L- and T-waves can be calculated by using the Foldy-type dispersion relation, which was extended to elastic waves in [3] and may be stated as

$$K_Z^2 = \chi_Z^2 + \varepsilon a^{-3} F_Z^*, \quad Z = L, T \tag{24}$$

In Eq. (24), ε is the density parameter of the inclusions, εa^{-3} corresponds to the number density of inclusions of the same radius, i.e. the number of inclusions per unit volume, F_Z^* is the average forward scattering amplitude of the corresponding wave mode by a single inclusion. For randomly oriented inclusions of variable mass the averages should be taken both over all possible inclusion orientations and masses. It should be noted here that the average over all inclusion orientations is the same as the average over all directions of the wave incidence (to avoid the additional average over all wave polarizations for an incoming *T*-wave, we assume that the normal to the inclusions lie in the plane of incidence of the incoming *TV*- or *TH*-wave). Hence, in the case of parallel inclusions the expressions for the average forward scattering amplitudes of corresponding wave mode are

$$F_Z^*(\theta_0) = \int_\tau^\beta F_Z(\theta_0,0,\gamma)g(\gamma)d\gamma, \qquad Z = L,TV,TH, \tag{25}$$

and in the case of randomly oriented inclusion they become the form

$$F_Z^* = \frac{1}{2}\int_0^\pi \int_\tau^\beta F_Z(\theta_0,0,\gamma)g(\gamma)\sin\theta_0 d\gamma d\theta_0, \qquad Z = L,TV,TH. \tag{26}$$

Here θ_0 is the angle characterizing the direction of the wave incidence, the parameters τ and β characterize the minimal and maximal masses, respectively, in the system of distributed inclusions with the density function g of inclusion mass, and $F_Z(\theta_0,0,\gamma)$ are the forward scattering amplitudes given by Eq. (19). The density distribution function of inclusion mass should satisfy the normalization condition

$$\int_\tau^\beta g(\gamma)d\gamma = 1 \tag{27}$$

A suitable set of inclusion mass variations, which corresponds to aligned, normal and uniform distributions, is defined in Table 1.

Distribution type	$g(\gamma)$
Aligned	$\delta(\gamma - \gamma_0)$
Normal	$3\gamma^2/(\beta^3 - \tau^3)$
Uniform	$1/(\beta - \tau)$

Table 1. Various types of density distribution functions of inclusion mass (δ is the Dirac delta function).

The approximation for the complex wave number (24) can be considered as a special case of the multiple wave scattering models of higher orders [4,5], and it involves only the first order in the inclusion density and is thus only valid for a dilute or small inclusion density. In the case of a large density or high concentration of inclusions, more sophisticated models such as the self-consistent approach or the multiple scattering models should be applied, to take the mutual dynamic interactions between individual inclusions into account.

Once the complex effective wave numbers K_Z have been determined via Eq. (24), the effective wave velocities c_Z^* and the attenuation coefficients α_Z of the plane L- and T-waves can be obtained by considering the definition (22). This results in

$$c_z^*(\omega) = \frac{\omega}{\mathrm{Re}\left[K_z(\omega)\right]}, \quad \alpha_z(\omega) = \mathrm{Im}\left[K_z(\omega)\right], \quad Z = L, TV, TH \tag{28}$$

It should be remarked here that Foldy's theory was derived for isotropic wave scattering, which is appropriate macroscopically for the configuration of randomly oriented inclusions. A composite solid with aligned (parallel) disk-shaped inclusions exhibits a macroscopic anisotropy, namely a transversal anisotropy. When an incident plane wave propagates in an arbitrary direction, this gives rise to a coupling between the L- and T-waves, and thus a change in the effective polarization vector. However, it is reasonable to apply Foldy's theory, when the wave propagation is along the principal axes because of the decoupling of the L- and T-waves. In this special case, wave propagation can be treated like in the isotropic case.

4. Numerical analysis of global dynamic parameters of a composite

The method presented in the previous sections is used to calculate the effective dynamic parameters of a composite elastic solid with both parallel and randomly oriented rigid disk-shaped inclusions of variable mass for the propagation of time-harmonic plane L- or TV-waves. For numerical discretization of the inclusion surface, the domain $\tilde{S}$ is divided into 264 rectangular elements of length $\Delta y_1 = \pi/22$ and $\Delta y_2 = \pi/12$ in the y_1- and y_2-directions, the domain $\tilde{S}_y^0$ is chosen as $\tilde{S}_y^0 = \tilde{S} \setminus \Delta\tilde{S}_y$, where $\Delta\tilde{S}_y$ is the boundary element with the point y in the center of the element. Poisson's ratio is selected as 0.3, the radii of inertia of the circular inclusion are defined as $i_1 = i_2 = a/2, i_3 = a/\sqrt{2}$. In the numerical examples, the radius a of the inclusions is assumed to be equal, while their masses are varied by the distribution laws defined in Table 1 from the minimal value $\tau = 5$ to the maximal value $\beta = 20$. The inclusion density parameter is fixed as $\varepsilon = 0.01$ in accordance to the assumption of low concentration of inclusions in a matrix.

For comparison purpose, normalized effective wave velocities and normalized attenuation coefficients are introduced as $\bar{c}_Z = c_Z^*/c_Z$, $\bar{\alpha}_Z = 2a \cdot \alpha_Z/(\pi\varepsilon)$, where the subscript $Z = L, TV$ stands for L- and TV-waves, respectively.

For parallel or aligned disk-shaped inclusions, the macroscopic dynamic behavior of the composite materials is transversely isotropic. Thus, the effective wave velocities and the attenuation coefficient are dependent on the direction of the wave incidence. In this analysis, only two wave incidence directions are considered, namely normal incidence $(\theta_0 = 0°)$ of L- and T-waves and grazing incidence $(\theta_0 = 90°)$ of L- wave. This choice provides the vanishing of the dynamic torque on the inclusions and, therefore, their zero-rotations.

For normal incidence of a plane L-wave, the normalized attenuation coefficient $\bar{\alpha}_L$ and the normalized effective wave velocity $\bar{c}_L$ versus the dimensionless wave number ϑ are presented in Figure 3. Three separate parametrical studies are performed to show the effects of different inclusion mass distributions: aligned with $\gamma_0 = \beta = 20$, normal and uniform. To check the accuracy of the implemented BEM, the present numerical results are compared with the analytical low-frequency solutions given in [9]. Here and hereafter a very good agreement between both results is observed in the frequency range $0 \le \vartheta \le 0.3$.

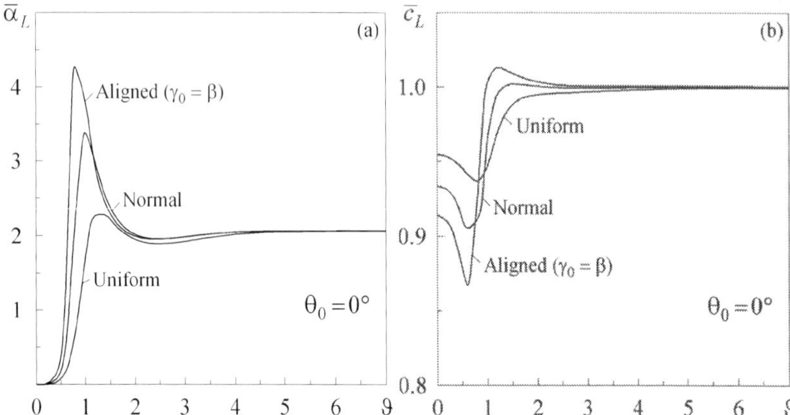

Figure 3. Effects of the inclusion mass distribution on the normalized attenuation coefficient (a) and the normalized effective wave velocity (b) as the functions of dimensionless wave number for parallel inclusions and normal L-wave incidence.

In the low frequency range, the normalized attenuation coefficient $\bar{\alpha}_L$ increases rapidly with increasing ϑ, after reaching a maximum it then decreases and approaches its high frequency limit (see Figure 3(a)). The peak value of $\bar{\alpha}_L$ increases and is shifted to a smaller value of the dimensionless frequency with changing of inclusion mass distribution from uniform to normal and then to aligned. The normalized attenuation coefficient $\bar{\alpha}_L$ in the high-frequency or short-wave limit does not depend on the frequency and the inclusion mass. Also, the inclusions are unmovable in the high-frequency limit and the normalized attenuation coefficient $\bar{\alpha}_L$ can be obtained by using the Kirchhoff approximation for short waves [19]. At low frequencies, the normalized effective wave velocity $\bar{c}_L$ in the composite is smaller than that in the homogeneous matrix material (see Figure 3(b)). Then the normal inclusion mass distribution is characterized by the bigger (smaller) value of $\bar{c}_L$ in comparison with the aligned (uniform) situation. An opposite tendency is observed in the range of higher frequencies. In addition, the normalized effective wave velocity $\bar{c}_L$ in the composite can be bigger than that in the homogeneous matrix material, for instance for the considered inclusions of aligned mass and $\vartheta > 1$. The high-frequency limit $\bar{c}_L \to 1$ at $\vartheta \to \infty$ means that the velocity of the L-wave in the short-wave limit coincides with that in the matrix. The explanation of this high-frequency limit follows from the geometrical optical interpretation of the wave field. The wave field at high frequencies may be considered as a set of independent beams propagating through the medium. Because of the existing continuous matrix material, the effective wave velocity should coincide with the wave velocity in the matrix in the high-frequency limit.

The corresponding numerical results for grazing incidence of a plane L-wave are presented in Figure 4. As followed from Figure 4(a), the normalized attenuation coefficient $\bar{\alpha}_L$ shows a similar dependence on the dimensionless wave number ϑ and the inclusion mass distribution. For the same density function g of inclusion mass, the peaks of $\bar{\alpha}_L$ are larger than that for normal incidence as depicted in Figure 3(a), but the high-frequency limit of $\bar{\alpha}_L$

in the case with $\theta_0 = 90°$ is smaller than that in the case with $\theta_0 = 0°$. The normalized effective wave velocity $\bar{c}_L$ in Figure 4(b) for grazing incidence of a plane L-wave is also very similar to Figure 3(b). Compared to the normalized effective wave velocity $\bar{c}_L$ for a normal incidence of plane L-wave, the maximum effective wave velocity $\bar{c}_L$ is increased while its minimum value is decreased. Also in contrast to Figure 3(b), the normalized effective wave velocity $\bar{c}_L$ for all considered functions g is larger than that of the matrix material at high frequencies. In general, the quite tangled effects of the inclusions' stiffness and mass on the average phase velocity are observed at different frequencies.

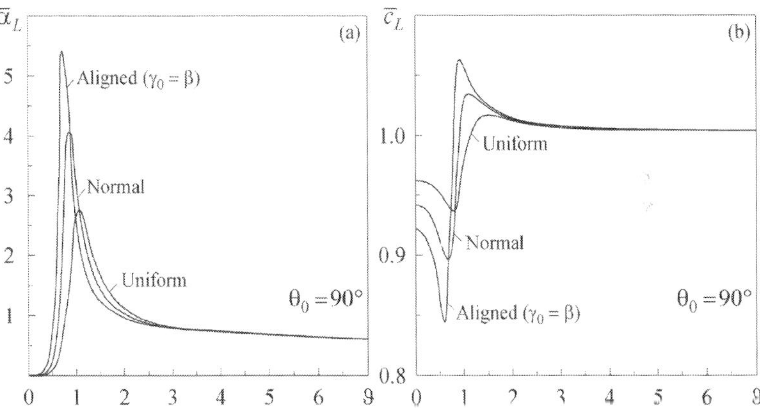

Figure 4. Effects of the inclusion mass distribution on the normalized attenuation coefficient (a) and the normalized effective wave velocity (b) as the functions of dimensionless wave number for parallel inclusions and grazing L-wave incidence.

For normal incidence of a plane T-wave (then TV- and TH-wave are the same), the numerical results for the normalized attenuation coefficient $\bar{\alpha}_T$ and the normalized effective wave velocity $\bar{c}_T$ are presented in Figure 5 versus the dimensionless wave number ϑ. The global behavior of the normalized attenuation coefficient $\bar{\alpha}_T$ and the normalized effective wave velocity $\bar{c}_T$ is very similar to that for grazing incidence (i.e. $\theta_0 = 90°$) of plane L-wave as presented in Figure 4. In comparison to the peak values of $\bar{\alpha}_L$ for normal and grazing incidence of a plane L-wave, the peak values of $\bar{\alpha}_T$ for an normal incidence of a plane T-wave are increased, what follows from Figures 3(a), 4(a) and 5(a). Comparison of Figures 3(b), 4(b) and 5(b) shows, that the minimum values of the normalized effective wave velocity $\bar{c}_T$ are reduced compared to $\bar{c}_L$ for normal and grazing incidence of a plane L-wave for the same inclusion mass distribution, while the maximum values of $\bar{c}_T$ lie between the $\bar{c}_L$ -values for normal and grazing incidence of a plane L-wave.

Next numerical examples concern the randomly oriented micro-inclusions, when the macroscopic dynamic behavior of the composite material is isotropic. It means that the effective wave velocity and the attenuation coefficient do not depend on the direction of the wave incidence. Both the translations and the rotations of the inclusions are exhibited in this case.

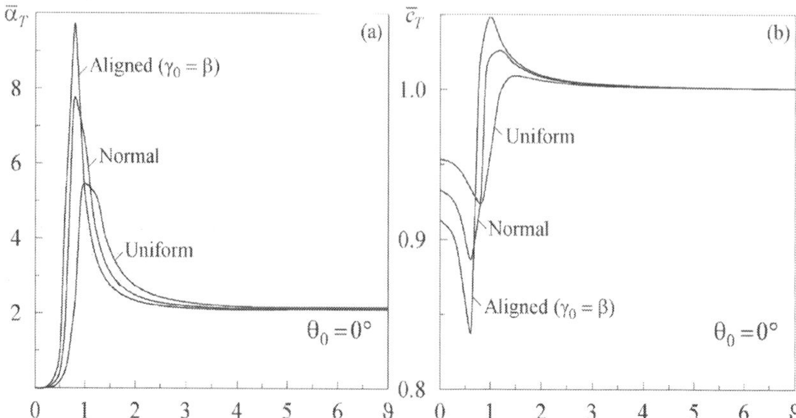

Figure 5. Effects of the inclusion mass distribution on the normalized attenuation coefficient (a) and the normalized effective wave velocity (b) as the functions of dimensionless wave number for parallel inclusions and normal T-wave incidence.

For an incident plane L-wave, the normalized attenuation coefficient $\bar{\alpha}_L$ and the normalized effective wave velocity $\bar{c}_L$ are presented in Figure 6 versus the dimensionless wave number ϑ. A comparison of Figure 6 with Figures 3 and 4 for parallel inclusions shows a similar dependence of the $\bar{\alpha}_L$ and $\bar{c}_L$ on the dimensionless wave number ϑ and the inclusion mass distribution. As expected, the peak values of $\bar{\alpha}_L$ and $\bar{c}_L$ for randomly oriented penny-shaped inclusions are larger than that for parallel inclusions with $\theta_0 = 0°$ but smaller than that for parallel inclusions with $\theta_0 = 90°$.

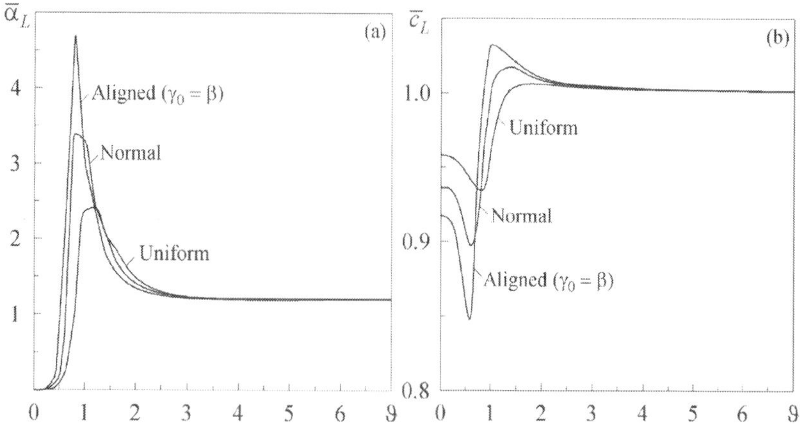

Figure 6. Effects of the inclusion mass distribution on the normalized attenuation coefficient (a) and the normalized effective wave velocity (b) as the functions of dimensionless wave number for randomly oriented inclusions and L-wave incidence.

Figure 7 demonstrates the corresponding results for the normalized attenuation coefficient $\bar{\alpha}_T$ and the normalized effective wave velocity $\bar{c}_T$ for an incident plane TV-wave. In contrast to parallel inclusions (see Figure 5), now the variations of $\bar{\alpha}_T$ and $\bar{c}_T$ with the dimensionless wave number ϑ are rather complicated at low frequencies. For instance, the normalized attenuation coefficient $\bar{\alpha}_T$ for aligned distribution of inclusion mass in Figure 7(a) shows two distinct peaks, which are not observed in the case of parallel inclusions. The normalized attenuation coefficient $\bar{\alpha}_T$ for randomly oriented inclusions is smaller than that for parallel inclusions with $\theta_0 = 0°$, while the normalized effective wave velocity $\bar{c}_T$ could be larger or smaller than that for parallel inclusions depending on the dimensionless wave number ϑ and the density function g of inclusion mass (see Figure 7(b)).

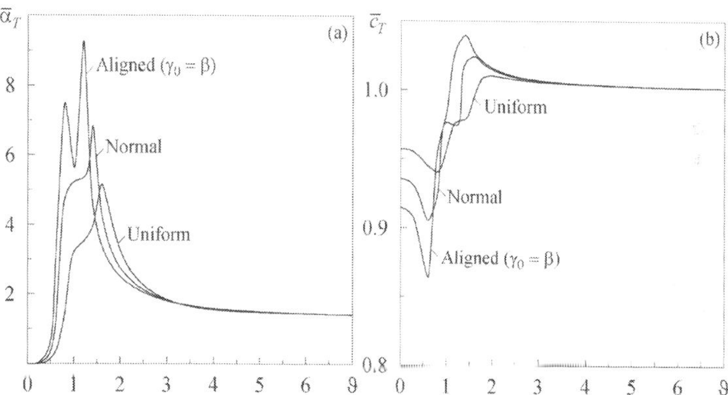

Figure 7. Effects of the inclusion mass distribution on the normalized attenuation coefficient (a) and the normalized effective wave velocity (b) as the functions of dimensionless wave number for randomly oriented inclusions and TV-wave incidence.

5. Numerical analysis of local dynamic parameters of a composite

Description of macroscopic dynamic response of a composite to elastic wave propagation by Eqs. (23) and (28) allows us the extension of analysis on the near-field quantities connected with each inclusion. Special attention should be paid to the dynamic stress intensity factors as the most important fracture parameters. Taking in mind the assumptions of neglecting the inclusions interaction, the relations (21) are applied for the estimation of the mode-I, II, and III dynamic stress intensity factors K_I, K_{II} and K_{III} in the vicinity of separate inclusion. We suppose the impinge on the inclusion of plane longitudinal L-wave with constant amplitude U_0, which should be corrected for each inclusion in accordance to its localization in a composite with the considering of attenuation and dispersion law (23). The value $K_* = U_0 G / \left[4(1-v)\sqrt{a} \right]$ is chosen as the normalizing factor for the amplitudes of dynamic stress intensity factors, so that $\bar{K}_{I,II,III} = \left| K_{I,II,III} \right| / K_*$. All material and discretization parameters are the same as it is fixed in the previous Section. Different inclusion masses are involved into analysis by the changing of parameter γ_0.

At normal L-wave incidence on the inclusion (antisymmetric problem) $K_I = K_{III} = 0$, and the mode-II dynamic stress intensity factor K_{II} does not vary along the inclusion contour (see

Figure 8). It follows from the Figure 8 that in the initial range of wave numbers $\bar{K}_{II}$-factor rapidly increases from a zero value, what is more pronounced for the inclusions of large mass. A further increase in ϑ in the case of an inclusion with the mass characteristic $\gamma_0 = 20$ leads to a local maximum of $\bar{K}_{II}$. For higher wave numbers, a regularity is the approaching of $\bar{K}_{II}$-factors for the inclusions of different mass, in addition, this approaching is from above as the mass of inclusion increases. Subsequently, a linear relationship between $\bar{K}_{II}$ and ϑ is reached in accordance to the increasing order of stresses in an incident wave.

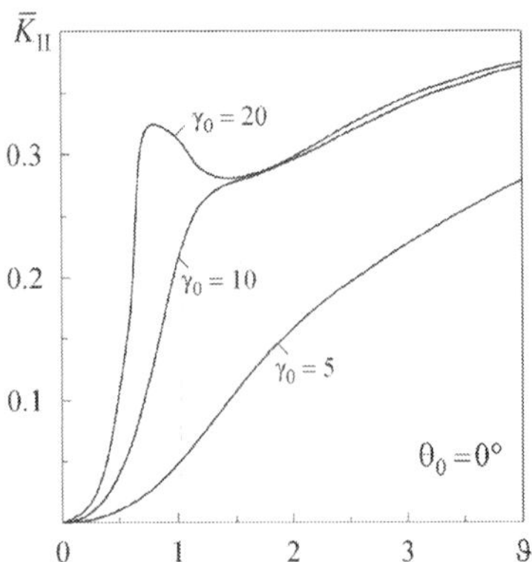

Figure 8. Effects of the inclusion mass on the normalized amplitude of the mode-II dynamic stress intensity factor as the function of dimensionless wave number for normal L-wave incidence on an inclusion.

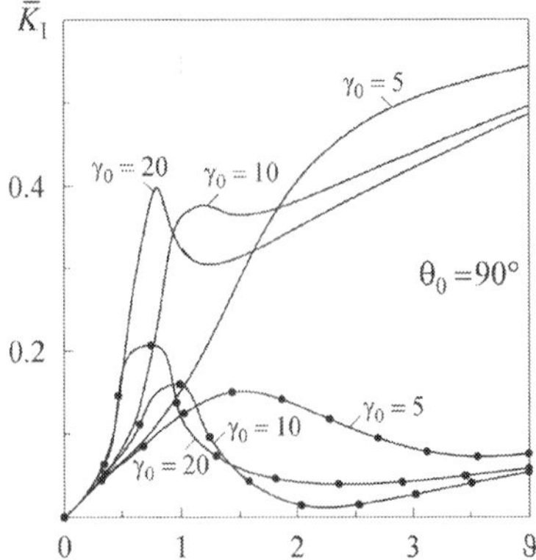

Figure 9. Effects of the inclusion mass on the normalized amplitude of the mode-I dynamic stress intensity factor as the function of dimensionless wave number for grazing L-wave incidence on an inclusion: solid lines correspond to the contour point where the wave runs on the inclusion; marked lines correspond to the contour point where the wave runs down from the inclusion.

At grazing L-wave incidence on the inclusion (symmetric problem) $K_{II} = 0$, and the mode-I and mode-III dynamic stress intensity factor K_I and K_{III} depend on the angular coordinate ϕ of the inclusion contour point. Hence, in Figure 9 $\bar{K}_I$-factor at the most representative points of the inclusion contour is showed, where $K_{III} = 0$ due to the symmetry conditions. A similar frequency behaviour (with a difference in the quantitative values) as in the previous antisymmetric problem is observed here for $\bar{K}_1$-factor at the point, where the wave runs on the inclusion. It should be mentioned that $\bar{K}_1$-factor at this point exceed one at the point, where the wave runs down from the inclusion. In the range of high wave numbers $\bar{K}_1$-factors for the inclusions of different mass approach also, but now from below as the mass of inclusion increases.

6. Conclusion

Attenuation and dispersion of time-harmonic elastic waves, as well as dynamic stress concentration in 3D composite materials consisting of a linear elastic matrix and rigid disk-shaped inclusions of variable mass is simulated numerically. Translations and rotations of the inclusions in the matrix are taken into account in the analysis. Wave scattering by a single disk-shaped inclusion is investigated by a boundary element method to obtain the stress jumps across the inclusion surfaces. Then, far-field scattering amplitudes of elastic waves are computed by using the stress jumps. To describe the average macroscopic dynamic properties of the composite materials with a random distribution of disk-shaped micro-inclusions, complex wave numbers are computed by the Foldy-type dispersion relations, from which the effective wave velocities and the wave attenuation can be obtained. The present analysis concerns a dilute distribution of micro-inclusions, when the mutual inclusion interactions and the multiple scattering effects are approximately neglected. Numerical examples involve:

- both longitudinal and transversal waves propagation in a composite material;
- parallel and randomly oriented rigid disk-shaped inclusions;
- aligned, normal and uniform distributions of inclusion mass;
- frequency-domain analysis of global dynamic parameters, such as the wave attenuation coefficients and effective wave velocities;
- frequency-domain analysis of local dynamic parameters, such as the dynamic stress intensity factors in the inclusion vicinities

As shown, particular dynamic properties of composite materials can be varied by controlled changes in the microstructure.

Author details

Viktor Mykhas'kiv
Department of Computational Mechanics of Deformable Systems,
Pidstryhach Institute for Applied Problems of Mechanics and Mathematics NASU, Lviv, Ukraine

Acknowledgement

This work is sponsored by the State Foundation for Fundamental Researches of Ukraine (Project No. 40.1/018), which is gratefully acknowledged.

7. References

[1] Foldy LL (1945) Multiple Scattering Theory of Waves. *Physical Review* 67: 107-119.

[2] Lax M (1952) Multiple Scattering of Waves. *Physical Review* 85: 621-629.

[3] Gubernatis JE, Domany E (1984) Effects of Microstructure on the Speed and Attenuation of Elastic Waves in Porous Materials. *Wave Motion* 6: 579-589.

[4] Martin PA (2006) *Multiple Scattering: Interaction of Time-harmonic Waves with N Obstacles.* Cambridge: Cambridge University Press 437 p.

[5] Conoir JM, Norris AN (2010) Effective Wavenumbers and Reflection Coefficients for an Elastic Medium Containing Random Configurations of Cylindrical Scatterers. *Wave Motion* 47: 183-197.

[6] Kerr FH (1992) The Scattering of a Plane Elastic Wave by Spherical Elastic Inclusions. *International Journal of Engineering Science* 30: 169-186.

[7] Kanaun SK, Levin VM (2007) Propagation of Longitudinal Elastic waves in Composites with a Random Set of Spherical Inclusions. *Archive of Applied Mechanics* 77: 627-651.

[8] Sabina FJ, Smyshlaev VP, Willis JR (1993) Self Consistent Analysis of Waves in a Matrix-Inclusion Composite I. Randomly Oriented Spheroidal Inclusions. *Journal of the Mechanics and Physics of Solids* 41: 1573-1588.

[9] Kanaun SK, Levin VM (2008) *Self-Consistent Methods for Composites. Volume 2 — Wave Propagation in Heterogeneous Materials.* Heidelberg: Springer 294 p.

[10] LevinV, Markov M, Kanaun S (2008) Propagation of Long Elastic Waves in Porous Rocks with Crack-Like Inclusions. *International Journal of Engineering Science* 46: 620-638.

[11] Eriksson AS, Boström A, Datta SK (1995) Ultrasonic Wave Propagation through a Cracked Solid. *Wave Motion* 22: 297-310.

[12] Zhang Ch, Gross D (1998) *On Wave Propagation in Elastic Solids with Cracks.* Southampton: Computational Mechanics Publications 248 p.

[13] Sato H, Shindo Y (2002) Influence of Microstructure on Scattering of Plane Elastic Waves by a Distribution of Partially Debonded Elliptical Inclusions. *Mechanics of Materials* 34: 401-409.

[14] Maurel A, Mercier JF, Lund F (2004) Elastic Wave Propagation through a Random Array of Dislocations. *Physical Review B* 70: 024303 (1-15).

[15] Mykhas'kiv VV, Khay OM, Zhang Ch, Boström A (2010) Effective Dynamic Properties of 3D Composite Materials Containing Rigid Penny-Shaped Inclusions. *Waves in Random and Complex Media* 20: 491-510.

[16] Kit HS, Mykhas'skiv VV, Khay OM (2002) Analysis of the Steady Oscillations of a Plane Absolutely Rigid Inclusion in a Three-Dimensional Elastic Body by the Boundary Element Method. *Journal of Applied Mathematics and Mechanics* 66: 817-824.

[17] Mykhas'kiv VV (2005) Transient Response of a Plane Rigid Inclusion to an Incident Wave in an Elastic Solid. *Wave Motion* 41: 133-144.

[18] Mykhas'kiv VV, Khay OM (2009) Interaction between Rigid-Disc Inclusion and Penny-Shaped Crack under Elastic Time-Harmonic Wave Incidence. *International Journal of Solids and Structures* 46: 602-616.

[19] Achenbach JD (1973) *Wave Propagation in Elastic Solids.* Amsterdam: North-Holland Publishing Company 425 p.

[20] Khaj MV (1993) *Two-Dimensional Integral Equations of Newton Potential Type and Their Applications.* Kyiv: Naukova Dumka 253 p. (in Russian).

[21] Kassir MK, Sih GC (1968) Some Three-Dimensional Inclusion Problems in Elasticity. *International Journal of Solids and Structures* 4: 225-241.

Chapter 3

Structural Health Monitoring for Composite Materials

Jian Cai, Lei Qiu, Shenfang Yuan, Lihua Shi, PeiPei Liu and Dong Liang

Additional information is available at the end of the chapter

http://dx.doi.org/10.5772/48215

1. Introduction

1.1. Composite materials

A composite material can be defined as a combination of two or more distinct materials at a macroscopic level to attain new properties that can't be achieved by those of individual components acting alone. Different from metallic alloys, each material keeps its own chemical, physical and mechanical properties [1]. Composite materials have reinforcing and matrix phases. The reinforcing phase with higher strength and stiffness is usually fibers, flakes or particles while the matrix phase can be polymers, ceramics or metals. Composite materials are commonly classified into four types, i.e., fibrous composite materials, laminated composite materials, particulate composite materials and the others [2].

Compared with traditional metallic materials, the main advantages of composites are: a) low destiny and high specific strength and stiffness, which are help for weight savings; b) good vibration damping ability, long fatigue life and high wear, creep, corrosion and temperature resistances; b) strong tailor ability in both microstructures and properties make them easily designed to satisfy different application needs; c) since detail accessories can be combined into a single cured assembly, the number of required fasteners and the amount of assembly labor can be significantly reduced [1].

The above advantages make composite materials wildly used in various fields. In aeronautic structures, composite materials are increasingly utilized to decrease weight for payload and radius purposes. The percentages by weight of composites in USA fighters rise from 2% in F-15E to 35.2% in F-35/CV. The overall structure of Eurofighter Typhoon is composed of 40% carbon-fiber composite materials. For commercial aircrafts, the usage percentages of fiber-reinforced composite materials in latest Boeing B787 and newly-designed Airbus A350-XWB reach 50% and 52%, respectively. To meet the performance and fuel efficiency

requirements, the consumption of composites in automobile industry is growing. The blades of wind turbines are normally made of composites to improve electrical energy harvest efficiency [1]. In ships or infrastructures, the composite materials with high corrosion resistance have received wild acceptance. The brake and engine parts working in high temperature are often fabricated from metal or ceramic matrix composites. In addition, the sports and recreation market is also one of the primary consumers of composites [3].

1.2. Problems of composite materials

Despite having the great advantages and applicability, composite materials are not exempt from some problems. As multiphase materials, composites exhibit distinct anisotropic properties. Their material capabilities, largely relating to manufacturing processes, are dispersive. Furthermore, the mechanisms of flaw initiating, spreading over the composite volume and leading to the ultimate failure are very complicated. So far, clearly description for the damage evolution and fracture behavior in composites remains a challenge work. Both the complex mechanical and damage characteristics can also make the optimization design for composites very difficult. Because of lacking enough data cumulation and available standards, the composite design efficiency usually depends on the designer's experiences and the final structures are easily prone to be over-designed [4].

Another important problem for composites is that they are susceptible to impact damages due to the lack of reinforcement in the out-of-plane direction. In a high energy impact, only small total penetration appears in composites. While in the low or medium energy impact, matrix crack will occur and interact, inducing delamination process. Fibre breakage would also happen on the opposite side to the impact [5]. Moreover, damages can be induced in composites by incorrect operations during manufacture and assembly, aging or service condition.

1.3. Requirements of SHM for composite materials

The common damages in composites are fibre breakage, matrix cracking, fibre-matrix debonding and delamination between plies, most of which occur beneath the top surfaces and are barely visible. They can severely degrade the performance of composites and should be identified in time to avoid catastrophic structural failures.

The conventional non-destructive testing (NDT) methods, such as ultrasonic, X-ray, thermography and eddy current methods can be adopted for detecting damages in composites. However, these NDT methods, merely allowing the off-line testing in a local manner with complicated and heavy equipments, are labor-extensive and time-consuming especially for large-scale structures. Meanwhile, disassembling the tested structures may be required to ensure the inspection area accessible, which can increase the maintenance costs [6-7].

Structural health monitoring (SHM), an emerging technique developed from NDT, combines advanced sensor technology with intelligent algorithms to interrogate the structural 'health' condition [8]. Different from NDT, the real-time and on-line damage detection via in-situ sensors can be achieved in SHM. Now, the needs of SHM for composites have been continuously increased. The potential benefits of SHM include improving reliability and safety, reducing lifecycle costs and helping design of composite materials.

In the following, after the sensors commonly applied in SHM are presented, some typical SHM methods for composites are reviewed. Hereafter arranged are the two SHM examples on composite structures. Summary and conclusions are given at last.

2. Sensors of SHM

In SHM, various sensors are integrated with target structures to obtain different structural information, such as temperature, stress, strain, vibration and so on. The familiar SHM sensors are resistance strain gages, fibre optic sensors, piezoelectric sensors, eddy current sensors, and microelectromechanical systems (MEMS) sensors.

2.1. Resistance strain gages

Resistance strain gage is a traditional strain sensor element. The gage mainly consists of a resistance grid of thin wire or foil, connector and encapsulation layer, as shown in figure 1(a). With the strain-resistance effect, the grid senses the structure's strain as its resistance value, which can be finally converted to the voltage signal with Wheatstone bridge circuit (seen in figure 1(b)). Resistance strain gage, of very small thickness but high sensitivity, can be easily bonded onto the structures and applicable in high temperature or pressure conditions.

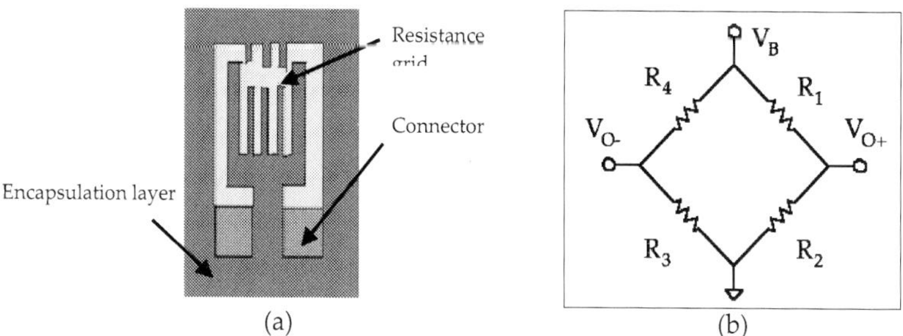

Figure 1. (a) Configuration of a resistance strain gage (b) Wheatstone bridge circuit

2.2. Fibre optic sensors

Fibre optic sensors (FOSs) are competitive candidates for SHM applications because of their unique advantages of light weight, high stability and reliability, long life cycle, low power utilization, EMI immunity, high bandwidth, compatibility with optical date transmission and processing, etc. According to the sensing range, FOSs can be categorized into local, quasi-distributed and distributed sensors [9]. The most-commonly used local FOSs are interferometric sensors, such as Mach-Zehnder, Michelson and Fabry-Perot FOSs. These sensors can measure strains and deformations at local sites by detecting the phase shifts of relative optical waves.

Fiber Bragg grating (FBG) sensors, with multiplexing capacity, are a kind of typical quasi-distributed FOSs. FBG is formed by inducing a periodic modulation of the refractive index in the core of a single mode optical fiber [10]. When light within a fiber passes through a FBG, constructive interference between the forward and contra-propagating light waves happens and leads to the narrowband back-reflection of light with Bragg wavelength λ_B. Any local changes along with FBG can be manifested as that of λ_B and therefore, from the measurement of the transmitted or reflected spectrum, as shown in figure 2, it is possible to monitor any strain-resulting parameters from temperatures to stress waves [9-11]. The major advantage of the sensor is that an array of wavelength-multiplexed FBGs can be deployed in a single fiber for quasi-distributed measurement. To further increase the number of FBGs, both spatial-division multiplexing and time-division multiplexing can be implemented.

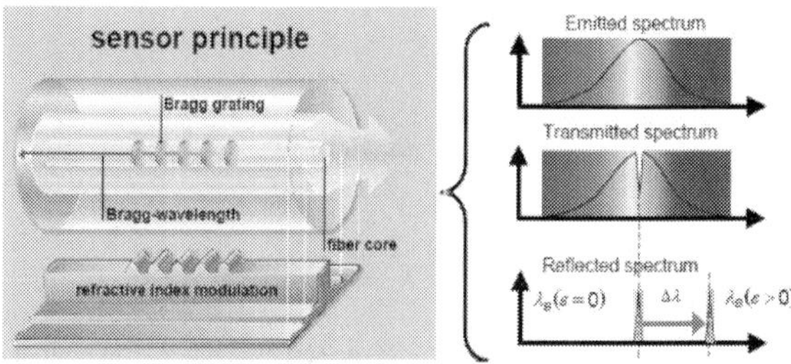

Figure 2. Fiber Bragg grating principle [11]

With all segments of an optical fiber acting as sensors, distributed FOSs can fulfill the real distributed measurement, which is very attractive for SHM of large structures. The sensors are based on the modulation of light intensity in a fiber. The optical time domain reflectometry (OTDR) and Brillouin scattering are the two main distributed sensor methodologies, in which Rayleigh and Fresnel scatterings and Doppler shift in light frequency are used for measuring, respectively [9].

2.3. Piezoelectric sensors

Piezoelectric sensors are frequently used for measuring low or high frequency vibrations, such as Lamb waves or acoustic emission. Compared with conventional acoustic probes, e.g., wedge or comb Lamb wave transducers, piezoelectric sensors are more desired for SHM in view of their weights, sizes and costs. The sensors, made of piezoelectric materials, operate on piezoelectric principles. With direct piezoelectric effect, the sensors in a stress field can generate charge response and vice verse, an external electric field applied to the sensors can result in an induced strain field through inverse piezoelectric effect. Consequently, piezoelectric sensors can be employed both as actuators and sensors [12].

Lead zirconium titanate ceramics (PZT) wafers and polyvinylidene fluoride (PVDF) films are the two common piezoelectric elements, as shown in figure 3. PZT wafers with high piezoelectric constant possess both excellent sensitivities as sensors and strong driving abilities as actuators, whereas the wafers are quite brittle due to ceramic inherent nature. In contrast, PVDF films have the advantages of high flexibility, low mass and cost and high internal damping [13]. However, because of the poor inverse piezoelectric properties and large compliance, PVDF films are usually preferred to be sensors [14]. To overcome the disadvantage of high brittleness of PZT wafers, piezoelectric composites, such as piezoelectric rubbers and piezoelectric paints, have been developed.

Figure 3. PZT and PVDF of various sizes and shapes [14]

2.4. Eddy current sensors

The concept of using eddy currents for damage detection stems from the electromagnetic induction [15], with which the eddy current can be induced to the tested conductive structure and sensed by the identical or different windings of eddy current sensors. The main application of eddy current sensors is crack or corrosion detection for metallic parts even through coatings or layers which may be non-conducting. This makes the sensors useful for such the composite structures as parent metal materials with composite doubler repairs and metal-matrix composites.

Due to the winding configurations, the conventional eddy current sensors with obtrusive size are hard to be integrated. Fortunately, with the development of micro-fabrication technique, the windings can be adhered or directly printed to a conformable substrate. As shown in figure 4, eddy current foil sensors, even in an array style with multiple sensing elements of various shapes, have been produced [16-17]. The new sensors are so thin and flexible that they can be easily surface-mounted or embedded between layers, offering the potential for on-line and continuous monitoring.

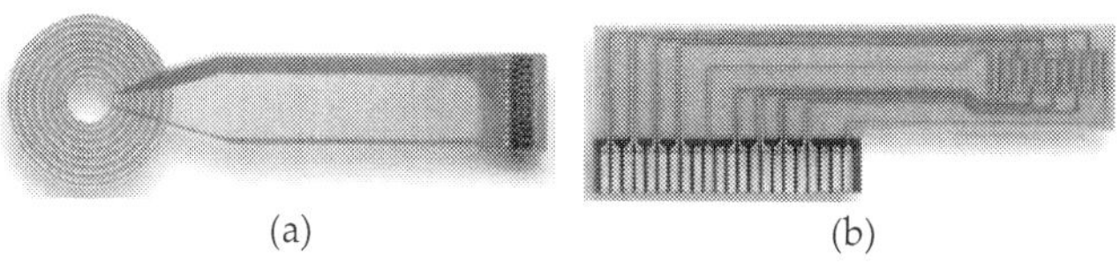

(a)   (b)

Figure 4. (a) 4-element rosette eddy current array and (b) 9-element linear eddy current array [18]

2.5. MEMS sensors

With the aid of advanced integrated circuit (IC) fabrication processes, MEMS is developed by co-fabricating microsensors, actuators and control functions in one silicon slice. MEMS is an intelligent system which can sense the circumstances and do some reactions by the microcircuit control [18]. At present, many MEMS sensors, such as MEMS accelerometers and pressure sensors, can be purchased commercially. Due to the extremely small size and large-scale integration degree, the sensors have the remarkable characteristics of light weight, flexibility in design, low power consumption and noise level, short response time, high reliability and economy, etc. To avoid the lengthy cables, the wireless communication capability can be added to the sensors with transmitter chips equipped.

Besides Comparative Vacuum Monitoring (CVM) sensors which will be later introduced together with CVM method, there are other sensors, such as laser scanners and microwave sensors, could be applied in SHM.

With the advances in SHM requirements for both monitoring area and damage quantity, a great number of same or different sensors are arranged to form large sensor arrays to the monitored structure [19], leading to the appearance of various sensor-array layers. Similar to the above eddy current arrays shown in figure 4, the layers are generally made by encapsulating sensor elements with thin and flexible dielectric films in desired configurations. The benefits from the layers are [20]: a) rapidly and consistently arranging a large number of sensors is allowed; b) connecting wires are avoid to reduce EMI; c) the layers can be surface-mounted on existing structures or embeded as extra layers in composites during manufacturing. Besides the PZT-array layer, known as SMART Layer [21], the PZT-FOS hybrid array layer [6] and HELP (Hybrid Electromagnetic Performing) layer [22] have also emerged, as shown in figure 5.

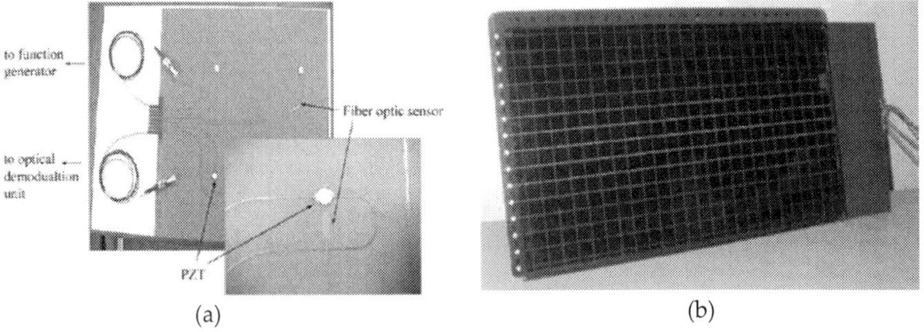

Figure 5. (a) PZT-FOS hybrid array layer [22] (b) HELP layer [23]

3. Typical SHM methods for composite materials

As illustrated in figure 6, SHM can be performed in either passive or active ways depending on whether actuators are used [23]. In passive SHM, various operational parameters, such as loads, stress, acoustic emission and circumstance condition, mainly concerned to infer the structural states in conjunction with signal and information processing technologies, mechanical modeling analysis or priori-knowledge. Passive SHM only 'listens' to the structures but does not interact with them, as figure 6(a) illustrates. While in active SHM, structures are firstly excited with actuators in prescribed manners and interrogated by analyzing the received structural responses. Though both actuators and sensors are required, as shown in figure 6(b), active SHM can be carried out whenever necessary.

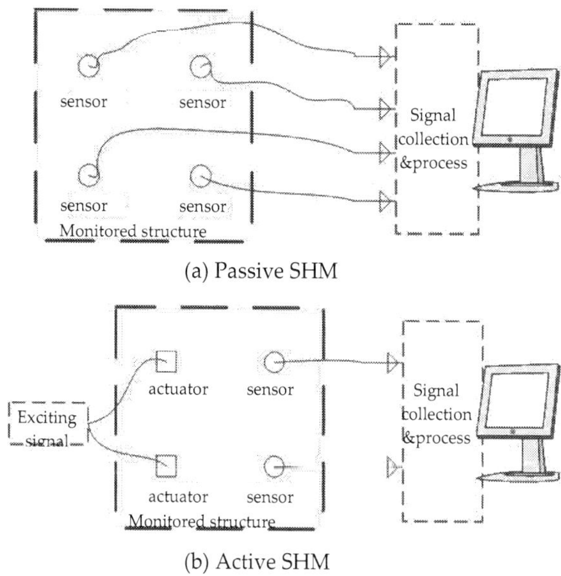

Figure 6. Schemes of passive SHM and active SHM

For composite materials, the common active SHM methods include Lamb wave, Electro-mechanical (E/M) impedance and active vibration-based methods. Acoustic emission, strain-based method and CVM are the typical passive approaches.

3.1. Lamb wave method

Lamb waves, first theoretically predicted by Horace Lamb in 1917, are a kind of guided ultrasonic waves existing in thin-wall structures. Because of the ability of long-distance transmission and high sensitivity to both the surface and the internal defects, Lamb waves are widely used as a promising tool for active SHM.

Lamb waves are usually excited and received by PZT wafers. FOS and PVDF can be also used as the wave sensors. Due to the multi-mode and dispersion characteristics, the propagation of Lamb waves is very complicated. In practical applications, a windowed toneburst is usually selected to generate the fundamental symmetric (S_0) and anti-symmetric (A_0) modes with the excitation frequency below the cut-off frequency of A_1 mode. To achieve single S_0 or A_0 mode generation, frequency tuning or double-side generation methods can be utilized [24-25].

Since defects in composites can bring about changes of geometric and mechanical boundary conditions, the phenomena of reflecting, scattering and energy attenuation could occur when propagating Lamb waves encounter the defects. Characteristic parameters can be then extracted from Lamb wave signals for damage monitoring.

Damage location is usually based on the time of flight (TOF) of Lamb wave signals. In ellipse location method [26], if the TOF of the damage scattered signal acquired by a transducer pair as well as the propagation velocity is known, an ellipse with the transducer pair at its foci can be determined to indicate the possible flaw locus. In order to identify the exact damage location, more ellipses are required to be constructed with other scattered signals from different transducer pairs and their intersection corresponds to the flaw site. Theoretically, a minimum of three ellipses can unambiguously locate the damage. When mode conversation severely takes place during damage scattering, the TOF of the damage-induced mode signals can be also used to estimate the defect point [27].

Time reversal (TR), based on spatial reciprocity and time invariance of linear wave equations, has been advocated as a baseline-free damage detection method. The presence of damage can induce nonlinearity and break down the reconstruction procedure of TR [28], resulting in divergence between the original and reconstructed waveforms. From the waveform difference, damage index can be then computed, in which original waveform rather than reference signal is involved in.

Damage imaging based on sensor arrays is often performed in Lamb wave monitoring to directly give a display of damage positions and intensities. The familiar imaging methods include delay-and-sum, phased array, and tomography methods.

In delay-and-sum imaging method [29-30], every point of the tested structure is considered as a potential flaw. As shown in figure 7, the traveling time $t_{ij}(x,y)$ of Lamb waves from actuator i at (x_i,y_i) to an imaging point O at (x,y) and then to sensor j at (x_j,y_j) is computed assuming that only one Lamb wave mode exists

$$t_{ij}(x,y) = t_{off} + \sqrt{(x_i - x)^2 + (y_i - y)^2} \Big/ c_i + \sqrt{(x_j - x)^2 + (y_j - y)^2} \Big/ c_j \qquad (1)$$

where t_{off} is the reference time, c_i and c_j are the group velocities for the wave mode propagating from i to O and from O and j, respectively.

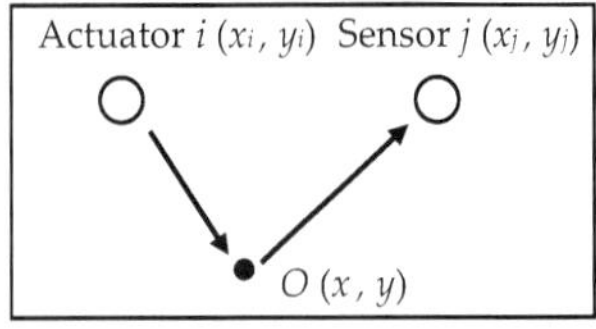

Figure 7. Illustration of delay-and-sum imaging

According to equation (1), the damage scattered signals $s_{ij}(t)$ measured by all transducer pairs can be time-shifted and summarized to get the pixel value at O as

$$E(x,y) = \left[\frac{2}{N(N-1)} \sum_{i=1}^{N} \sum_{j=i+1}^{N} s_{ij}(t_{ij}(x,y)) \right]^2 \tag{2}$$

where N is the transducer number.

An image is gained with pixel values at all points calculated. Though delay-and-sum imaging method is very similar to the above ellipse location technique in theory, the special calibration for TOF of every damage scattered signal is not required. Furthermore, those methods [31-32] based on TR focusing are essentially identical to the imaging method.

Phased arrays are generally compact transducer arrays in linear, circular or other patterns. Every array element, commonly PZT wafer, is individually used as actuator and sensor in a round-robin fashion such that a group of sensor signals are collected. Based on the synthetic-beam principles [33], all the signals, supplied with different phase delays, can be combined into one synthetic beamforming signal at one given steering angle, which can be implemented in either time or wavenumber domain [34]. Through the similar processes for all angles, the virtual scanning can be achieved without any physical manipulation of the array. Since constructive interference for the damage scattered signals is actually realized during beamforming, the signal-to-noise (SNR) of diagnostic signals and inspection distance can be largely improved [35]. An image of the scanned area is finally generated by directly mapping all the synthetic signals with the known velocity. Note that phased arrays work in pulse echo mode.

In tomography imaging, an array of transducers should be arranged around the tested area and used for Lamb wave exciting and receiving in pitch-catch mode. A tomographic image can be reconstructed by using wave speed, waveform or amplitude as flaw-relevant features. The standard parallel projection, fan-beam or crosshole schemes are usually adopted in tomography technique [36]. The crosshole scheme with iterative nature and great flexibly is more suited for any geometry and incomplete data set. To increase the sensitivity of tomography with sparse arrays, reconstruction algorithm for probabilistic inspection of defects (RAPID) is introduced [37]. In RAPID, the probabilities of defect occurrence at a point can be estimated from the changing severity of the signal of each transducer pair and its relative position to the pair.

3.2. E/M impedance method

The structural mechanical impendence, defined as the ratio of the applied force to the resulting velocity, can be easily affected by damages, such as cracks, disbonds and delaminations. However, direct measurement for the mechanical impendence is very hard. With PZT transducers, mechanical impendence is indirectly measured as E/M one for damage detection in E/M impedance method [38].

The method utilizes PZT transducers as both actuators and sensors to acquire structural dynamic responses. The electro-mechanical coupling model between the transducer and the structure is shown in figure 8. In the model, the PZT wafer is axially connected to a single degree-of-freedom spring-mass-damper system represented for the structural impendence. Through the mechanical coupling between the transducer and the tested structure and electro-mechanical transduction inside the transducer, the structural impedance gets reflected in the electric one at the transducer terminals as [39-40]

$$Z(\omega) = \left[i\omega a \left(\bar{\varepsilon}^{T}_{33} - \frac{Z_s(\omega)}{Z_s(\omega) + Z_a(\omega)} d^2_{3x} \hat{Y}^E_{xx} \right) \right]^{-1} \tag{3}$$

where $Z(\omega)$ is the electric impedance computed as the ratio between the input voltage and the output current of the PZT wafer. $Z_s(\omega)$ and $Z_a(\omega)$ are the structure and PZT wafer mechanical impedances, respectively. a, $\bar{\varepsilon}^{T}_{33}$, d_{3x} and $\hat{Y}^E_{xx}$ are the geometry constant, the complex dielectric constant of the PZT wafers at zero stress, the piezoelectric coupling constant and Young's modulus, respectively.

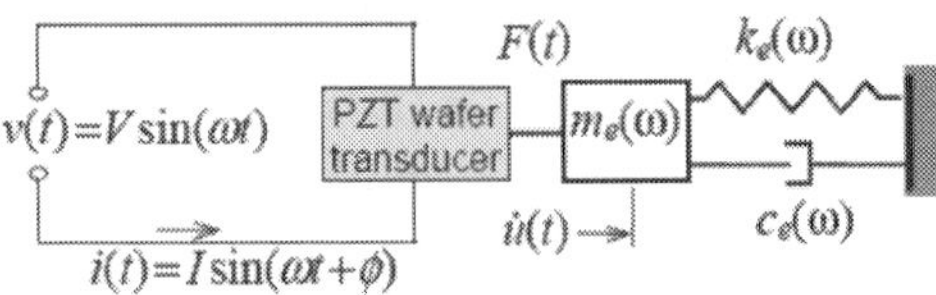

Figure 8. Electro-mechanical coupling model between the PZT transducer and the structure [39]

Equation (3) shows that, as long as the mechanical properties of PZT wafers keep invariable, $Z(\omega)$ is uniquely determined by $Z_s(\omega)$. Thus, the variations of $Z(\omega)$ can be mainly attributed to those of structural integrity. In E/M impedance monitoring, $Z(\omega)$ over a specified bandwidth, i.e., the complex impedance spectrum, is obtained by driving the transducer with sinusoid voltage sweeping and compared with its baseline. Usually, the existence of flaw exhibits as the resonance frequency or amplitude modification in the spectrum. Since the imaginary part of $Z(\omega)$ is temperature-sensitive due to $\bar{\varepsilon}^{T}_{33}$ in equation (3), the real part is more reactive to defects and can be considered for damage assessment. For instance, a damage index is computed as the Euclidean norm of the real portion of the spectrum [39], i.e.,

$$DI = \sqrt{\frac{\sum_{N} \left[R_e(Z^1_i) - R_e(Z^0_i) \right]^2}{\sum_{N} \left[R_e(Z^0_i) \right]^2}} \tag{4}$$

where N is the number of sampling points in the spectrum. Z^1_i and Z^0_i are the electric impedances measured in current and health states at frequency sampling point i, respectively.

Note that the impedance measurement is often performed in ultrasonic frequency range. At such high frequencies, the dynamic response is dominated in local modes and the excitation wavelength is small enough to ensure the high sensitivity to incipient local flaws [38-40].

3.3. Active vibration-based method

Active vibration-based method is a classical SHM technique. The basic idea behind the method is that structural dynamic characteristics are functions of the physical properties, such as mass, stiffness and damping [5, 41]. Therefore, damages, arising with physical property changes, can cause detectable differences in vibration responses. The dynamic characteristic parameters commonly used in the method include frequency, mode shape, power spectrum, mode curvature, frequency response function (FRF), mode flexibility matrix, energy transfer rate (ETR), etc.

A typical SHM system based on active vibrations is shown in figure 9.

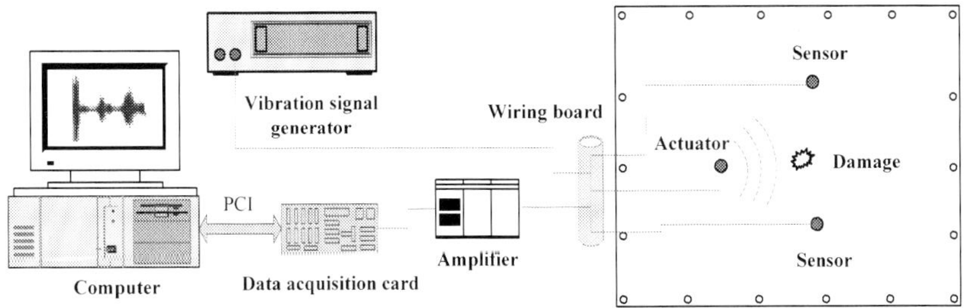

Figure 9. A typical active vibration-based SHM system

The method can be either model based or non-model based. The model based methods undertake structural model analysis and use model characteristic parameters to identify defects, while non-model based ones permit damage detection relying on the vibration characteristics independent of structural model.

Model-based methods are applied much more in practical SHM applications and can be roughly classified into three groups. One group of the methods, regarded as the forward problems, consists in calibrating model parameters in various known damage cases and defects can be then determined by comparing the measured parameters to the predicted ones. The main challenge of the method is how to obtain the sufficient and accurate characteristic parameters related to all structural circumstances. Particularly for large structures, experimental measurement is unpractical and finite element method (FEM) may a better choice.

In the second group of the methods, criteria or indicators are defined to examine model parameter variations for damage identifying. The ordinary natural frequency criteria are Cawley–Adams criterion and damage location assurance criterion (DLAC). Multiple damage location assurance criterion (MDLAC) is an extension of DLAC to detection multiple flaw sites [42]. Frequency response assurance criterion (FRAC), frequency domain assurance criterion

(FDAC) [43], global shape correlation (GSC) function and global amplitude correlation (GAC) function are the criteria of FRFs [44]. Modal assurance criterion (MAC) can quantify the correlation between measured and analytical mode shapes in a scalar number from zero to unity. The co-ordinate MAC (COMAC) and partial MAC (PMAC) are the developed forms of MAC [5]. The discrepancies of the other model properties, such as mode shape curvature and dynamic flexibility, can be also computed as damage indicators.

The last group of the methods is based on the structural model modification. The discrepancy between the original and modified models can provide the damage information. Mathematically, model modification is a constrained optimization problem based on the structural equations of motion, the nominal model and the measured data [45].

Note that compared with the aforementioned active Lamb wave or E/M impedance methods, the vibration frequencies in the active vibration-based method are generally much lower.

3.4. Acoustic emission method

Acoustic emission (AE) can be defined as the sudden release of localized strain energy in the form of transient elastic wave, due to a distortion or change in the structural integrity of material [46]. Many AEs arise during damage processes within structures. These AEs are referred to as primary ones while the secondary AEs are the others induced from external sources, such as impacts [47].

AE phenomena could appear evidently even when a structure is in microscopic-level damage status, which provides the possibility for defect forecasting and real-time monitoring. Generally, the procedure of AE testing can be summarized as: a) AE waves originate from AE source and propagate to the sensors; b) AE waves are captured by the sensors and converted to electrical signals; c) The AE signals are processed and interpreted to evaluate structural condition. Since only sensors are used to passively detect AE signals, AE method is a passive SHM technique.

From an AE signal, the parameters of AE event, ring-down count, count rate and total count are traditionally extracted to describe the damage mechanisms. The extracting procedures for AE event and ring-down count are illustrated in figures 10 and 11, respectively. As figure 10 shows, providing the envelope picked up from an AE signal with a proper voltage threshold V_1, a square impulse is obtained and related to an AE event. The impulse number over unit time and the accumulative impulse number are respectively defined as event count rate and total event count. If a threshold is directly set to the AE waveform, the ring-down count is gotten by quantitatively recording the resultant ring impulses, as shown in figure 11. Ring-down count per event is the so-called AE rate. The other signal features including amplitude, duration, rise time and energy can be also correlated with the defect characteristics [48]. Additionally, because different damages could result in different frequency contents, the spectrum of the AE signal can be calculated for damage discrimination [49].

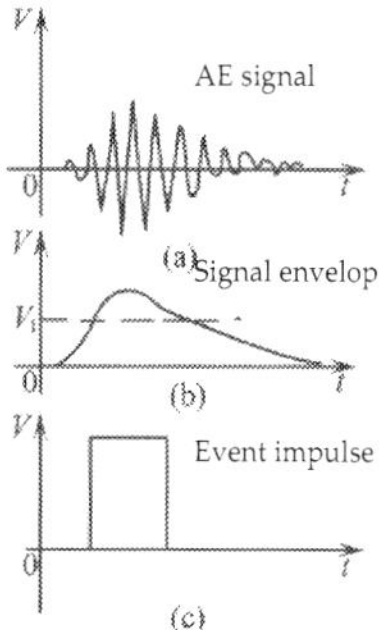

Figure 10. Extracting procedure for AE event

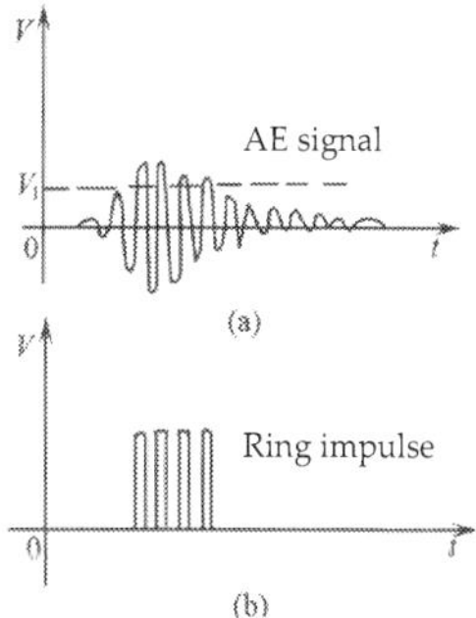

Figure 11. Extracting procedure for ring-down count

Using the time information of AE signals, AE source location can be realized. Taking the triangulation method for example, the location principle is illustrated in figure 12. At least three sensors, P_1, P_2 and P_3, should be used to decide a triangle $\Delta P_1 P_2 P_3$. As figure 12 shows, supposing AE source S is inside $\Delta P_1 P_2 P_3$ and the angles between S and the three sensors are θ_1, θ_2 and θ_3, respectively. According to the geometric relationship, a nonlinear equation can be finally derived as

$$\begin{cases} C_{g_1} C_{g_2} \Delta t_{12} \sin(\theta_1 + \theta_2) - L_{12}(C_{g_2} \sin\theta_2 - C_{g_1} \sin\theta_1) = 0 \\ C_{g_2} C_{g_3} \Delta t_{23} \sin(\theta_3 + \hat{S}_2 - \theta_2) - L_{23}\left(C_{g_3} \sin\theta_3 - C_{g_2} \sin(\hat{S}_2 - \theta_2)\right) = 0 \\ L_{12} \sin\theta_1 \sin(\theta_3 + \hat{S}_2 - \theta_2) - L_{23} \sin\theta_3 \sin(\theta_1 + \theta_2) = 0 \end{cases} \tag{3}$$

where L_{12}, L_{13} and L_{23} are the lengths of three sides of $\Delta P_1 P_2 P_3$. $\hat{S}_2$ is the internal angle $\angle P_1 P_2 P_3$. C_{g_1}, C_{g_2} and C_{g_3} are the velocities for the AE waves propagating from S to P_1, P_2 and P_3, respectively. Δt_{12} and Δt_{23} are the arrival time differences between P_1 and P_2, and between P_2 and P_3, respectively.

Note that equation (3) can be also applicable when S is outside $\Delta P_1 P_2 P_3$. By solving the equation (3), θ_1, θ_2 and θ_3 can be gained to locate S.

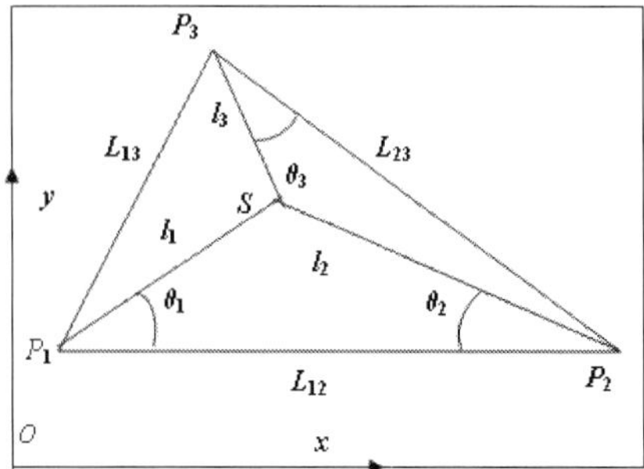

Figure 12. Illustration of triangulation method

For plate-like structures, modal AE (MAE) [50-52] is often presented based on elastic wave theory. In MAE, AE signal is analyzed in terms of different propagating wave modes, among which the basic A_0 and S_0 modes are mostly concerned.

3.5. Strain-based method

Strain-based method is an effective passive SHM method, because the presence of damage in the structure under normal operational loads can alter the local strain distribution due to the changing load path [53]. Besides the resistance strain gages, FOSs are usually applied in the method to measure the distributed strains. Figure 13 gives a typical strain distribution measurement system based on an array of multiplexed FBG sensors.

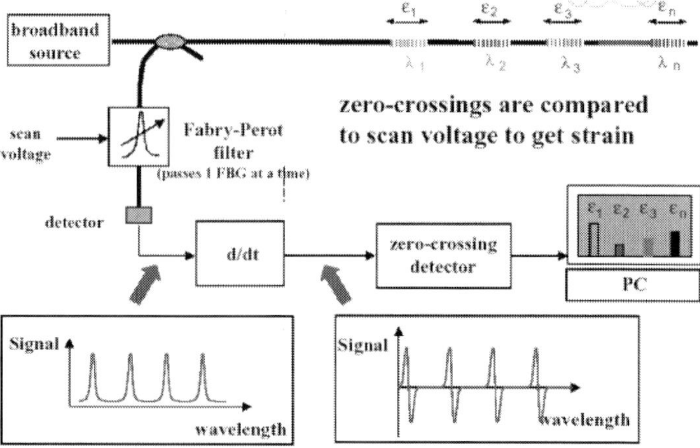

Figure 13. A strain distribution measurement system based on FBG sensors

In practical applications, the strain-based method can be performed in two ways. In one way, the strain distribution of the intact structure is measured as the baseline in advance. Damage can be then detected when the current strain measurement significantly diverges from the baseline. In the other way, a theoretical model for the structure is established and analyzed to acquire the strain data corresponding to various structural states. Comparing the data to the actually acquired ones directly or with criterions, the structural integrity is evaluated. The key issue lies in this methodology is how to make the model exact enough especially for the complex real-life structure.

3.6. CVM method

CVM method is a very mature technique and is ready for deployment onto operational platforms [54]. CVM has been developed by Structural Monitoring System Ltd. (SMS), with the original patents being granted in 1995. The CVM system has three primary components: a CVM sensor, fluid flow meter and stable low vacuum source [55]. The CVM sensor is directly adhered to the surface of the monitored structure to form a series of long and narrow galleries, which are alternately placed in the low vacuum or atmosphere states, as figure 14 illustrates. With the stable vacuum reference provided by the vacuum source, the air pressure of vacuum galleries is measured by the flow meter. If no flaw presents, the galleries remain sealed and there should be no leaks and pressure changes happening. However, if a flaw develops and breaks the galleries, air will flow along the breakage passage from the atmosphere to the vacuum galleries, increasing the pressure [54, 56]. Furthermore, the rate of pressure rising can be the indication of damage size.

Obviously, the sensitivity of the CVM sensor is determined by the gallery wall thickness. Now, the commercially available sensor can have a sensitivity down to $250\mu m$ with an accuracy better than 4% [55]. Since the exposed structural surface becomes one part of the galleries, CVM method is very suitable for surface crack or corrosion detection in metals. Using embedded CVM sensor, the applicability of the method for monitoring crack growth, debonding or delamination in composite structures has been also demonstrated.

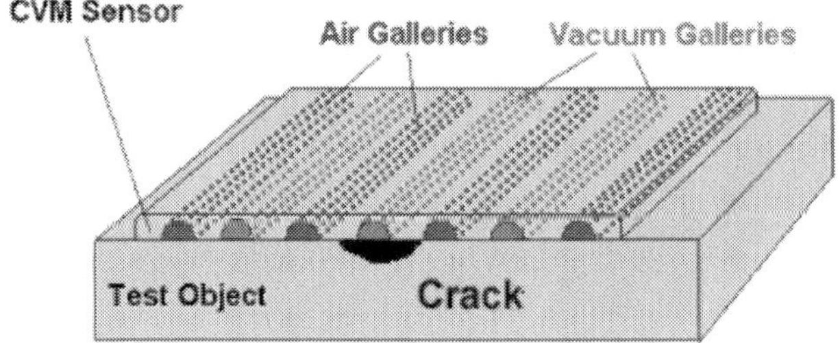

Figure 14. Measuring principle of CVM [61]

All the above methods can be summarized in table 1.

Methods	Used sensors	Monitoring objects					Characteristics
		Strain	Impact	Debonding	Delamination	Crack	
Lamb Wave method	Piezoelectric sensor	D	D	E	E	E	Global monitoring, high sensitivity, on-line & off-line
	FOS	D	D	E	E	E	Global monitoring, requiring PZT actuators, limited by high-frequency modulation
E/M impedance method	Piezoelectric sensor	D	D	E	E	E	Local monitoring, off-line
Active vibration-based method	Piezoelectric sensor & accelerometer	E	D	E	E	E	On-line & off-line, medium and high frequency vibration and acceleration monitoring
	FOS	E	D	E	E	E	On-line & off-line, low frequency (<1kHZ) vibration monitoring
Strain-based method	Resistance strain gauge	E	D	E	E	E	On-line, relying on loads
	FOS	E	D	E	E	E	Distribution measurement, on-line, rely on loads
Acoustic emission	Piezoelectric sensor & AE sensor	D	E	D	E	E	On-line
CVM	CVM sensor	D	D	E	E	E	Local monitoring, mature method

Table 1. Typical SHM methods for composites

4. SHM examples on composite materials

To verify the SHM methods, two examples of Lamb wave imaging and impact location for composite structures are arranged as the representations of active and passive methods, respectively.

4.1. Lamb wave imaging

The tested specimen is a quasi-isotropic epoxy glass-fiber composite plate with the dimension of 600mm×600mm×2mm. Eight PZT wafers P_1~P_8 are mounted on the plate to form a sparse PZT array, as shown in figure 15. The diameter of each PZT is 8mm and its thickness is 0.48 mm. Two identical hexagonal hollow screws, denoted as D_1 and D_2, are bonded on the plate to simulate damages. The positions of PZT wafers and damages are listed in table 2. The overall experimental setup, including Lamb wave detection system, matrix switch, power amplifier and the specimen, is shown in figure 16. Lamb wave detection system is built based on an industrial computer, in which a LAI200-ISA arbitrary wave generator (up to 50MHz DAC clock, ±5V output scale, 12 bit resolution), a charge amplifier and a PCI-9812 analog input card (10MHz sampling rate, ±5V sampling scale and 12 bit acquisition resolution) are integrated to generate Lamb wave signals, amplify and collect sensor signals. Matrix switch controls the working sequence of all PZT pairs and power amplifier is used to amplify the excitation signal to enlarge the monitoring area in the plate.

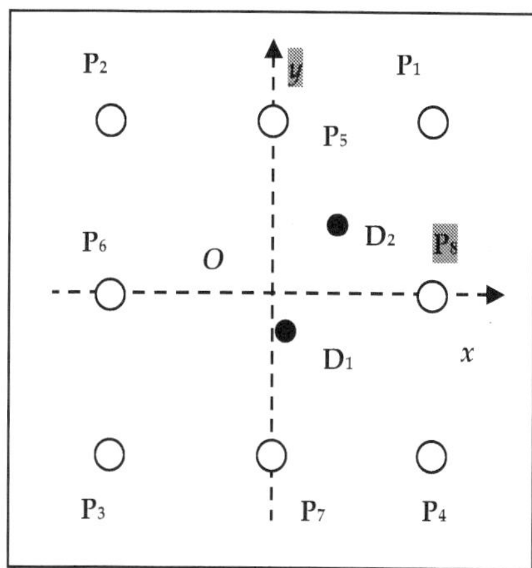

Figure 15. Configuration of the specimen of Lamb wave imaging

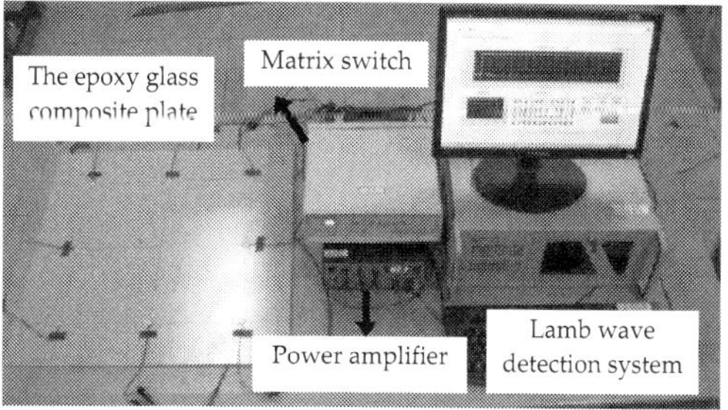

Figure 16. Figure 16 Experiment setup of Lamb wave imaging

	(x, y)/(mm)		(x, y)/(mm)
P_1	(200 , 200)	P_6	(-200 , 0)
P_2	(-200 , 200)	P_7	(0 , -200)
P_3	(-200 , -200)	P_8	(200 , 0)
P_4	(200 , -200)	D_1	(60 , 7 0)
P_5	(0 , 200)	D_2	(10 , -30)

Table 2. The coordinates (x, y) of PZT wafers and damages in the epoxy glass composite plate

As shown in figure 17, a symmetrical modulated 5-cycle sine burst with the central frequency of 50 kHz is adopted to excite diagnostic waves of single A_0 mode into the composite plate. The damage scattered signals can be obtained by subtracting the baseline response of the undamaged plate from the response of the damaged plate under the sin burst excitation. Figure 18 shows the damage scattered signals measured by P_1-P_5 pair. The wavepacket scattered from D_1 can be observed from figure 18(a). When D_1 and D_2 exist, the two damage scattered wavepackts appear in figure 18(b).

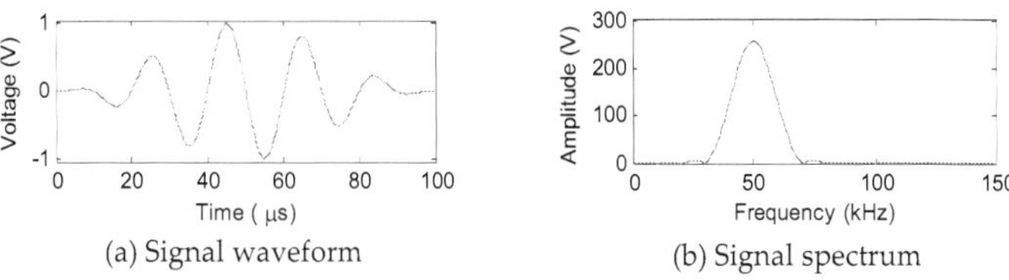

(a) Signal waveform   (b) Signal spectrum

Figure 17. Excitation signal

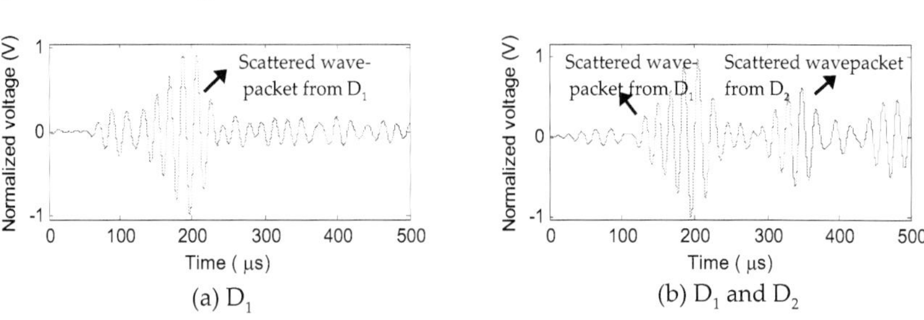

(a) D_1 (b) D_1 and D_2

Figure 18. Damage scattered signals measured by P_1-P_5 pair

After the group velocity of the A_0 mode at 50 kHz is measured as 1331.4m/s in the composite plate, the damage images can be constructed by using the envelopes of the twenty-eight scattered signals acquired by all the PZT pairs in the sparse array based on equation (2). The imaging results are shown in figure 19 where the symbol 'X' denotes the actual damage location. As displayed in figure19 (a) and (b), each flaw point is clearly and accurately represented by a bright focalized spot.

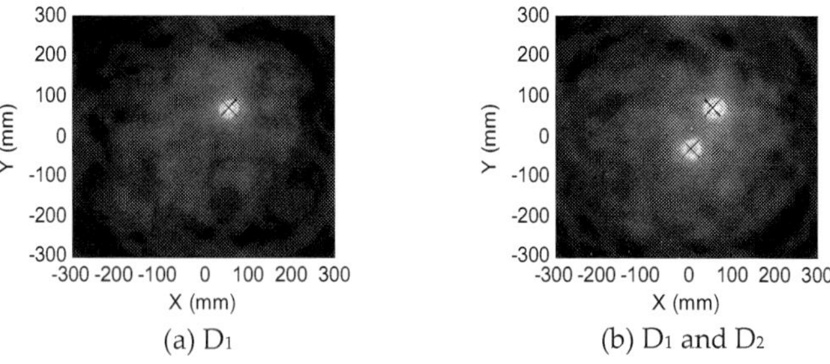

(a) D_1 (b) D_1 and D_2

Figure 19. Imaging results

4.2. Impact location

The experiment system for impact location, mainly composed of an aircraft wing box and an integrated SHM system, is shown in figure 20. The size of the wing box is 1000mm×1800mm× 200mm. The top panel is made of carbon fiber composite material and the bottom panel is made of aluminum. The panels are fastened to steel box frame. There are in total six T-shaped stiffeners with a distance of 130mm between each other on the panels. Vertical to the stiffeners there are five rows of bolt holes with a distance of 280mm. The experiment system is built on the top panel of the carbon fiber composite material. An array of smart layers [52] with three PZT wafers is attached on the inner surface of the top panel. Four impacts are produced to the plate using a hammer (seen in figure 20). The detailed positions of the impacts and the used PZT wafers (P_1, P_3, P_4, P_7, P_{16}, P_{19}, P_{20} and P_{22}) are shown in figure 21 and table 3.

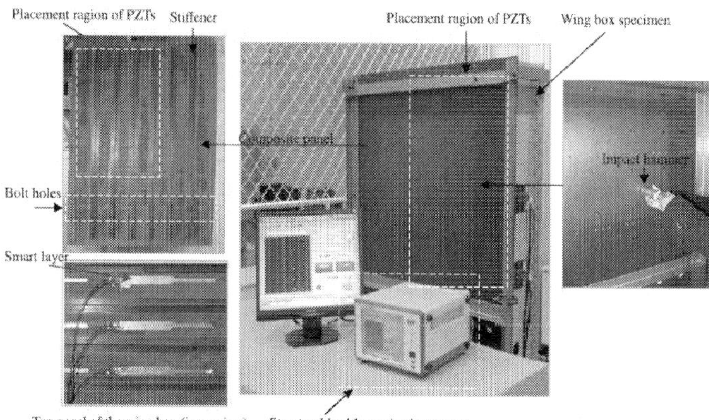

Figure 20. Experiment setup of impact location [52]

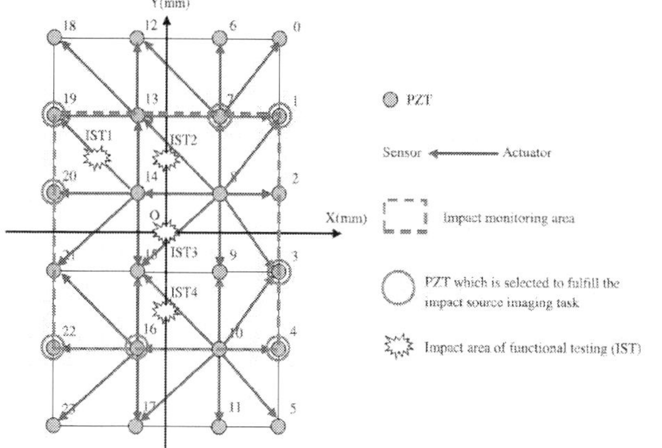

Figure 21. Configuration of the specimen of impact location [52]

	(x, y)/(mm)		(x, y)/(mm)
P_1	(-210 , 225)	P_{20}	(-200 , 75)
P_3	(210 , -75)	P_{22}	(-200 , -225)
P_4	(210 , -225)	IST_1	(-120 , 150)
P_7	(-100 , 225)	IST_2	(0 , 150)
P_{16}	(-50 , -225)	IST_3	(0 , 0)
P_{19}	(-200 , 225)	IST_4	(0 , -150)

Table 3. The coordinates (x, y) of PZT wafers and impacts in the top panel

The impact responses are fed into the integrated SHM system. Figure 22 shows a waterfall plot of normalized responses produced by the impact at IST_1. Here, the narrow-band components with central frequency of 100kHz are extracted from all the responses and their envelopes are then computed for arrival time determination, which can be performed with the complex wavelet transform [57]. After the arrival times are decided by the first peaks in the obtained envelopes, the impact can be located based on equation (3). Note that the anisotropic properties in the composite panel should be considered during impact location. The location result is given in table 4, in which the location error is defined as the spatial interval between the actual and the estimated impact sites.

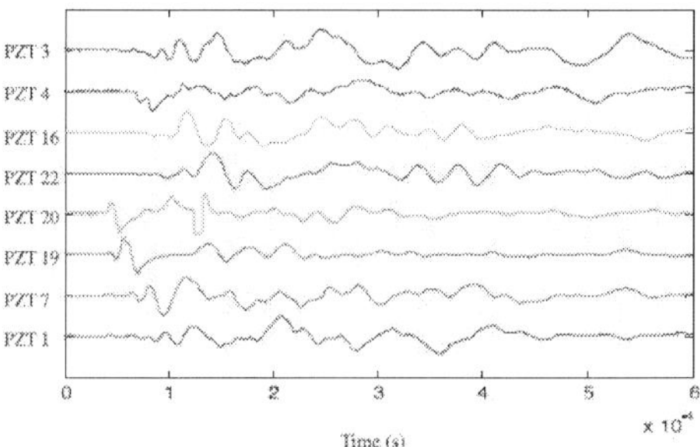

Figure 22. Acquired impact responses [52]

Impact Number	Impact location result		
	Actual site	Estimated site	Location error
IST_1	(-120mm, 150mm)	(-130mm, 125mm)	27mm
IST_2	(0mm, 150mm)	(15mm, 120mm)	16mm
IST_3	(0mm, 0mm)	(0mm, -20mm)	20mm
IST_4	(0mm, -150mm)	(-20mm, -120mm)	36mm

Table 4. Impact location result in the carbon-fiber composite panel

5. Summary and conclusions

SHM for composite materials is briefly described in the chapter. Firstly, an introduction involving advantages, problems and SHM requirements of composites is made. The common sensors used in SHM as well as typical SHM methods for composite materials are then introduced. Two examples, Lamb wave imaging for a glass-fiber composite plate and impact location in an aircraft composite wing box, are also arranged.

Though much development of SHM has been achieved, a great deal of work is still required for the further practical SHM applications in composite materials.

Author details

Jian Cai, Lei Qiu, Shenfang Yuan, PeiPei Liu and Dong Liang
The Aeronautic Key Lab for Smart Materials and Structures,
Nanjing University of Aeronautics and Astronautics, Nanjing, Jiangsu Province, P.R. China

Lihua Shi
National Key Lab of Electromagnetic Environmental Effect and Electro-optical Engineering, Nanjing, Jiangsu Province, P.R. China

Acknowledgement

This work is supported by EU-FP7 SICA programme (Grant No.FP7-PEOPLE-2010-IRSES-269202), Natural Science Foundation of China (Grant No.50830201).

6. References

[1] Campbell FC. Structural Composite Materials. Ohio: ASM International; 2010.

[2] Jones RM. Mechanics of composite materials (2nd edition). Philadelphia PA: Taylor & Francis; 1999.

[3] Miracle DB, Donaldson SL. Introduction to Composites. In: Miracle DB, Donaldson SL (ed.) ASM Handbook Volume 21: Composites. ASM International; 2001. p3-17

[4] Awad ZK, Aravinthan T, Zhuge Y, et al. A review of optimization techniques used in the design of fibre composite structures for civil engineering applications. Materials and Design 2012; 33(1) 534-544.

[5] Montalvao D, Maia NMM and Ribeiro AMR. A Review of Vibration-based Structural Health Monitoring with Special Emphasis on Composite Materials. The Shock and Vibration Digest 2006; 38(4) 295-324.

[6] Qing XL, Kumar A, Zhang C. A hybrid piezoelectric/fiber optic diagnostic system for structural health monitoring. Smart Materials and Structures 2005; 14(3) S98-S103.

[7] Qing XL, Chan HL, Beard SJ, et al. An Active Diagnostic System for Structural Health Monitoring of Rocket Engines. Journal of intelligent material systems and structures 2006; 17(7): 619-628.

[8] Ihn JB, Chang FK. Pitch-catch Active Sensing Methods in Structural Health Monitoring for Aircraft Structures. Structural Health Monitoring 2008; 7(1): 5-19.

[9] Lia HN, Lia DS and Song GB. Recent applications of fiber optic sensors to health monitoring in civil engineering. Engineering Structures 2004; 26(11): 1647-1657.

[10] Loayssa A. Optical Fiber Sensors for Structural Health Monitoring. In: Mukhopadhyay S. (ed.) New Developments in Sensing Technology for Structural Health Monitoring. Heidelberg: Springer-Verlag; 2011. p335-358.

[11] Speckmann H, Henrich R. Structural health monitoring (SHM) - overview on technologies under development. In: Proceedings of the 16th world conference on NDT (WCNDT). Montreal, Canada, August 30-September 3; 2004.

[12] Lin B, Giurgiutiu V. Modeling and testing of PZT and PVDF piezoelectric wafer active sensors. Smart Materials and Structures 2006; 15(4): 1085-1093.

[13] Boller C. Next generation structural health monitoring and its integration into aircraft design. International Journal of Systems Science 2000; 31(11): 1333-1349.

[14] Raghavan A, Cesnik CES. Review of Guided-wave Structural Health Monitoring. The Shock and Vibration Digest 2007; 39(2): 91-114.

[15] Sodano HA. Development of an Automated Eddy Current Structural Health Monitoring Technique with an Extended Sensing Region for Corrosion Detection. Structural Health Monitoring 2007; 6(2): 111-119.

[16] Zilberstein V, Walrath K, Grundy D, et al. MWM eddy-current arrays for crack initiation and growth monitoring. International Journal of Fatigue 2003; 25(9-11): 1147-1155.

[17] Goldfine N, Zilberstein V, Schlicker D, et al., 2001, Surface Mounted Periodic Field Eddy Current Sensors for Structural Health Monitoring, in Proc. Of SPIE, Advanced Nondestructive Evaluation for Structural and Biological Health Monitoring, ed. T. Kundu, vol. 4335.

[18] Varadan VK, Varadan VV. Microsensors, microelectromechanical systems (MEMS), and electronics for smart structures and systems. Smart Materials and Structures 2000; 9(6): 953-972.

[19] Yuan SF, Liang DK, Shi LH, et al. Recent Progress on Distributed Structural Health Monitoring Research at NUAA. Journal of Intelligent Material Systems and Structures 2008; 19(3): 373-386.

[20] Qing XL, Beard SJ, Kumar A, et al. Advances in the development of built-in diagnostic system for filament wound composite structures. Composites Science and Technology 2006; 66(11-12): 1694-1702.

[21] Lin M, Chang FG. 1998. Design and Fabrication of Built-in Diagnostic for Composite Structures. In: 12th American Society of Composites Technical Conference, Baltimore, USA.

[22] Lemistre MB, Balageas DL. A Hybrid Electromagnetic Acousto-ultrasonic Method for SHM of Carbon/epoxy Structures. Structural Health Monitoring 2003; 2(2): 153-160.

[23] Giurgiutiu V. Structural Health Monitoring with Piezoelectric Wafer Active Sensors. Boston: Elsevier Academic Press; 2008.

[24] Bottai GS, Chrysochoidis NA, Giurgiutiu V, et al. Analytical and experimental evaluation of piezoelectric wafer active sensors performances for Lamb waves based structural health monitoring in composite laminates, 2007, Proceedings of SPIE.

[25] Su Z, Ye L. Selective generation of Lamb wave modes and their propagation characteristics in defective composite laminates. Journal of Materials: Design and Applications 2004; 218(2): 95-110.

[26] Quek ST, Tua PS, Jin J. Comparison of Plain Piezoceramics and Inter-digital Transducer for Crack Detection in Plates. Journal of Intelligent Material Systems and Structures 2007; 18(9): 949-961.

[27] Su ZQ, Ye L, Bu XZ. A damage identification technique for CF/EP composite laminates using distributed piezoelectric transducers. Composite Structures 2002; 57(1-4): 465-471.

[28] Park HW, Sohn H, Law, KH, et al. Time reversal active sensing for health monitoring of a composite plate. Journal of Sound and Vibration 2007; 302(1-2): 50-66.

[29] Michaels JE. Detection, localization and characterization of damage in plates with an in situ array of spatially distributed ultrasonic sensors. Smart Materials and Structures 2008; 17(3): 1-15.

[30] Cai J, Shi LH, Yuan SF, et al. High spatial resolution imaging for structural health monitoring based on virtual time reversal 2011; 20(5): 055018-55028.

[31] Wang HC, Rose JT, Chang FG. A synthetic time-reversal imaging method for structural health monitoring. Smart Materials and Structures 2004; 13(2): 415-423.

[32] Wang Q, Yuan SF. Baseline-free Imaging Method based on New PZT Sensor Arrangements. Journal of Intelligent Material Systems and Structures 2009; 20(14): 1663-1673.

[33] Giurgiutiu V, Bao J. Embedded-ultrasonics Structural Radar for In Situ Structural Health Monitoring of Thin-wall Structures. Structural Health Monitoring 2004; 3(2): 121-140.

[34] Wilcox PD. Omni-directional guided wave transducer arrays for the rapid inspection of large areas of plate structures. IEEE Transactions on Ultrasonics, Ferroelectrics and Frequency Control 2003; 50(6): 699-709.

[35] Yan F, Royer RL, Rose JL. Ultrasonic Guided Wave Imaging Techniques in Structural Health Monitoring. Journal of Intelligent Material Systems and Structures 2010; 21(3): 377-384.

[36] Leonard KR, Malyarenko EV, Hinders MK. Ultrasonic Lamb wave tomography. Inverse Problems 2002; 18(6): 1795-1808.

[37] Zhao XL, Gao HD, Zhang GF, et al. Active health monitoring of an aircraft wing with embedded piezoelectric sensor actuator network. Smart Materials and Structures 2007; 16(4): 1208-1217.

[38] Ayres JW, Lalande F, Chaudhry Z, et al. Qualitative impedance-based health monitoring of civil infrastructures. Smart Materials and Structures 1998; 7(5): 599-605.

[39] Giurgiutiu V, Rogers C A. 1998 Recent advancements in the electro-mechanical (E/M) impedance method for structural health monitoring and NDE Proc. Conf. on Smart Structures and Materials (San Diego, CA) vol 3329 (SPIE) pp 536–47.

[40] Gyuhae P, Sohn H, Farrar CR, et al. Overview of Piezoelectric Impedance-Based Health Monitoring and Path Forward. The Shock and Vibration Digest 2003; 35(6): 451-463.

[41] Farrar CR, Doebling SW, Nix DA. Vibration-based structural damage identification. Philosophical transactions of the Royal Society. Mathematical, physical, and engineering sciences 2001; 359(1778): 131-149.

[42] Messina A, Williams EJ, Contursi T. Structural damage detection by a sensitivity and statistical-based method. Journal of Sound and Vibration 1998; 216(5): 791-808.

[43] Pascual R, Golinval JC, Razeto M. A frequency domain correlation technique for model correlation and updating. International Modal Analysis Conference (IMAC), 15th, Orlando, FL; UNITED STATES; 3-6 Feb. 1997. pp. 587-592. 1997.

[44] Zang C, Friswell MI, Imregun M. Structural Health Monitoring and Damage Assessment Using Measured FRFs from Multiple Sensors, Part I: The Indicator of Correlation Criteria. Key Engineering Materials 2003; 245-246(131): 131-140.

[45] Doebling SW, Farrar CR, Prime MB. Review of Vibration-Based Damage Identification Methods. Shock and Vibration Digest 1998; 30(2): 91-104.

[46] Gostautas RS, Ramirez G, Peterman RJ. Acoustic Emission Monitoring and Analysis of Glass Fiber-Reinforced Composites Bridge Decks. Journal of Bridge Engineering 2005; 10(6): 713-721.

[47] Nair A, Cai CS. Acoustic emission monitoring of bridges: Review and case studies. Engineering Structures 2010; 32(6): 1704-1714.

[48] Huang M, Jiang L, Liaw PK. Using Acoustic Emission in Fatigue and Fracture Materials Research. JOM-e 1998; 50(11): 1-14.
http://www.tms.org/pubs/journals/JOM/9811/Huang/Huang-9811.html

[49] Groot PJ, Wijnen PA, Janssen RB. Real-time frequency determination of acoustic emission for different fracture mechanisms in carbon epoxy composites. Composites Science and Technology 1995; 55(4): 405-412.

[50] Morscher GN, Modal acoustic emission of damage accumulation in a woven SiC/SiC composite. Composites Science and Technology 1999; 59(5): 687-697.

[51] Surgeon M, Wevers M. Modal analysis of acoustic emission signals from CFRP laminates. NDT&E International 1999; 32(6): 311-322.

[52] Qiu L, Yuan SF, Zhang XY, et al. A time reversal focusing based impact imaging method and its evaluation on complex composite structures. Smart Materials and Structures 2011; 20(10): 105014.

[53] Silva-Munoz RA, Lopez-Anido RA. Structural health monitoring of marine composite structural joints using embedded fiber Bragg grating strain sensors. Composite Structures 2009; 89(2): 224-234.

[54] Mrad N. State of Development of Advanced Sensory Systems for Structural Health Monitoring Applications. Proceedings of the NATO RTO AVT-144 Workshop on Enhanced Aircraft Platform Availability Through Advanced Maintenance Concepts and Technologies, Vilnius, Lithuania, 3-5 October 2006 (DRDC Atlantic SL-2008-260).

[55] Wishaw M, Barton DP. Comparative Vacuum Monitoring: a New Method of In-Situ Real-Time Crack Detection and Monitoring. In: Proceding of 10th Asia-Pacific Conference On Nondestructive Testing, 2001. Brisbane.

[56] Stehmeier H, Speckmann H. Comparative Vacuum Monitoring (CVM) Monitoring of fatigue cracking in aircraft structures. 2nd European Workshop on Structural Health Monitoring July 7-9, 2004 Amazeum Conference Centre at Deutsches Museum, Munich, Germany.

[57] Jeong H. Analysis of plate wave propagation in anisotropic laminates using a wavelet transform. NDT&E International 2001; 34(3): 185-190.

Chapter 4

Acoustic Emission of Composite Vessel

Hyun-Sup Jee and Jong-O Lee

Additional information is available at the end of the chapter

http://dx.doi.org/10.5772/47877

1. Introduction

There are about 10 million vehicles run on natural gas in the world. There are about 1.7 million low-pressure LPG vehicles [1], and 17,000 high-pressure CNG vehicles running in Korea and the number is increasing.[2]

Generally for CNG vehicles, type II vessel, where the increase in used pressure and lighter weight are achieved through fiber-reinforced composite material, which is wrapped in hoop-direction on the steel liner is used. Since 1984, the U.S. experienced more than 80 cases of vehicle fuel tank-related accidents [3] and Korea also experienced 8 cases in which the CNG tank exploded; thus, there is a need for the development of inspection technology for high-pressure fuel tanks. In the case of the U.S., the inspection technology of high-pressure fuel tanks were developed by DOD and NASA as an inspection technology for missile fuel tanks[4] but as the use of high-pressure fuel tanks for transport increased DOT executed a research on inspection technology for vehicles based on the research results of NASA and reported that among several NDT technology, Acoustic Emission(AE) has a possibility of being used as an inspection technology for vehicles[5,6]. The gas vessel, which is made of fiber-reinforced composite material, is unlike vessel made of only steel materials in that when the damage increases the acoustic generation activity increases but when the degree of damage increases even more, the acoustic generation activity rather decreases [7]. A study of defect detection and failure analysis for composite Materials using acoustic emission is progressing steadily [9-11,14].

2. Experiment

2.1. Experimental vessel

Experiment vessel used in this research is a 64 Liter CNG fuel tank used in vehicles. The thickness of the liner in the shell is about 6 mm and was made using the DDI (Deep Drawing Ironing) method[12,13] using 34CrMo4 steel plate, and is a type-II vessel in which glass fiber is hoop-wrapped on the shell of the liner.

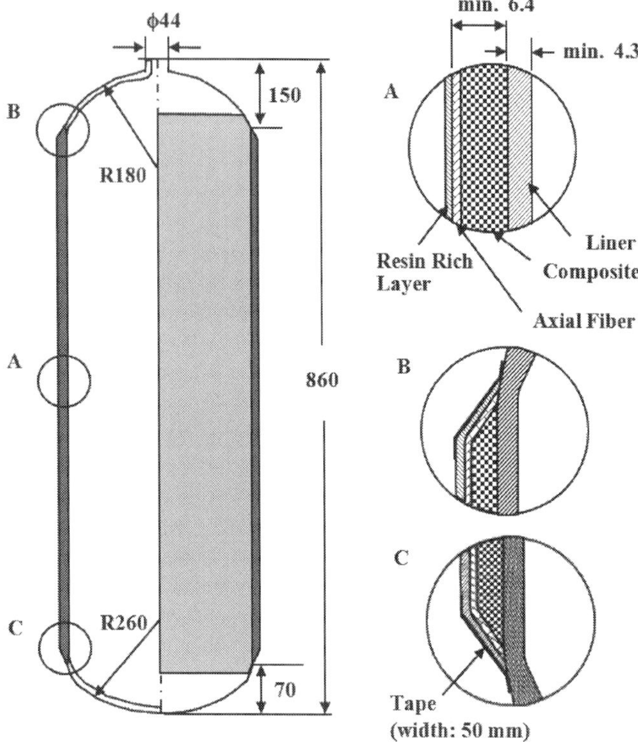

Figure 1. Schematic diagram of experimental vessel

	C	Mn	Si	P	S	Cr	Mo
Max	0.38	1.00	0.40	0.015	0.010	1.20	0.40
Min	0.25	0.40	0.10	-	-	0.80	0.15
P+S	≤0.020						

Table 1. Chemical composition of vessel liner (unit : wt.%)

2.2. Method of experiment

For the test, the acoustic emission sensor is R15I (PAC) with the resonance frequency of 150 kHz and cable of RG58A/U (10m) is put on the middle of shell using the vacuum grease.

The detected AE signal is put into the DiSP-52 Acoustic emission workstation (PAC) for processing. In addition, the water was used as medium for burst test. The threshold value of test was set at 45dB. The source of simulated sound was the destruction of the 2H Pentel pencil lead. The average sensitivity of sensor was 98dB within 1 inch from sensor.

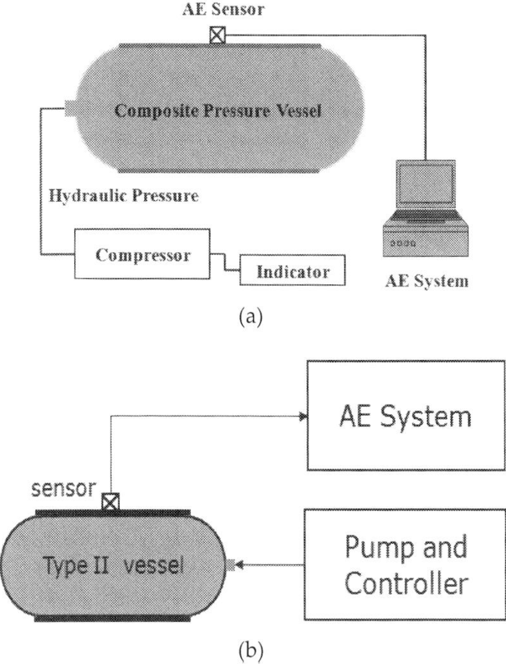

Figure 2. Diagram of Experimentation : (a) Burst and (b) fatigue test of the composite vessel

The burst pressure was estimated to be 600 bar under the pressured conditions. By raising the pressure to 30%, 50%, 60%, 70%, 80% and 90% of estimated burst pressure and keeping each pressure stage for 10 minutes as in Figure 3(a), we acquired the AE signal from each stage. As can be seen in the figure, pressure was put on the vessel with a pump to control the pressure and acoustic emission signals were detected using acoustic emission sensors attached to the vessel. And the signal were processed and analyzed after fed into AE equipment. The fatigue test repeated 20000 cycles between 0 and 207 bar, afterwards which has a used pressure of 0 and afterwards, the pressure was continuously increased and the burst test was carried out. Figure 3(b) shows the conditions for pressurization.

3. Result and research

3.1. Burst test

3.1.1. Damage mechanism of composite vessel

First, the composite materials wrapped in the metal liner were separated due to the matrix crack and then each layer was separated from each other. Then, some section of the fuel tank was weakened due to the cutting of some reinforced fiber, causing the destruction of metal liner of that part and finally destroying the vessel.

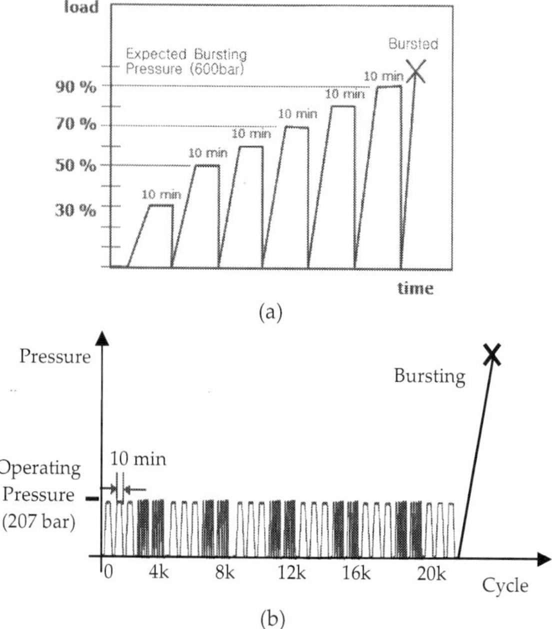

Figure 3. Loading sequence : (a) Burst and (b) fatigue test

Figure 4. Composite vessel after burst test

3.1.2. AE signal generated during burst test

In this test, considering the flow noise during the initial pressure, I excluded the signal during the first two minutes from the data but included the remaining 8 minutes data for evaluation. Generally, for the evaluation of the soundness of vessel, a tester imposes pressure on the vessel and keeps the pressure for some time until some sound emits from it. This is called the analysis of creep effect. The values in Figure 5 show that the pressure up to 360 bar, or 60% of the expected burst press is weak, showing that the vessel is not likely to receive significant damages. But after 70% of the expected pressure, the signals of 60dB or more are often shown, which may mean that there is a lot of creep effect and that the significant damages have been done to the vessel.

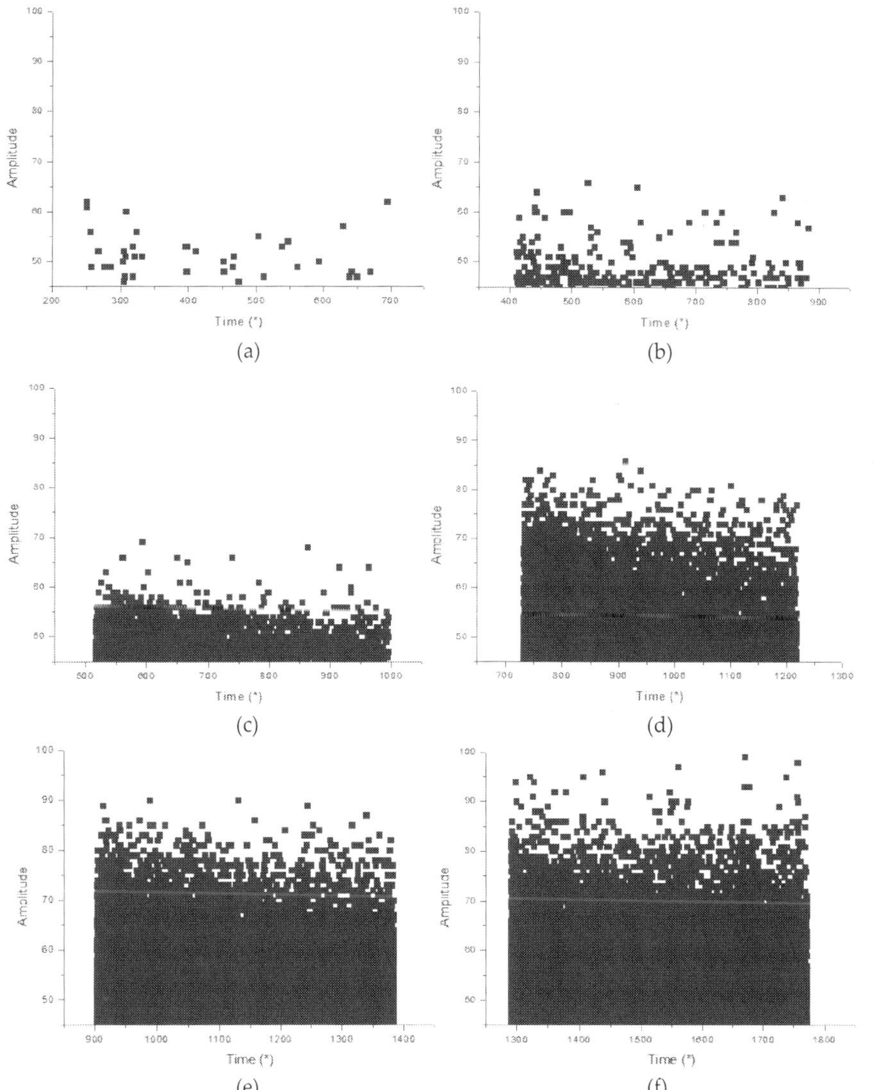

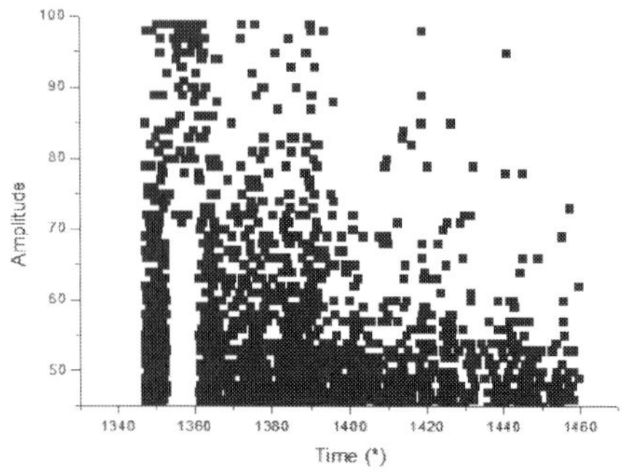

(g)

Figure 5. Amplitude vs Time plot during load holding : (a)30% (b)50% (c)60% (d)70% (e)80% (f)90% and (g)fractured

The pressure rise to the 80 % and 90 % of the expected burst pressure shows a lot of signals with high amplitudes. At the pressure stage of 540 bar or higher, where the burst is likely to occur, the signals with the amplitude of 80-100 dB continue to occur while giving the continuous leak signal from 1352 seconds to 1361 seconds. The burst occured when the pressure rapidly went down to 400 bar. Figure 6 shows the total hits from 2 minutes after the start of each load. There is a rapid increase in the number of hits when the load pressure goes over 70 % of the expected burst pressure where the significant damage is likely to happen. In the final destruction, the number of total hits is low because the time to destruction was short, considering the felicity ratio and the pressure rising up to the fracture.

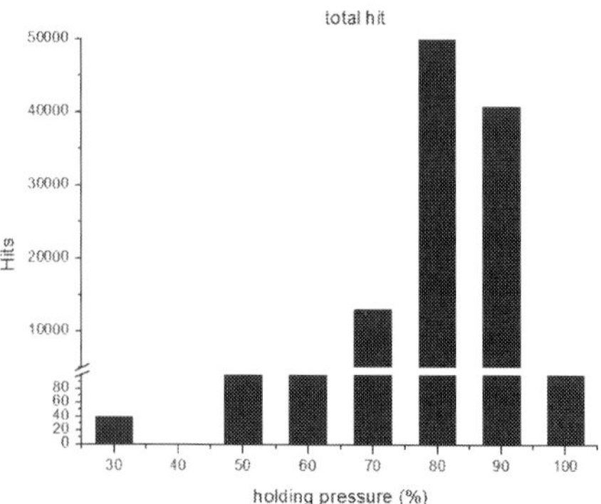

Figure 6. Total hits during load holding

Figure 7 and 8 show that the average rise time and duration for the signals occurred during the holding time of each load. The rise time and duration become shorter up to loads up to 70 % while those from time of 80 % to the time of burst get longer, during which the damage is likely to be greater.

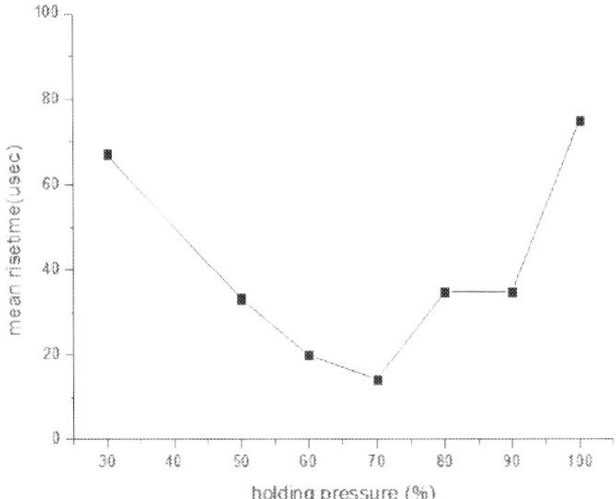

Figure 7. Mean rise time during load holding

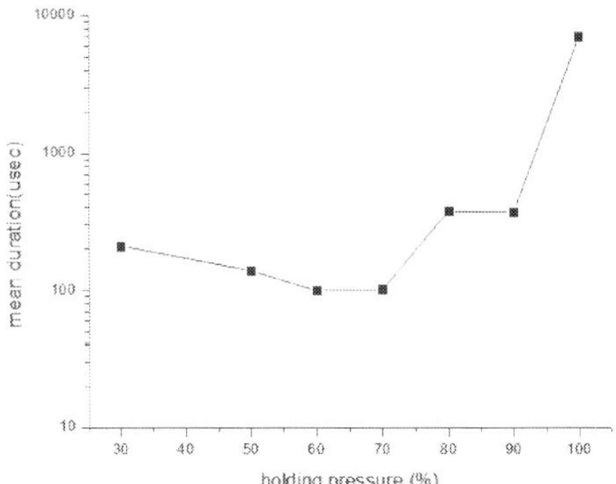

Figure 8. Mean duration during load holding

Unlike the metal vessel, the pressure vessel that is made with composite materials has various damage mechanisms including the matrix crack, crack growth, separation between layers, cutting of fiber, and destruction of metal liner. Figure 9 shows that at the load of 30 % stage, the amplitude of signal rises due to the increased sound emission from matrix crack of composite materials. It also shows that at the load of 50-60 % of the estimated burst pressure, the sound emission is weakened and thus the amplitude of the signal is also reduced. At the load of 70 % pressure, the interlayer separation and the cut of reinforced fiber start and the amplitude of the signal increases again. Then, more reinforcement fibers become severed and the metal liner begins to burst, showing a little increase in amplitude.

As shown below, the composite material pressure vessel has many damage mechanisms in it. The initial damage mechanism contains the matrix crack due to the stress as well as the crack grown. As shown in Figure 10 (a), the rise time appears at the range of 10 μs and 100 μs. This shows the creation of matrix crack and the growth of this crack. The signal from around 10 μs is related to the creation of cracks while the signals from 100 μs look to be related to the growth of the cracks.

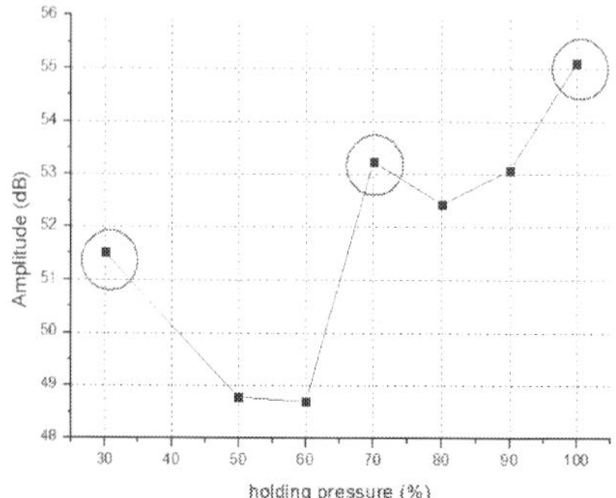

Figure 9. Mean amplitude during load holding

AE signals at the load of 50% and 60% show that the matrix crack and growth occur in this range instead of the additional causes for damages. At the load of 70%, a lot of the rise time of AE signals come from 500 μs appear. It is likely to indicate that the new damage elements such as cut of reinforcement fiber appear. The rapid increase in AE signals around 10 μs shows that the growth of matrix cracks such as growth of existing cracks and interlayer separation occur at a fast pace. Then, at the load of 80~90%, it is estimated that rapid damages such as growth of the previous matrix cracks and the cutting of reinforcement fiber occur, making much sound from cutting of reinforcement fiber. The 800 μs of rise time at this stage is likely to be caused by the cutting of several lines of reinforcement fibers. Then, at 100% load, a composite of damage mechanisms occur, increasing the damage of vessel and then destroying the last metal liner.

Figure 11 shows the total count of sound emission signals which occur during the 2 minutes of holding time. These variables are better indications than mean rise time in Figure 7, mean duration of Figure 8 or mean amplitude of Figure 9. It was discovered that it was difficult to assess the damage against the vessel with variables such as mean amplitude, total hit, mean rise time, or mean duration but the total count and the total signal strength of the sound emitted when the load is pressed on, activation and vessel's damage were better indications for damages on the vessel.

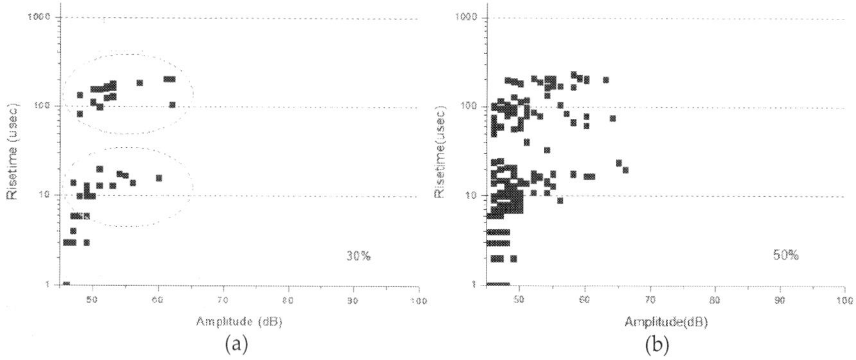

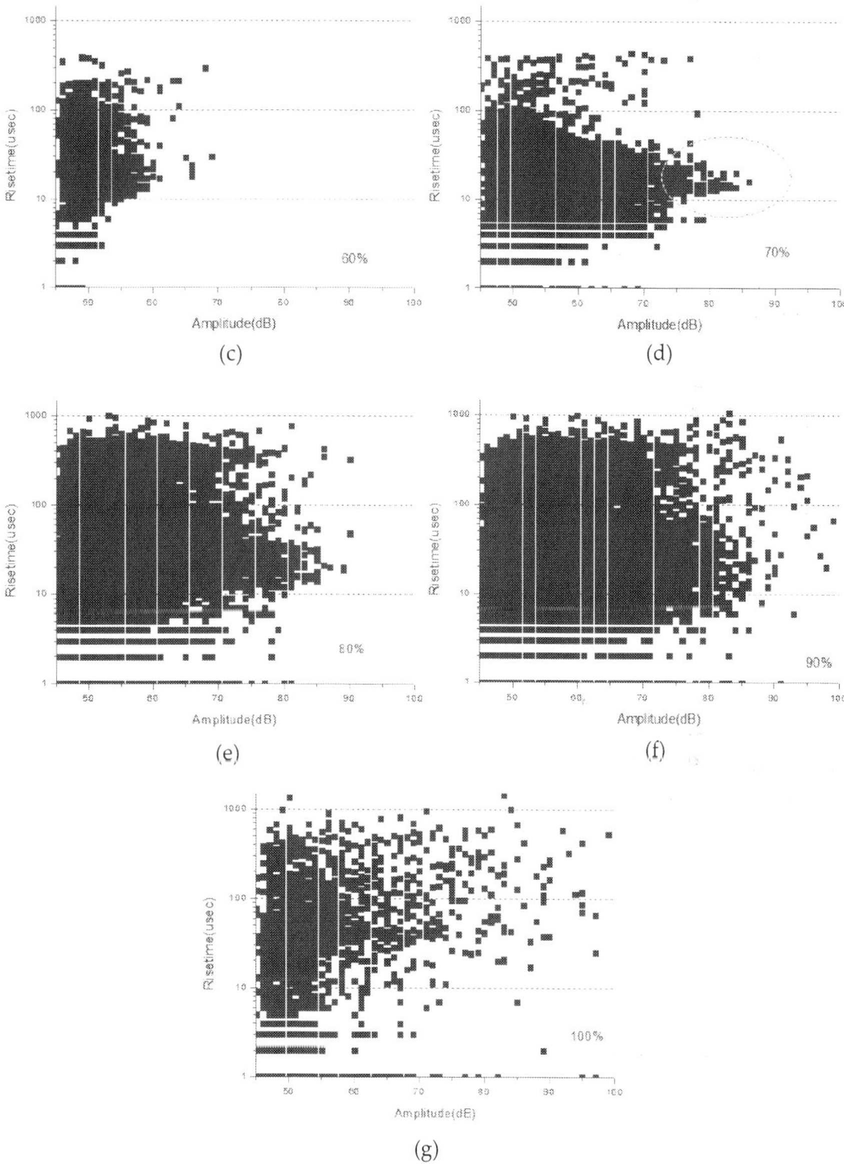

Figure 10. Rise time vs amplitude plot during load holding : (a)30% (b)50% (c)60% (d)70% (e)80% (f)90% and (g)fractured

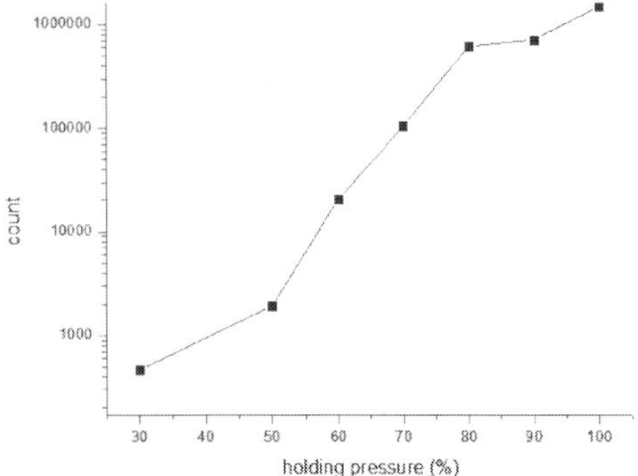

Figure 11. Total count during load holding

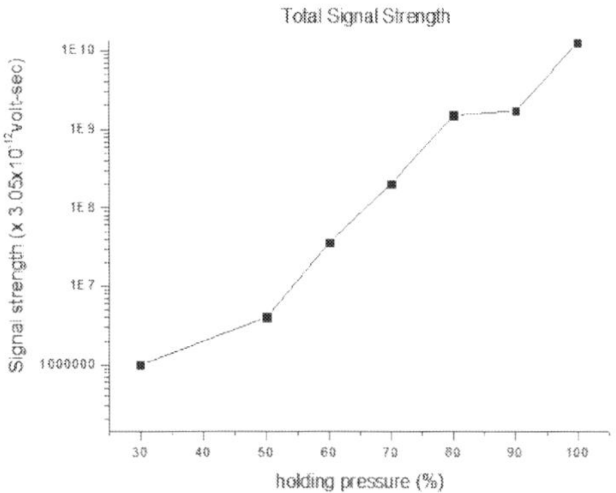

Figure 12. Total signal strength during load holding

3.1.3. Distribution of AE signal during the holding time

I did not use the signals during the initial 2 minutes among the 10 minutes to minimize the flow noise which often includes the initial stages of pressing in the pressure test. After that, I observed the number of hits and creep effect of the sound emission signals which occur during the holding time. Even though there were creation and growth of cracks on substrate under the load of 60 % of the estimated burst pressure, they did not cause much damage to the vessel. But it was found that at the load of 70 % pressure, the vessel was quickly destroyed.[4] Figure 13 shows the distribution of amplitude of signals occurring after 2 minutes of the pressure holding at each load. At the load of 60 % or less, there were not a lot of signals of 60 dB or higher but less than 7 0dB. At this time, I could estimate that at the initial stage of the vessel damage, the composite material substrate wrapping around the vessel were showing some cracks and their growth. I could observe the cracks with my own eyes at the 2000th cycle of fatigue test when the pressure was at 207 bar. The used pressure was more than 30 % of the estimated burst pressure (180 bar). During the burst test, at the load of 30 % to 60 %, even though there was increase in the number of signals, there was not a big increase in amplitude. Accordingly, we now that up to 60 %, there is not much of a creep affect but just the low amplitude of signals. I think that this is mainly caused by the creation and growth of cracks of substrate. After pressing, there were the creep effects, which show the significant increase in AE signal and the signals with the amplitude of 85 dB. In addition, as the distribution of amplitude appears different from the previous one, I think that the damages on the vessel are significant. It seems that there are more creations and growths of cracks in the composite materials and the cutting of reinforcement fiber and peeling of substrate cause the damage mechanism. In addition, the slope of the distribution of amplitude is known to be related to the mechanism of the signals[8], In the load of 70 % or higher, the slop looks similar, meaning that the damage is caused by the similar damage mechanism.

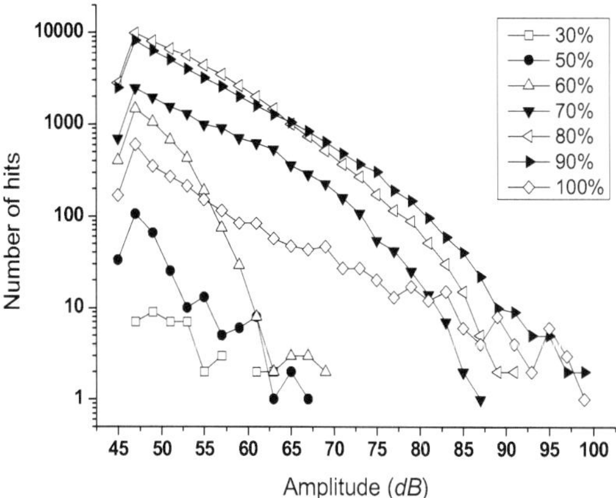

Figure 13. Amplitude distribution during load holding

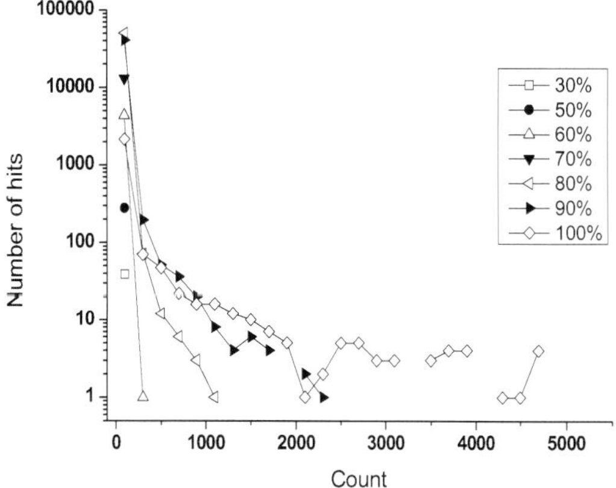

Figure 14. Count distribution during load holding

Figure 14 shows the distribution of counts of AE signals coming from each load of pressure. When the signal with counts of 500 or less occurred under the load of 60 %, the damage was caused by the creation of growth of substrate crack under the load of 60 % pressure. At the pressure of 80-90 % where the significant damage is done to vessel, the number of counts was 1000 or more. At the burst stage the number of counts was 2,500 or more.

Figure 15 shows that the duration of signal occurs with the range of 500 µs at the load of up to 70 % and that the signal of over 10000 µs started to appear at the load of 80 % or higher. At the load of 90 %, the 20000 µs appears during the final burst stage where the signals of 25000 ~ 50000 µs appeared.

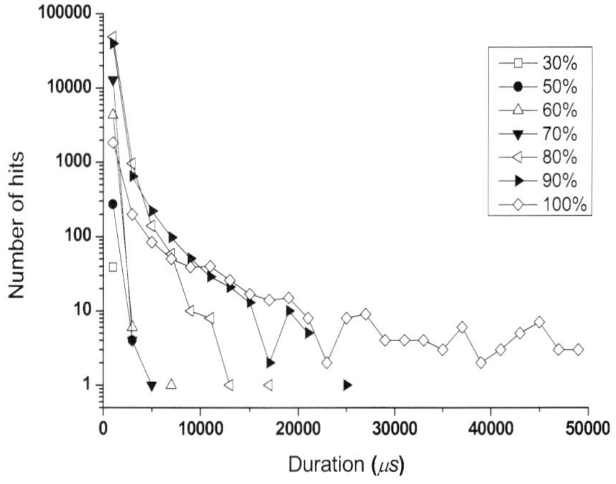

Figure 15. Duration distribution during load holding

Figure 16 shows the rise time of 100 µs or less and the signal of 100-200 µs at the load of up to the initial 30 %. As we understand the distribution of the amplitude from the fatigue test of 20,000 cycles, there may be no mechanism other than the creation and growth of cracks at the load of 30 % pressure. It means that the signal of 200 µs or less is caused by the creation and growth of cracks of the vessel materials. Even though there were 1 and 4 signals of 300 µs or higher at the load of 50-60 %, they are negligible considering the total number of signals. It seems that the damage was mainly caused by creation and growth of cracks on composite materials rather than by new damage mechanisms. At the load of 70 % pressure, the signal of 250-450 µs, which is longer than at previous load stage appeared, meaning that there may be new damage mechanism. It seems that it was caused by the additional damage mechanism such as the cutting of reinforcement fibers and interlayer peeling due to the increase in internal pressures. At the load of 80-90 %, the rise time of 500 µs or more is observed. It may be affected by the composite damage mechanism such as the growth of existing cracks, cutting of reinforcement fiber and inter-layer separation happening at the neighboring location. This trend is true of the distribution of counts and durations as specified above. The observation of the sound emission signals coming out of the neighboring location through the composite damage mechanism would be a good tool to assess the vessel.

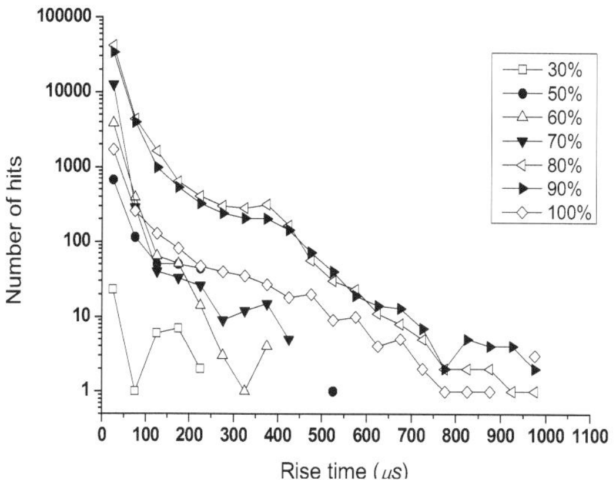

Figure 16. Rise time distribution during load holding

3.1.4. Mean frequency of AE signal during the holding time

To estimate the damage with the sound emission signal, the signal of 60 dB or higher is used. Figure 17 shows the distribution of amplitude for hits of 60 dB or higher for each load of pressure. AE time becomes lower until the pressure of 70 % but at 80 %, it becomes faster. As the pressure rises at constant pace during the test, at 70 % pressure, there is no AE signal until 60 % pressures, showing the Kaiser effect. Accordingly, up to 60 % pressure, it looks like that there was no significant damage to the vessel. As the pressure rises up to 70 % that may have damaged the vessel, it looks like that AE happened due to the Felicity effect when the pressure goes up to 80 % as the vessel was damaged during the rise up to 80 %. At the 90 and 100 % pressure, there is the Felicity effect. Particularly, in the pressure range of up to 100 %, which experienced the load of 90 % or higher, there is AE signals of 60 dB or more from the beginning, showing that the damage is significant. In the actual burst test, the damage mechanism of Type II vessel, or the matrix crack in the direction of initial hoop occurs and then the creation of cracks and its growth and the interlayer separation (which belongs to matrix crack but has different damage mode) occur [4]. At the load of 60 %, the matrix crack in the direction of hoop was shown. But, the matrix crack in the direction of hoop did not affect the burst pressure of the vessel [8].

Figure 18 shows the count of hits where sound exceeds 60 dB at each stage of burst test. The rate decreases from 7 % to 1 % during the pressure of 30 % to 60 %. Then, the signals of 60 dB or higher rapidly go up to 2 0% at the load of 70 %. The rate goes up to 30 % at the load of 100 %. The signals of 60 dB or higher at an early stage may be caused by the creation of matrix crack on the composite materials. The reduction thereafter may be caused by the low amplitude (60 dB or less) rather than the creation of crack. At the load of 70 % or higher, there are not only the matrix cracks but also the damages by the new damage mechanism such as the cut of fibers, increasing the signals of 60 dB or higher. The measurement of the count of hits which are 60 dB or higher is a good tool to assess the damage.

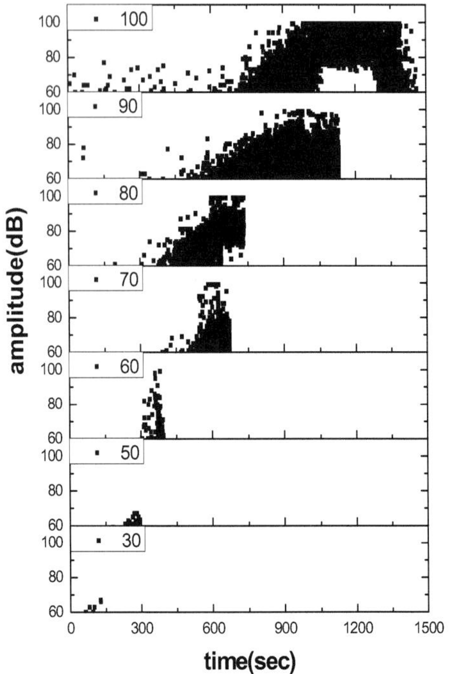

Figure 17. Amplitude distribution during loading at each loading stage (over 60 dB)

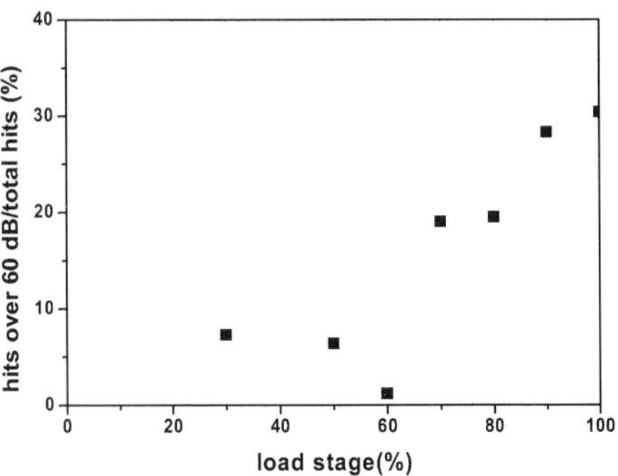

Figure 18. Ratio of hits over 60 dB / total hits during each loading stage

Figure 19-21 show the mean initial, reverberation and average frequency of AE signal at each load stage. The frequency is obtained not from the analysis of wave form but from the duration and counts. The frequency shows the difference between above and below 70 % load. As for the initial frequency, it did not show a big change at the load of up to 60 % or 100 kHz but it goes up at the load of 70 % until going down a little. This is likely to be related to the source mechanism. It shows the creation and growth of matrix crack at the load of up to 60 % and the additional damage mechanism such as the cut of fiber at the load of 70 % or higher. The reverberation frequency shows a little difference between below and above 70% but is constant in most stages while the average frequency was also constant with a little rise at the load of initial 30 % pressure.

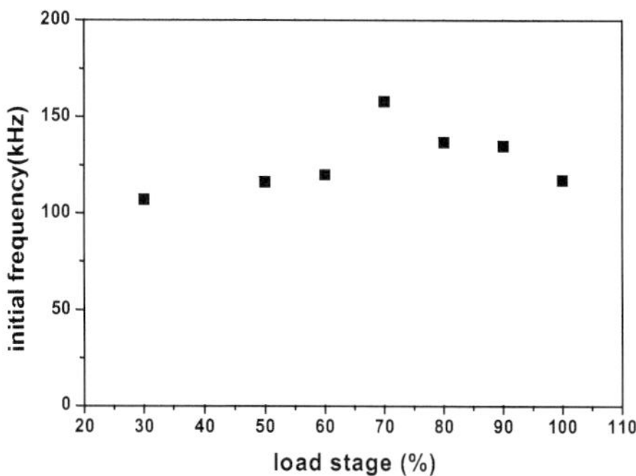

Figure 19. Mean initial frequency during each loading stage

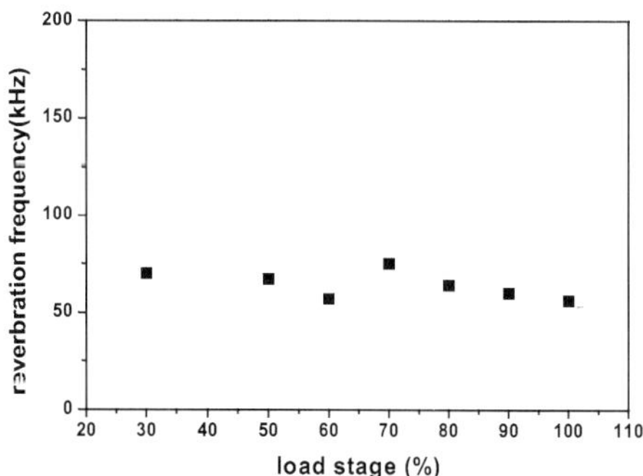

Figure 20. Mean reverberation frequency during each loading stage

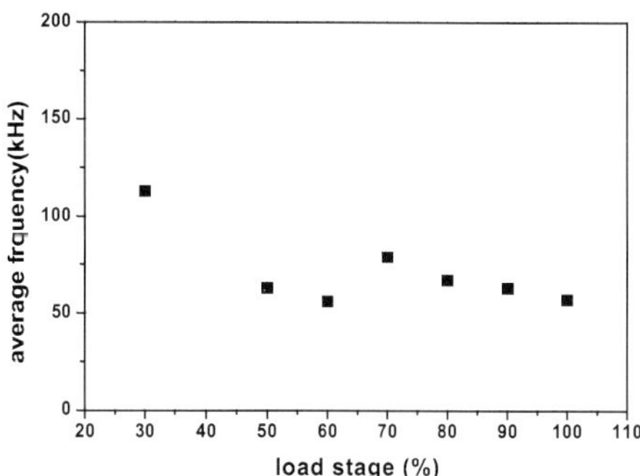

Figure 21. Mean average frequency during each loading stage

3.2. Fatigue test

3.2.1. Identification and Verification of Acoustic Emission Location

Figure 22 shows the results of measuring the elastic wave speed of the artificial acoustic emission source for the acoustic emission location on the fiber-wrapped composite material. As shown on the picture, the speed of longitudinal elastic wave was 4512 m/sec and the elastic wave speed of the wrapped direction was 5689 m/sec showing the characteristic of anisotropy. Thus, the location was identified using anisotropic vessel source location that uses the difference of time for the acoustic emission signal is to be reached.

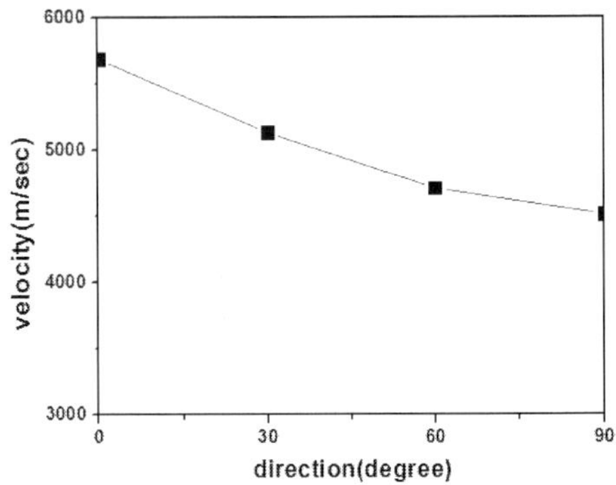

Figure 22. The elastic wave velocity with degree between propagation and wrapping direction

Figure 23 is the location of the sensor and the result of the identification test of the location using the artificial acoustic emission source. The four sensors were attached in staggered locations in channel 5, 6, 7, 8 and channel 7 refers to the backside of 8. The acoustic emission source is located diagonally in equal intervals between channel 5 and 8 and between channel 6 and 8.

Figure 23. Confirm of Source location

3.2.2. Fatigue Test and Location of Artificial Defect

Figure 24 shows the size of the artificial defect realized on the composite material which is wrapped and the length is 50 mm, width 3 mm, and the depth is 3 mm which is 50 % of the thickness of the wrapping composite material. The direction of the realized defect was longitudinal and transverse, two types of artificial defects.

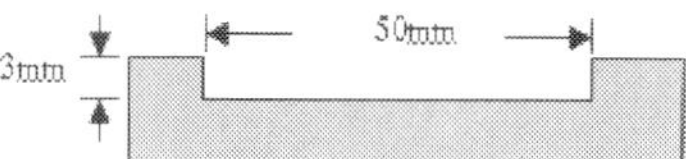

Figure 24. Schematic diagram of artificial defect

Figure 25 shows the average number of hit per sensor during the fatigue test of vessel. The vessel with the artificial defect has more number of hits compared to the sound vessel and when you see the tendency in increase and decrease, the number of hits for transverse defect vessel and sound vessel decreases as the number of cycle increases; however, the number of hits for longitudinal defect vessel decreases in 4000 cycle and then increases and then decreases in 8000 and 12000 cycles.

The reason the number of hit in the early stages is big is related to the initiation and growth of the matrix rupture in the comparatively weak areas within the vessel. The number of hits decrease and the initiation and growth of ruptures are comparatively slowed until sufficient amount of resilience is stored to bring about new initiation and growth.

In the case of longitudinal defects vessel, the amount of resilience needed to bring about progress in the defect (growth of rupture) is comparatively smaller than other vessel, so in the 8000th and 12000th fatigue test, the number of hits is bigger than in other vessel and because there is a need for another accumulation of resilience, the number of hits decrease afterwards. Such phenomena are clearly distinct in the composite material as mentioned in the introduction and in the early stages of damage, resilience increases like the number of hits and when there are some increases in damage, the resilience afterwards is comparatively low.

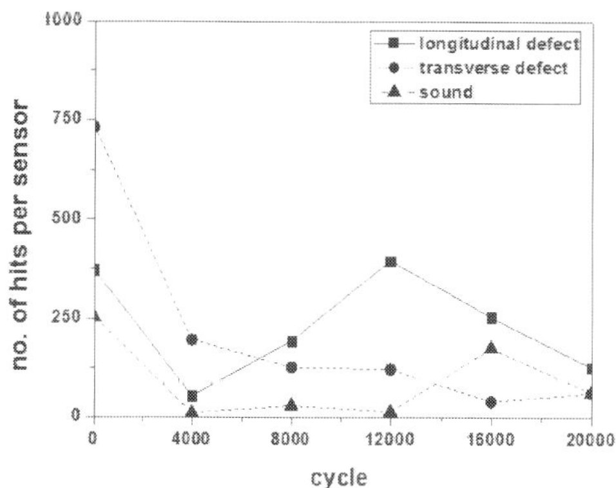

Figure 25. The number of hits per channel during fatigue test for artificial defect and sound vessel

Figure 26 shows the number of events extracted during the fatigue test on longitudinal defect and transverse defect vessel. An event shows the number of sources calculating the location of the acoustic emission source within vessel using the acoustic emission signal that hit the sensor and in the case transverse defect vessel which includes a transverse defect, the number of events is noticeably decreased as the number of cycle increases, but in the case of longitudinal defect vessel which has longitudinal defect, the number of events increases a lot in 8000th cycle and does not change much as the number of cycle increases.

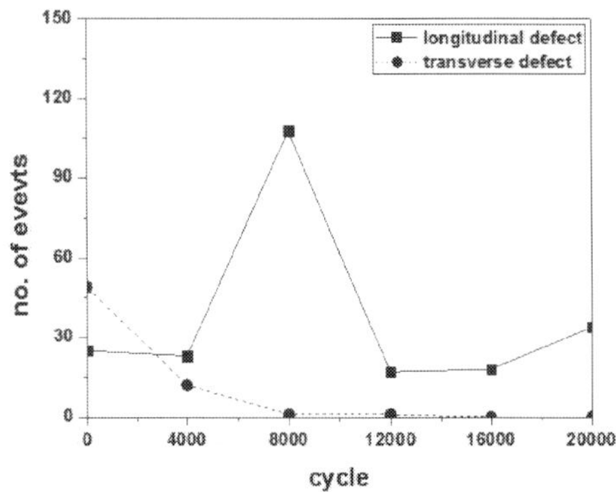

Figure 26. The number of events during fatigue test for artificial defect vessel

Figure 27 compares the number of hit per sensor and the number of events using Figure 25 and 26 and shows the event/hit ratio according to the increase in the number of cycle. This is a figure that shows the % of the number of hits that can precisely show the source among the total number of hits. In the case of longitudinal defect vessel, it was 41.8 % in the 4000th cycle, 55.7 % in the 8000th cycle. An event can be calculated using the difference in time that the resilience from the source (acoustic emission signal) sent through the walls of the vessel reaches the sensor and has to be extracted from at least 3 sensors. Hits that do not go by such standards cannot be used to calculate events and if the location was identified by simulation or if the signal is weak or is static, it cannot be recorded as an event. In the case of longitudinal defect vessel, the number of hits is small in the 4000th and 8000th cycle but the growth of the rupture is relatively easy and it sends hits to at least 3 sensors with signals with sufficient amplitude so the number of event/hit rate increases.

Figure 28 is the result of the location due to the acoustic emission test performed during the fatigue test on longitudinal defect vessel. Figure 28 (a) shows the location of 25 events during the first 3 fatigue test cycles. It does not show the location of artificial defects but shows the overall looks on the whole vessel. Such results are not shown in pictures but also in transverse defects vessel, 49 events in Figure 26 shows looks like a) on the whole vessel and it seems that signals were created on the weakest areas of the whole vessel when the first pressure was put. In the case of b), after the 4,000th cycle, in the 3rd fatigue test, 23 events were shown to be clustered around the artificial defect. As for the vessel with transverse defects, after the 4,000th fatigue test cycle, in the 3rd fatigue test, 12 events were created but they were all over the whole vessel so we could not show the location of the artificial defect.

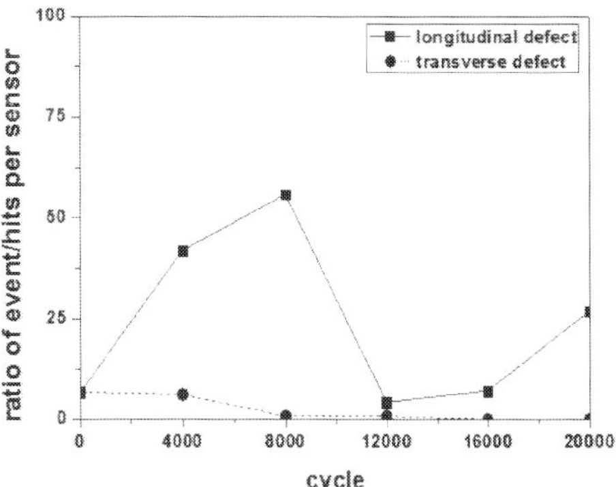

Figure 27. The ratio of events/ hits per sensor during fatigue test for artificial defect vessel

In the case of c), in the 8000th fatigue cycle, 108 events were clustered around the artificial defect and as mentioned in the explanation in Figure 27 more than 50% of the occurred hits were signals related to the artificial defect. Afterwards, events that occurred in d) ~f) were those that mostly occurred near the artificial defect so we can acknowledge that the damage on the composite material is progressed around the artificial defect. Figure 29 shows the surroundings of the longitudinal artificial defect after the 20000th cycle and shows that at the end of the defect, there is a matrix rupture progressing in a hoop-direction and although not clear in the picture, at the end of the depth in the artificial defect, delaminating was observed on the overall defect. In the case of vessel with transverse defects, after the 8,000th fatigue test, less than 1 event was created and the location of the artificial defect could not be shown clearly.

On the other hand, after the 20000th fatigue test, in the burst test, the location of the burst is marked in c) of Figure 28 and events are also observed in a), d), e), and f). The source of the acoustic emission signal is assumed to be the fatigue rupture in the weak areas of the steel liner rather than in the composite material. The final burst in the case of the longitudinal defect accompanies matrix rupture and delaminating as mentioned above during the fatigue test and burst test so the whole vessel area, which is the length of the defect, is thinner in terms of thickness like the depth of the defect and thus is weaker than other areas and it is thought that it burst at the final burst location in which the fatigue rupture occurred in the steel liner.

3.2.3. AE parameters during fatigue test

Figure 30 shows the relationship between the amplitude of the signal occurring during the early three cycles and the rise time, and can be clearly distinguished as around 10 µs and over 100 µs and the grey mark shows the rise time while holding the load during the three cycles and also at 90 %, which is the highest, has a rise time of about 10 µs. Generally while load holding, it can be inferred that there is likely to be growth of an existent crack rather

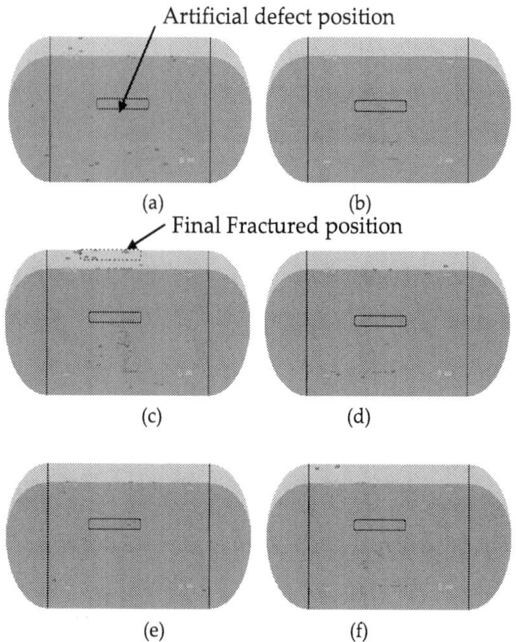

Figure 28. The result of source location with cycle for longitudinal defect : a)0, b)4000, c)8000, d)12000, e)16000, f)20000

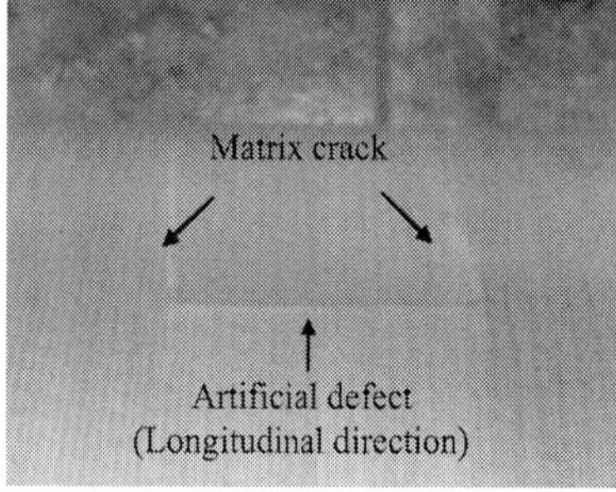

Figure 29. Longitudinal defect and matrix crack after 20000 cycle fatigue test

than initiation of a new matrix crack and thus can be said that it is a growth of crack around 10 μs. And the rise time of AE signal which occurs during the initiation of a matrix crack can be said to be more than 100μs. This accord to the result that the rise time of the AE signal occurring during the initiation of matrix crack during the burst test is around 100 μs and that the rise time of the AE signal occurring during the growth of the crack is around 10 μs. [4]

On the other hand, in the case of a sound cylinder, after the 4000th fatigue test, the average rise time noticeably decreased and increased little up to the 20000th test. After the 4000th fatigue test, there are not many initiations of new matrix cracks and you can see the growth of the cracks created in early stages. On the other hand, in the case of vessel with defects, the average rise time of related hits for events occurred during the first three fatigue test was 56 μs and afterwards decreased to 34 μs after the 8000th cycle and increased to 82 μs about after the 12000th cycle and then decreased again.

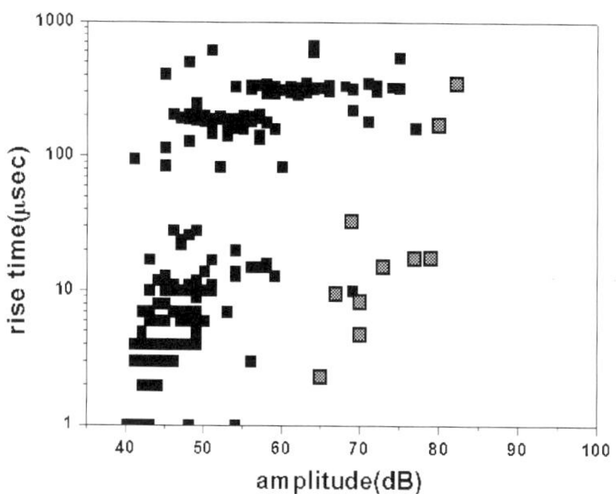

Figure 30. The rise time distribution during initial 3 cycles for sound vessel

Figure 31 a) shows the relationship between the amplitude of the signals of events occurring during the first three cycles of the vessel and related hits and rise time and is clustered around 10 μs in the figure and is also scattered around 100-800 μs. It is judged that the initiation and growth of matrix crack will happen in weak areas of the overall vessel. Especially, a higher rise time can be observed when the source forms a cluster around the defect as shown in b), c), e), f) of Figure 31 than when the source is scattered overall on the vessel like a) or d) so the growth of cracks around the defect has a lower rise time than the initiation of cracks.

On the other hand, after the 4000th fatigue cycle, the average rise time of the signal occurring in defect vessel are shown to be higher than the average rise time of signals occurring in sound vessel.

In the case of a sound vessel, the growth of individual cracks occurs far away in other cracks. But, in the case of vessel with defects, the growth of multiple cracks occurs around the defects because the elastic energy needed for the growth of matrix crack in the vessel with defect is smaller than that of sound vessel. And generated signals overlap with other signals. Therefore, the rise time of signal is extended.

Generally, in order to analysis of frequency, we use wideband sensors to detect signals, and analyze by the Fourier transform of detected signals. However, because the sensitivity of wideband sensors fall behind resonant sensors, signals are also detected using resonant sensors and the frequency is calculated by rise time, duration, and count. Frequency by calculation was used and analyzed in this research because resonant sensors were used for the source location.

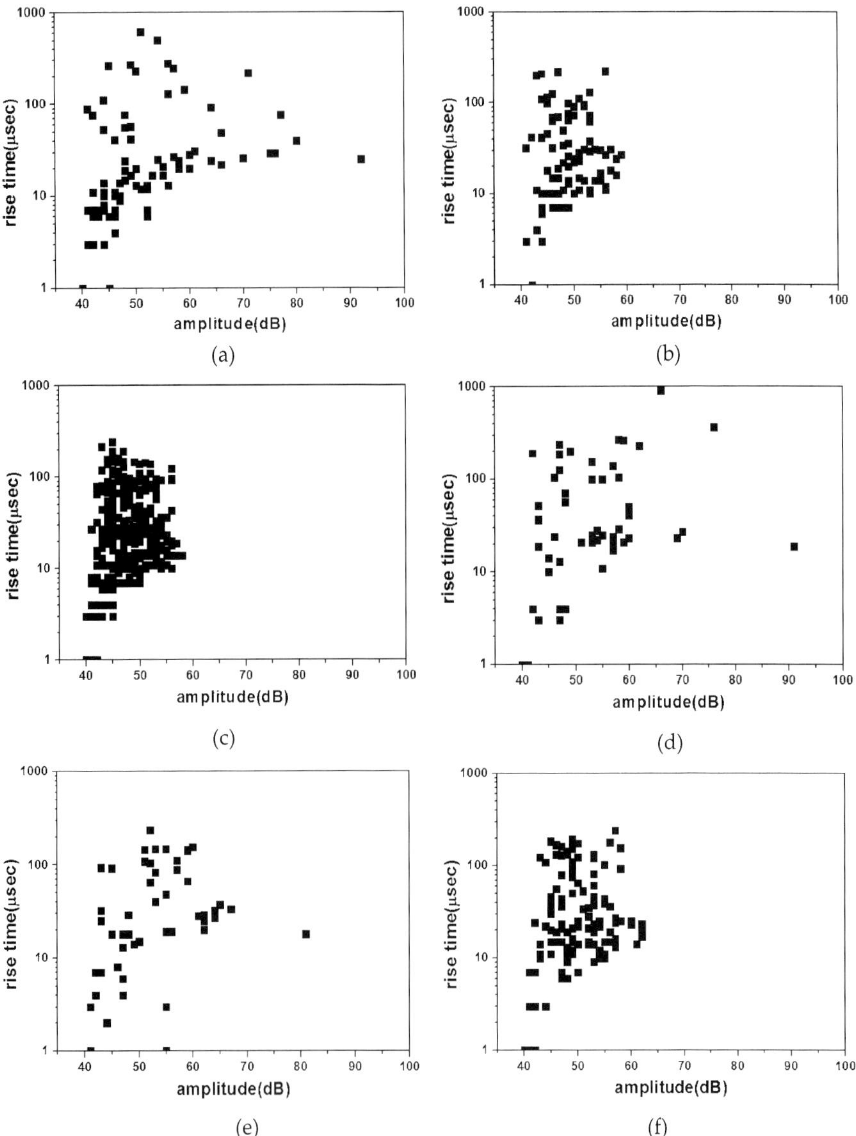

Figure 31. The rise time distribution of hits connected with event during fatigue cycle for longitudinal defect vessel: a) 0, b) 4000, c) 8000, d) 12000, e) 16000, f) 20000 cycles

Figure 32 shows the distribution of initial frequency versus the amplitude with fatigue cycle but it is dispersed around 100-200 kHz, which is unrelated to the fatigue cycle. It is in accord with the result that the average frequency of signal which is generated due to initiation and growth of matrix crack is around 100 kHz in the burst test using resonant type sensors [7].

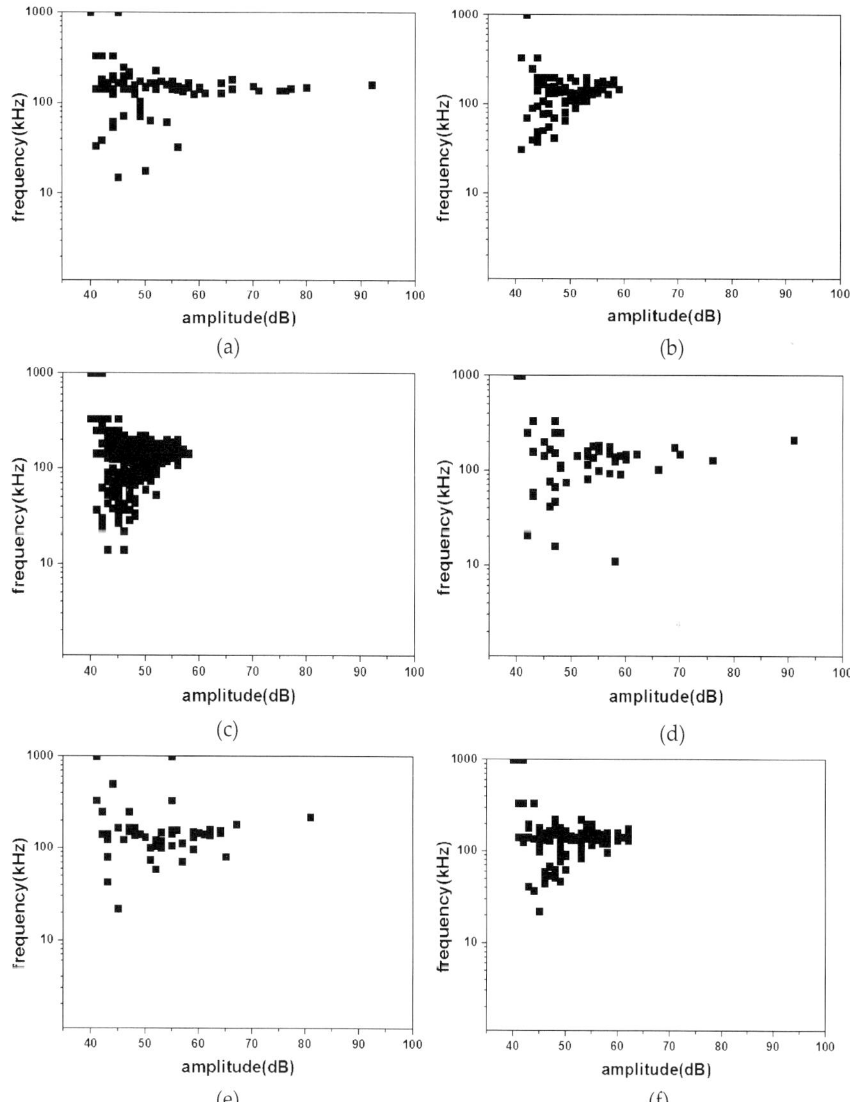

Figure 32. The initial frequency distribution of hits connected with event during fatigue cycle for longitudinal defect vessel: a) 0, b) 4000, c) 8000, d) 12000, e) 16000, f) 20000 cycles

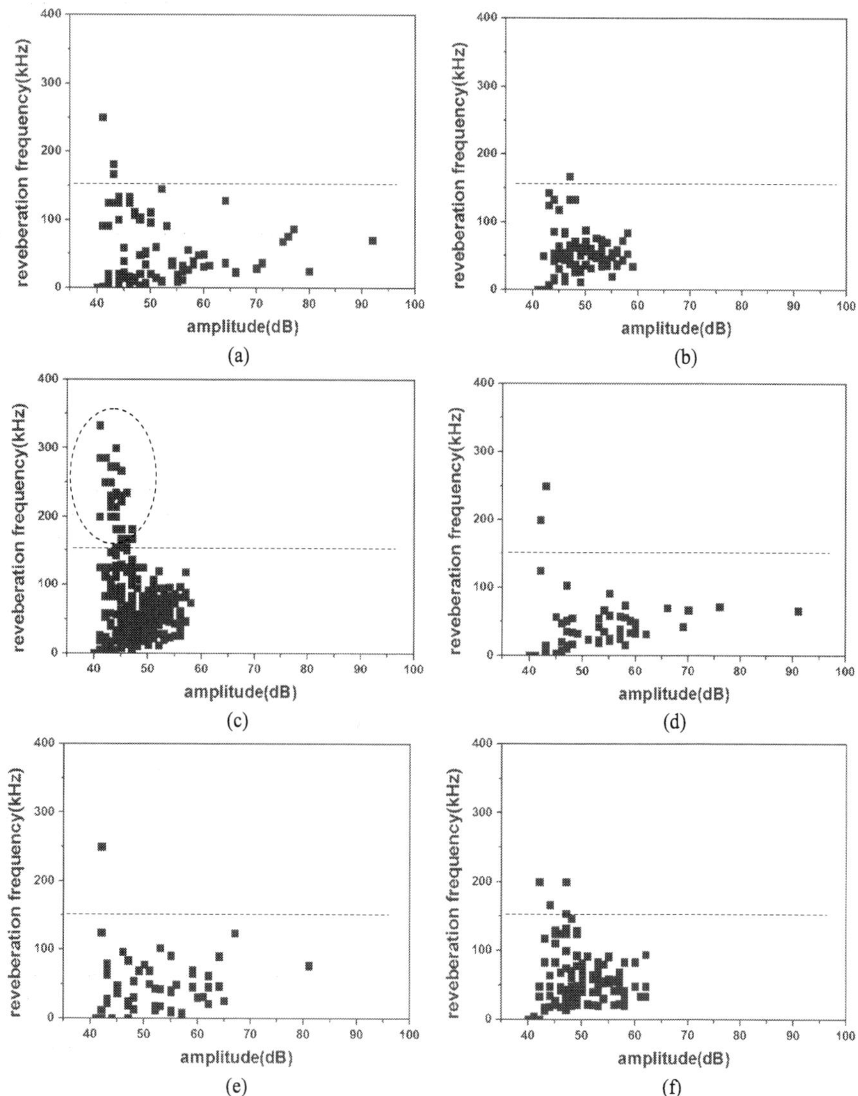

Figure 33. The reverberation frequency distribution of hits connected with event during fatigue cycle for longitudinal defect vessel: a) 0, b) 4000, c) 8000, d) 12000, e) 16000, f) 20000 cycles

Figure 33 shows the distribution of reverberation frequency on amplitude according to the number of fatigue tests but besides the oval area in Figure 33 c) which is the result of the 8000th fatigue test, most are dispersed below 150 kHz. The average reverberation frequency on signals relating to all events of Figure 33 c) was 73 kHz and these accords to the fact that the average reverberation frequency is between 50 - 75 kHz in all stage of burst which includes the damage mechanism of composite materials in the cast of burst test.[7] On the other hand, the reverberation frequency during the burst test of the same vessel was known to occur in all stage of the burst test in the range of 150 - 350 kHz[8], and in such case, because it includes the mechanism of all damage, it is hard to differentiate damage mechanism as a frequency. There were 23 event signals that had a reverberation frequency in the 150 - 350 kHz area in Figure 33 c) and 16 of them, which are 70 %, occurred in the matrix crack area of Figure 29, which is an observation of artificial defects after the 20000th fatigue test. More than 90 % of the related 44 hits(150 - 350 kHz) were signals with rise time lower than 100 μs and average rise time of 31 μs. As mentioned in 3.2.1, it is assumed that it is due to the growth of matrix crack.

3.2.4. Amplitude distribution slop during fatigue test

Figure 34 shows the amplitude distribution of accumulated hits according to the number of fatigue test and its slope has two types of shapes. Generally, the slope in the amplitude distribution of accumulated hit is known to be related to the mechanism of the source [8] and in vessel with defects as used in the experiment, as mentioned in the previous chapter, it includes mechanisms such as the initiation of matrix crack, growth of the created cracks (including delaminating), and the initiation and growth of liner fatigue cracks.

Initiation and growth of liner fatigue crack will be mentioned in the following chapter and in the view of the estimated result of the damage mechanism according to the number of fatigue explained in the previous chapter, the case of the initiation of matrix cracks is estimated to have a slope of ①(0.04) and the growth of cracks, a slope of ②(0.12).

In order to analyze the characteristics of acoustic emission signals that occurred in the final burst position, event signals observed around the final burst location within 200 mm hoop direction in terms of length as marked in Figure 29 were analyzed. 18 events were occurred during the 20000th fatigue cycle and the number of related hits was 59. Figure 35 shows accumulated amplitude distribution and there are only 3 that are over 60 dB and 49 of them are below 50 dB. You can see the slope as ③ but if you observe closely, it is possible to observe that the same slope exists as ① in Figure 34 and also that it include the initiation of matrix crack. Seeing it as showing the slope of ③, as a signal according to a sole damage mechanism, it is estimated to be related to liner damage and the size is 0.06

Figure 36 shows the distribution of rise time on the amplitude of the signals occurred in the final burst location. There is almost no rise time above 100 μs and is dispersed around 10 μs. It can be thought that the rise time which occurs during the growth of steel liner fatigue crack is similar to that during the growth of matrix crack.

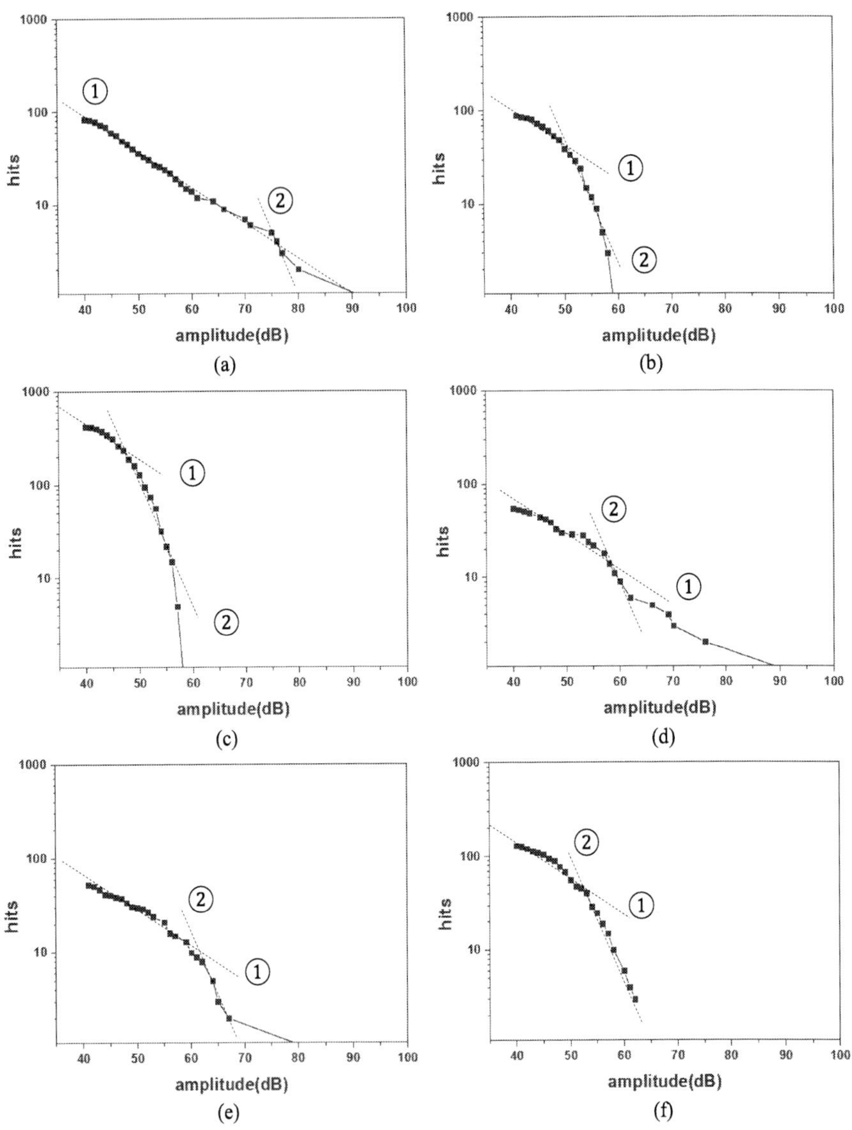

Figure 34. The amplitude distribution of accumulated hits connected with event during fatigue cycle for longitudinal defect vessel: a) 0, b) 4000, c) 8000, d) 12000, e) 16000, f) 20000 cycles

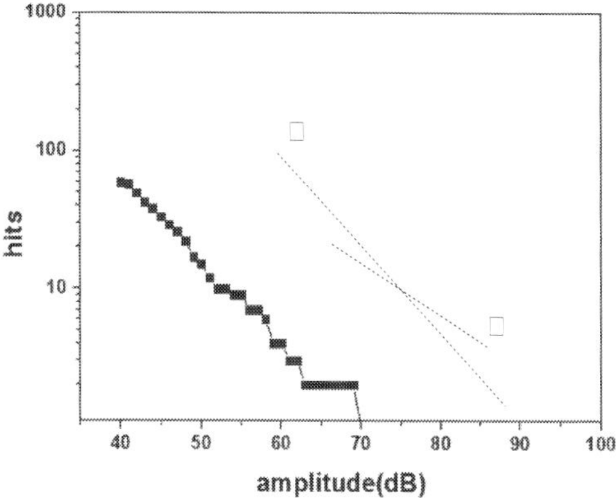

Figure 35. The amplitude distribution of accumulated hits connected with event during fatigue cycle for longitudinal defect vessel

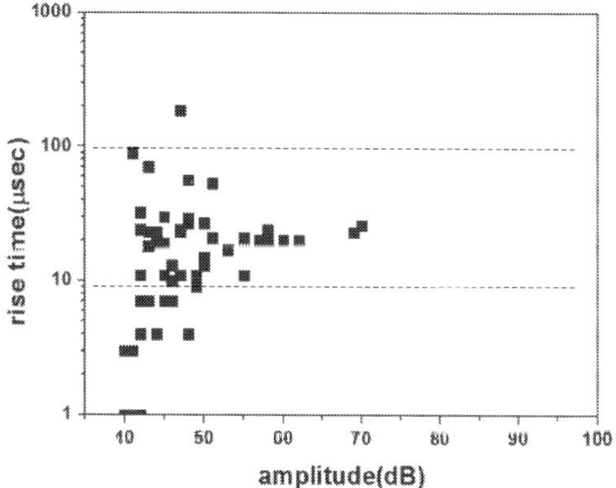

Figure 36. Distribution of rise time on the amplitude of the signals

3.2.5. Burst test after 20000 cycles fatigue test

We burst the vessel with two types of artificial defect after carrying out 20000 cycles fatigue test on a sound vessel and continuously increasing the pressure where the burst pressure is 590~615 bar with the difference in the pressure of the vessel were within 5 % and thus was irrelevant to the existence of defects.

Generally, 20000 cycles of fatigue is equivalent to a vessel used for more than 50 years if you are to put pressure on the vessel once a day although, of course, in the case of a real gas vessel, gas is used as a pressure medium so it may be different from the case in which machine oil is used as a pressure medium, but if the vessel with artificial defects and sound vessel were tested in the same conditions, the burst pressure is shown to be almost the same. Thus, in the case of the size of artificial defect used in this research, the direction of defect is shown to have almost no effect on the life of the vessel.

Figure 37 is a picture that shows the burst location in the vessel with the artificial defect. As shown in the picture, we can see that in the case of a vessel with a transverse defect, the final burst location is in the general burst location (cylinder and head area) for a well-constructed type II vessel. However, in the case of a vessel with a longitudinal defect, in both vessels, the final burst location was within the transverse vessel in which the defect was located. We think that this is because in the case of the longitudinal defect, the wrapped fiber is cut in 3 mm depth and in 50 mm length so the effect in which the thickness of the composite material is big, but in the case of the transverse defect, the fiber is cut only in 3 mm depth and in 3 mm length so the effect is small.

In this research, we cannot precisely know how much the depth of the defect has to be in order for the final burst pressure to change; however, the direction of the defect and the final burst location do have a correlation and we can infer that the longitudinal defect has a bigger effect on the final burst location.

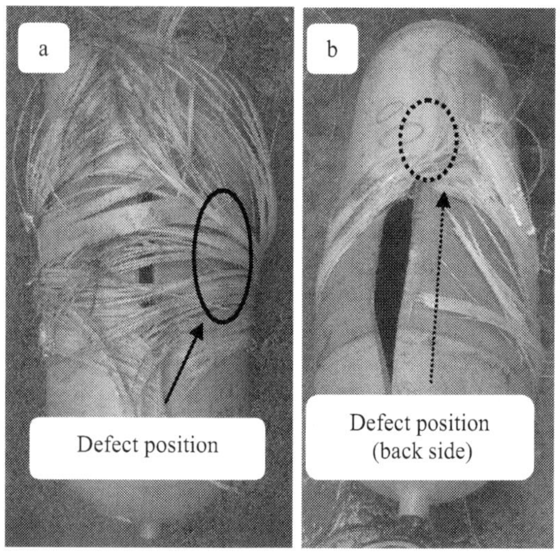

Figure 37. Position of artificial defect and burst location : a) longitudinal, b) transverse

4. Conclusion

4.1. Burst test

By increasing the loading of pressure up to the expected burst pressure, I obtained the sound emission signal during 10 minutes holding time after loading. Considering that there would be flow noises during the initial 2 minutes, I obtained data for the remaining 8 minutes except the initial 2 minutes to analyze the AE variables.

1. Up to 360 bar, or 60 % of the estimated burst pressure, which is equivalent to 1.8 times of the usage pressure, it seems that there is little creep effect as there is little damage to

the vessel. If the pressure is over 420 bar, or 70 % of the estimated burst pressure, the damage to the vessel becomes greater, meaning that the creep effect becomes larger.

2. The sound emission signal variables such as mean amplitude, mean rise time, means duration, and rise time amplitude correlation can be obtained when the vessel is damaged at each stage of pressure load. Though the variables were not enough to evaluate but were effective to estimate the damage mechanism.

3. It was discovered that the total count, and total signal strength at the pressure holding stage were the sound emission variables, which represent the degree of damages on the vessel.

4. The rate of number of hits with 60dB or higher in amplitudes in the number of total hits is likely to indicate a damage of vessel.

5. We can estimate the damage mechanism through mean rise time, mean amplitude and frequency analysis.

4.2. Fatigue test

After manufacturing a sound vessel and a vessel with artificial defect for the composite vessel, we executed acoustic emission test during fatigue test and came up with the following conclusion.

1. Vessel with two types of artificial defects (longitudinal and transverse) and a sound vessel was put in 20000 fatigue test and the pressure was continuously increased and then was burst, the burst pressure was 590~615 bar and the differences in pressure on the vessel were less than 5 % and was not relevant to the existence of defects.

2. There is a correlation between the direction of defect and the final burst location and the longitudinal defect had a greater effect on the final burst location of the vessel rather than the transverse defect.

3. Acoustic Emission Signal, which occurs during the fatigue test, occurred more in vessel with defects rather than in sound vessel, and as the number of fatigue test accumulated, the number of hits increased more in vessel with longitudinal defects than in those with transverse defects.

4. In the case of vessel with longitudinal defects, events were clustered around the artificial defect and more than 50 % of the occurred hits were signals that were related to artificial defects and the source location was precisely found on the defect location but in the case of vessel with transverse defect, events rarely occurred and even if they occurred, the source location relevant to a defect did not match.

5. Longitudinal defect of the vessel created matrix rupture and delaminating of the composite material during the fatigue test and the burst test and the thickness of the whole vessel area became thinner like the length of the defect and thus was weaker than other areas of the vessel. And the final burst was at the location in which the fatigue rupture of the steel liner occurred.

6. The position of the longitudinal defect was shown well using the identification of acoustic emission location during the fatigue test and the average rise time of acoustic emission signal related to events occurring here was about 30-90 µs and signals with a shorter rise time can be observed more in the growth rather than in the initiation of matrix cracks.

7. The initial frequency is distributed around 100-200 kHz and signals with reverberation frequency higher than 150 kHz are related to the growth of matrix cracks.

8. The slope of accumulated amplitude distribution is about 0.04 during the initiation of matrix cracks and is about 0.12 during the growth. However, signals estimated to be liner fatigue crack growth have a slope of 0.06 and a rise time similar to that of during the growth of matrix cracks.

Author details

Hyun-Sup Jee and Jong-O Lee
Korea Institute of Materials Science, South Korea

5. References

[1] Statics of korean Association for Natural Gas Vehicles (2009)

[2] Mark Toughiry (2002) Examination Of The Nondestructive Evaluation Of Composite Gas Cylinders, United States Department of Transportation, NTIAC/A7621-18:CRC-CD8.1, 10

[3] General Motors Corporation (1997) Development of Inspection Technology for NGV Fuel Tanks, FaAA-SF-R-97-05-04

[4] H. S. Jee, J. O. Lee, N. H. Ju and J. K. Lee (2011) Study of acoustic emission parameters during a burst test for CNG vehicle fuel tank, Journal of KSME, 35(9), 1131

[5] J. O. Lee, J. S. Lee, U. H. Yoon and S. H. Lee (1996) Evaluation of adhesive bonding quality by Acoustic emission, Journal of KSNT, 16(2), 79

[6] H. S. Jee, J. O. Lee, N. H. Ju, J. K. Lee and C. H. So (2011) Development of in-service inspection for type-ll gas cylinder, Proceeding for Spring Conference of KIGAS

[7] H. S. Jee, J. O. Lee, N. H. Ju, J. K. Lee and C. H. So (2011) Damage Evaluation for High Pressure Fuel Tank by Analysis of AE Parameters Journal of KSCM, 24(4), 25

[8] S. Yuyama, T. Kishi and Y. Hisamatsu (1982) Detection and Analysis of Crevice Corrosion-SCC Process by the Use of AE Technique, The Iron and Steel Institute of Japan (ISIJ), 64(14),2019-2028

[9] S. H. Paik, S. H. Park and S. J. Kim (2001) Three dimensional FE analysis of acoustic emission of composite plate, Journal of KSCM, 18(5), 15-20

[10] J. S. Park, K. S. Kim and H. S. Lee (2003) A study on the acoustic emission characteristics of laminated composite structures, Journal of KSCM, 16(6), 16.

[11] L. Dong and J. Mistry (1998) Acoustic emission monitoring of composite cylinder, Composite Structures, 40(2), 149-158

[12] J. C. Choi, J. S. Jung, C. Kim, Y. Choi and J. H. Yoon (2002) A study on the Development of computer-aided Process planning system for the deep drawing & Ironing of high pressure gas cylinder, Journal of KSPE, 19(2), 177-186

[13] Y. Choi, J. H. Yooh, Y. S. Park and J. C. Choi (2004) A study on the Die Design for manufacturing of High Pressure Gas cylinder, Journal of KSPE, 21(7), 153-162

[14] A. Bussiba, M. Kupiec, S. Ifergane, R. Piat and T. Bohlke (2008) Damage evaluation and fracture events sequence in various composites by acoustic emission technique, Composite Science and Technology, 68, 1144-1155

Locating Delamination in Composite Laminated Beams Using the Zero-Order Mode of Lamb Waves

Yaolu Liu, Alamusi, Jinhua Li, Huiming Ning,
Liangke Wu, Weifeng Yuan, Bin Gu and Ning Hu

Additional information is available at the end of the chapter

http://dx.doi.org/10.5772/49991

1. Introduction

To improve the safety and reliability of various engineering structure, it is essential to develop efficient techniques for non-destructive damage detection or structural health monitoring. Lamb wave can travel a long distance in plate-like and shell-like structures made of materials even with high attenuation ratio (e.g. Carbon Fibre/Epoxy Polymer composites). To take this advantage, many researchers have recently explored the possibility of using Lame waves for damage identification [1]. To date, many developed Lamb wave-based techniques are generally based on so called two stage prediction models by which the difference in the signals between a defective structure and a benchmark (intact structure) can be evaluated. Then, the residual error is easy to be defined no matter what information extracted from the signals is used, such as the information in time domain [2, 3] or frequency domain [4, 5]. Therefore, a benchmark or baseline signal is essential for the detection, which is very reliable and suitable for monitoring the propagation of damage. Also, tremendous efforts have been put to the delamination identification, which could be treated as a problem of inverse pattern recognition using calibrated numerical methods such as artificial neural network [6]. The interaction between Lamb waves and delamination has also been investigated numerically and theoretically [7-9]. However, the complex wave scattering phenomenon in a delamination area has not been clearly understood in these studies.

In this chapter, a technique for delamination identification in laminated composites using the zero-order mode of Lamb waves, i.e., S0 mode and A0 mode, without referring to the baseline data, is described. Through measuring the propagation speed of a wave and the traveling time of a reflected wave from the delamination, the delamination position can be

accurately identified. Moreover, to understand the complex interaction of Lamb waves with a long delamination damage, the numerical simulations have been carried out.

2. Detective technique without the baseline data

2.1. Experiments and numerical analyses

In general, compared with S0 mode, the attenuation of A0 mode in structures is more severe. Here, S0 mode corresponds to an axial deformation mode while A0 mode corresponds to a flexural deformation mode. Therefore, it should be paid more attention to the detective capability of both modes for a delamination damage in the different interfaces along the thickness direction. Moreover, the experimental setup of S0 mode is slightly different with that of A0 mode.

2.1.1. Materials and experimental procedure

For S0 mode, as shown in Fig. 1, a CFRP laminated composite beam of stack sequence of $[0_{10}/90_{12}/0_{10}]$ was used. An artificial delamination damage with different lengths for several cases, i.e., 30 mm, 20 mm and 10 mm, respectively, was intentionally created at the interface between the 10th and 11th plies by inserting a Teflon film with a thickness of 25 μm. A PZT actuator was attached on the top surface of the left end of the beam. The PZT actuator had a diameter of 10 mm and a thickness of 0.5 mm. In this case, the generated waves from the actuator merged into the reflected waves from the left end of beam, which leads to a simpler signal. And the same PZT unit was used as a sensor to pick up the reflected wave from the delamination damage. Because the propagation speed of S0 mode is much higher than that of A0 mode, confirmed as 4 times higher in the used composites from our testing, it is expected that in a specified time domain, no A0 mode will be collected due to its slow speed, making the analysis of S0 modes much easier.

For the A0 mode, a similar composite beam was used, as shown in Fig. 2. Two kinds of stack sequence of the laminated beam were used, i.e., $[0_{10}/90_{12}/0_{10}]$ and $[0_{12}/0_4/0_4/0_{12}]$. Note that $[0_{12}/0_4/0_4/0_{12}]$ is a unidirectional laminate, and it can be simply expressed as $[0_{32}]$. To simplify the description of delamination later, this expression is used. Like the case of S0 mode, a delamination damage with different lengths, i.e., 30 mm, 20 mm and 10 mm, was intentionally created at the interface between two plies by inserting a Teflon film with a thickness of 25 μm. The distance between the center of the delamination and the left end of beam is 790 mm. To generate the A0 mode, two actuators were attached on the top and bottom surfaces of the beam with applied out-of-phase voltages because the difference of signals in the two actuators can produce the pure A_0 mode, as shown in Fig. 2. Although this arrangement can generate a relatively pure A0 mode, there are still reflections in S0 mode due to the mode change caused by the scattering between the Lamb waves and the delamination. To pick up the pure A_0 mode, two PZT sensors were attached. We know that two components of A0 mode in the two sensor signals are of the electrical charges of opposite signs, and two components of S0 mode in the two sensor signals are of the electrical charges of same signs. Naturally, the difference of two signals can yield the pure A0 mode. In experiments, an excitation signal in the following Eq. (1), was adopted to generate S0 or A0 mode.

$$P(t) = \begin{cases} 0.5[1 - \cos(2\pi ft / N)\sin(2\pi ft)], & t \leq N / f \\ 0, & t > N / f \end{cases} \tag{1}$$

where f is the central frequency in Hz and N is the number of sinusoidal cycles within a pulse. In this experiment, the signal with f=100 kHz (Fig. 1 for S0 mode) or f=50 kHz (Fig. 2 for A0 mode) and N=5 was used and the electrical voltage on the actuator was 10 V. The material properties of PZT are listed in Table 1.

Without the baseline data, it is still easy to estimate the arrival of the reflections from the beam boundaries, provided that the wave propagation speed and the dimensions of the beam are known. Then, if the reflected signal from the delamination is not overlapped with the reflections from the boundaries, it can be simply collected by the sensor. Therefore, it is still a technical challenge when the delamination is located close to the beam boundaries and the reflected signal of the delamination is completely overlapped with the reflections from the boundaries.

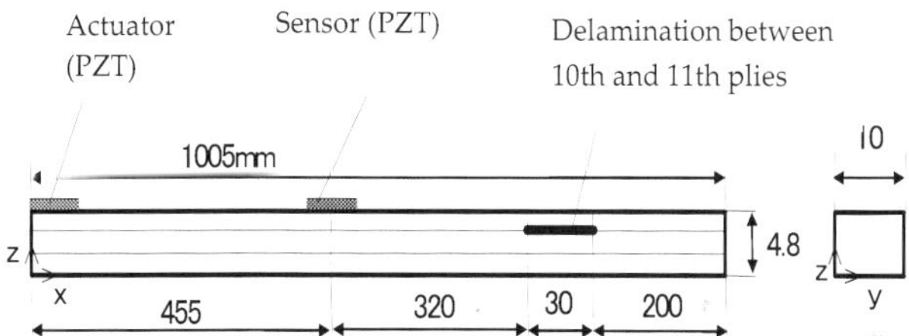

Figure 1. Schematic view of models for experiments using S0 mode (unit: mm)

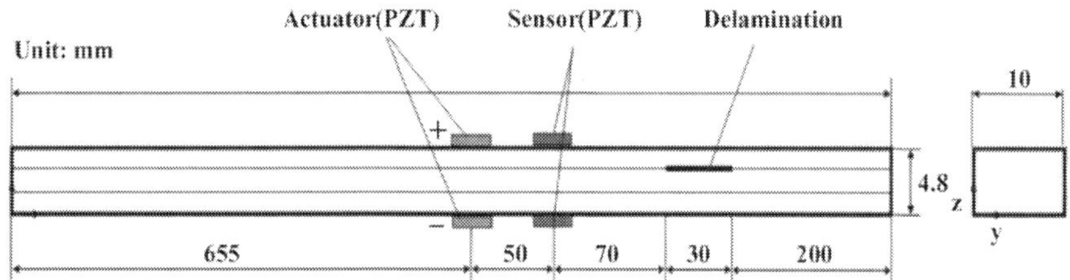

Figure 2. Schematic view of models for experiments using A0 mode (unit: mm)

	Material properties
PZT	E_{11}=62 GPa, E_{33}=49 GPa, d_{33}=472 pC/N, d_{31}=-210 pC/N, ρ=7500 kg/m³
CFRP lamina	E_{11}=115 GPa, E_{22}=E_{33}=9 GPa, G_{12}=G_{23}=5.5 GPa, G_{13}=3 GPa, ν_{12}=ν_{13}=0.3, ν_{23}=0.45, ρ=1600 kg/m³

Table 1. Material properties of PZT and CFRP lamina

2.1.2. Finite element analysis procedure

To explore the wave propagation in laminates with a long delamination case, a three-dimensional 8-noded brick hybrid element proposed by the authors [10] and the explicit time integration algorithm were used in the finite element simulations without considering the dynamic contact effects like those in some previous studies [11-13]. For the delamination area, one node used on the intact interface along the through-thickness direction, and double nodes were used on the delamination interface, i.e., one belongs to the elements of upper delaminated portion, and the other belongs to the elements of lower delaminated portion. The contact effects in the delamination area were neglected. The same signal stated in Eq. (1) was also used here. Furthermore, the actuator and the sensor were discretized using 3D brick elements. If an electrical field is applied on the actuator, the tensional or compressive strains will be generated in it from a proper relation, which connects the applied voltage and the generated internal strains. The induced stresses from the generated strains can be used to calculate the elemental nodal forces, which form axial force and bending moment in the beam simultaneously. Then, both S0 and A0 modes can be automatically generated. The material properties of CFRP are shown in Table 1.

2.2. Results of S0 mode

2.2.1. Experimental analysis

Firstly, to evaluate the wave propagation speed, an intact beam with two sensors attached, was employed. The distance between the two sensors was 420 mm. By using the wavelet transformation technique [14], the arrival times of the incident waves to the two sensors were determined and the wave propagation speed of S0 mode at f=100 kHz was estimated as 6210 m/s. To verify the experimental results, the theoretical wave speed of the S0 mode was estimated based on the transfer matrix method [15]. The calculated speed was 6467 m/s if the material properties of CFRP were taken as the values of Table 1. On the other hand, the corresponding propagation speed of A0 mode was estimated as 1506 m/s, much lower that the S0 mode. The higher theoretical wave speed of S0 mode may be due to the transfer matrix method used here, which is actually for an infinite plate. For a beam with a finite width, the influences of Poisson's ratio and boundary conditions are expected.

A comparison of the signals from an intact beam and a delaminated one (a 30 mm delamination damage) for the case in Fig. 1 is shown in Fig. 3. It is clear that there is a reflected wave in the signals from the delaminated beam, which is located between the incident and the reflected waves. However, no clear reflected wave could be detected when the delamination length was reduced to 10 mm. No A0 mode can be observed in Fig. 3, since its incident wave is not fast enough to reach the sensor within 300 μs. For a signal with f=100 kHz and N=5, the minimum detectable length of the delamination was 20 mm, which is about 1/3 of the wavelength of the S0 mode at 100 kHz. It is reasonable to assume that a higher excitation frequency may have the benefit of detecting smaller delamination cases. Once the arrival time of a reflected wave from the delamination is determined, the difference between the arrival times of the incident and reflected waves can be used to detect the delamination position, as shown in Fig. 4. For the 20 mm long delamination, a slightly higher error in the delamination position is observed. The reason is the overlapping between the reflected signals from the delamination and those from the right end of the beam, causing an increased error in determination of the arrival time of a reflected wave from the delamination.

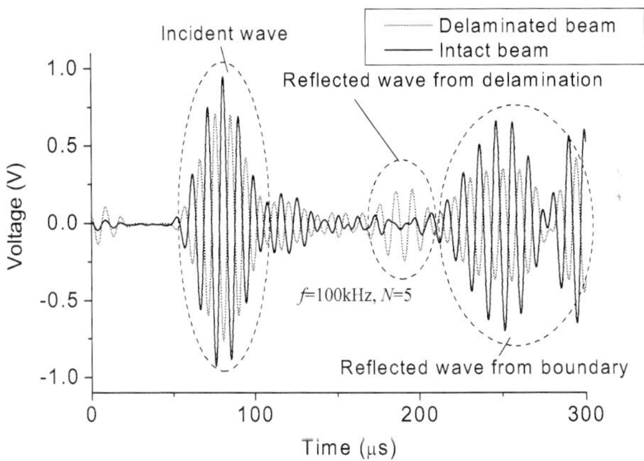

Figure 3. Comparison between signals of delaminated and intact beams (a 30 mm delamination damage between the 10th and the 11th plies)

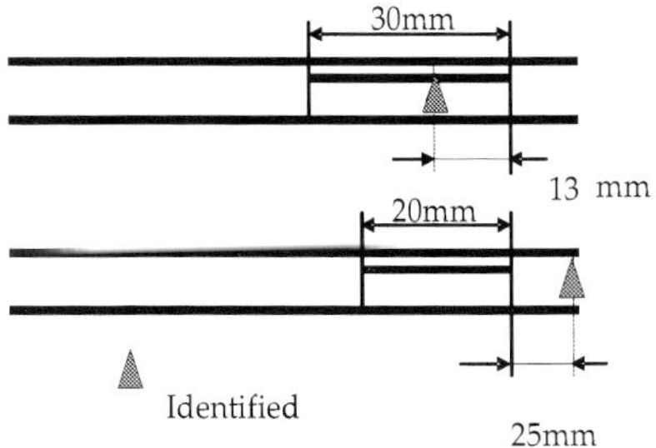

Figure 4. Delamination positions identified experimentally

2.2.2. Numerical simulation

A comparison between the experimental result and numerical simulation for the case of Fig. 1 is shown in Fig.5. A good agreement between two results is observed. For convenience, the amplitude of the first arrival S0 mode in the simulated waves was calibrated using the experimental data. The attenuation coefficients of the CFRP in the numerical model were determined by matching the amplitudes of the reflected waves from the right end of the beam to the experimental data. When the actual sensor thickness, i.e., 0.5 mm, was used in numerical simulations, a small reflected wave from the sensor could be observed immediately next to the incident wave. This was also confirmed by the experimental result. Corresponding to a 0.05 mm sensor thickness, however, no obvious reflection from the sensor could be observed in the numerical simulation. Therefore, the sensor thickness of 0.05 mm was eventually chosen in the simulations. Note that the selection of sensor thickness does not affect the simulation results as the calculated amplitudes will be calibrated using the experimental data. Similar to the experimental results, the numerical simulation showed that the reflected wave from the 10 mm delamination was very weak.

To understand the effect of delamination position along the through-thickness direction on the propagation of a Lamb wave, a 30 mm delamination damage was created in the mid-plane of the laminated beam, i.e., between the 16th and the 17th plies. With the signal of f=80 kHz and N=5, the comparison between the numerical and experimental results is shown in Fig.6. In Fig.6, no obvious reflected waves from the delamination can be observed. This finding is consistent with the work of Guo et al. [7]. They showed that the delamination at the positions of zero shear stress along the through-thickness direction, such as the mid-plane of a beam, had no effect on Lamb wave propagation in a S0 mode.

To examine the capability of the numerical simulation to identify the location of delamination, the length of delamination L was increased from 30 to 90 mm. The length of the beam was also increased up to 1500 mm. Under a condition of f=100 kHz and N=5, the propagation speed of S0 mode was firstly calculated in an intact beam with two sensors. The estimated propagation speed was 6260m/s, which is very close to the experimental result. Based on this speed, the numerically identified positions for various delamination lengths are shown in Fig. 7(a). It is surprising to note that all predicted positions are beyond the right end of the delamination. It has been demonstrated that a higher propagation speed of S0 mode in the delaminated 0°layer can make the predicted delamination positions behind the actual delamination [16, 17]. From the FEM simulations, the estimated speed of S0 mode in the 0°delaminated layer was 8483 m/s. If we use this speed in the delaminated region only (in the intact region, we still use 6260 m/s), with the known left end of the delamination, the predicted delamination positions or the reflected positions of S0 mode are shown in Fig. 7(b). Compare it with Fig. 7(a), we can find that there is no obvious difference in the delamination positions for short delamination cases, e.g. 30 and 50 mm. However, for a longer delamination case (e.g., 70 and 90 mm), an increased deviation from the actual delamination is observed when 8483 m/s was used in the prediction. It implies that the use of the actual wave speed of the delaminated layer cannot improve the prediction and a reflected wave with observable intensity is normally from the right end of the delamination.

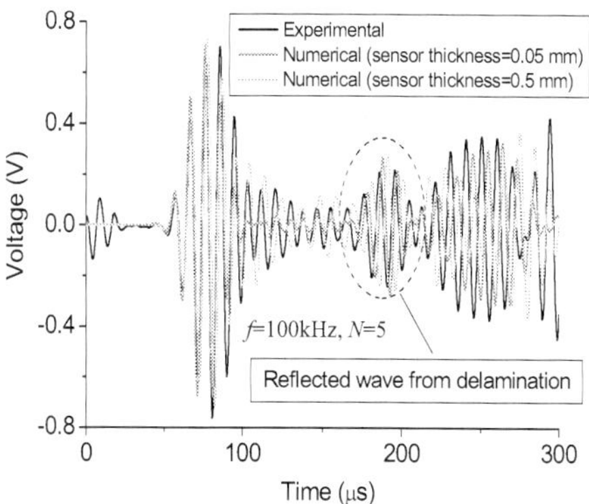

Figure 5. Comparison of experimental and numerical results for a 30 mm delamination damage between the 10th and the 11th plies

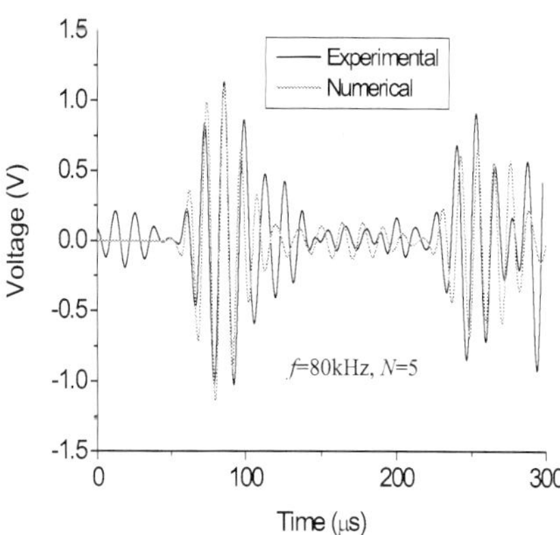

Figure 6. Comparison of experimental and numerical results for a 30 mm delamination damage at the mid-plane between the 16th and the 17th plies

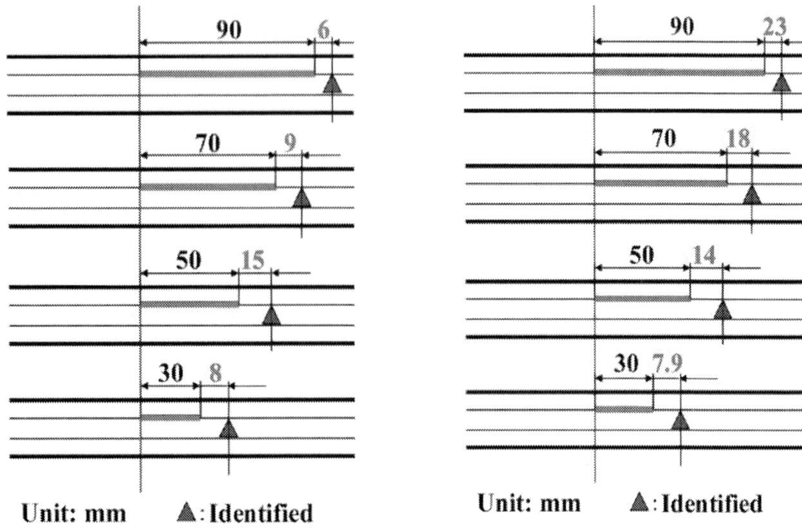

(a) Results using 6260 m/s   (b) Results using 8483 m/s in 0° delaminated layer

Figure 7. Numerical identification of positions of various delamination cases

2.3. Results of A0 mode

To confirm the effectiveness of present excitation and sensing techniques for A0 mode (Fig. 2), numerical simulation was firstly conducted for the laminate with $[0_{10}//90_6/90_6/0_{10}]$ and 30 mm delamination under a condition of f=50 kHz and N=5. Note that "//" denotes the position of the delamination. The signals from upper and lower sensors are plotted in Fig. 8(a). From the first wave packet, it can be observed that the pure A0 mode can be excited by two actuators with the out-of-phase applied voltages and two signals with a mutual phase difference of 180° are generated. Also, after the incident wave, the direct reflected signal from the delamination can be clearly observed. As mentioned before, the reflected signals from the delamination may include two modes, i.e., S0 mode and A0 mode as marked in Fig. 8(a). It can be explained by the wave mode change when the incident A0 mode interacts with the delamination. In Fig. 8(a), the reflections of the transmitted waves from the boundary can be identified, which also contain the S0 mode and A0 mode. The difference of two signals between the two sensors in Fig. 8(a) is shown in Fig. 8(b). It is clear that the S0 mode has been completely removed, and only pure A0 mode exists.

The results of the laminate $[0_{10}//90_6/90_6/0_{10}]$ with 30 mm and 10 mm delamination damages are shown in Fig 9. Sampling time for both numerical and experimental results was 1.0×10^{-4} ms. For the delamination of the lengths of 30 mm and 10mm, the reflections from the delamination can be clearly identified. In Figs. 9(a) and 9(b), it is interesting to note that the reflection signal from the 10 mm delamination is stronger than that from the 30 mm delamination. The reason may be that the reflections from the two ends of delamination are more easily overlapped in a shorter delamination case, e.g., the 10 mm one, resulting in a higher total reflection. Therefore, the intensity of the reflection from the delamination does not certainly depend on the length of the delamination. When the delamination is located on the mid-plane of the laminates, i.e., $[0_{10}/90_6//90_6/0_{10}]$, the results are shown in Fig. 10. Similar to Fig. 9, the reflection can be identified clearly. Also, the reflection from 10 mm delamination is still stronger than that of 30 mm. For the short delamination, e.g., smaller than 30 mm, reflections from the two ends of the delamination should be overlapped if we consider the duration time of a 5 cycle signal (N/f with the number of cycles N and wave central frequency f), the length of delamination and the travelling speed of A0 mode. By comparing Figs. 10(a) and 10(b), it can found that the duration of the reflected signal from the 30 mm delamination is obviously longer than that of the 10 mm delamination although its amplitude is smaller. It implies that the overlapping degree of two reflections from the two ends of the 30 mm delamination is lower than that of the 10 mm delamination, which leads to the lower intensity reflection from the 30 mm one. For the laminates with $[0_{12}/0_4//0_4/0_{12}]$ and $[0_{12}//0_4/0_4/0_{12}]$, the anti-symmetric A0 mode still works very well and the reflection from the delamination can be clearly identified. Only the difference of two sensor signals for the case of $[0_{12}/0_4//0_4/0_{12}]$ with 10 mm delamination is illustrated in Fig. 11.

From Fig. 9 to Fig. 11, a good agreement between experimental and numerical results is observed. For convenience, the amplitude of the first arrival A0 mode in the simulated waves was calibrated using the experimental data. The attenuation coefficients of the CFRP in the numerical model were determined by matching the amplitude of the reflected waves from the right end of the beam to that of experimental data. Naturally, the material properties of CFRP in Table 1 were also adjusted slightly within a reasonable range to match the numerical results to the experimental ones.

To determine the wave propagation speed of A0 mode, two sets of distant sensor pairs were attached on an intact laminated beam. The wave speed was determined from the distance and the difference of arrival times between the two sets of sensor pairs. For the laminate with $[0_{10}/90_6/90_6/0_{10}]$, the experimental and numerical wave speeds were 1555 m/s and 1470 m/s, respectively. For the case of $[0_{12}/0_4/0_4/0_{12}]$, the experimental and numerical wave speeds were 1723 m/s and 1616 m/s, respectively. From these results, it can be found that the wave speed of $[0_{12}/0_4/0_4/0_{12}]$ is only slightly higher than that of $[0_{10}/90_6/90_6/0_{10}]$. The reason is that the wave speed of A0 mode is determined by the bending stiffness of laminates, which is dominated by the fibre orientation of outer layers near top and bottom surfaces of laminates. In both cases, $0°$ degree layers are located near the two surfaces of laminates, which results in the comparatively small difference of wave speeds in both cases. For the case of $[0_{10}/90_6/90_6/0_{10}]$, if $90°$ degree layers were located near the two surfaces of laminates, it would be a completely different story.

With the knowledge of the wave speeds, the delamination positions for various situations were evaluated, as shown in Table 2. The distances between identified positions and the actual centers of delamination are listed. In this table, the negative values denote that the identified positions are located on the left side of the delamination centers, and positive values represent that the identified positions are on the right side of the delamination centers. In this table, it is clear that the delamination positions in almost all laminates have been successfully detected compared with the beam length. For the laminate with $[0_{10}//90_6/90_6/0_{10}]$ and 20 mm delamination, the reflection is not very clear due to strong noises in experiments. As a result, only numerical estimation is presented in this table. The similar case can be found in the laminate with $[0_{12}/0_4//0_4/0_{12}]$ and 30 mm delamination, as shown in this table.

In summary, unlike S0 mode, A0 mode can be used for different applications. However, the amplitude of reflection from delamination may not depend on the length of the delamination. The amplitude of reflection from short delamination is higher than that from long delamination due to overlapping of reflections from the ends of short delamination. The disadvantage of A0 mode is the short travelling distance due to its high attenuation in CFRP materials. Therefore, if possible, both the actuators and sensors should be placed close to the potential damage sites.

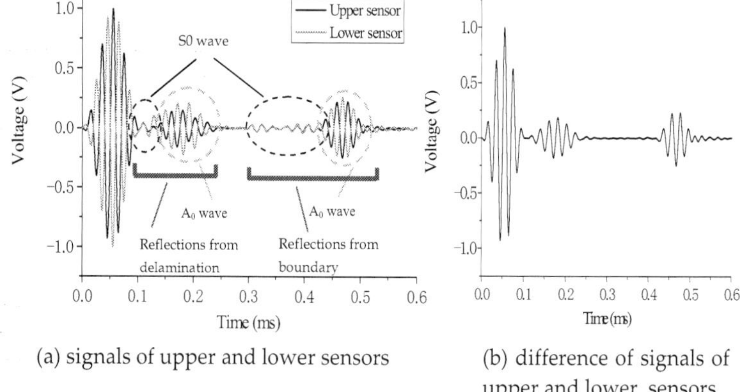

(a) signals of upper and lower sensors

(b) difference of signals of upper and lower sensors

Figure 8. FEM numerical results for $[0_{10}//90_6/90_6/0_{10}]$ (delamination length: 30 mm)

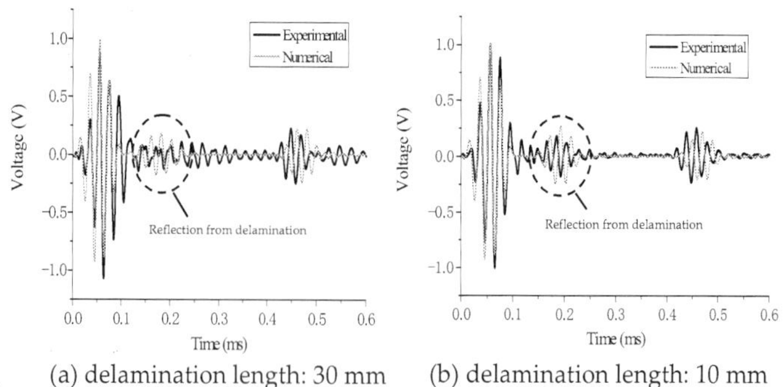

(a) delamination length: 30 mm

(b) delamination length: 10 mm

Figure 9. Comparison of numerical and experimental results for $[0_{10}//90_6/90_6/0_{10}]$

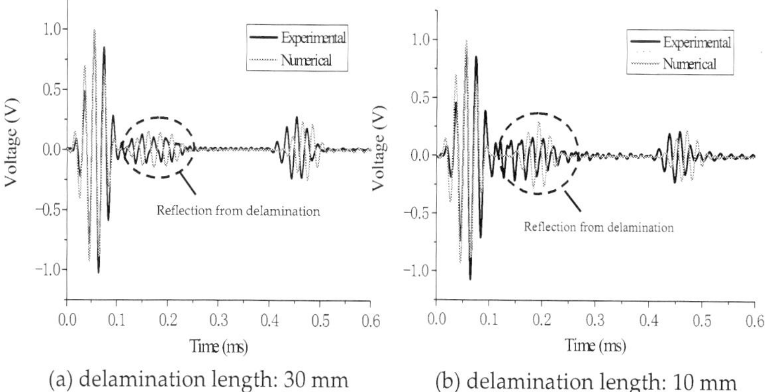

(a) delamination length: 30 mm   (b) delamination length: 10 mm

Figure 10. Comparison of numerical and experimental results for $[0_{10}/90_6//90_6/0_{10}]$

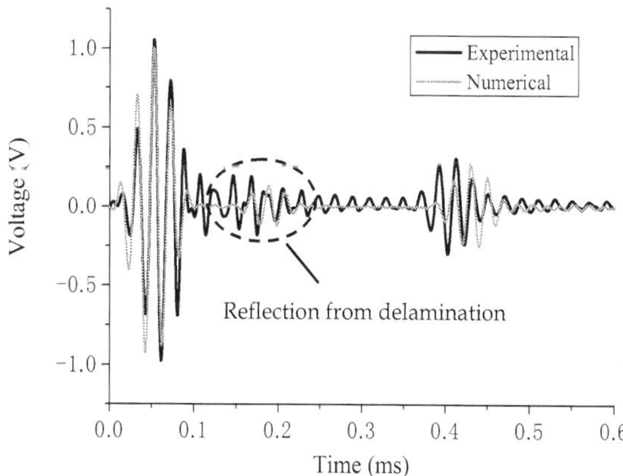

Figure 11. Comparison between experimental and numerical results for a 10 mm delamination ($[0_{12}/0_4//0_4/0_{12}]$)

Unit (mm)		[0/90//90/0]	[0//90/90/0]	[0/0//0/0]	[0//0/0/0]
30mm	Num.	-11.0	4.0	0.0	18.0
	Exp.	-19.0	39.0		-11.0
20mm	Num.	-12.0	-5.0	-14.0	18.0
	Exp.	9.0		-11.0	4.0
10mm	Num.	2.0	2.0	1.0	1.0
	Exp.	2.0	-2.0	10.0	-17.0

Table 2. Identified delamination positions for various cases

3. Interaction between Lamb waves and delamination

3.1. S0 mode

To gain a better understanding of this complex interaction, a beam with a length of 1500 mm and a 200 mm long delamination damage was examined using the numerical simulation under a condition of f=100 kHz and N=5. The very long delamination was chosen for obtaining the separated reflected signals from the two ends of the delamination. The transverse deflections at various mesh points on the top surface of the beam were used and the subtraction between the wave signals from the intact and the delaminated beams was done to amplify the scattered wave signals from the delamination. Fig. 12 shows the detail information of the scattered wave signals from the delamination at different time domains. The typical sensor signal for accurately evaluating the arrival times of various waves is shown in Fig. 13.

At the time domain of 139 μs, the S0 mode arrives at the left end of the delamination and then reflects from or transmits from the left end, as shown in Fig. 12(a). At 182 μs, the S0 mode passes through the right end of the delamination. Fig. 12(b) shows two separate reflected modes, i.e., A0 and S0 from the left end of the delamination. In principle, it is easy to distinguish a S0 from an A0 mode by the propagation speed and wave length. However, the amplitude of the reflected S0 mode is smaller than that of the A0 mode. This can be explained by the fact that the bending deformation in a laminated beam is much higher as compared to the axial deformation. Also, when the S0 mode travels through the left end of delamination, a new transmitted A0 mode is generated as shown in Fig. 12(b). The reflected waves from the right end of the delamination, consisted both A0 and S0 modes, can be also observed in Fig. 12(b). Corresponding to 236 μs, the incident A0 mode from the actuator does not reach the sensor, and the reflected S0 mode from the left side of the delamination has already passed through the sensor. However, at the same time point in Fig. 13, this reflected S0 mode cannot be detected by the sensor due to its small amplitude. Also, the reflected S0 mode from the right end of the delamination, which is surpassing the reflected A0 mode from the left end of the delamination, can be observed, as shown in Fig. 12(c). A new transmitted A0 mode and the transmitted S0 mode from the right end of the delamination can be clearly observed, Fig. 12(c). At 303 μs, the incident A0 mode does not arrive at the sensor. As indicated in Fig. 12(d), the reflected S0 mode from the right end of the delamination completely surpasses the reflected A0 mode from its left end. This reflected S0 mode has already passed through the sensor position. From Fig. 13, the arrival time of the reflected wave from the delamination is identified as around 275 μs. The reflected wave actually arrives at the sensor between the time points shown in Figs. 12(c) and 12(d), i.e., 236 μs and 303 μs. Therefore, the reflected wave signal detected clearly by the sensor is considered to be the reflected S0 mode from the right end rather than the left end of the beam, due to its higher intensity. Based on the numerical simulations above, the very complex interaction between the different signal modes and the boundaries of delamination can be identified. When only a single S0 mode passes through the delamination, four modes will be generated at one end of the delamination, including two transmitted A0 and S0

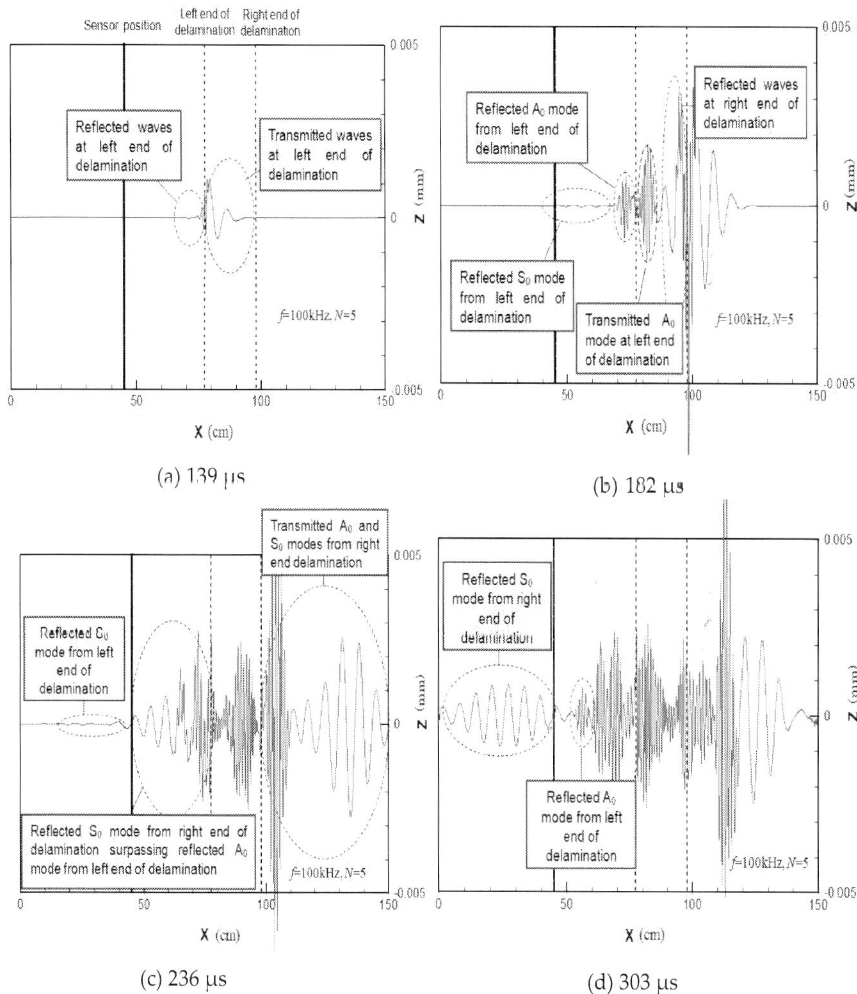

Figure 12. Wave signal differences of intact and delaminated beams(S0 mode)

modes, and two reflected A0 and S0 modes. Further study is required to understand why the stronger reflected A0 and S0 modes. However, the reflected S0 mode received by a sensor is from the reflections are from the right end of delamination. One possible explanation is that the delaminated region is of lower bending stiffness, which can be considered to be a softer region. When a wave propagates from a harder or intact region into a softer region, the reflections become weak. In contrast, when the wave propagates from a softer region into a harder one, a stronger reflection is expected.

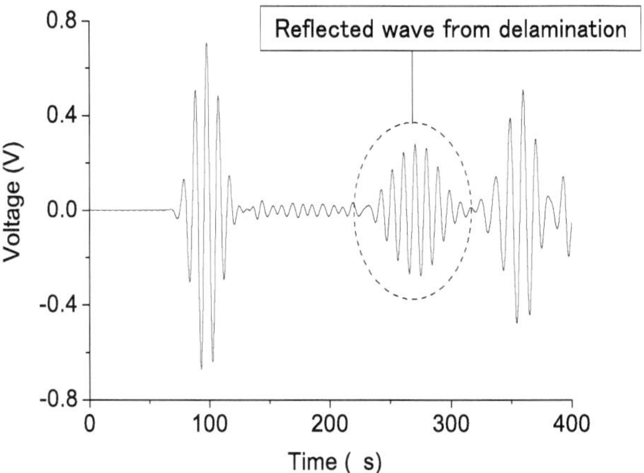

Figure 13. Sensor signal for a delamination of length of 200 mm

3.2. A0 mode

Like S0 mode, a beam with a length of 1200 mm and a 200 mm long delamination case was examined using the numerical simulation under a condition of f=50 kHz and N=5. The transverse deflections (out-of-plane deformation) at various mesh points on the top surface of the beam were used. Fig. 14 also shows the subtraction between the wave signals from the intact and the delaminated beams, which can amplify the scattered wave signals from the delamination. The typical sensor signal for accurately evaluating the arrival times of various waves is shown in Fig. 15 with an enlarged picture for reflected waves from the delamination.

Firstly, for the laminate with $[0_{10}//90_6/90_6/0_{10}]$, at the time domain of 318 µs, two A0 mode waves were excited by the actuator pair, which propagates to the left and right directions, independently. The A0 mode wave propagating to the right direction arrives at the left end of the delamination and then reflects from or transmits from the left end, as shown in Fig. 14(a), respectively. At 433 µs, the transmitted A0 mode from the left end of delamination passes through the right end. Fig. 14(b) shows the reflected wave from the right end of delamination. By comparing Fig. 14(b) and Fig. 15 at the same time domains, we can identify that the reflected A0 mode from the left end of delamination passes through the sensor. The reflected S0 mode due to mode change can also be identified. In principle, it is easy to

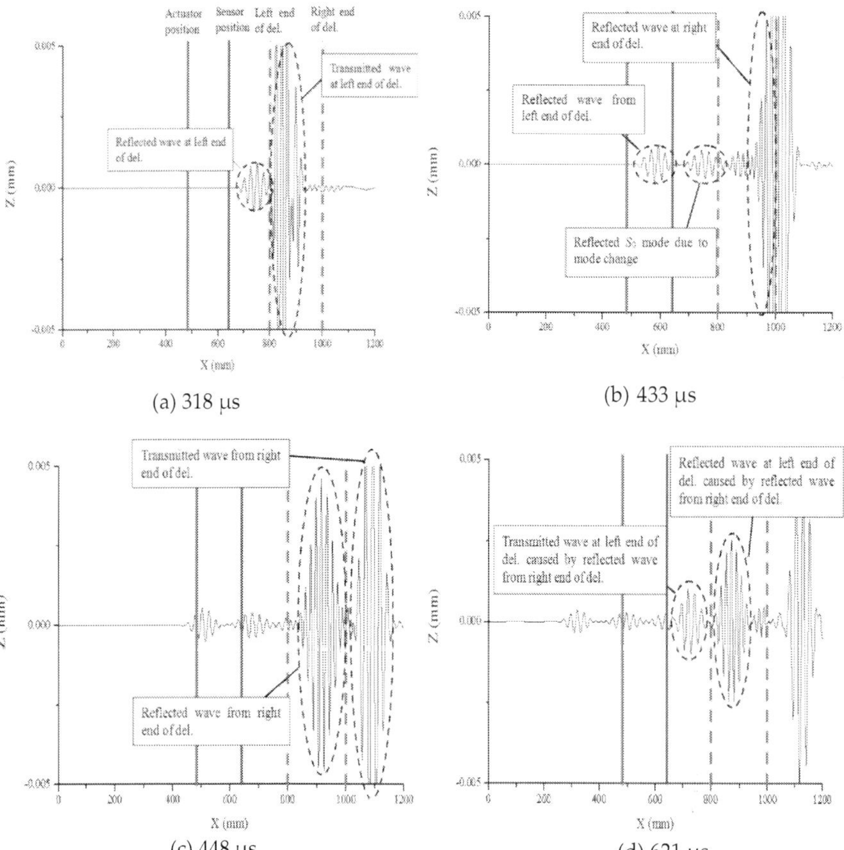

(a) 318 μs

(b) 433 μs

(c) 448 μs

(d) 621 μs

Figure 14. Wave signal differences of intact and delaminated beams(A0 mode)

distinguish a S0 from an A0 mode by using phase information of signals from the two sensors. At 488 μs, the reflected and transmitted waves at the right end of the delamination separate to each other, as shown in Figs. 14(c). Also, the amplitude of the reflected wave is much higher than that of the reflected wave at the left end of the delamination, confirmed in Fig. 14(b). This phenomenon is similar to that obtained for S0 mode. Fig. 14(d) shows the transmitted and reflected waves at the left end of the delamination when the reflected wave from the right end of the delamination arrives at the left end at 621 μs. It is worthwhile mentioning that although the amplitude of reflected wave from the right end of delamination is very strong, the reflection at another end (left end) of the delamination can significantly reduce the amplitude of the transmitted wave, which is expected to be monitored by the sensors. Therefore, wave mode change and multiple reflections may create complicated interactions between the A0 mode and the delamination.

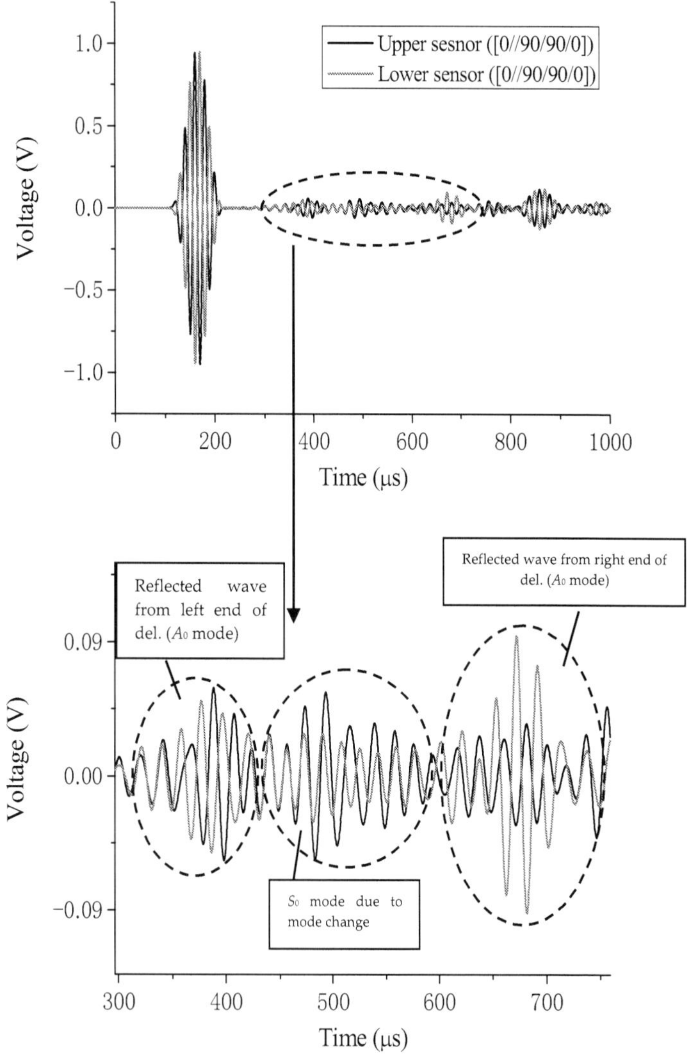

Figure 15. Sensor signal for delamination of 200 mm in length ($[0_{10}//90_6/90_6/0_{10}]$)

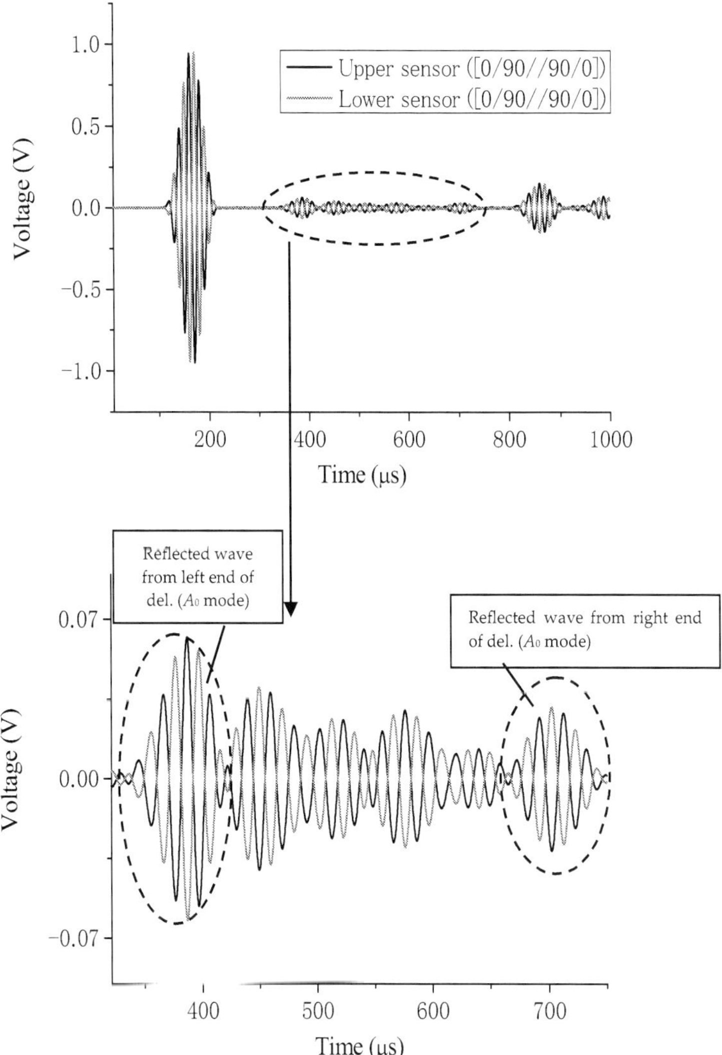

Figure 16. Sensor signal for delamination of 200 mm in length ($[0_{10}/90_6//90_6/0_{10}]$)

In Fig. 15, the amplitude of the reflected wave from the right end of the delamination is still higher than that of the reflected wave from the left end. Similar results can be observed in the laminate with $[0_{12}//0_4/0_4/0_{12}]$. In consequence, as shown in Table 2, more identified delamination positions are close to the right end of delamination for the laminates with $[0_{10}//90_6/90_6/0_{10}]$ and $[0_{12}//0_4/0_4/0_{12}]$. This phenomenon is similar to the result obtained for S0 mode in the case of $[0_{10}//90_6/90_6/0_{10}]$.

For the case of $[0_{10}/90_6//90_6/0_{10}]$, the signals of two sensors are plotted in Fig. 16 with the enlarged picture for the reflection from the delamination. In Fig. 16, there is no S0 mode and the mode change does not occur as the delamination is located on the mid-plane of the laminates. In contrast to the cases of $[0_{10}//90_6/90_6/0_{10}]$ and $[0_{12}//0_4/0_4/0_{12}]$, it is interesting to note that the amplitude of the reflected wave from the left end of delamination is higher than that of the right end. Similar results are obtained in the case of $[0_{12}/0_4//0_4/0_{12}]$. Corresponding to Table 2, more identified delamination positions are close to the left end of the delamination. The physical meaning of why the stronger reflection occurs at the different ends of delamination for different delamination cases in Figs 15 and 16 is not clear. The position of the stronger reflection seems to mainly depend on the following two factors: bending stiffness of upper and lower delaminated layers and their difference, and Lamb wave deformation mode.

4. Conclusion

A technique for identifying a delamination damage in cross-ply laminated composite beams has been developed by using the zero-order mode of Lame waves. The delamination position can be identified if the arrival time of the arrival time of a reflected wave from a delamination and the wave propagation speed of Lame waves are known. One of the advantages of this approach is the baseline data acquired from intact beams are not required.

And, extensive finite element simulations have been conducted to investigate the interaction of Lame waves with various delamination cases. For S0 mode, when a single S0 mode propagates into a delamination, mode change happens and reflected S0 mode with high amplitude is generated from the end point of the delamination where the waves propagate out; for A0 mode, it has been found that the situation is more complex than S0 mode. For the laminates with stack sequence of $[0_{10}//90_6/90_6/0_{10}]$ and $[0_{12}//0_4/0_4/0_{12},]$, mode change also happens and the reflected wave from the right end of delamination is slight higher than that from the left end. However, for the case of $[0_{10}/90_6//0_6/0_{10}]$ and $[0_{12}/0_4//0_4/0_{12}]$, there is no mode change when a single A0 mode passes through delamination, and the stronger reflection from the delamination occurs at the left end of the delamination.

Author details

Yaolu Liu, Alamusi, Jinhua Li, Huiming Ning, Liangke Wu and Ning Hu

Department of Mechanical Engineering, Chiba University, Yayoi-cho, Inage-ku, Chiba, Japan

Weifeng Yuan and Bin Gu
School of Manufacturing Science and Engineering,
Southwest University of Science and Technology, Mianyang, P.R.China

5. References

[1] Su ZQ, Ye L, Lu Y. Guided Lamb waves for identification of damage in composite structures: A review. *J Sound & Vibr* 2006; 295: 753-80.

[2] Valdes SHD, Soutis C. A structural health monitoring system for laminated composite. *Proceedings of DETC, Pittsburgh, PA, USA*; 2001.

[3] Sohn H, Farrar CR. Damage diagnosis using time series analysis of vibration signals. *Smart Mater and Struct* 2001; 10: 1-6.

[4] Ihn JB, Chang FK. Built-in diagnostics for monitoring crack growth in aircraft structures. *In: Chang Fu-Kuo (ed.), Proceedings of the international workshop on structural health monitoring. Stanford University, USA*; 2001.

[5] Hurlesbaus S, Niethammer M, Jacobs LJ, Valle C. Automated methodology to locate notches with Lamb waves. *Acoustics Research Letters Online* 2001; 2: 97-102.

[6] Su Z, Yang C, Pan N, Ye L, Zhou LM. Assessment of delamination in composite beams using shear horizontal (SH) wave mode. *Compos Sci Technol* 2007; 67: 244-51.

[7] Guo N, Cawley P. The interaction of Lamb waves with delaminations in composite laminates. *J Acoust Soc Am* 1993; 94: 2240-6.

[8] Wang CH, Rose LRF. Wave reflection and transmission in beams containing delamination and inhomogeneity. *J Sound & Vibration* 2003; 264: 851-72.

[9] Mahapatra DR, Gopalakrishnan S. A spectral finite element model for analysis of axial–flexural–shear coupled wave propagation in laminated composite beams. *Compos Struct* 2003; 59: 67-88.

[10] Cao YP, Hu N, Lu J, Fukunaga H, Yao ZH. A 3D brick element based on Hu-Washizu variational principle for mesh distorsion. *Int J Numer Meth Engrg* 2002; 53: 2529-48.

[11] Hu N, Sekine H, Fukunaga H, Yao ZH. Impact analysis of composite laminates with multiple delaminations. *Int J Impact Eng* 1999; 22: 633-48.

[12] Hu N, A solution method for dynamics contact problems. *Comput & Struct* 1997; 63: 1053-63.

[13] Li CF, Hu N, Cheng JG, Sekine H, Fukunaga H. Low-velocity impact-induced damage of continuous fiber-reinforced composite laminates. Part II. verification and numerical investigation, *Compos part A* 2002; 33: 1063-72.

[14] Jeong H, Jang YS. Wavelet analysis of plate wave propagation in composite laminates. *Compos Struct* 2000; 49: 443-50.

[15] Nayfeh AH, Chimenti DE. Elastic wave propagation in fluid-loaded multi-axial anisotropic media. *J Acoust Soc Am* 1991; 89: 542-9.

[16] Toyama N, Noda J, Okabe T. Quantitative damage detection in cross-ply laminates using Lamb wave method. *Compos Sci and Technol* 2003; 63: 1473-79.

[17] Takeda N, Okabe Y, Kuwahara J, Kojima S, Ogisu T. Development of smart composite structures with small-diameter fiber Bragg grating sensors for damage detection: Quantitative evaluation of delamination length in CFRP laminates using Lamb wave sensing. *Compos Sci Technol* 2005; 65: 2575-87.

Bio-Medical Composites and Their Applications

Chapter 6

Biocomposite Materials

Khaled R. Mohamed

Additional information is available at the end of the chapter

http://dx.doi.org/10.5772/48302

1. Introduction

Composite materials may be restricted to emphasize those materials that contain a continuous matrix constituent that binds together and provides form to an array of a stronger, stiffer reinforcement constituent. The resulting composite material has a balance of structural properties that is superior to either constituent material alone. Combining the advantages of inorganic and organic components such as HA/organic composites that show good biocompatibility and favorable bonding ability with surrounding host tissues inherent from IIA. Besides, the problems associated with HA ceramic, such as its intrinsic brittleness, poor formability and migration of HA particles from the implanted sites, can be circumvented by the integration of HA ceramic with biopolymers. In order to achieve controlled bioactivity and biodegradability, polymer-ceramic composites have been proposed. The composites of ceramics with natural degradable polymers have attracted much interest as bone filler. Several particle composites based on degradable biopolymers such as chitosan and gelatin with inorganic powders, were developed as bone filler.

2. Hydroxyapatite (HA)

Hydroxyapatite (HA) (Ca_{10} $(PO_4)_6(OH)_2$, is widely used in musculoskeletal procedures due to its chemical and crystallographic similarity to the carbonated apatite in human bones and teeth (Suchanek and Yoshimura, 1998). While sintered HA can be machined and used in pre-fabricated forms, several formulations of calcium phosphate cements can be molded as pastes and harden in situ. HA is the main component of teeth and bones in vertebrates. Good mechanical properties with superior biocompatibility of sintered HA make it well preferred bone and tooth implant material (Kim et al., 2008). Calcium phosphates, especially HA, are excellent candidates for bone repair and regeneration and have been used in bone tissue engineering for two decades. Although HA is bioactive and osteoconductive, its mechanical properties are inadequate, making it unable to be used as a load bearing implant.

2.1. Properties

HA power (HAp), a major inorganic component of bone, has been used extensively for biomedical implant applications and bone regeneration due to its bioactive, biodegradable and osteoconductive properties (Gomez-Vega et al., 2000). HA is available in market in many forms like solids blocks, micro-porous blocks and as granules (Murgan and Ramakrishna, 2005). Nano-hydroxyapatite (n-HA) has been proven to be of great biological efficacy. nHA precipitates may have higher solubility and therefore affect the biological responses. It was able to promote the attachment and growth of human osteoblast-like cells (Huang et al., 2004). Clinical trials have shown that HA cement is both biocompatible and resistant to infection and that the HA coating improves the success rate of implants. It has also been demonstrated that HA ceramics support mesenchymal stem cell (MSC) attachment, proliferation, and differentiation (Zhao et al., 2006). Nano scaled HA with extraordinary properties such as high surface area to volume ratio and ultra fine structure similar to that of biological apatite, which is of great effect on cell-biomaterial interaction, has been reported to be used for the treatment of bone defects, and it could bond to living bone in implanted areas (Liuyun et al., 2008). HAp has been used in bone regeneration and as a substitute of bone and teeth because it is a biocompatible, bioactive, non-inflammatory, non-toxic, osteoconductive and non-immunogenic material (Grande et al., 2009).

Instability of the particulate nHA is often encountered when the particles are mixed with saline or patient's blood and hence migrate from the implanted site into surrounding tissues and causing damage to health tissue (Miyamato and Shikawa, 1998). Also, HA ceramic is difficult to shape in specific forms required for bone substitution, due to its hardness and brittleness. Therefore, a composites of HAp and organic polymers have become of great interest to compensate the weak mechanical points of HAp (Furukawa et al., 2000). Mohamed and Mostafa, (2008) reported that HA has low fracture toughness, hardness and brittleness, therefore, HA cannot serve as a bulk implant material under the high physiological loading conditions traditionally associated with implants.

Since the natural bone is a composite mainly consisted of nano-sized, needle-like HAp crystals and collagen fibers, many efforts have been made to modify HAp by polymers, such as poly-lactic acid (Kasuga et al., 2001), chitosan (Viala et al., 1998), polyethylene (Wang and Bonfield, 2001) to compensate the weak mechanical points of HAp. Yamaguchi et al., (2003) reported that chitosan is flexible and has a high resistance upon heating due to the intra-molecular hydrogen bonds formed between hydroxyl and amino groups. A composite biomaterial of HAp and chitosan is, therefore, expected to show an increased osteoconductivity and biodegradation together with a sufficient mechanical strength for orthopedic use. A major disadvantage of current orthopedic implant materials such as sintered hydroxyapatite is that they exist in a hardened form, requiring the surgeon to drill the surgical site around the implant or to carve the graft to the desired shape. This can lead to increases in bone loss, trauma, and surgical time. Hence, the moldable, self-setting calcium phosphate cement (CPC)-chitosan composite is desirable for dental, craniofacial and orthopedic repairs, especially where shaping and contouring for esthetics are needed. The

CPC powder can be mixed with the chitosan liquid to form a paste that can be applied in surgery via minimally invasive techniques such as injection, with fast-setting and anti-washout capabilities to form a scaffold in situ (Moreau and Xu, 2009).

2.2. Applications

Bone defects that are generated by tumor resection, trauma, and congenital abnormality have been clinically treated by the implantation of bioceramics or autogenous and allogenous bone grafts. Although autografting is a popular procedure for reconstructive surgery, it has several disadvantages, such as the shortage of donor supply, the persistence of pain, the nerve damage, fracture, and cosmetic disability at the donor site. On the other hand, there are no donor site problems for allografting, while allografting has some clinical risks including disease transmission and immunological reaction (Takahashi et al., 2005). HA has been incorporated into a wide variety of biomedical devices including dental implants, coatings on Ti based hip implants, biodegradable scaffolds, and other types of orthopedic implants (Wilson and Hullb, 2007). Synthetic HA, is a bioactive material that is chemically similar to biological apatite HA, has been used as a bioactive phase in the composites, coating on metal implants, and granular filler for direct incorporation into human tissues (Rehman et al., 1995). The characteristics of an ideal ceramic composite for bone tissue engineering should comply to the following parameters:(1) A biodegradability for bone remodeling, (2) A macroporosity for in-growth into the composite, (3) Mechanical stability/ease of handling, (4) Osteoconductivity to guide bone around/inside the implant, (5) Carrier for growth factors/cells. The use of these materials for tissue engineering purposes is still explored. Most researchers are aware that HA has low resorbability of sintered Ca P-ceramics. Because of the positive influence of ceramics on cell differentiation/proliferation, it is not surprising that bone forming cells are introduced into these ceramics to speed-up tissue in-growth. The surface of sintered ceramics is chemically stable and therefore a good substrate for seeding cells. In the field of bone tissue engineering most scaffolds are ceramic or ceramic-derivatives. Especially HA-based calcium phosphate compound is regarded as high-potential scaffolds due to their osteoconductive properties. Next to the scaffold material, in bone tissue engineering, cells and growth factors are also introduced to speed-up tissue in-growth. In most ceramic scaffolds, however, difficulties arise when cells are added, most probably due to a limited supply of nutrition at the inside of the implant or less than optimal cell-cell interactions. Growth factors like bone morphogenic protein-2 (BMP-2), transforming growth factor (TGF-β), basic fibroblast growth factor (b FGF) and vascular endothelial growth factor (VEGF) are commonly introduced into these scaffolds due to their osteoinductive properties and vascularization (Habrakenb et al., (2007). HA has a composition and structure very close to natural bone mineral and therefore has been considered to be the ideal material to build bone tissue engineering scaffold due to its osteoconductivity and osteoinductivity (Wang et al., 2007). Some in vitro studies demonstrated that nano-phase HA (67 nm grain size) significantly enhanced osteoblast adhesion and strikingly inhibited competitive fibroblast adhesion compared to conventional, 179 nm grain size HA, after just 4 h of culture. Researchers believe they know why they have elucidated the highest adsorption of vitronectin (a protein well known to promote osteoblast adhesion) on nanophase ceramics, which may explain the subsequent enhanced osteoblast adhesion on these materials. In addition, enhanced osteoclast-like cell functions (such as the synthesis of tartrate-resistant acid phosphatase (TRAP) and the formation of resorption pits) have also been observed on nano-HA compared to conventional HA, nano-porous or nano-fibrous polymer matrices can be fabricated via electrospinning, phase separation, particulate leaching, chemical etching and 3-D printing techniques (Zhang et al., 2008).

3. Chitosan polymer (C)

3.1. Structure

Brimacombe and Webber, (1964) reported that the chitosan consists of repeating units of beta (1-4) 2-amino-2-deoxy-D-glucopyranon (D-glucosamine) (Fig.1). The primary unit of chitin is 2-acetamido-2-deoxy-D-glucose, while that of chitosan is 2-amino-2-deoxy-D-glucose with beta, 1-4 glucosidic linkages (Muzarelli, 1985). Chitosan is a natural polysaccharide derived from chitin by its deacetylation (Dang and Leong, 2006). Chitosan shares a number of chemical and structural similarities with collagen. These similarities form the basis for the development of HAp/chitosan nano-composites for use in bone tissue engineering (Wilson et al., 2007

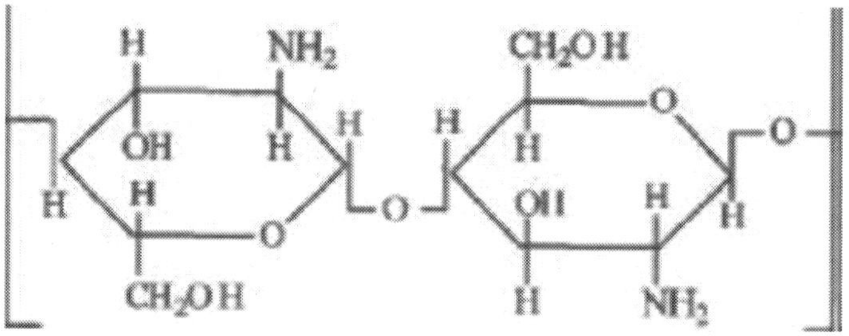

Figure 1. Structure of chitosan

3.2. Origin

The history of chitosan dates back to the 19 th century, when Rouget, 1859 discussed the deacetylated form of chitosan. Chitin, the source material for chitosan, is one of the most abundant organic materials, being second only to cellulose in the amount produced annually by biosynthesis. It is an important constituent of the exoskeleton in animals, especially in crustacean, molluscs and insects. It is also the principal fibrillar polymer in the cell wall of certain fungi (Eugene and Lee, 2003). Chitosan is one of the most abundant naturally occurring polysaccharides, primarily obtained as a sub-product of seafood, containing amino and hydroxyl groups (Muzarelli, 1985 and Manjubala et al., 2006).

3.3. Properties

3.3.1. Physical and chemical properties

The size of chitosan molecule plays an important role in drug delivery system where the release rate was reported to be inversely proportional to the molecular weight of the reservoir. Chitosan is insoluble in water, alkali and many organic solvents but soluble in many dilute aqueous solutions of organic acids, of which most commonly used are formic acid and acetic acid. Tomihata and Ikada, (1997) concluded that molecular weight of chitosan affect the crystal size and morphological character of its cast film. Gorbunoff, (1984) concluded that the chemical interaction between NH^{3+} and HPO_4^{2-} control the adsorption of chitosan on octacalcium phosphate (OCP) crystal surface which is similar to the binding of protein to the surface of apatite crystals. Chitosan permeability increased in the amorphous region of the membrane (Brine, 1984). Chitosan forms a polycation in aqueous acid solutions like acetic acid and hydrochloric acid via protonation of amine functions (Muzarelli, 1985).

Blair et al., (1987) reported that chitosan prepared from crab shell chitin has a lower molecular weight than that prepared from prawn shells and tensile strength versus elongation decrease at the break with prolonged treatment in alkali solutions and both increase proportional to higher molecular weight. Also, crystallinity of membrane increased with decreasing molecular weight of chitosan and it is hydrophilic i.e it contains a large number of OH groups which are easy to combine with another group and form new bond (Ogawa et al., 1992). The larger molecular weight of chitosan used, the higher tensile strength and higher tensile elongation of membrane were obtained (Chen and Hwa, 1995). The degree of deacetylation affects the physical properties of chitosan membrane and it does not change during ultrasonic degradation on chitosan. The molecular weight of the prepared chitosan depends on the severity of the deacetylation process.

Chitosan is flexible and has a high resistance upon heating due to the intra-molecular hydrogen bonds formed between hydroxyl and amino groups (Lee et al., 1999). The degradation rate is inversely related to the degree of crystallinity which is controlled mainly by the degree of deacetylation (DD). Highly deacetylated forms (85 %) exhibit relatively a low degradation rate and may take several months in vivo, whereas, the forms with lower DD degrade more rapidly. The degradation rates also inherently affect both the mechanical and solubility properties (Kamiyama et al., 1999).

The cationic nature of chitosan also allows for pH-dependent electrostatic interactions with anionic glycosaminoglycans (GAG) and proteoglycans distributed widely throughout the body and other negatively charged species. This property is one of the important elements for tissue engineering applications because numbers of cytokines/growth factors are known to be bound and modulated by GAG including heparin and heparin sulfate (Nishikawa et al., 2000). Chitosan is easy to handle for its resistive nature to heating due to intra-molecular hydrogen bonds between hydroxyl and amino groups (Itoh et al., 2003). Chitosan is a very versatile biopolymer with film, fiber, and micro/nanoparticle forming properties. It is biocompatible, biodegradable and non-toxic (Yilmaz, 2004). Deacetylation of chitin with a degree of deacetylation more than 50 % gives chitosan, which is soluble in organic acids such as acetic or formic acid, and has been more widely used than chitin as films, membranes, fibres and particles (Kang et al., 2006). Chitosan has three types of reactive functional groups, an amino group as well as both primary and secondary hydroxyl groups at the C(2), C(3), and C(6) positions, respectively. These groups allow modification of chitosan like graft copolymerization for specific applications, which can produce various useful scaffolds for tissue engineering applications. The chemical nature of chitosan in turn provides many possibilities for covalent and ionic modifications that allow extensive adjustment of mechanical and biological properties (Kim et al., 2008). When the temperature is 30°C, the crystallinity of chitosan membrane is relatively high and its crystal particles are much small, which makes the mechanical properties poor. When the temperature is as high as 90°C, the temperature will cause the change of chitosan properties and the chitosan membrane is nearly not a crystal structure at this moment, resulting in the fall of mechanical properties (Xianmiao et al., 2009).

3.3.2. Biological properties

Chitosan potentiates the differentiation of osteoprogenitor cells and may facilitate the formation of bone. Rao and Sharma, (1997) concluded that chitosan is an ideal non-toxic biopolymer and the cell binding and cell activating properties of chitosan play a crucial role in its potential action. Chitosan degrades in the body to non-harmful and non-toxic compounds and has been used in various fields such as nutrition, metal recovery and biomaterials (Muzzarelli et al., 2001). It has gained much attention as a biomaterial in diverse tissue engineering applications due to its low cost, large-scale availability, anti-microbial activity, and biocompatibility (Khora and Limb, 2003). Chitosan was suggested as an alternative polymer for use in orthopedic applications to provide temporary mechanical support to the regeneration of bone cell in-growth due to its good biocompatible (Khora and Limb, 2003), non-toxic, biodegradable and inherent wound healing characteristics (Eugene and Lee 2003). Chitosan had been used in various forms such as zero dimension microspheres, two-dimension membrane, three-dimension pin or rod (Hu et al., 2003). Therefore, much attention has been paid to chitosan-based biomedical materials, for instance, as a drug delivery carrier or a wound-healing agent. Chitosan is structurally similar to glycosaminoglycan (GAG) and has many desirable properties as tissue engineering scaffolds (Kuma, et al., 2004). It was reported as being neither antigenic in mamalian test system nor thrombogenic and chitosan reported to improve hemostasis, decreased fibroplasias with enhanced tissue organization as well as normal bone formation (Mohamed, 2004). Chitosan marginally supports biological activity of diverse cell types (Sarasam and Madihally, 2005).

A number of natural and synthetic polymers have been studied for overcoming the weak points as bone substitutes. Chitosan has been found in a broad spectrum of applications along with unique biological properties including biocompatibility, biodegradability to harmless products, non-toxicity, physiological inertness, remarkable affinity to proteins, antibacterial, haemostatic, fungi-static, anti-tumoral and anti-cholesteremic properties (Kim et al., 2008). Chitosan has been shown to degrade in vivo, which is mainly by enzymatic hydrolysis. The degradability of a scaffold plays a crucial role on the long-term performance of tissue-engineered cell/material construct because it affect many cellular process, including cell growth, tissue regeneration, and host response. If a scaffold is used for tissue engineering of skeletal system, degradation of the scaffold biomaterial should be relatively slow, as it has to maintain the mechanical strength until tissue regeneration is almost completed. Lysozyme is the primary enzyme responsible for in vivo degradation of chitosan, which appears to target acetylated residues (Kim et al., 2008).

3.4. Applications

Chitosan of biopolymer have been used as blood coagulant, in artificial kidney membrane, digestive sutures, hypercholesterolemic agents, media for the slow release of drugs and hemostatic agent (Mohamed, 2004). It is a good candidate for biomedical applications such as for wound healing, vaccine delivery, as well as tissue regeneration (Yilmaz, 2004). It has been extensively investigated in biotechnological, biomedical, and environmental fields (Dang and Leong, 2006). Chitosan polymer is used in dentistry, because it prevents the formation of plaque and tooth decay. Since chitosan can regenerate the connective tissue that covers the teeth near the gums, it offers possibilities for treating periodontal diseases such as gingivitis and periodontitis (Elizalde-Pen et al., 2007).

3.4.1. Membrane

Aiba et al., (1986) used chitosan in membrane separation, chemical engineering, medicine and biotechnology areas. It was also found that the water adsorption and the mechanical properties of fibroin membrane were improved by blending chitosan (Chen et al., 1998).

3.4.2. Skin

Muzzarelli et al., (1988) used chitosan as artificial skin substitute and they reported that no adverse effect after implantation in tissue. In general, these materials have been found to evoke a minimal foreign body reaction, with little or no fibrous encapsulation. It observed the typical course of healing with formation of normal granulation tissue, often with accelerated angiogenesis (Suh and Matthew, 2000). Also, chitosan has many advantages for wound healing such as hemostasis, accelerating the tissue regeneration and stimulating the fibroblast synthesis of collagen (Mi et al., 2001). Chitosan possesses the properties favorable for promoting rapid dermal regeneration and accelerate wound healing suitable for applications extending from simple wound coverings to sophisticated artificial skin matrices (Kim et al., 2008).

3.4.3. Bone substitutes

Sapelli et al., (1986) used chitosan powder to promote healing of periodontal pockets, palatal wounds and extraction sites. Malette et al., (1986) proved enhanced leg bone regeneration in dogs using chitosan. It was reported to accelerate wound healing and was applied for bone wound repair in dogs (Borah et al., 1992) as well as bone growth in critical size metacarpal fibular defects. Klokkevold et al., (1992) concluded that chitosan solution may enhance the formation of bone. Later, it was reported to improve osseous healing of defects in femoral coundyl of sheep and stimulated cell proliferation and organized the hystoarchitectural tissue structure (Muzzarelli et al., 1994). Chitosan has been also extensively used in bone tissue engineering since after exploring its capacity to promote growth and mineral rich matrix deposition by osteoblasts in culture. Also, chitosan is biocompatible (additional minimizes local inflammation), biodegradable, and can be molded into porous structures (allows osteoconduction) (Martino et al., 2005 and Kim et al., 2008).

3.4.4. Drug delivery system (DDS)

Chitosan as an inert and hydrophilic material, its gel is suitable for application as matrices for enzyme/cell immobilization (Roberts, 1992) and for separation processes (Li et al., 1992). Felt et al., (1998) discussed the use of chitosan to manufacture sustained release systems deliverable by other routes such as nasal, ophthalmic, transdermal and implantable devices. Tarsi et al., (1998) suggested that low molecular weight chitosan may be very interesting as potential antidental caries agents. Chitosan has been reported to enhance drug delivery across the nasal or mucosal layer without damage (van der Lubben et al., 2001). The selectivity of membrane is a critical parameter in membrane separation and several factors are affecting its selectivity such as pore size, thus it is very suitable for the use as DDS and in artificial kidney (Mohamed, 2004). The cationic properties of chitosan offer valuable properties for drug delivery systems, gene delivery systems, and tissue engineering, that is, the formation of ion complexes between chitosan and anionic drugs or DNA can be used as a delivery vehicle (Kim et al., 2007).

3.4.5. Anti-bacterial

The experiments of antibacterial activity of chitosan-graft-polyethylene terephthalate (PET) against S. aureus showed a high growth inhibition in the range of 75–86% and still maintained a 48–58% bacterial growth inhibition after laundering (Hu et al., 2003). Chitosan and chito-oligosaccharides grafted membranes showed antibacterial activity against Escherichia coli, Pseudomonas aeruginosa, methicilin-resistant Staphylococcus aureus (MRSA), and S. aureus (Hu et al., 2003). Chitosan is a biomaterial with antiseptic property and the influence of the release or positive migration of protonated glucosamine fractions from the biopolymer into the microbial culture is the responsible event for the antimicrobial performance of the biopolymer (Beherei et al., 2009).

3.4.6. In-vitro application

Chitosan and collagen have intrinsic properties that support growth and differentiation of osteoblasts. Collagen was combined with chitosan and cross-linked to improve the biological stability and strength of chitosan–collagen composite sponges to reach the demand of an application in bone tissue engineering. The incorporation of chitosan into a collagen scaffold increases the mechanical strength of the scaffold and reduces the biodegradation rate against collagenase (Arpornmaeklong et al., 2007).

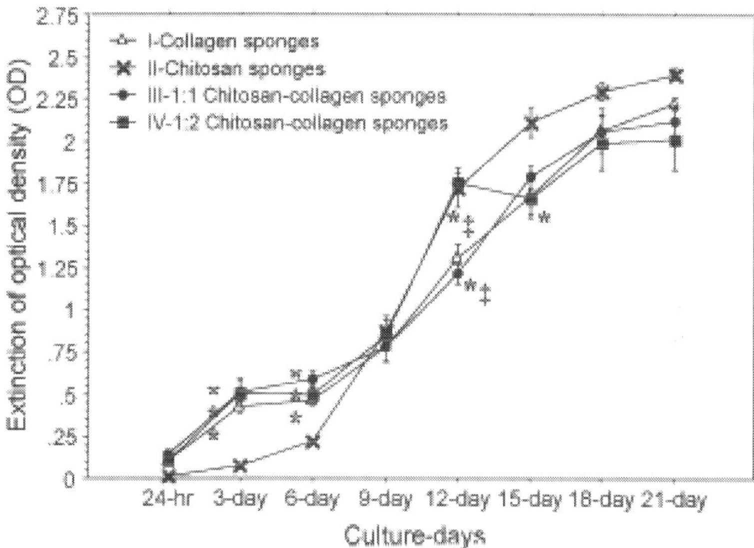

Figure 2. Cell viability of osteoblasts cultured for 21 days on the sponges (Arpornmaeklong et al., 2007).

The collagen, chitosan and chitosan–collagen sponges were biocompatible. All sponges supported growth of cells on three-dimensional structures in a similar manner. Chitosan sponges had a tendency to promote growth of cells to a greater extent than the other groups (Fig.2). It is postulated that strong attraction between positive charges on the chitosan surface and negative charges on the cell surface enhanced the metabolic activity of cells on chitosan sponges (Mi et al., 2001). In vivo, chitosan enhances angiogenesis and wound healing, and supports growth and differentiation of osteoblasts (Lee et al., 2004).

3.4.7. Other applications

a) Blood vessels:The chitosan fibers offer the potential of being fabricated into blood vessels and their blood compatibility results demonstrated for applications where hemocompatibility is required (Khora and Limb, 2003). Kim et al., (2008) reported that the effort has made to overcome both incomplete endothelialization and smooth muscle cell hyperplasia, which are two of the problems contributing to the poor performance of existing small-diameter (4 mm) vascular grafts, through complexation of GAGs with porous chitosan scaffolds. GAG-based material should promise because of their growth inhibitory effects on vascular smooth muscle cells and their anti-coagulant activity. However, few data regarding chitosan as a scaffold of tissue engineered blood vessels have been reported. Chitosan itself was documented to promote migration of endothelial cells and fibroblasts so as to accelerate wound healing. b) Nerve: Chitosan has been studied as a candidate material for nerve regeneration due to its properties such as antitumor, antibacterial activity, biodegradability and biocompatibility. Neurons that were cultured on the chitosan membrane can grow well and that chitosan tube can greatly promote the repair of the peripheral nervous system, also the chitosan fibers supported the adhesion, migration and proliferation of Schwann cells (SCs), which provide a similar guide for regenerating axons to Büngner bands in the nervous system (Yuan et al., 2004). c) Liver: Chitosan as a promising biomaterial can be applied in liver tissue engineering due to its various properties such as its structure that is similar to glycosamineglycans (GAGs), which are components of the liver extracellular matrix (ECM) (Li et al., 2003). d) Cartilages: Chitosan is one of the most abundant polysaccharides and thus shares some bioactivities with various glycosaminoglycans and hyaluronic acid present in articular cartilage (Suh and Matthew, 2000). Lu et al., 1999 has demonstrated that the chitosan solution injected into the knee articular cavity of rats lead to a significant increase in the density of chondrocytes in the knee articular cartilage, indicating that chitosan could be potentially beneficial to the wound healing of articular cartilage (Lee et al., 2004).

4. Bone structure

Bone is a specialized tissue comprising mineral substances, organic tissue, and water (Otto et al., 1997). Cortical bone is largely a composite of collagen, fiber and biological apatite (Fricain et al., 1998). The inorganic component of bone (bone mineral) is calcium phosphate that contains up to 8 wt% carbonate. Substitution of carbonate or other ions HA can occur in two distinct atomic sites in the lattice (Suchanek et al., 2002). These ions can partially substituted in the lattice for hydroxyl ions (OH), known as the A site, and/or for phosphate ions, known as the B site (Gibson and Bonfield, 2002). Skeletal bone is of two types: cortical bone and trabecular bone. Cortical bone is the outermost mineralized cortex. It is compact, strong, and densely packed as an intricate calcium matrix. Cortical bone comprises 85% of the skeleton, specifically 75% in the femoral neck, 75% in the distal radius, and 95% in the midradius. Cortical bone has no contact with marrow. Trabecular bone is the inner spongy structure composed of the sturdy collagen matrix. It comprises 15% of the skeletal mass and has structural rigidity and elasticity to withstand mechanical stress. Trabecular bone contains hematopoietic tissue in its central cavity. The bone of diaphysis consists of cancellous bone covered with a shell of cortical bone (Fig.3). The flat bones of the skull have a middle layer of cancellous bone sandwiched between two relatively thick layers of cortical bone (Liu et al., 2009).

Bone remodeling occurs continuously throughout the lifetime, although the process slows with age. The balance of osteoclastic and osteoblastic activity results in breakdown and reconstruction, which ensures skeletal integrity and maintains mineral homeostasis. Osteoclasts and osteoblasts are interconnected and influence the activity of each other. Osteoclasts are large multinuclear cells that develop from monocyte-macrophage precursors. They are imbedded in the bone matrix at or near the site of bone resorption and dissolve first the calcium and then the organic matrix of the bone. Osteoblasts arise from mesenchymal cells and are found layered over the bone. They deposit calcium into the matrix that is building up cortical bone and produce collagen and other proteins to synthesize the bone matrix. Osteoclasts that are in close proximity can increase sensitivity of osteoblasts to growth factors. When a bone is broken there is usually bleeding into the space between the bone and the periosteum. This produces a hematoma, swelling due to blood. The osteoblasts, bone-producing cells, near the hematoma invade it along with small blood vessels from the bone. In a short time, the hematoma is replaced by bone tissue produced by the osteoblasts. In general, this bone is cancellous, but the added bone makes the broken junction much thicker than it was. The swollen bone is called the callus. A remarkable process follows. Where there is strain, the newly formed cancellous bone condenses into compact bone. Where strain is absent, the cancellous bone disappears through the activity of osteoclasts, bone destroying cells. This process seems to determine the development of the skeleton in normal growth (Roodman, 2004).

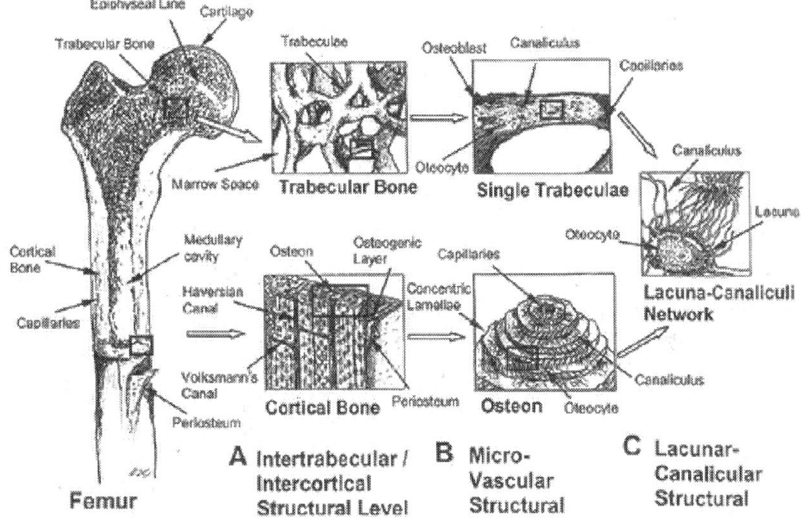

Figure 3. Bone structure

5. Hydroxyapatite/chitosan biocomposites: Introduction

For the increase of bioactivity and mechanical property, some composites of polymer and bioactive ceramics have been developed for bone tissue engineering. These composites fulfill the mechanical properties required for their function as skeleton, teeth and cells of organisms. Among these composites, HAp/polymer composites have attracted much attention since such composites may have osteconductivity due to the presence of HAp, which has a similar chemical composition and structure to the mineral phase of human bones and hard tissues. Thus, HA/polymer composite scaffolds are of interest for biomedical applications (Jin et al.,

2008). Kikuchi et al., (2004) prepared a self organized HA-collagen nano-composite by a biomimetic co-precipitation method. It was reported that the composite had similar microstructure to native bone and showed osteoclastic resorption and good osteoconductivity. However, a major concern over HA-collagen composite is the high cost of collagen, which limits its clinical application in healing bone defect to its insouciant formability and flexibility. Polymers such as chitosan have a higher degradation rate than bioceramics. Incorporation of HA into a chitosan polymer matrix has been shown to increase osteoconductivity and biodegradability with significant enhancement of mechanical strength (Yamaguchi et al., 2001). Chitosan's primary attractive features including its biocompatibility, biodegradability, flexibility, adhesiveness and anti-infectivity, make it as a feasible wound healing agent and an ideal polymeric matrix for HA ceramic (Rusu et al., 2005).

5.1. Structure

The free amino groups of chitosan ($C-NH_2$) was protonated to $C-NH_3^+$, when chitosan was dissolved in acetic acid (HAc) solution, which was shown as follows:

$$C-NH_2 + HAc \longrightarrow C-NH_3^+ + Ac \qquad pH = 4.2$$

The presence of calcium and phosphate ions in chitosan solutions leads to formation of HA\chitosan composites through electrostatic interactions between $C-NH_3^+$ and Ca^{2+} and\or PO_4^{3-} ions to form C-Ca and C-PO$_4$ complexes. There is also an interaction between OH of chitosan and OH of HA via hydrogen bond (Hu et al., 2004).

5.2. Properties

Chitosan can be utilized in combination with other bioactive inorganic ceramics, especially HA to further enhance tissue regenerative efficacy and osteoconductivity. Incorporation of HA with chitosan, the mineral component of bone, could improve the bioactivity and the bone bonding ability of the chitosan/HA composites (Wang et al., 2002). Chitosan just plays in a role of adhesive to dissolve the problem of difficulty of HA specific shape and migration of HA powder when implanted. HA/chitosan nano-composites are prepared by the precipitation. It is proposed that the nano-structure of HA/chitosan composite will have the best biomedical properties in the biomaterials applications (Chen et al., 2002). Also, it has been demonstrated that chitosan-hydroxyapatite composite induces osteoconductivity in osseous defects and could act as drug vehicle. It is important to be able to load these composites with short-time life and controlled action anti-inflammatory to reduce or eliminate undesirable inflammatory processes (Larena et al., 2004).

It is desirable to develop a composite material with favorable properties of chitosan and HA. The designed composites are expected to have an optimal mechanical performance and a controllable degradation rate as well as eminent bioactivity and this will be of great importance for bone remodeling and growth (Zhang et al., 2005). It must be emphasized at this point that the successful design of a bone substitute material requires an appreciation of the structure of bone. Thus, the use of a hybrid composite that makes up of chitosan and calcium phosphate resembles the morphology and properties of natural bone. This may be one-way to solve the problem of calcium phosphates brittleness, besides possessing good biocompatibility, high bioactivity and great bone-bonding properties (Ding, 2007). The mechanical strength of the chitosan/calcium phosphate composite fiber with core-shell structure increased with an increased concentration of chitosan solution (Matsuda et al., 2004). Among the composites studied, the 30/70 chitosan/n HA exhibits the maximum value of compressive strength, about 120 MPa, which is strong enough to be used in load-bearing sites of bone tissue. In contrast the compressive strength of pure HA compact prepared by the similar method has been reported as 6.5 MPa about one twentieth of the maximum value of the composite. In general, the proper stress transfer occurring between the reinforcement and the matrix governs the mechanical characteristics of filled polymers. The chemical and mechanical interlocking between n-HA and chitosan are accounts for the efficient stress-transfer in the composite system. Besides, the interactions such as hydrogen bonding and chelation between the two phases, also contribute to the good mechanical properties of chitosan/n-HA composite (Zhang et al., 2005). The biodegradable composites based on chitosan and calcium phosphate have been prepared using a simple mixing and heating method. The detrimental effects of the simulated physiological environment on mechanical properties of the hybrid composites resulted in the significantly decrease in strength and modulus. The chitosan/calcium phosphate composites containing 10 wt/v % might an optimal material in terms of initial strength and degradation behavior. Although susceptibility to solution attack, this type of chitosan/calcium phosphate composites with high initial strength might be acceptable for use in bone tissue repair (Ding, 2007).

Three-dimensional biodegradable chitosan/nHA composite scaffolds were characterized by superior mechanical, physicochemical, and biological properties compared to pure chitosan scaffolds for bone tissue engineering. The nanocomposite scaffolds were characterized by a highly porous structure and the pore size was similar for scaffolds with varying n-HA content. The nano-composite scaffolds exhibited greater compression modulus, slower degradation rate and reduced water uptake, but the water retention ability was similar to that of pure chitosan scaffolds. Favorable biological response of pre-osteoblast on nanocomposite scaffolds included improved cell adhesion, higher proliferation, and well spreading morphology in relation to pure chitosan scaffold (Thein-Han and Misra, 2009).

Mechanical properties of biocomposites: hydroxyapatite/chitosan (HA/CS) nano-composite rods were reinforced via a covalently cross-linking method. The bending strength and bending modulus of the cross-linked HA/CS (5/100, wt/wt) rods could arrive at 178 MPa and 5.2 GPa, respectively, increased by 107% and 52.9% compared with uncross-linked HA/CS (5/100, wt/wt) rods (Takagi et al., 2003). The presence of HA-DBM filler into the grafted chitosan copolymer matrix resulted in compressive strength properties are quite close to those of cancellous bone (2–12 MPa). This result is due to effect of the presence of demineralized bone matrix (DBM) powder and pMMA having bone cement formation within this composite (Mohamed et al., 2007). The E-modulus and compressive strength for the three composites HA/, 90%HA-10Ti/, and 70%HA-30%Ti/grafted chitosan copolymer composites recorded comparable values compared to the cancellous bone. Therefore, the presence of HA filler or HA filler containing titania content up to 30% into the copolymer resulted in compressive strength properties that are quite close to those of cancellous bone (2–12 MPa) (Mohamed et a., 2008).

Collagen/apatite composite membranes exhibited significantly improved mechanical properties compared with their pure collagen equivalent; their mechanical properties were still lower than those of natural bone (Teng et al., 2009). It was notified that hardness of calcium pyrophosphate (CPP)/ chitosan composite (66.80) was increased compared to chitosan copolymer (60.88) and the hardness of CPP/chitosan-grafted composite (68.23) was also increased compared to chitosan–gelatin copolymer (84.12) proving the polymer/filler interaction and adhesion. Also, the compressive strength of CPP/chitosan composite (6.53 MPa) was increased compared to chitosan copolymer (5.11 MPa). These values of compressive strength were comparable to those of human cancellous bone (Kokubo et al., 2003). As a result, CPP filler powder into the chitosan copolymer matrix containing chitosan or chitosan–gelatin polymer resulted in more effective reinforcement of the composite, then, stiffer composite (El-Kady et al., 2009). The CPC–chitosan composites were more stable in water than conventional calcium phosphate cement (CPC). They did not disintegrate even when placed in water immediately after mixing. The CPC–chitosan paste hardened within 10 min in all cases. The authors demonstrated that CPC–chitosan composites are stable in a wet environment and have acceptable mechanical strengths for clinical applications (Wang et al., 2010).

5.3. Preparation

Although powder ceramics remain the form of choice for filling small irregular defects, the therapeutic effect of the filling implant was lost by migration of particles from the defect site. Furthermore, it was difficult to be handled and kept in place compactly for convenient fabrication and operation of block-type ceramics (Lin et al., 1998). Thus, it is necessary to mix a suitable binder with the granular material to overcome these problems. Presently, the approaches to obtain chitosan/hydroxyapatite (HAp) composite materials are based either on mixing or co-precipitation methods. Yamaguchi et al., (2001) have developed one of the co-precipitation methods that lead to a type of chitosan/HAp composites. In this approach, the composite was co-precipitated in one step, by dropping a chitosan solution containing phosphoric acid into a calcium hydroxide suspension. Other approaches employ either the biomineralization of chitosan in a solid form (especially as membranes) in simulated body fluids (SBF) (Beppu and Santana, 2001) or by mixing of a chitosan solution with different calcium phosphate fillers followed by their precipitation as hydrogel composite. The use of these approaches leads to incorporation of inorganic fillers into the structure of composites (Schwarz and Epple, 1998), either as nano-sized or micro-sized particles.

Different preparation methods of HAp/chitosan composites have been reported, such as mechanical mixing of HAp powder in a chitosan solution, coating of HAp particles onto a chitosan sheet, coating of HAp crystals onto a tendon chitosan (Yamaguchi et al., 2003) and a co-precipitation method. The chitosan/HAp hybrid fiber has also been reported by Chung and Korean, (2002). However, these materials were shown in the macroscopically homogeneous. The conventional method to fabricate chitosan/HA composite is that HA powder was mixed with chitosan, dissolved in 2% acetic acid solution, then the mixture was impressed into mold, finally was freezing-dried to make sponge composite. Surface modification and polyblend methods can be used to change the physicochemical properties of chitosan–gelatin membranes or scaffolds by incorporating hyaluronic acid. Adding hyaluronic acid can improve the mechanical, biological and anti-degradation properties of the membranes or the scaffolds (Mao et al., 2003).

Hydroxyapatite (HAp) was prepared by precipitation method, while the biphasic hydroxyapatite/tricalcium phosphate (HA/B-TCP) was prepared by heating the prepared HAp at 900°C for 5 hours in air. To improve bioactivity both HAp and HAp/TCP fillers were loaded onto chitosan grafted with two monomers, hydroxyethylmethacrylate (HEMA) and methylmethacrylate (MMA) during copolymerization process (Hashem and Mohamed, 2007). Also, biocomposites containing HA-DBM mixture powder loaded onto the copolymer matrix containing the grafted chitosan with poly methylmethacrylate (pMMA) and its derivative during copolymerization were fabricated (Mohamed et al, 2007).

The chitosan mineralization in the case of using a stepwise co-precipitation approach involves the following stages: First, chitosan chains change their conformation as a function of environmental parameters, such as pH. Starting from extended conformations such as worm like, as seen in Fig.4 (1), adopted at low pH (until 3–3.5), chitosan turns to the more compact conformations such as extended random coil Fig.4(2) and even to more compacted random coil conformations, at higher pH values (from 3.5 until 6). Between 5.5 (the pH at which brushite is precipitated) and 6.5–6.7 (the pH at which chitosan is precipitated) is the pH range that leads to formation of an interconnected three-dimensional net work between chitosan and brushite that can be approximated as a dendritic-like structure, as shown in Fig.4(3). This structural model is characterized by highly dense irregular shaped cores which are linked by the chitosan bridging segments. In the dendritic core, the chitosan chains are randomly packed as amorphous regions while in the bridging regions more extended conformations are found, in some parts parallel oriented chains domains are presented. Since the dendritic core is in the range between 200 and 600nm hence it is formed by the many compact random coil chitosan units that are approaching one another chitosan chain segments can interact with the CaPs phases (i.e. seeds of HAp already formed at pH 5.5 and identified by its XRD pattern, brushite and some other ACP). In this way, they achieve of so called "anchoring regions" by different specific interactions such as ion-dipole or/and through the complexation of Ca with chitosan. Considering that a dendritic-like structure is formed in the earlier stage of composite formation, one can explain the complex bimodal distribution of HAp nano-crystallites in the chitosan matrix. Inside the dendritic core, the chitosan chains density is much greater than outside of them, shown as inter-connection blob regions. We assume that the probability to find a certain number of HAp crystallite seeds per a chitosan chain unit is the same, inside and outside the dendritic core regions. Since the chitosan chain density is much greater inside the cores, this leads to the conclusion that the higher density of HAp crystallite seeds is located inside the dendritic cores. Consequently inside the core favorite the formation of the "small HAp nanocrystallites (their growth is spatially limited) whereas outside the cores, along the interconnection regions where the space constraint is not that much limited, "large" HAp nanocrystallites are favorized to be formed. Finally, the cluster-like and scattered-like size domains are generated in this way, as seen in Fig.4(4) This theoretical model is supported by the experimental data, since we could demonstrate that the amount of chitosan in the composite can be used to control the HAp nano-crystallite size, otherwise no influence should be observed (Rusu et al., 2005).

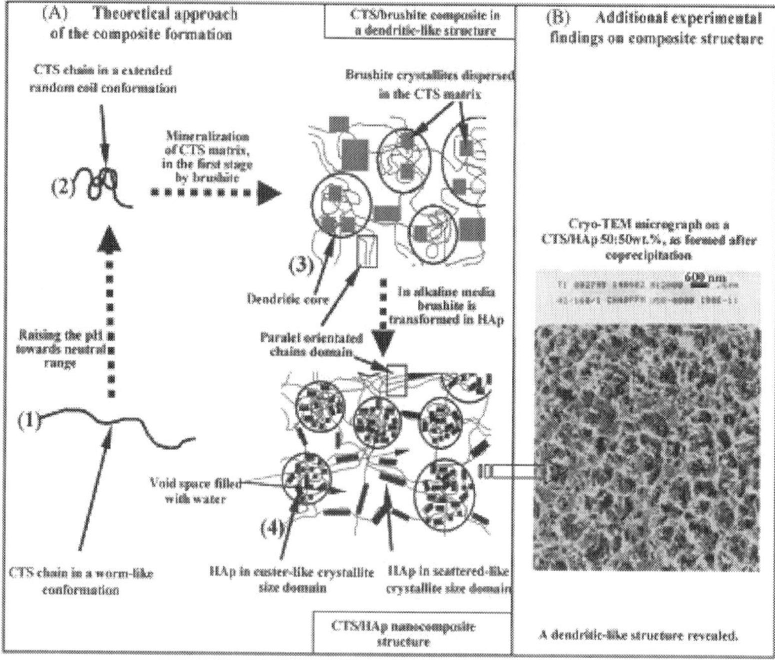

Figure 4. Sketch of chitosan (CTS) mineralization through nanosized HAp. The main stages in the formation of the composite structure are outlined by the theoretical approach, as shown on the left side (A). On the right side (B), we present a TEM micrograph a CTS/HAp 50:50wt% composite, which emphasizes the dendritic-like structure of sample at this stage of formation (in gel-like form). Furthermore, when the excess water is released a solid, rigid composite is obtained.

An interesting approach was reported by Hu et al., (2004) in which the chitosan hydrogel is mineralized via in situ hybridization by the ionic diffusion processes in a controlled manner. It should be mentioned that, each approach of those cited above leads to a particular type of chitosan/HAp composite materials with respect to their structure and properties. In order to prepare such types of composites, we have reported a stepwise co-precipitation method in which the pH of the chitosan solution is gradually increased in a stepwise fashion (Ng et al.,

2008). Biodegradable hydroxyapatite/chitosan-gelatin polymeric biocomposites were fabricated by using HA powder and HA filler containing titania powder (10 and 30%) with a chitosan and gelatin grafted co-polymeric matrix during copolymerization process (Mohamed and Mostafa, 2008). Preparation of a model HAp/CTS (30:70 in mass ratio) nanocomposite nanofibers using a two-step method, which involves firstly preparing HAp/CTS nanocomposites by a co-precipitation synthesis approach and then fabricating the resultant HAp/CTS nanocomposites, aided with a fiber-forming additive–ultrahigh molecular weight poly(ethylene oxide), into nanofibers via the electrospinning process (Zhang et al., 2008).

5.4. Characterization

The FT-IR spectra show that the two characteristic bands of amide I (1655 cm^{-1}) and amide II (1599 cm^{-1}) for chitosan shift to lower wavenumber after being compounded, which suggests that interaction must take place between chitosan and n-HA, including hydrogen bonds between -NH$_2$ and-OH of n-HA as well as the chelation between -NH$_2$ and Ca^{++}. The more shift of these bands to lower wave number, the stronger the hydrogen bonds between these groups and also the stronger the interaction between these molecules (Zhang et al., 2005). The XRD pattern of precipitated HA shows un-differentiated broad peaks with poor crystallinity around the characteristic region. However, the crystallographic structure of precipitated HA nano-crystals is more identical with natural bone mineral (biological apatite). Hence, the prepared HA nano-crystallites in this investigation have more similarity with natural bone mineral in terms of degree of crystallinity and structural morphology. The calcined HA exhibited all the characteristic diffracted peaks of stoichiometric HA with higher degree of crystallinity. Rising in the calcination temperature shapes the diffracted peaks more sharper, which is a good sign for the improvement of crystallinity of precipitated HA. The obtained results did not show any peaks corresponding to calcium carbonate and calcium oxide and hence suggesting that the ingredients were reacted completely and produced a homogeneous HA. The XRD analysis of HA/chitosan composites proved that, a broad peak assigned to chitosan at 20° becomes wider and weaker with increase of n-HA. It suggests that the addition of n-HA obviously affects the crystallinity of chitosan. The characteristic peaks at 25.8° (002) and 39.6° (310) are used to calculate the n-HA crystal sizes (Xianmiao et al., 2009). The TGA mass loss of HA\C composite increased from 3.3–6.5 mass% as the chitosan concentration increased from 0–2.5 mass%. The amount of chitosan that adsorbed on HA was 2.8–3.1 mass% based on Carbon–Hydrogen–Nitrogen (CHN) analysis. The specific surface area of HA increased after aging in chitosan acetate gel solutions and attained a high value of 160 m^2/g in comparison to 85 m^2/g for untreated HA (Wilson and Hull, 2008).

The SEM micrograph of precipitated HA (Fig.5a) exhibited nano-sized crystals with almost uniform particles size. The HA particles prepared were not only stoichiometric but also mono-dispersive and roughly particles were not fused together with other crystals. It can be inferred that majority of the particles were of single crystals, regular shape and cleaner contours with no agglomeration which are highly beneficial for coating of nano HA onto biomedical implants. On the other hand, composites bone paste (Figs.5b and c) showed heterogeneous phases with complete fusion of HA crystallites into chitosan matrix. Major changes in the crystal size of composites were monitored as compared to single phase HA due to the presence of chitosan macromolecules. The physical appearances of the composites are quite different from the starting material and also apparent that ultra-fine particles of HA are found to aggregate into large clusters and precipitate in the chitosan matrix. One of the possible reason may be due to some of the HA nano-particles might have partially dissolved in the acidic chitosan solution that permitting the HA particles more easily to penetrate into the chitosan matrix. The particles of composites showed a high tendency to agglomerate and hence it can have capability to prevent the particle mobilization after post-implantation. Both the SEM pictures (b and c) of composites exhibited porous surfaces, but the pores were not uniform. The average pore size was found to 105 and 80mm for 5 and 10 wt chitosan composites, respectively. The composites containing porous structure on their surfaces will be more beneficial for tissue in-growth (Murugan and Ramakrishna, 2004). Also, the TEM micrographs of HA\chitosan =100\5 (wt\wt) proved that the HA particle size was 100 nm in length and 20-50 nm in width which dispersed well in chitosan matrix homogenously (Hu et al., 2004)

Figure 5. SEM photographs of (a) precipitated nano HA; (b) composite with 5% chitosan sol; and (c) composite with 10% chitosan sol.

5.5. Applications

One of the present trends in implantable applications requires materials that are derived from nature. The impetus is twofold. First, such "natural" materials have been shown to better promote healing at a faster rate and are expected to exhibit greater compatibility with humans. Second, new concepts in implantable medical devices especially tissue engineering derived from a combination of biomaterial onto which cells are seeded, require "temporal" features dictating the biomaterial to matrix was progressively resorbed. Therefore, HA or other calcium containing materials incorporated into chitosan has been a primary research area where orthopedic or bone substitution and periodontal applications were the focus (Khor and Lim, 2003).

5.5.1. Bone substitutes

Maruyama and Ito, 1996 reported that the strength of chitosan-HA hardened composite was comparable to that the cancellous bone derived from tibial eminentia. Pal et al., (1997) prepared different varieties of HA in conjunction with chitosan, as binder, to know its unique biological behavior in bone bonding. These hybrid materials displayed good blood compatibility (Chen et al., 1998). Apatite cement (AC) was almost completely surrounded by mature bone at eight weeks. No promotion or production of osteoconductivity was observed by chitosan even though it is considered to promote bone formation. Then, they concluded that there is enhancement of bone formation (Takechi et al., 2001).

Chitosan and silk fobroin (SF) together as a complex organic matrix for HA granules were employed attempting to obtain a novel composite HA/chitosan–silk fobroin (HA/CTS–SF) with good osteoconductivity, enhanced mechanical strength and sufficient formability and flexibility. Additionally, chitosan and SF are easily derived from naturally abundant chitin and silk cocoon, respectively, which offers a great promise for the potential use of HA/CTS–SF composite as bone scaffold material. HA/CTS–SF composite was obtained via a simple co-precipitation method at room temperature with chitosan and SF serving as a complex organic matrix. The inorganic component in the composite is identified as mono-phase poorly crystalline HA containing carbonate ions. The chemical interactions between the inorganic and organic constituents in the composite, probably take place via the chemical bonding between Ca^{2+} and the amino group of chitosan or the amide bands of SF. The involvement of chitosan and SF endows the composite with higher compressive strengths compared to pure HA. These findings suggest that HA/ CTS–SF composite may be a promising biomaterial for bone in-growth and implant fixation (Wang and Li, 2007).

A novel bone repair material can be obtained by incorporating carboxymethyl cellulose (CMC) into n-HA/CS system. Not only did it compound uniformity by chemical interactions and resembled natural bone apatite in composition morphology and size, but also it improved the compressive strength compared with n-HA/CS composite and had controllable degradation rate via adjusting the CS/CMC weight ratio (Liuyun et al., 2008). HA can promote the formation of bone-like apatite on its surface. Polymers combined with HA are capable of promoting osteoblast adhesion, migration, differentiation and proliferation, especially useful for potential applications in bone repair and regeneration. HA particles have been incorporated into chitosan matrices to enhance the bioactivity of tissue engineering scaffolds for hard tissue regeneration. Therefore, composite membrane of HA and chitosan is expected to be a good degradable barrier membrane for guided bone regeneration (GBR) technique (Xianmiao et al., 2009).

5.5.2. Bone tissue engineering

The scaffold is a key component of tissue engineering (Langer and Vacanti, 1993). The study of inorganic crystal assembly in or on an organic polymer matrix is an important focus of bio-mineralization to produce nano composites, which can mimic natural bone. The 3D macro porous scaffolds play an important role in the formation of new tissues and provide a temporary scaffold to guide new tissue in-growth and regeneration (Nikalson and Langer, 1997). Chitosan has been proposed to serve as a non-protein matrix for 3D tissue growth. Chitosan could provide the biological primer for cell-tissue proliferation and reconstruction. One of the most promising features of chitosan is its excellent ability to be processed into porous structures for use in cell transplantation and tissue regeneration. In tissue engineering, the porous structure of chitosan provides a scaffold for bone cells to grow in and seed new bone regeneration. For rapid cell growth, the scaffold must have optimal micro architecture such as pore size, shape and specific surface area (Madihally and Matthew, 1999).

In bone tissue engineering, the biodegradable substitutes act as a temporary skeleton inserted into the defective sites of skeleton or lost bone sites, in order to support and stimulate bone tissue regeneration while they gradually degrade and are replaced by new bone tissue (Service, 2000). Chitosan-based scaffolds possess some special properties for use in tissue engineering. The major goal in fabricating scaffolds for bone tissue engineering is to accurately control pore size and porosity. Porous chitosan structures can be formed by freezing and lyophilizing chitosan–acetic acid solutions in suitable moulds (Chow and Khor, 2000). Bone regeneration research needed to deal with various clinical bone diseases such as bone infections, bone tumors and bone loss by trauma (Braddock et al., 2001). To combine the osteoconductivity of calcium phosphate and good biodegradability of polymers, composites have been developed for bone tissue engineering either by directly mixing the components or by a biomimetic approach (Wei and Ma, 2004). Polymer-ceramic composite scaffolds are expected to mimic natural bone, in the way that natural bone is also a composite of inorganic compounds (calcium phosphates especially substituted carbonated hydroxyapatite) and organic compounds (collagen, protein matrix, etc.). The hydroxyapatite–chitosan–alginate porous network has been reported and demonstrated to be suitable for bone tissue engineering applications using osteoblast cells (Zhao et al., 2003).

Research advances in bone regeneration in tissue engineering have focused on the development of three dimensional (3D) porous scaffolds that can serve as a support, reinforce and in some cases organize the tissue regeneration or replacement in a natural way (Sachlos et al., 2003). Several studies have been focused on chitosan–calcium phosphates (CP) composites for this purpose in bone tissue engineering. Beta-tricalcium phosphate (β-TCP) and hydroxyapatite (HA) of CP bioceramics are excellent candidates for bone repair and regeneration because of their similarity in chemical composition with inorganic components of bone (Zhang et al., 2003). Tissue engineering is regarded as an ultimately ideal medical treatment for diseases that have been too difficult to be cured by existing methods. This biomedical engineering is designed to repair injured body parts and restore their functions by using laboratory-grown tissues, materials and artificial implants. For regeneration of failed tissues, this biomedical engineering utilizes three fundamental tools: living cell, signal molecules, and scaffold. The choice of chitosan as a tissue support material is governed among others by multiple ways by which its biological, physical and chemical properties can be controlled and engineered under mild conditions (Krajewska, 2005). The 3D macro porous scaffolds play an important role in the formation of new tissues and provide a temporary scaffold to guide new tissue in-growth and regeneration. The fabrication of biodegradable and osteoconductive scaffolds with a 3D interconnected porous network has been a formidable challenge. The feasibility of producing cost effective organic–inorganic scaffolds for tissue engineering to mimic bone by the diffusion method was performed. The porous structure of chitosan scaffold was homogeneously mineralized using this technique of apatite formation at room temperature. The mineralized scaffolds were found to be non-cytotoxic and better for cell proliferation and growth, as indicated by the enzyme activity and protein levels, than un-mineralized scaffold. This suggested that it could be used for further osteoconductivity studies. A biodegradable matrix with sufficient mechanical strength, optimized architecture and suitable degradation rate, which could finally be replaced by newly formed bone, is most desirable (Manjubala et al., 2006). Although the chitosan based composite biomaterials need to improve their mechanical properties for bone tissue engineering, no doubt that chitosan is a promising candidate scaffold material in clinical practice due to the worthiest ability to bind anionic molecules such as growth factors, GAG and DNA. Especially, the ability to link chitosan to DNA may render this material a good potential as a substrate for gene activated matrices in gene therapy application in orthopedics (Kim et al., 2008).

Shen et al., (2007) performed that with the increase of pH after the addition of ammonia, carboxyl groups of citric acid may begin to act as nucleation center for calcium phosphate formation. These negatively charged carboxyls in the reaction system can bond Ca^{2+} strongly and thus forms a large scale of local super saturation microenvironment, and strong electric field resulted from high concentration of negatively charged carboxyls are favor of the interaction that with the most positively charged crystalline plane, so there are many nucleation sites in the network of hydrogel template, each point of nucleation can result in microcrystal. Here, to our attention, biocompatible citric acid took the place of acetic acid in this work because three carboxyl of citric acid could provide more nucleation sites which were appropriate to formation of ultra fine nano-sized carbonate apatite. And it has been conjectured that appropriate increase of citrate ions can benefit the bone resorption and ossification through the formation of dissociated calcium citrate complexes in the surrounding body fluid (Rhee and Tanaka, 1999). Each citric acid molecule can provide three negatively charged carboxyls which act as nucleation center for calcium phosphate formation. Increase of nucleation center can be appropriate to fine crystallites. Furthermore, cross-linking chitosan hydrogel was provided with three dimensional network microstructures, its compartment effect limited the growth of inorganic mineral particles, so the inorganic nano-particles were limited to aggregate in the compartment of the chitosan hydrogel template according to orientation of preferential growth of crystal plane. This multiple-order template effect based on multiple-point nucleation of citric acid and compartment of hydrogel network had a very obvious mediation in the formation process of homogeneous composites (Shen et al., 2007) (Fig.6). In bone tissue engineering, the biodegradable scaffold is a temporary template introduced at the defective site or lost bone to initiate bone tissue regeneration, while it gradually degrades and is replaced by newly formed bone tissue. Finally, an ideal scaffold is characterized by excellent biocompatibility, controllable biodegradability, cytocompatibility, suitable microstructure (pore size and porosity) and mechanical properties. Additionally, it must be capable of promoting cell adhesion and retaining the metabolic functions of attached cells (Thein-Han and Misra, 2009).

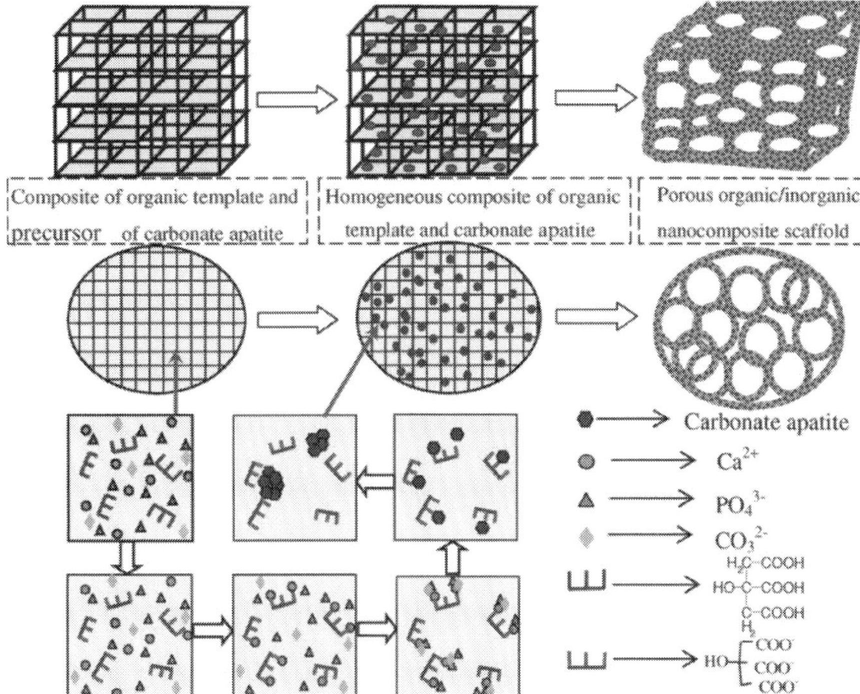

Figure 6. The scheme of formation of homogeneous chitosan/carbonate apatite composite and 3D nanocomposite scaffold (Shen et al., 2007).

5.5.3. In vitro applications

The process of apatite formation on the bioactive materials in living body could be reproduced in simulated body fluid (SBF), which means that in vivo bone bioactivity of a material can be predicted by assessing apatite formation on its surface in SBF. They confirmed that there are two types of material which inserted into living body. One of them was able to have apatite form on its surface in SBF, and consequently has apatite produced on its surface in the living body, and bonds to living bone through this apatite layer. The other type was directly bond to living bone without the formation of apatite on their surfaces, so examination of apatite formation on the surface of a material in SBF is useful for predicting the in vivo bone bioactivity of the material, not only qualitatively but also quantitatively (Kokubo and Takadama, 2006).

5.5.3.1. In simulated body fluid (SBF)

From SEM photos (Fig.7), it can be known that chitosan in the chitosan /nHA composite gradually degraded during the soaking in SBF solution, which resulted in plenty of macro- and micropores on the surface of and inside the specimens. At the same time, a lot of tiny apatite crystals deposited on the surface of the specimens, and till the 8th week, a thin layer of bone-like apatite, being highly bioactive was formed. At the first 4 weeks, the degradation rate of chitosan was higher than the deposition rate of apatite on the surface of specimens, which corresponding to a continuous increase of the rate of weight loss. After that, the deposition of apatite is prior to the degradation of chitosan, so the rate of weight loss decreased. This was also confirmed by the rate of water adsorption with the degradation of chitosan during the specimen's soaking in SBF solution, a more sponge-like structure was formed, which can hold more water. However, with more apatite crystals deposition, some of these pores were filled or covered, so water adsorption decreased (Zhang et al., 2005). Kong et al., (2006) reported that chitosan/nano-hydroxyapatite composite scaffolds analysis showed that after incubation in simulated body fluid on both of the scaffolds (the apatite-coated composite scaffolds and apatite-coated chitosan scaffolds), carbonate hydroxyapatite was formed. With increasing nano-hydroxyapatite content in the composite, the quantity of the apatite formed on the scaffolds increased. Compared with pure chitosan, the composite with nano-hydroxyapatite could form apatite more readily during the biomimetic process, which suggests that the composite possessed better mineralization activity (Kong et al., 2006).

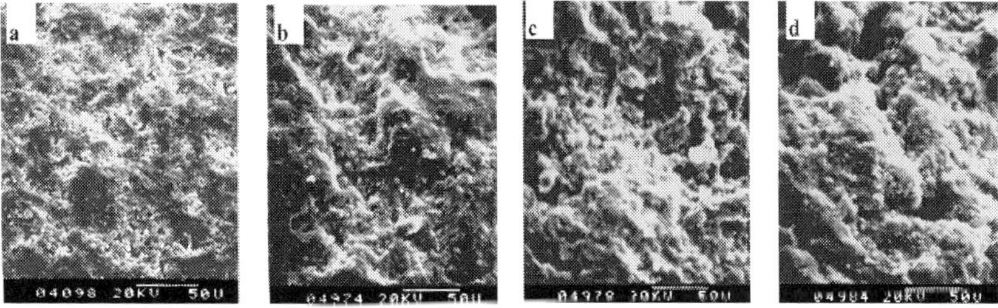

Figure 7. The SEM images of chitosan/nHA composites after soaking in SBF solutions for (a) 0 week, (b) 1week, (c) 4 weeks and (d) 8 weeks, Magn. × 400.

The swelling properties and degradation behavior proved the stability of hydroxyapatite-titania/chitosan-gelatin polymeric biocomposites into the media. In-vitro test behavior confirmed that the prepared composite enhanced the deposition of Ca^{++} and P ions onto the surface that is in the favor of the formation of apatite layer. FT-IR and SEM-EDAX of copolymer and three composites post-immersion verified the formation of spherical apatite particles onto the copolymer surface; therefore, it was expected to enhance the apatite nucleation onto the filler composite surface especially hydroxyapatite-titania/chitosan-gelatin (AK1) composite containing 10% content of titania (Mohamed and Mostafa, 2008).

5.5.3.2. Bone Tissue engineering (Scaffold)

Anti-washout scaffold paste could be directly applied to fit complex shapes of bone defects, without involving machining as in the case of sintered hydroxyapatite. The synergistic use of a reinforcing agent (e.g., chitosan) and a pore-forming agent (e.g., mannitol) in a bone graft may be applicable to other tissue engineering materials. In developing strong and macro-porous calcium phosphate cement (CPC) scaffolds by incorporating chitosan and water-soluble mannitol. The new CPC–chitosan formulation was biocompatible and supported the adhesion, spreading, proliferation and viability of osteoblast cells. The cells were observed to infiltrate into the pores of the scaffold and establish cell–cell interactions. The increased strength and macroporosity of the new apatite scaffold may help facilitate bone ingrowth, implant fixation, and more rapid new bone formation (Fig.8) (Hockin et al., 2005).

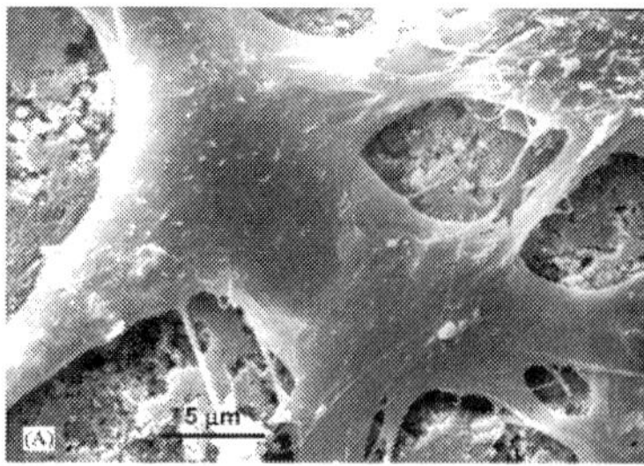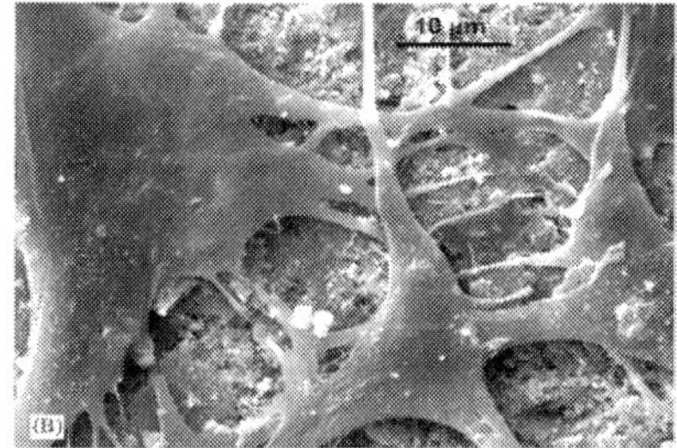

Figure 8. The SEM of cell attachment on (A) CPC control and (B) CPC chitosan composite. The cells developed cytoplasmic processes with lengths ranging from approximately 20 to 50 mm, and the materials exhibited similar cell attachment and cytoplasmic processes development (Hockin et al., 2005).

Kong et al., (2006) reported that pre-osteoblast cells cultured on the apatite-coated scaffolds showed different behavior. On the apatite-coated chitosan/nano-hydroxyapatite composite scaffolds cells presented better proliferation than on apatite-coated chitosan scaffolds. The cells on composite scaffolds showed a higher alkaline phosphatase activity which suggested a higher differentiation level. The results indicated that the addition of nano-hydroxyapatite improved the bioactivity of chitosan/nano-hydroxyapatite composite scaffolds. MSCs do not appear to be rejected by the immune system, allowing for large-scale production, appropriate characterization and testing, and the subsequent ready availability of allogeneic tissue repair enhancing cellular therapeutics. Overall it can be said that, for now, MSCs present more advantages than other cells and have already been widely used in bone tissue engineering. All the superiorities of MSCs encourage us to introduce MSCs into n-HA composite scaffolds for tissue engineering application (Wang et al., 2007).

The morphology and behavior of bone marrow stem cell (BMSCs) cultured in-vitro with the n-HA/chitosan (CS) composite membranes are observed under phase-contrast microscope. Fig.9 shows representative phase-contrast micrographs of cell attachment on the membrane with a n-HA/CS ratio of 4:6 after culture for 1 day, 7 days and 11 days. At the first day, only a few BMSCs are present with the elongated fusing form shape. At 7 days, a large amount of cells proliferate and form cell colony. At 11 days, the population of cells increases manifestly and cells fully attach to the membrane. Obviously, the n-HA/CS composite membrane has no negative effect on the cell morphology, viability and proliferation (Xianmiao et al., 2009).

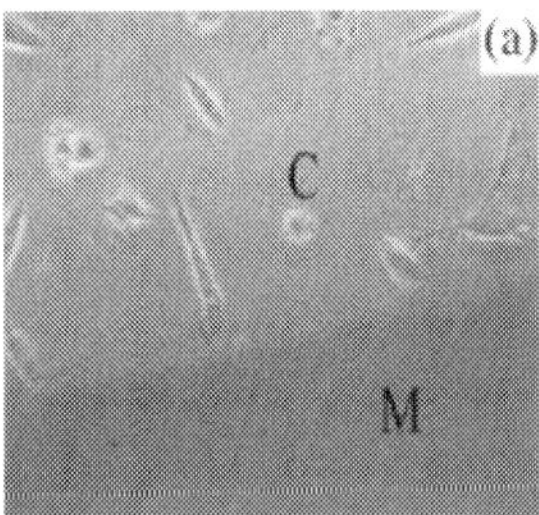

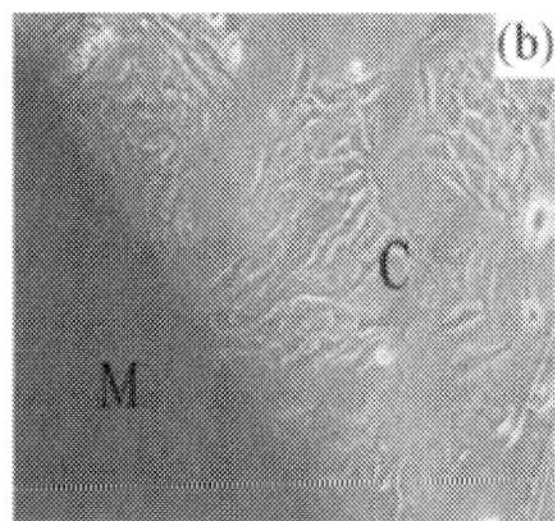

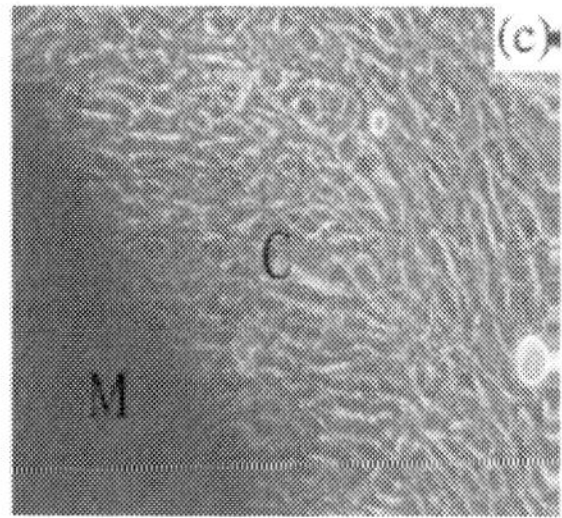

Figure 9. Phase-contrast micrographs of the BMSCs (denoted as C) attached to n-HA/CS (4:6) membrane (denoted as M) after in vitro culture for 1d (a), 7d (b) and 11 d (c).

In chitosan–nHA scaffolds, the presence of extensive filopodia, flat morphology, and excellent spreading in and around the interconnected porous structure, indicated strong cellular adhesion and growth (Fig.10e–h). Furthermore, cell density, cell–cell contact, sheet-like structure, and formation of extracellular matrix and cytoplasmic extensions were more pronounced on the chitosan–nHA surface than on pure chitosan. They believe based on Fig.10e–h that the steps involved in the development of sheet-like morphology involves clustering of cells and bridge formation between the pore walls with consequent formation of a multilayer structure. These steps occurred during early stages in the nano-composite constructs in relation to chitosan scaffolds, suggesting that pre-osteoblasts have high affinity to the surface of chitosan–nHA composite, which is attributed to its increase surface area and composition. Chitosan–nanocrystalline calcium phosphate scaffolds characterized by a relatively rough surface and approximately 20 times greater area/unit mass than chitosan scaffold indicated increase adsorption of fibronectin and improved cell attachment (Thein-Han and Misra, 2009).

5.5.4. In-vivo application

The extracellular matrix (ECM) is a powerful regulator of cell adhesion and indeed cells respond to the ECM by means of integrins, which couple the component of the ECM with the actin cytoskeleton. This structure thus mediates adhesion to the ECM and therefore to the implant material. In this respect, the possibility of bonding osteoinductive polymers such as modified chitosan to the ceramic substrate could enhance cell proliferation and consequently anchorage to the implant (Mattiolibelmonte et al., 1998). Sections from chitosan coated HA implants exhibited an evident mesenchymal reaction between bone and implant with several features of osteoinduction. Bone trabeculae penetrating the HA implant were also observed.

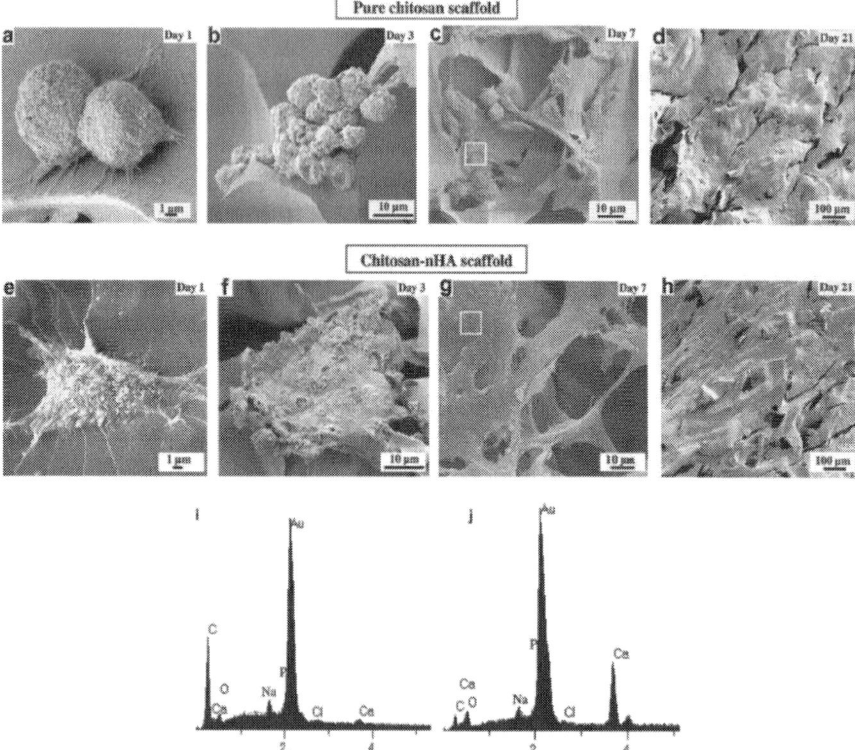

Figure 10. Scanning electron micrographs illustrating morphology of pre-osteoblasts seeded on high-MW chitosan (CH) and chitosan–nHA (CH1) scaffolds (Saggital section). Pre-osteoblasts on chitosan surface after (a) day 1, (b) day 3, (c) day 7 and (d) day 21 of cell culture; and on chitosan–nHA surface after (e) day 1, (f) day 3, (g) day 7 and (h) day 21 of cell culture. EDS spectra for the boxed region in (c) and (g) are presented in (i) and (j), respectively, showing the presence of Ca and P. The P peak is merged with the Au peak, which is due to conductive gold coating on the sample.

Sunny et al., (2002) have reported the preparation of HA-chitosan microspheres as potential bone and periodontal filling materials. HA powder was mixed with chitosan solution followed by paraffin oil, hexane and a surfactant and the microsphere production process commenced. Subsequently, glutaraldehyde was added to crosslink chitosan to give spherical particles ranging from 125 to 1000 mm. When the chitosan/n-HA composite implanted in body using as tissue scaffold, the degradation of chitosan makes room for the growth of new bone and then is substituted by new bone completely. It has been reported that chitosan can promote nucleation and growth of apatite and calcite crystals as well. Moreover, the surface of chitosan is hydrophilic, which can facilitate cell adhesion, proliferation and differentiation (Fig. 11). So, the chitosan/n-HA composite, used as bone substitutes, are hopeful to activate the regeneration and remodeling of bone tissue (Zhang et al., 2005).

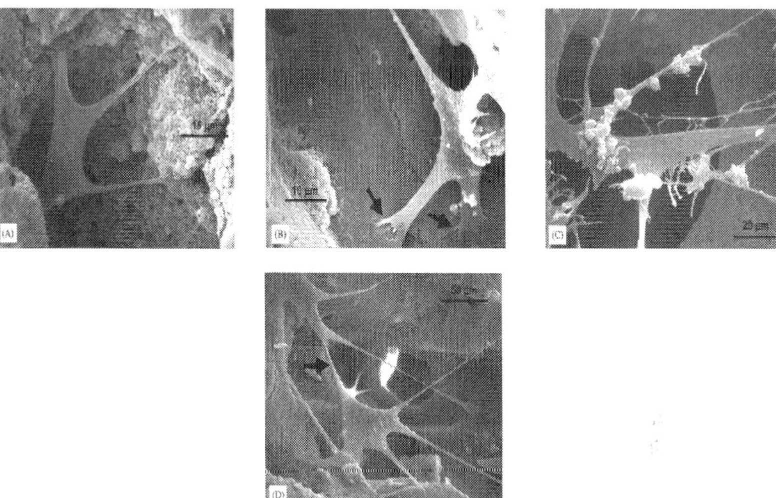

Figure 11. (A) SEM of cell infiltration into a macropore. (B) Cell attachment (arrows) to the bottom of a pore. (C) Cells inside a large pore near an opening at the bottom of the pore. (D) Cell-cell interactions inside a pore (arrow indicates a cell-cell junction).

The development of suitable three-dimensional scaffold for the maintenance of cellular viability and differentiation is critical for applications in periodontal tissue engineering. The different ratios of porous nanohydroxyapatite/chitosan (HA/chitosan) scaffolds were prepared through a freeze-drying process. The results indicated that the porosity and pore diameter of the HA/chitosan scaffolds were lower than those of pure chitosan scaffold. The HA/chitosan scaffold containing 1% HA exhibited better cytocompatibility than the pure chitosan scaffold. These scaffolds are evaluated in vitro by the analysis of microscopic structure, porosity, and cytocompatibility. The expression of type I collagen and alkaline phosphatase (ALP) activity are detected with real-time polymerase chain reaction (RT-PCR). Human periodontal ligament cells (HPLCs) transfected with enhanced green fluorescence protein (EGFP) are seeded onto the scaffolds, and then these scaffolds are implanted subcutaneously into athymic mice after implanted in vivo, EGFP transfected with HPLCs) not only proliferate but also recruit surrounding tissue to grow in the scaffold. The degradation of the scaffold significantly decreased in the presence of HA. This study demonstrated the potential of HA/ chitosan scaffold as a good substrate candidate in periodontal tissue engineering (Zhang et al., 2007).

Author details

Khaled R. Mohamed

Biomaterials Department, National Research Centre, Cairo, Egypt

6. References

Aiba I., Izume S., Minoura M.N. and Fujiwara Y. In chitin in Nature and Technology Ed. R.A.A. Muzzarelli, C. Jeuniaux and G.M. Goodey, Plenum Press, New York, (1986) 396-398.

Arpornmaeklong P., Suwatwirote N., Pripatnanont P., Oungbho K. Growth and differentiation of mouse osteoblasts on chitosan–collagen sponges. Int. J. Oral Maxillofac. Surg. 36 (2007) 328–337.

Beherei H.H., Mohamed K.R., Mahmoud A. I. Egyptian J. of Chemistry, xxx (2009) xxx-xxx. Under press.

Beppu M.M. and Santana C.C. In vitro biomineralization of chitosan. Key Eng Mater (2001) 192–195:31–4.

Blair H. S., Guthrie J., Law T. and Turkington p. Chitosan and modified membranes. App. Polymer. Sci., Vol. 33 (1987) 641 – 656.

Borah G., Scott G. and Wortham K. Bone induction by chitosan in endochondral bones of the extremities" Brine, C., Sanford P.A., Zikakis JP, 5th Int. Conf. Chitin and Chitosan , 1991, Princeton , N.J, London : Elsevier Applied Science, (1992) 47- 53.

Braddock M., Houston P., Campbell C., Aschroft P. Born again bone: tissue engineering for bone repair. News Physiol Sci 16 (2001) 208–13.

Brine C.J. Introduction: Chitin: Accomplishments and Perspectives. In Zikakis JP, ed chitin, chitosan and related enzymes. London: Academic Press (1984).

Chen H., Tian, X. and Zou, H. Preparation and blood compatibility of new silica-chitosan hybrid biomaterials. Artif. Cells Blood Substit. Immobil. Biotechnol., 26(4) (1998) 431-6.

Chen F., Wang Z., Lin C. Preparation and characterization of nano-sized hydroxyapatite particles and hydroxyapatite/chitosan nano-composite for use in biomedical materials Materials Letters 57 (2002) 858–861.

Chen R.H. and Hwa, H.D. Effect of molecular weight of chitosan with the same degree of deacetylation on the thermal, mechanical, and permeability properties of the prepared membrane. J. Carbohydrate Polymers, 29 (1995) 353-358.

Chow K.S. and Khor E. Novel fabrication of open-pore chitin matrixes. Biomacromolecules 1 (2000) 61–7.

Chung Y.S. and Korean J. Fiber Soc. 39 (5) (2002) 532– 536.

Dang, J. M.; Leong, K. W. Adv. Drug Delivery Rev. 58 (2006) 487–499.

Ding S. Biodegradation behavior of chitosan/calcium phosphate composites. J. Non-Crystalline Solids 353 (2007) 2367–2373.

Elizalde-Pen E.A., Flores-Ramirez N., Luna-Barcenas G., Va´squez-Garcı S.R., Ara´mbula-Villa G., Garcı´a-Gaita B., Rutiaga-Quinones J.G., Gonza´lez-Herna´ndeze J. Synthesis and characterization of chitosan-g-glycidyl methacrylate with methyl methacrylate. European Polymer Journal 43 (2007) 3963–3969.

Eugene K. and Lee Y.L. Implantable applications of chitin and chitosan. Biomaterials 24 (2003) 339-49.

El-Kady A.M., Mohamed K.R., El-Bassyouni G.T, Fabrication, characterization and bioactivity evaluation of calcium pyrophosphate/ polymeric biocomposites. Ceramics International 35 (2009) 2933–2942.

Felt O., Buri P. and Gurny R. Chitosan: a unique polysaccharide for drug delivery. J. Drug Dev. Ind. Pharm, 24 (11) (1998) 979- 93.

Fricain J.C., Bareille R., Ulysse F., Dupuy B., Amedee J. J. Biomed. Mater.Res. 42 (1998) 96-102.

Furukawa T., Matsusue Y., Yasunaga T., Shikinami Y., Okuno M., Nakamura T., Biomaterials 21 (2000) 889.

Gibson I.R. and Bonfield W. J. Biomed. Mater. Res. 59 (2002) 697-708.

Gomez-Vega J.M., Saiz E., et al., Biomaterials 21 (2000) 105.

Gorbunoff J. M. J. Interaction of proteins with hydroxyapatite. Anal. Biochem., 136 (1984) 425-445.

Grande C. J, Torres F. G, Gomez C. M, Bano M. C. Nanocomposites of bacterial cellulose/hydroxyapatite3 for biomedical application. Acta Biomaterialia xxx (2009) xxx–xxx.

Habraken W.J.E.M., Wolke J.G.C., Jansen J.A. Ceramic composites as matrices and scaffolds for drug delivery in tissue engineering, Advanced Drug Delivery Reviews 59 (2007) 234–248.

Hashem A.H and Mohamed K.R. Chitosan graft copolymer-HA/DBM biocomposites: Preparation, characterization and in-vitro evaluation Chitosan graft copolymer-HA/DBM biocomposites: Preparation, characterization and in-vitro evaluation. Egyptian J. of Chemistry, 50 (5) (2007) 625-644.

Hockin H., Xua K., Carl G., Simon Jrb. Fast setting calcium phosphate-chitosan scaffold: Mechanical properties, biocompatibility and Biomaterials 26 (2005) 1337–1348.

Hockin H.K., Xua C.G., Simon J. Fast setting calcium phosphate–chitosan scaffold: mechanical properties and biocompatibility Biomaterials 26 (2005)1337 1348.

Hu Q., Li B., Wang M., Shen J. Preparation and characterization of biodegradable chitosan/hydroxyapatite nanocomposite rods via in situ hybridization: a potential material as internal fixation of bone fracture. Biomaterials 25(5) (2004) 779–85

Hu Q.L., Qian XZ., Li BQ., Shen J.C. Study on chitosan rods prepared by in situ precipitation method, Chinese. J Chem Univ 24(3) (2003) 528-31.

Hu S. G., Jou C. H., & Yang M. C. J. Applied Polymer Science 88 (12) (2003) 2797–2803.

Huang J., Best S.M., Bonfieu W., Brooks R.A., et al. In vitro assessment of the biological response to nano-size Hydroxyapatite. J Mater Sci-Mater Med 15 (2004) 441–5.

Itoh S., Yamaguchi I., Suzuki M., Ichinose S., Takakuda K., Kobayashi H., Shinomiyag K., Tanaka J. Hydroxyapatite-coated tendon chitosan tubes with adsorbed laminin peptides facilitate nerve regeneration In vivo Brain Research 993 (2003) 111 – 123.

Kamiyama K., Onishi H., Machida Y. Biodisposition characteristics of N-succinyl–chitosan and glycol–chitosan in normal and tumor bearing mice. Biol Pharm Bull 22(2) (1999)179–86.

Kang H., Cai Y and Liu P. Synthesis, characterization and thermal sensitivity of chitosan-based graft copolymers. Carbohydrate Research 341 (2006) 2851–2857.

Kasuga T., Ota Y., et al., Biomaterials 22 (2001) 19.

Khora E. and Limb L.Y. Implantable applications of chitin and chitosan Biomaterials 24 (2003) 2339–2349.

Kikuchi, M., Matsumoto H. N., Yamada T., Koyama Y., Takakuda K., Tanaka J. Glutaraldhyde cross-linked hydroxyapatite/collagen self-organization nanocomposites. Biomaterials, 25 (2004) 63–69.

Kim D. G., Jeong Y. I., Nah J. W. J. Appl. Polym. Sci. 105 (2007) 3246–3254.

Kim S.B., Y J Kim Y.J., Yoon T. L., Park S.A., Cho I. H., Kim E.J, Kim I.A., Shin J.W. The characteristics of a hydroxyapatite–chitosan–PMMA bone cement Biomaterials 25 (2004) 5715–5723.

Kim Y., Seo S., Moon H., Yoo M., Park I., Kim B., Cho C. Chitosan and its derivatives for tissue engineering applications. Biotechnology Advances 26 (2008) 1-21

Kokubo T., Takadama H. How useful is SBF in predicting in vivo bone bioactivity? Biomater 27(15) (2006) 2907–15.

Kokubo T., Kim H., Kawashita M., Novel bioactive materials with different mechanical properties, Biomaterials 24 (2003) 2161–2175.

Kong L., Gao Y., Lu G., Gong Y., Zhao N., Zhang X. A study on the bioactivity of chitosan/nanohydroxyapatite composite scaffolds for bone tissue engineering. European Polymer Journal 42 (2006) 3171–3179.

Krajewska B. Membrane-based processes performed with use of chitin/chitosan materials. Sep Purif Technol 41 (2005) 305–12.

Langer R. and Vacanti J.P. Tissue engineering. Science 260 (1993) 920–6.

Larena A., Caceres D. A., Vicario C., Fuentes A. Release of a chitosan–hydroxyapatite composite loaded with ibuprofen and acetyl-salicylic acid submitted to different sterilization treatments. Applied Surface Science 238 (2004) 518–522

Lee S.B., Kim Y.H., Chong M.S., Lee Y.M. Preparation and characteristics of hybrid scaffolds composed of beta-chitin and collagen. Biomaterials 25 (2004)2309-2317.

Lee Y.L., Khor E., Ling C.E., J. Biomed. Mater. Res. 48 (1999) 111.

Li G., Dunn E.T., Grandmaison E.W. and Goosen, M. F.A. Applications and properties of chitosan. J. Bioact. Compat. Polyss, 7 (1992) 370-397.

Li J., Pan J., Zhang L., Guo X., Yu Y. Culture of primary rat hepatocytes within porous chitosan scaffolds. J Biomed Mater Res A 67 (2003a) 938-43

Li X., TsushimaY., Morimoto M., Saimoto H., OkamotoY., Minami S., et al. Biological activity of chitosan–sugar hybrids: specific interaction with lectin. Polym Adv Technol 11 (2000) 176–9.

Lin F.H., Yao C.H., Sun J.S., Liu H.C., Huang C.W. Biological effects and cytotoxicity of the composite composed by tricalcium phosphate and glutaraldehyde cross-linked gelatin. Biomaterials 19 (1998) 905–17.

Liu Z., Han J. and Czernuszka T. Gradient collagen/nano-hydroxyapatite composite scaffold: Development and characterization. Acta Biomaterialia 5 (2009) 661–669.

Liuyun J, Yubao L, Li Z, Jianguo L. Preparation and properties of a novel bone repair composite: nano-hydroxyapatite/chitosan/carboxymethyl cellulose. J Mater Sci: Mater Med (2008) 19:981–987.

Lu J.X., Prudhommeaux F., Meunier A., Sedel L., Guillemin G. Effects of chitosan on rat knee cartilages. Biomaterials 20 (1999) 1937–44.

Madihally S.V. and Matthew H.W.T. Porous chitosan scaffolds for tissue engineering. Biomaterials 20 (1999)1133–42.

Malette W.G., Quigley H.J., Adickes E.D. Chitin in nature and Technology Muzzarelli R, Jeuniauxc, Gooday, G.W, eds: Chitosan effect in vascular surgery Tissue culture and Tissue regenerations, New York: Plenum press, (1988) 435 - 442.

Manjubala I., Scheler S., Bossert J, Jandt K.D. Mineralization of chitosan scaffolds with nano-apatite formation by double diffusion technique. Acta Biomaterialia 2 (2006) 75–84.

Mao J.S., liu H. F., Yin Y. J., Yao K.D. The properties of chitosan–gelatin membranes and scaffolds modified with hyaluronic acid by different methods. J. Biomaterials 24 (2003) 1621–1629.

Martino A.D., Sittinger M., Risbud M.V. Chitosan: a versatile biopolymer for orthopaedic tissue-engineering. Biomaterials 26 (2005) 5983–90.

Matsuda A., Ikoma T., Kobayashi H., Tanaka J. Mater. Sci. Eng. C 24 (2004) 723.

Mattiolibelmonte M., De Benedittis A., Muzzarelli R. A. A., Mengucci P., et al

Bioactivity modulation of bioactive materials in view of their application in osteoporotic patients. J Mat. Sci: Mat. In Med. 9 (1998) 485-492.

Mi F.L., Shyu S.S., Wu Y.B., Lee S.T., Shyong J.Y., Huang R.N., Biomaterials 22 (2) (2001) 165.

Miyamato Y. and Shikawa K.I. Basic properties of calcium phosphate cement containing atelocollagen in its liquid or powder phases. Biomaterials 19 (1998) 707–15).

Mohamed K.R. Preparation of ceramic/ceramic and/or ceramic/biopolymer composites in the system of " Al_2O_3 – CaO – P_2O_5" and their characterization as bioceramic" Ph.D, Biophysics Dept., Faculty of Science, Cairo University.

Mohamed K.R and Mostafa A.A. Preparation and bioactivity evaluation of hydroxyapatite-titania/chitosan-gelatin polymeric biocomposites J. Materials Science and Engineering C, 28 (2008) 1087–1099.

Mohamed K.R., El Bassyouni G.E., Beheri H.H. Chitosan graft copolymer-HA/DBM biocomposites: Preparation, characterization and in-vitro evaluation. J. Applied Polymers Scienc, 105 (2007) 2553-2563.

Moreau J L. and Xu H.H.K. Mesenchymal stem cell proliferation and differentiation on an injectable calcium phosphate-chitosan composite scaffold. Biomaterials, xxx (2009) 1–8.

Murgan R. and Ramakrishna S. Crystallographic study of hydroxyapatite bioceramics derived from various sources: Cryst Growth Des 5 (2005)111-2).

Murugan R and Ramakrishna S. Bioresorbable composite bone paste using polysaccharide based nano hydroxyapatite. Biomaterials 25 (2004) 3829–3835

Muzarelli R.A. In: Aspinall GOA, editor. The polysaccharides. New York, NY: Academic Press; (1985) 417–25.

Muzzarelli R., Baldassarre V., Conti F., Ferrara P., Biagini G., Gazzanelli G. and Vasi V. Biological activity of chitosan : Ultrastructural study. Biomaterials, 9 (3) (1988) 247-52.

Muzzarelli R.A., Mattioli-Belmonte M., Tietz C., et al., Stimulatory effect on bone formation exerted by a modified chitosan. Biomat. J., 15 (13) (1994) 1075-1081.

Muzzarelli R.A.A., Biagini G., et al., Carbohydr. Polym. 45 (2001) 35.

Ng C H., Rusu V M., Peter M.G. Formation of chitosan hydroxyapatite composites in the presence of different organic acids. Adv Chitin Sci;7, (2008).

Nikalson L.E. and Langer R.S. Advances in tissue engineering of blood vessels and other tissues. Trans Immunol 5 (1997) 303–6.

Nishikawa H., Ueno A., Nishikawa S., Kido J., Ohishi M., Inoue H., et al. Sulfated glycosaminoglycan synthesis and its regulation by transforming growth factor-beta in rat clonal dental pulp cells. J Endod 26 (2000)169–7.

Ogawa K., Yui, T. and Miu M. J. Bioscince. Biotech. Biochem., 56 (6) (1992) 858.

Otto A. W., Klau J.W., Johnson J .M.,George S. M. Thin solid films 292 (1-2) (1997) 135-144.

Rao S.B. and Sharma, C.P. Use of chitosan as a biomaterial: Studies on its safety and hemostatic potential" Biomed Mat. Res., Vol. 34 (1) (1997) 21-28, Jan.

Rehman I., Smith R., Hench, L.L., Bonfield W. J. Biomed. Mater. Res. 29 (1995) 1287-1294.

Rhee S.H. and Tanaka J. Effect of citric acid on the nucleation of hydroxyapatite in a simulated body fluid. Biomaterials 20 (1999) 2155–60.

Roberts G.A.F., Chitin Chemistry, macmillan (1992)

Roodman G.D. Mechanisms of bone metastasis. NEJM. 350 (2004)1655-1664.

Rouget, (1859) : Book of chitin , Edited by Muzzarelli, R.A.A., (1977), Ancona, Italy, 60100, Pergamon Press.

Rusu V. M., Ng C. H., Wilke M., Tiersch B., Fratzl P., Peter, M. G. Size-controlled hydroxyapatite nanoparticles as self-organized organic–inorganic composite materials. Biomaterials, 26 (2005) 5414–5426.

Sachlos E., Reis N., Amsley C., Derby B., Czernuska J.T. Novel collagen scaffolds with predefined internal morphology made by solid free form fabrication. Biomaterials 4 (2003)1487–97.

Sapelli P.L., Baldassare V., Muzzarelli R.A.A. and Emanuelli M. Chitosan in dentistry. Chitin in Nature and Technology, (1986) 507-512.

Sarasam A. and Madihally S.V. Characterization of chitosan-polycaprolactone blends for tissue engineering applications. Biomaterials 26 (2005) 5500–8.

Schwarz K. and Epple M. Biomimetic crystallisation of apatite in a porous polymer matrix. Chem Eur J. 4(10) (1998) 1898–903.

Service R.F. Tissue engineers build new bone. Science 289 (2000)1498–500.

Shen X., Tong H., Jiang T., Zhu Z., Wan P., Hu J., Composites Science and Technology 67 (2007) 2238–2245

Suchanek W.L., Shuk P., Byrappa K., Riman R.E.,TenHuisen K.S., Janas V.F. Biomaterials 23 (2002) 699-710.

Suchanek W. and Yoshimura M., Processing and properties of hydroxyapatite-based biomaterials for use as hard tissue replacement implants. J Mater Res 13 (1998) 94–117.

Suh J.K.F and Matthew H.W.T., Application of chitosan-based polysaccharide biomaterials in cartilage tissue engineering: a review. Biomaterials 21(24) (2000) 2589–98.

Sunny M.C., Ramesh P.,Varma H.K. Microstructured microspheres of hydroxyapatite ceramic. J Mater Sci Mater Med 13 (2002) 623–32.

Takahashi Y., Yamamoto M., Tabata Y. Enhanced osteoinduction by controlled release of bone morphogenetic protein-2 from biodegradable sponge composed of gelatin and β-tricalcium phosphate. Biomaterials 26 (2005) 4856-5.

Takagi S., Chow L., Hirayama S., Eichmiller F., Properties of elastomeric calcium phosphate cement–chitosan composites, Dental Materials, 19 (8) (2003) 797-804.

Takechi M., Ishikawa K., Miyamoto Y., Nagayama M., Suzuki K. Tissue responses to anti-washout apatite cement using chitosan when implanted in the rat tibia JMat.Sci. In Medicine 12, (2001)597-602

Tarsi R., Corbin B., Pruzzo C. and Muzzarelli R.A. Effect of low molecular weight chitosan on the adhesive properties of oral streptococci. Oral Microbial Immunol., 13 (4) (1998) 217-24.

Teng S., Lee E., Yoon B., Shin D., Kim H., Oh J. Chitosan/nano-hydroxyapatite composite membranes via dynamic filtration for guided bone regeneration. J. Biomed. Mater. Res. Part A, 88A(3) (2009) 569–580.

Thein-Han W.W and Misra R.D.K. Biomimetic chitosan-nanohydroxyapatite composite scaffolds for bone tissue engineering. Acta Biomaterialia, 5(4), (2009) 1182-1197.

Tomihata K. and Ikada Y. In-vitro and in-vivo degradation of films of chitin and its deacetylated derivatives. J. Biomater., 18 (7) (1997) 567 – 75.

Van Der Lubben I. M., Verhoef J. C., Borchard G., Junginger H. E. Adv. Drug Delivery Rev. 52 (2001)136–144.

Viala S., Freche M., Lacout J.L., Ann. Chim. Sci. Mater. 23 (1998) 69.

Wang H., Li Y., Zuo Y., Li J., Ma S., Cheng I. Biocompatibility and osteogenesis of biomimetic nano-hydroxyapatite/polyamide composite scaffolds for bone tissue engineering. Biomaterials 28 (2007) 3338–3348.

Wang L. and Li C., Preparation and physicochemical properties of a novel hydroxyapatite/chitosan–silk fobroin composite, Carbohydrate Polymers 68 (2007) 740–745.

Wang M. and Bonfield W. Biomaterials 22 (2001)1311.

Wang X., Ma J., Wang Y., He B. Bone repair in radii and tibias of rabbits with phosphorylated chitosan reinforced calcium phosphate cements. Biomaterials 23 (2002) 4167–76.

Wang Z and Hu Q, Preparation and properties of three-dimensional hydroxyapatite/chitosan nano-composite rods. Biomed. Mater. (2010) 5 045007 doi:10.1088/1748-6041/5/4/045007.

Wei G. and Ma P.X. Structure and properties of nano-hydroxyapatite/ polymer composite scaffolds for bone tissue engineering. Biomaterials 25 (2004) 4749–57.

Wilson O.C., and Hull J.R. et al., (2008): "Surface modification of nanophase hydroxyapatite with chitosan" Materials Science and Engineering C 28 , 434–437.

Xianmiao C., Yubao L., Yi Z., Li Z., Jidong L., and Huanan W., "Properties and in vitro biological evaluation of nano-hydroxyapatite/chitosan membranes for bone guided regeneration" Materials Science and Engineering C 29, (2009) 29–35.

Yamaguchi I., Iizuka S., Osaka A., Monma H., Tanaka J. The effect of citric acid addition on chitosan/hydroxyapatite composites. Colloids and Surfaces A: Physicochem. Eng. Aspects 214 (2003) 111 -118.

Yamaguchi I., Itoh S., Suzuki M., Osaka A., Tanaka J., Biomaterials 24 (2003) 3285– 3292.

Yamaguchi I., Tokuchi K., Fukuzaki H., Koyama Y., Takakuda K., Monma H., et al. Preparation and microstructure analysis of chitosan/hydroxyapatite nanocomposites. J of Biomedical Materials Research, 55 (2001) 20–27.

Yilmaz E. Adv. Exp. Med. Biol. 553 (2004) 59.

Yuan Y., Zhang P., Yang Y., Wang X., Gu X. The interaction of Schwann cells with chitosan membranes and fibers in vitro. Biomaterial.25 (8) (2004) 4273.

Zhang Y., Ni M., Zhang M., Ratner B. Calcium phosphate chitosan composite scaffolds for bone tissue engineering. Tissue Eng 9 (2005) 337-45.

Zhang Y.F. ,Cheng X.R. , Chen Y., Shi B., Chen X-H . , Xu D-X., Ke J. Three dimensional Nanohydroxyapatite/chitosan scaffold as potential tissue engineered periodontal tissue J of Biomaterials Applications, Vol. 21, No. 4, (2007) 333-349

Zhang Y., Venugopal J. R., Adel El-Turki , Ramakrishna S., Bo Su , Lim C. T. Electrospun biomimetic nano-composite nano-fibers of hydroxyapatite/chitosan for bone tissue engineering. Biomaterials 29 (2008) 4314-4322.

Zhao A. F., Graysona W. L., Maa T., Bunnellb B., Luc W. W. Effects of hydroxyapatite in 3-D chitosan–gelatin polymer network on human mesenchymal stem cell construct development Biomaterials 27 (2006) 1859–1867.

Zhao F., Yin Y., Lu W.W., Leong C., Zhang W., Zhang J., et al. Preparation and histological evaluation of biomimetic three-dimensional hydroxyapatite/chitosan–gelatin network composite scaffolds. Biomaterials 23 (2003) 3227–34.

Non-Destructive Examination of Interfacial Debonding in Dental Composite Restorations Using Acoustic Emission

Haiyan Li, Jianying Li, Xiaozhou Liu and Alex Fok

Additional information is available at the end of the chapter

http://dx.doi.org/10.5772/51369

1. Introduction

Light-cured, dental resin composites are widely used to repair decayed or damaged teeth because of their superior esthetics, ease of use and ability to bond to tooth tissues. However, during setting or polymerization, the resin composite shrinks, producing shrinkage stresses within the tooth and composite [1]. If the internal shrinkage stress is high enough, debonding between the tooth and restoration will occur, causing problems such as reduced fracture resistance and increased micro-leakage. The latter will ultimately lead to secondary caries. Clinical studies have identified the loss of interfacial integrity as one of the main causes for replacement of composite restorations [2-4].

Unfortunately, interfacial debonding caused by the shrinkage of composite resins can be hard to avoid. Fig. 1 shows evidence of interfacial debonding by comparing the Micro-CT images of a restored tooth before and after curing the composite. It can be seen that an interfacial crack of a significant length appeared after curing the composite along the initially intact interface. On the other hand, in the example shown in Fig. 2, where a different resin composite was used, no clear interfacial debonding was observed from the Micro-CT images. Therefore, whether interfacial debonding will occur depends on the restorative composite material, specifically the level of shrinkage stress it produces. There are other factors that may affect the initial quality and subsequent degradation of bonding at the tooth-restoration interface, e.g. the adhesive/bonding materials, cavity geometry, restorative techniques, thermal and mechanical loading, etc. [5-8].

It is very difficult to predict whether interfacial debonding will take place in a particular composite restoration system. This is because the shrinkage stress within a real tooth restoration is difficult to predict or measure due to the small but complicated geometry and

rapidly changing material properties during curing. At the same time, the bond strength between the tooth and restoration is difficult to determine. Tensile and shear tests have been widely used for bond strength testing. However, the results are highly variable, being dependent on the test devices and specimen size used [9-11]. Therefore, they are not very predictive of the actual clinical performance.

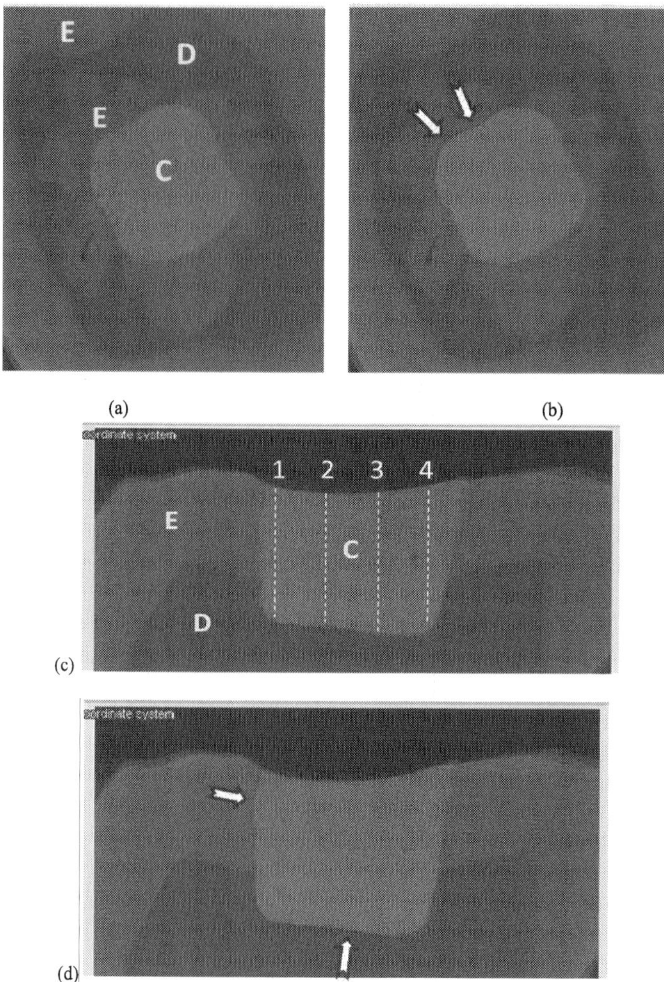

Figure 1. Micro-CT images of a tooth specimen restored with Z100: (a) horizontal cross-section before curing, (b) horizontal cross-section after curing, (c) vertical cross-section before curing and (d) vertical cross-section after curing. *(E-enamal, D-Dentin, C-composite. Lines 1-4 in (c) schematically indicate the locations of micro-hardness measurement described in Section 3.2.5 and Section 4.1. The arrows in (b) and (d) point out the debonding positions at the interface.)*

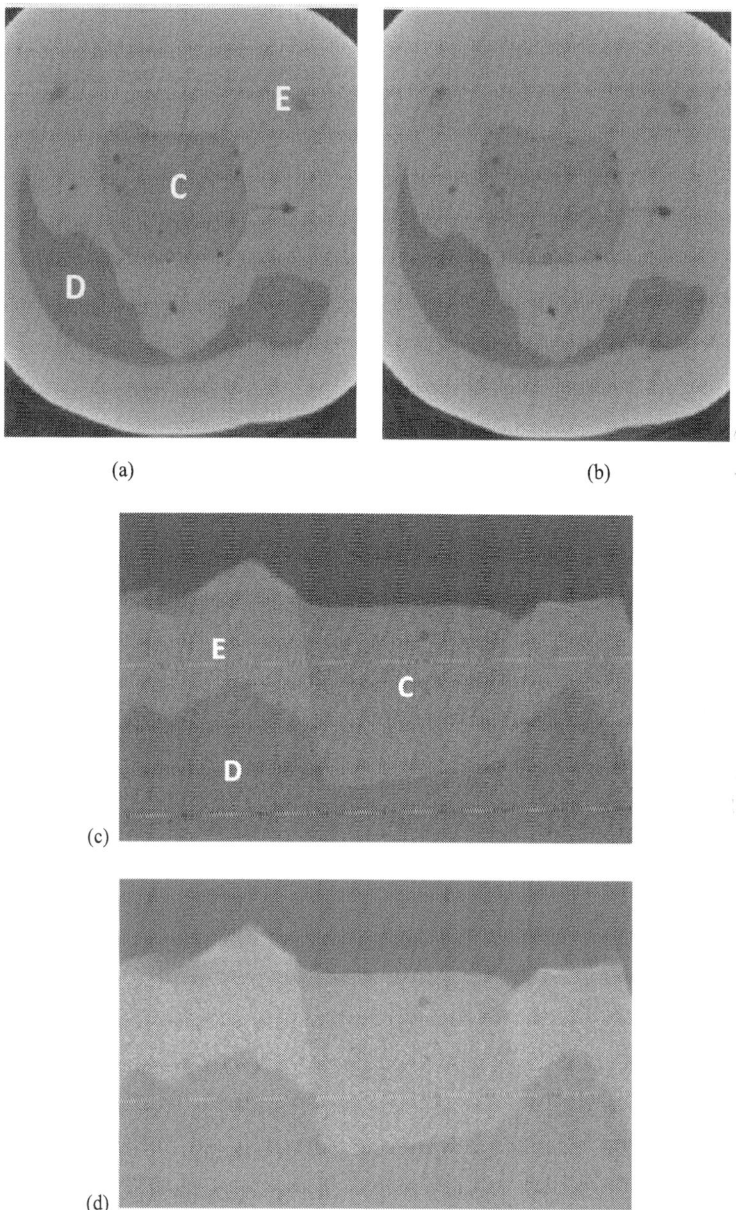

Figure 2. Micro-CT images of a tooth specimen restored with P90: (a) horizontal cross-section before curing, (b) horizontal cross-section after curing, (c) vertical cross-section before curing and (d) vertical cross-section after curing. *(Note: E-enamal, D-Dentin, C-composite.)*

Further details on the development of shrinkage stress in composite restorations and some current methods for assessing the resulting interfacial debonding are described in Section 2. The aim of this study is to develop a new method to evaluate the interfacial debonding of dental composite restorations. A non-destructive method, the acoustic emission (AE) technique, will be used to monitor *in-situ* the interfacial debonding of composite restorations during polymerization. The AE technique and the system used for monitoring interfacial debonding will be described in Section 3, where the capability of this new evaluation method will be verified by different tests. Then, in Section 4, the AE technique will be used to study the influences of several factors, including the composite material, C-factor (ratio between bonded and unbonded surfaces) and filling technique, on the debonding behavior of composite restorations.

2. Shrinkage stress and interfacial debonding in dental composite restorations

The dental resin composite is a mixture of organic monomer systems and inorganic filler particles. The monomer systems, which act as a matrix for the dental composite, are flowable before curing so that the composite can be easily packed into the tooth cavity and achieve good marginal adaptation with the tooth tissues. When polymerization is activated, the resin matrix will solidify, with its mechanical properties (Young's modulus, viscosity, hardness, strength, etc.) changing significantly and rapidly in the process. Volumetric shrinkage also occurs during polymerization due to the reduction of intermolecular separations in the monomers [12]. This is a very quick process which normally takes place within the first 30-50 seconds. The volumetric contraction that accompanies polymerization is typically on the order of 1.5–5% [1].

For most dental composite restorations, a bonding agent (adhesive) is used to create a strong bond at the tooth-composite interface. Thus, during the polymerization process, because the composite is constrained along the interfaces, shrinkage stresses are built up within the composite and tooth structure. These shrinkage stresses are difficult to predict or measure because of the small but complicated tooth geometry and rapidly changing material properties. Some experimental methods using simple specimens have been developed to estimate the polymerization shrinkage stress that develops in dental composites [1, 13, 14]. However, the measured shrinkage stresses are sensitive to the test configuration and procedures, e.g. the instrument's compliance, direction of curing light application and specimen shape. Also, because the material parameters (shrinkage strain, viscosity, Young's modulus and Poisson's ratio) of the composite that control the development of shrinkage stresses all change rapidly with time during polymerization, even with a suitable mathematical model, the accurate prediction of the shrinkage stresses is not trivial [15, 16].

If the shrinkage stress within a composite restoration is high enough, interfacial debonding or, more generally, failure between the tooth and restoration will occur. Interfacial failure can be adhesive, in which case failure occurs right in the adhesive layer; or it can be cohesive, in which case failure occurs in the tooth or composite materials near the interface.

Interfacial debonding has been identified as one of the main causes for replacement of composite restorations [2-4].

To date, there is no reliable tool to detect or monitor debonding between the composite and tooth during the curing process. Traditional methods for studying the interfacial integrity of dental restorations include optical microscopy, scanning electron microscopy (SEM) and transmission electron microscopy (TEM) [5-8, 17, 18]. The major disadvantage of these methods is that they are limited to essentially surface examination; internal debonding cannot be detected. To see whether debonding has occurred inside the restoration, these methods require destructive sectioning of the specimens which may introduce more uncertainties to the results due to possible machining damage. As an alternative, non-destructive methods, such as X-ray micro-computed tomography (micro-CT), have received much attention in recent years in the study of interfacial bonding/debonding of composite restorations [19]. Micro-CT is a computer-aided, 3D reconstruction of a structure or material that can be sliced virtually along any direction to gain accurate information on their internal geometric properties and structural parameters. Figures 1 and 2 show Micro-CT images used to study the interfacial debonding of composite restorations. Although Micro-CT can provide 3D examinations of the whole structure, its lower resolution means that it cannot detect debonding at a submicron level. Most of all, none of these imaging techniques can be used to monitor debonding as it happens.

3. AE technique and its application to interfacial debonding examination

3.1. The acoustic emission (AE) measurement technique

The acoustic emission (AE) measurement technique is also a non-destructive method. It is normally used to monitor the integrity of structures by providing real-time information of the fracture or damage process. It uses transducers or sensors to detect the high-frequency sound waves produced as a result of the sudden strain energy released within a material following fracture. Figure 3 shows schematically how an AE test system works. First, the cracking event within a material is captured by the AE sensor attached on the surface of the component. The raw signal is then passed through a preamplifier for pre-amplification and then to the acquisition system for acquisition and storage. Finally, the AE data can be displayed and analyzed by specially designed software.

AE technology has been widely used in research and industry to monitor the development of crack growth, wear, fiber-matrix debonding in composites, phase transformation, etc [20-22]. It has also been used to detect the fracture of different dental structures. For example, Ereifej et al. [23] used the AE technique to detect the initial fracture of ceramic crowns; Vallittu [24] used it to study the fracture of a composite veneer reinforced by woven glass fibres; and Kim and Okuno [25] used it to study the micro-fracture behavior of composite resins containing irregular-shaped fillers.

The interfacial debonding in composite restorations, either adhesive or cohesive, is actually a kind of fracture within the restored tooth structure. Therefore, it is possible to use the AE technique to capture the cracking events. Unlike some of the imaging methods mentioned above, the AE method does not need destructive sample preparation and is not affected by the curing light. Therefore, it can be used to measure interfacial debonding in real time during the curing process, which cannot be achieved by micro-CT despite it being a nondestructive method as well.

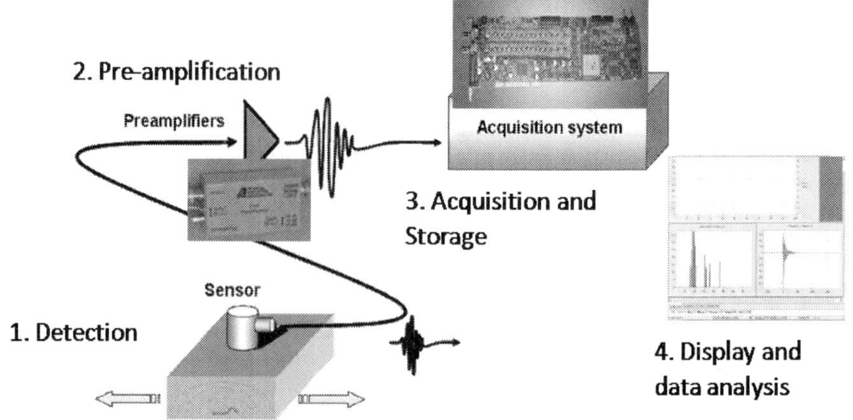

Figure 3. Schematic diagram of the AE test system

3.2. Application of AE technique to debonding measurement [26]

When using the AE technique to study interfacial debonding of dental restorations, it is necessary to prove that the captured AE signals indeed represent the cracking/micro-cracking events caused by interfacial debonding during curing of the composite. To this end, three groups of tests with different boundary constraints were designed and conducted, as described in this sub-section.

3.2.1. Specimens and materials

Figure 4 shows the specimens used in the three groups of tests: (a) free standing pea-size specimens of composite placed directly on the AE sensor, (b) ring specimens prepared from the root of a single bovine tooth and (c) intact human molars with a Class-I restoration. The composites in Group a were constrained least while those in Group c were constrained most. Each group had 4 specimens and Z100™ (3M ESPE) was the composite material used. More details of the specimens are given below.

The free-standing pea-size specimens of Group a were about 5mm in diameter. They were directly placed on the AE sensor without using any adhesive material, as shown in Figure 4(a). Each of these specimens was cured with a blue light (Elipar TriLight, 3M ESPE, US) at an intensity of $550mW/cm^2$ for 40s. Testing with these specimens helped to verify that free shrinkage of the composite resin itself did not induce any AE event.

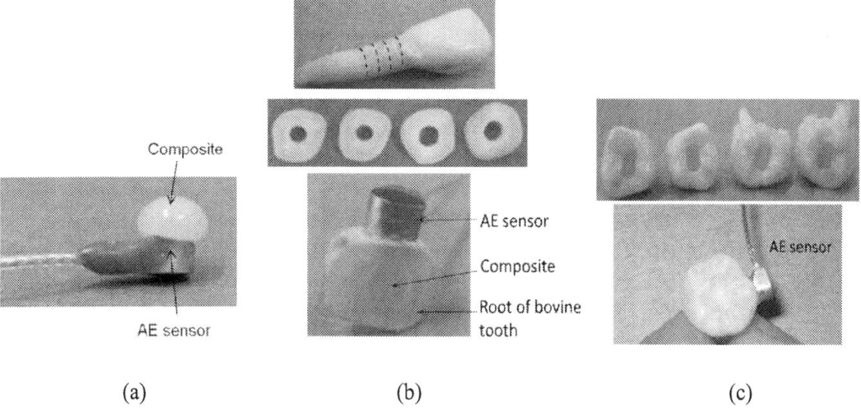

Figure 4. Samples: (a) pea-size composite specimen directly attached to the AE sensor, (b) rings prepared from the root of a bovine tooth and (c) human teeth with Class-I cavities

The four ring specimens in Group b were cut from the root of a single bovine tooth. They were, therefore, similar in material properties and structural anatomy; see Figure 4(b). The central holes of the specimens were originally the root canal, which was enlarged to 3mm in diameter with a high-speed handpiece. Compared with the whole tooth specimens in Group c, the composites in the ring specimens had larger free surface areas. This allowed the effect of the so-called C-factor (ratio between the bonded and unbonded surfaces) on interfacial debonding to be investigated.

For Group c, 4 intact human molars with similar dimensions were selected; they had been extracted and stored in saturated thymol solution at 4°C for less than one month. Standard Class-I cavities were prepared on these teeth by a single operator following clinical procedures with a high-speed handpiece and dental cutting burs; see Figure 4(c). The whole tooth specimens were considered to be the most representative of those in real clinical situations.

Each of the ring and tooth specimens was first treated with a bonding agent (Adper™ Scotchbond™ SE Self-Etch, 3M ESPE, US) to the cavity surface and then restored with the composite resin Z100™ (3M ESPE, US). Again, the composite was cured with a blue light (Elipar TriLight, 3M ESPE, US) at 550mW/cm² for 40s.

3.2.2. Shrinkage stress measurement

To explain the different levels of interfacial debonding measured by the AE method for different composite materials, the development of shrinkage stress for the composite materials during curing were measured using a tensometer (American Dental Association Foundation) [27]. This device is based on the basic engineering cantilever beam bending theory. The tensile force generated by the shrinking composite was calculated from the beam deflection using a previously obtained calibration constant. Further details of the shrinkage stress measuring method using the tensometer are given in Ref. [27]. Before placing the composite material between the two glass-rod holders at the free end of the cantilever beam, the end surfaces of the glass rods were first polished with 600-grit sandpaper, silanized with a porcelain primer (Bisco Inc., Schaumburg, IL, USA), and then applied with a layer of adhesive (Scotchbond Multi-purpose, 3M, St. Paul, MN, USA). The dimensions of the composite specimens were 6mm in diameter and 2mm in height. The temporal developments of shrinkage stress for Z100™ (3M ESPE, US) and Filtek P90 (3M, St. Paul, MN, USA) are plotted in Figure 5. As can be seen, the shrinkage stresses developed rapidly and reached their maximum values within the first 50s, with Z100 producing a much higher shrinkage stress than Filtek P90.

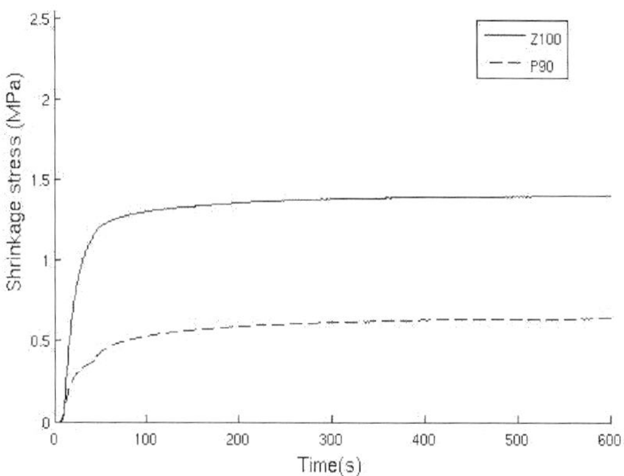

Figure 5. Shrinkage stress against time for Z100 and P90

3.2.3. AE measurement

A 2-channel AE system (PCI-2, Physical Acoustic Corporation, USA) was used in this study for AE data acquisition and digital signal processing. The AE sensor/transducer used for detecting interfacial debonding was S9225 (Physical Acoustic Corporation, USA), which had a resonance frequency of 250kHz. For the whole tooth and ring specimens, the AE sensor was attached to their outer surfaces using cyanoacrylate adhesive (Super Bond, Staples Inc, USA). The signals acquired with the sensor were amplified by a preamplifier with 20/40/60 dB gains. The parameters selected for the signal acquisition were: a 40dB gain for the preamplifier, a 100kHz-2MHz band pass and a 32dB threshold. These parameters were selected through many trial tests, with the aim of maximizing the system sensitivity while minimizing the background noises.

Figure 6 shows the procedures of a typical AE test on composite restoration debonding: (1) Prepare the cavity on a tooth sample; (2) Apply adhesive (bond agent) to cavity walls, fill the cavity with composite resin, and attach the AE sensor onto the tooth surface; (3) Turn on the AE system and blue curing light simultaneously. Cure the composite resin for 40s and record AE data continuously for 10 minutes. (4) Analyze the AE data. Use the curves of instantaneous and accumulated AE events against time to study the curing behavior of the different specimens. During the AE tests, the teeth were wrapped by wet paper tissue to avoid cracking through dehydration.

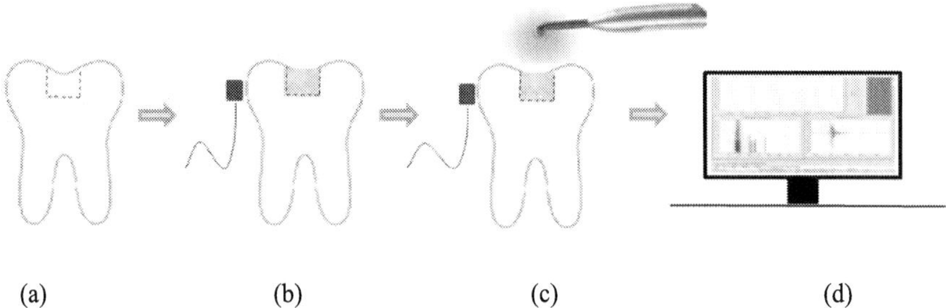

(a)   (b)   (c)   (d)

Figure 6. Procedure of an AE test on composite debonding: (a) sample preparation; (b) placement of adhesive, restoration, and AE sensor; (c) AE recording (10mins) & curing of composite resin (~40s); (d) AE data analysis

Figures 7 and 8 show the cumulative numbers of AE events against time for the ring specimens (Group b) and the human tooth specimens (Group c), respectively. The temporal development of the shrinkage stress of Z100, as shown in Figure 5, was also plotted for comparison. It was found that the temporal developments of the AE events followed roughly those of the shrinkage stress. It can also be seen that while some AE events occurred during the initial rapid polymerization of the composite resin, there were still some late events which took place a few minutes after the completion of light curing. In order to compare these two groups, the average temporal AE developments for Group b and Group c are plotted together in Figure 9, with the standard deviation of the cumulative number of AE events of each group being shown as red bars. No AE events were detected for the free-standing composite blob specimens (Group a) placed directly onto the AE sensor, illustrating that the free shrinkage of composite does not produce any AE event during curing. The mean and standard deviation of the total number of AE events for Group a, Group b and Group c were 0(-), 3.7(2.1) and 9.0(1.6), respectively, as summarized in Table 1.

3.2.4. Debonding evaluation using Micro-CT

The integrity of the tooth-restoration interface within the real tooth specimens (Group c) was examined further using a micro-CT machine (XT H 225, X-TEK Systems LTD). Two specimens were selected and they were first scanned immediately after placement of the composites to see how well they had adapted to the cavity walls. After curing and AE measurement, they were scanned for the second time to look for any detachment of the restoration from the cavity walls. In order to ensure the same position and orientation for the two scans to facilitate "same-slice" comparison, each of the specimens was mounted into a Teflon ring with positioning pins using an orthodontic resin (DENTSPLY International Inc., US). During scanning, the teeth were covered by wet paper tissue to avoid cracking through dehydration.

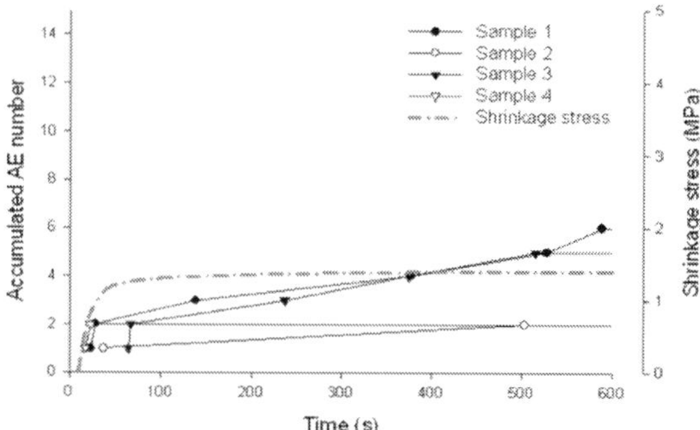

Figure 7. AE results for the ring specimens (Group-b)

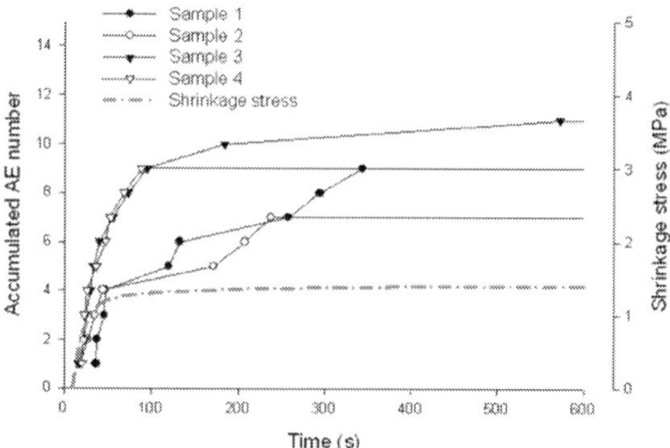

Figure 8. AE results for the real tooth specimens (Group-c)

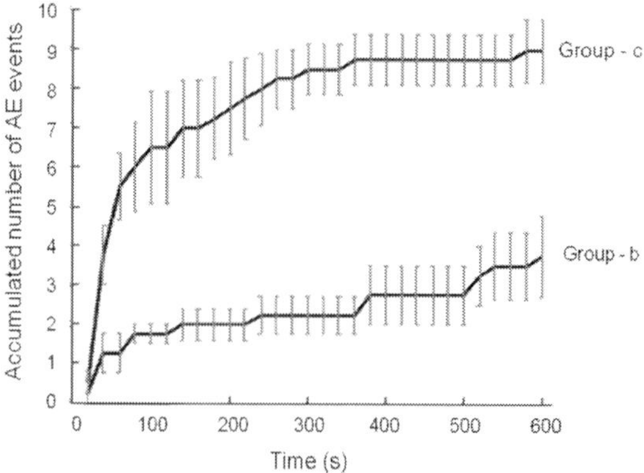

Figure 9. Average cumulative number of AE events for specimens in Groups b and c, with the standard deviation being plotted with red bars

	Group a	Group b	Group c
Mean	0	3.7	9.0
STD	-	2.1	1.6

Table 1. Total number of AE events: mean and standard deviation (STD)

The cross-sectional images obtained from the 3D Micro-CT reconstructions of one of the human tooth specimens restored with Z100 are shown in Figure 1. The micro-CT images clearly show the internal structures of the three main components of the restored teeth: enamel, dentin and composite. Before curing, as shown in Figures 1a and 1c, the composite can be seen to be perfectly in contact with the surrounding tooth tissues. After curing, however, the specimen showed clear interfacial debonding along the side walls and at the bottom of the cavity.

3.2.5. Micro-hardness measurement

To ensure that any lack of AE events/debonding was not due to uncured composite, the solidification of the composite within the tooth samples (Group c) after curing was assessed by using mico-hardness tests. After curing and AE measurement, one of the specimens was cut along a vertical central plane with a 102mm Dia. x 0.3mm thick diamond blade (Buehler, USA). The Vickers hardness was then measured with a micro-hardness testing machine (Micromet 5104, Buehler, USA). The indenter load was 100g and the load-holding time was 10s. The measurement points were located along 4 vertical lines within the restoration, which were schematically shown in Figure 1c. The measurement began from the top surface and continued until the bottom interface was reached.

Figure 10 shows the Vickers hardness of the cured composite (Z100) along its depth. It shows very uniform hardness distributions within the composite restoration, indicating that the polymerization of the composite resin was uniform and complete.

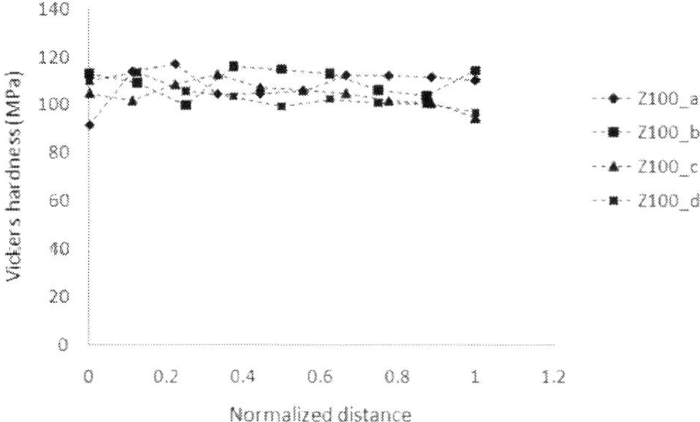

Figure 10. Vickers hardness along the depth of a Z100 restoration (see Figure 1c)

3.2.6. Discussion on the AE verification tests

From the AE results shown in Figures 7 and 8, it can be seen that the temporal developments of the AE events followed roughly those of the shrinkage stress. Also, polymerisation of the composite itself did not create any detectable AE events, as demonstrated by the negative results of the freestanding pea-size specimens placed directly on the sensor. All these indicate strongly that the cracking leading to the AE events in the restored samples were caused by the shrinkage stress produced by the polymerization of the composite resin.

Possible sources of AE events include debonding at the tooth-restoration interface and cohesive cracking in the tooth tissues or composite resin under the action of tensile shrinkage stresses. The only tensile stress in the ring specimens was the radial stress, which was maximum at the tooth-restoration interface. Therefore, cracking, either cohesive or adhesive, at the interface was the most possible reason causing those AE events in the ring specimens. The stress distributions within the restored tooth specimens were more complicated, due to the complex geometry involved. During shrinkage of the composite restoration, tensile stresses could also be induced in the axial direction, which would be maximum on the outer tooth surfaces. However, the micro-CT images of the tooth specimen, shown in Figure 1, confirm again that the AE events recorded were probably induced by adhesive debonding or cohesive cracking at the tooth-restoration interface.

Figure 9 shows that there were more AE events in the whole tooth specimens than there were in the ring specimens. This could be attributed to the larger restoration volume, and thus larger volumetric shrinkage, and/or the higher ratio of bonded-to-nonbonded surface area in the whole tooth specimens, i.e. the so-called C-factor. Also, the compliances of the two samples were different, with the ring specimens having a lower stiffness and thus a lower shrinkage stress, which led to fewer AE events. The more complicated geometry of the Class-I restorations also meant that local stress concentrations were more likely to exist in the tooth specimens which would lead to more debonding or AE events.

These results verified that the non-destructive AE measurement technique is an effective tool to detect and monitor *in-situ* the interfacial debonding of composite restorations during curing. It will allow quantitative studies to be carried out of the effects of factors such as composite material properties, cavity geometries, and restorative techniques on the integrity of the tooth-restoration interface.

4. Factors that affect interfacial debonding

There are many factors that can affect the initial quality and subsequent degradation of bonding at the tooth-restoration interface, e.g. properties of the composite and adhesive materials, cavity geometry, layering techniques, thermal and mechanical loading, etc. [5-8]. In this section, the influences of three factors on interfacial debonding will be studied using the AE technique described above. The three factors studied include: the composite material, the cavity configuration, and the filling technique.

4.1. Influence of the composite material

The interfacial debonding in composite restorations is mainly caused by the shrinkage stress produced during the polymerization of the composite. The material properties of the composite, i.e. the volume shrinkage, Young's modulus and strength, etc., will affect the internal shrinkage stress level and, thus, the degree of interfacial debonding. To investigate the influence of the composite resin on interfacial debonding, two commercial composite materials: Z100™ (3M ESPE, US) and Filtek™ P90 (3M ESPE, US) were chosen to conduct further AE debonding tests. The shrinkage stress curves for the two materials obtained using a tensometer (see Session 3.2.2) are shown in Figure 5. It can be seen that P90 produces a much lower shrinkage stress than Z100. This is because P90 is a low-shrinkage composite with special molecular structures that can open up to counter the shrinkage during polymerization.

8 intact human molars with similar dimensions were selected and randomly divided into 2 groups of 4. The specimen preparation and testing procedure and test equipments are the same as that in Section 3.2. The bonding agent used for all the specimens was Adper™ Scotchbond™ SE Self-Etch (3M ESPE, US).

Figure 11 shows the average temporal development of AE events during curing for the two groups, with the standard deviation being plotted with red bars. Just as it produced a lower shrinkage stress, Filtek P90 also produced a lower number of AE events than Z100, indicating that there was less interfacial debonding/cracking in the P90 specimens. In fact, two of the P90 specimens did not produce any detectable AE events at all.

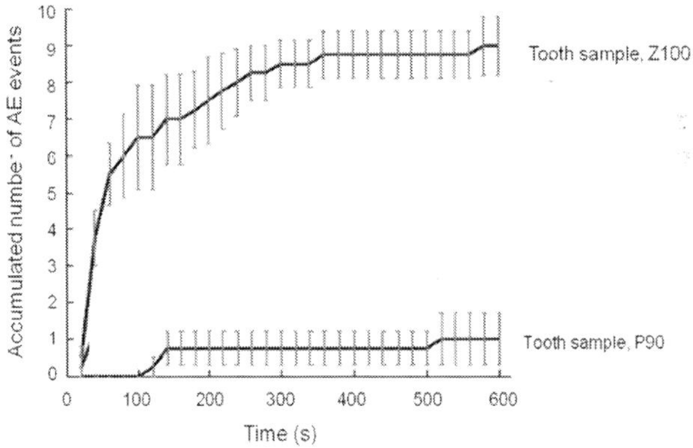

Figure 11. Average cumulative number of AE events for specimens restored with Z100 and P90 (the red bars show the standard deviation)

The above conclusion is supported by comparing the micro-CT images of the two specimens, as shown in Figures 1 and 2. After curing, the specimen restored with Z100 can be seen to have clear interfacial debonding along the walls and at the bottom of the cavity (Figures 1b and 1d), while that restored with P90 did not show any obvious debonding (Figures 2b and 2d).

Again, to discard the possibility that the low number of AE events in the P90 specimens was caused by incomplete polymerization, mico-hardness tests were also done on one of the specimens to assess the solidification of the composites. Figure 12 shows the Vickers hardness along the depth of the restoration for both Z100 and P90 specimens. It can be seen that, while Z100 was harder than P90, restorations of both materials had very uniform hardness distributions within them, indicating that the polymerization of all specimens were uniform and complete.

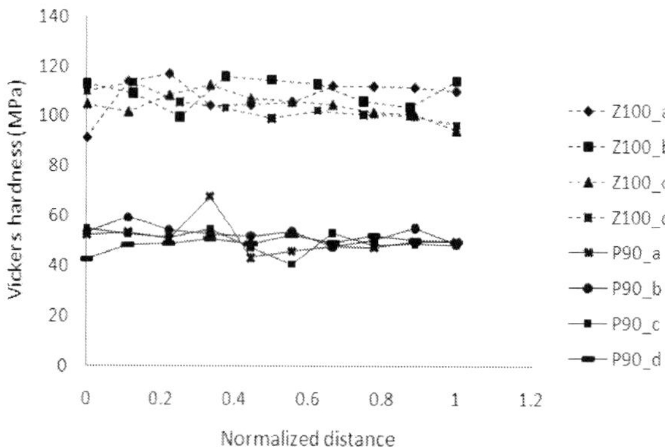

Figure 12. Vickers hardness along the depth of restoration (see Figure 1c) for specimens restored with P90. The results of Z100 specimens were plotted for comparison.

4.2. Influence of cavity configuration (C- factor) [28]

The C-factor of a restoration is defined as the ratio of the bonded areas to the unbonded areas, and it is normally used to characterize the configuration of the cavity [29]. Experimental studies [30-32] have shown that increasing the C-factor would reduce the interfacial bond strength and increase the incidence of interfacial failure. That was attributed to the higher shrinkage stress in restorations with a higher C-factor which had caused more interfacial debonding. In this section, three cavity configurations with different C-factors were used and their interfacial debonding was evaluated with the AE technique.

20 intact human molars with similar dimensions were selected and randomly divided into 4 groups of 5: Group 1 with large Class-I cavities, Group 2 with small Class-I cavities, Group 3 with small Class-II cavities and Group 4 with large Class-II cavities. The specimen preparation, testing procedure and test equipments were similar to those in Section 3.2 and Section 4.1. The bonding agent used for all the specimens was total-etch adhesive Adper™ Single Bond Plus (3M ESPE, USA) and the composite resin was Z100™ (3M ESPE, US). Figure 13 shows schematically the Class-I and Class-II cavities prepared and the dimensions of interest. The dimensions of each specimen were measured with a micrometer (Mitutoyo, Japan) with which the C-factor was calculated. The average C-factors of the four groups were 3.37, 2.90, 2.00 and 1.79, respectively, as summarized in Table 2.

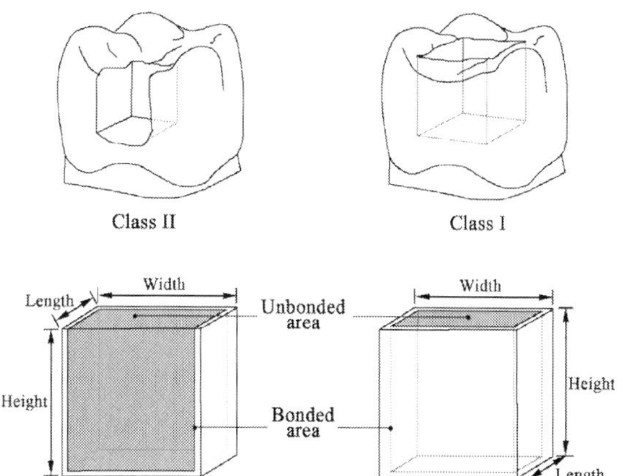

Figure 13. Schematic diagrams of the prepared tooth cavities with dimensions of interest. The shaded areas are the free, unbonded surface areas

Cavity type	Group	Length (mm)	Width (mm)	Depth (mm)	C-factor	Bond area (mm²)	Volume (mm³)
Class I	1	3.94	2.99	2.08	3.45	40.62	24.50
		3.76	3.68	1.88	3.02	41.81	26.01
		4.13	4.06	2.48	3.42	57.39	41.58
		4.00	3.79	2.02	3.08	46.63	30.62
		3.99	3.61	2.72	3.87	55.75	39.18
Mean (STD)		3.96(0.13)	3.63(0.39)	2.24(0.35)	3.37(0.34)	48.44(7.78)	30.68(7.72)
Class I	2	2.66	2.77	1.40	3.06	22.57	10.32
		3.29	3.04	1.80	3.28	32.79	18.00
		2.86	2.87	1.05	2.47	20.24	8.62
		3.59	2.70	1.10	2.43	23.53	10.66
		2.73	2.70	1.45	3.14	23.12	10.69
Mean (STD)		3.03(0.40)	2.82(0.14)	1.36(0.30)	2.88(0.40)	24.45(4.83)	11.66(3.65)
Class II	3	2.09	1.85	1.68	2.01	14.00	6.50
		2.35	1.92	1.42	1.92	13.91	6.41
		2.30	1.91	1.73	2.03	15.66	7.60
		2.39	1.94	2.02	2.13	18.21	9.37
		2.11	2.10	1.70	1.90	15.18	7.53
Mean (STD)		2.25(0.14)	1.94(0.09)	1.71(0.21)	2.00(0.09)	15.39(1.75)	7.48(1.19)
Class II	4	3.75	3.34	2.47	1.89	39.30	30.94
		4.18	3.39	2.72	1.97	46.13	38.54
		4.49	3.36	2.19	1.88	42.11	33.04
		3.63	3.51	1.58	1.63	29.76	20.13
		4.57	4.50	1.80	1.57	45.12	37.02
Mean (STD)		4.12(0.42)	3.62(0.50)	2.15(0.47)	1.79(0.18)	40.48(6.57)	31.93(7.26)

Table 2. Dimensions and geometrical factors of the specimens used in the experimental study on C-factor

Figure 14 shows the mean cumulative number of AE events against time for the four test groups, with the standard deviations shown as red bars. AE caused by interfacial debonding was first detected about 20s into the curing of the composite and developed rapidly thereafter. The mean and standard deviation of the total number of AE events for the four groups were 29.6±15.7, 10.0±5.8, 2.6±1.5, and 2.2±1.3 (Table 3), respectively, which showed an increase with an increasing C-factor. Table 4 shows the statistical significance (p-value) of the differences in the total number of AE events between the different groups. It can be seen that the differences were significant, except that between Groups 3 and 4 which had very similar C-factors.

In Figure 15(a), the total number of AE events for all the specimens were plotted against their individual C-factors. To account for the differences in the bonded area or volume of restorations with similar C-factors, the total number of AE events per unit bonded area and that per unit composite volume are plotted in Figure 15(b) and 15(c), respectively, for comparison. The mean and standard deviation of these normalized numbers for the 4 groups are also listed in Table 3. Despite the large variations in the number of AE events detected among specimens with similar C-factors, it can be clearly seen that the amount of debonding increased with an increase in the C-factor. There appeared to be a critical value for the C-factor associated with Z-100, i.e. 1.5, below which no AE events could be detected. The trend remained the same even when possible influence from the bonded area and restoration volume was taken into account.

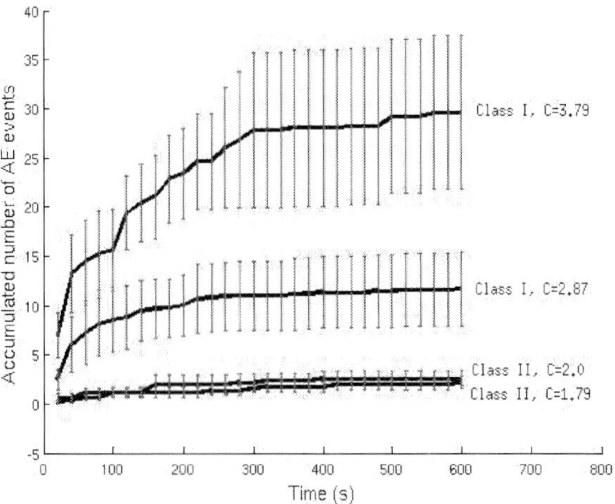

Figure 14. The cumulative number of AE events against time for the 4 test groups

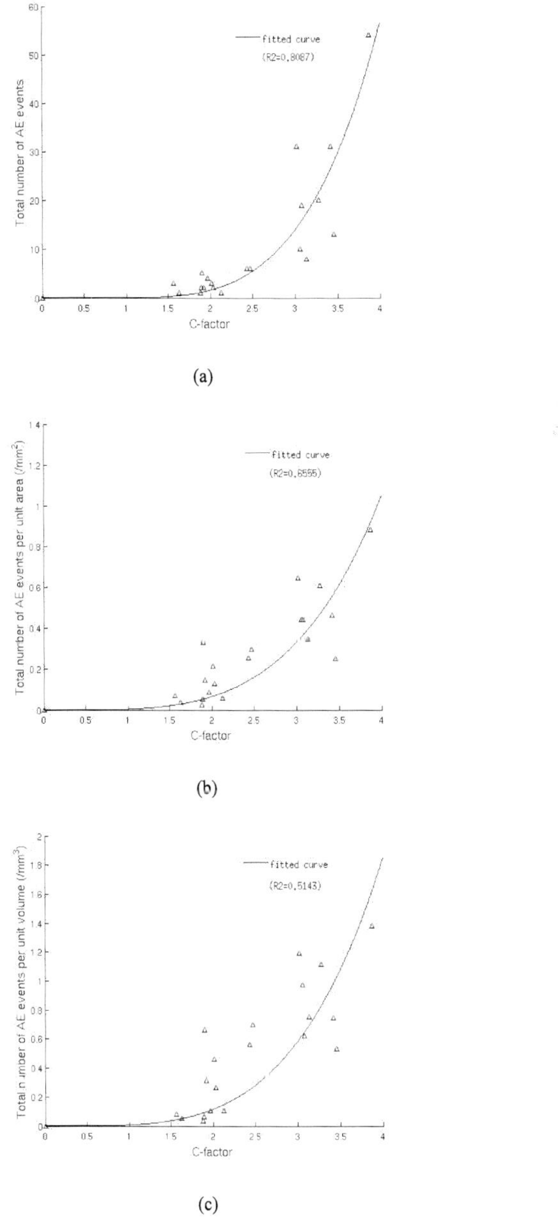

Figure 15. The total number of AE events as a function of the C-factor: (a) total number, (b) total number per unit bond area (/mm²), and (c) total number per unit composite volume (/mm³)

Group	Total number of AE events	Total number of AE events per area (/mm²)	Total number of AE events per volume (/mm³)
1	29.6 (15.7)	0.54(0.24)	0.89(0.37)
2	10.0 (5.8)	0.39(0.14)	0.81(0.22)
3	2.6 (1.5)	0.17(0.10)	0.36(0.21)
4	2.2 (1.3)	0.05(0.02)	0.07(0.03)

Table 3. Results from the AE tests: mean (standard deviation)

Group	2	3	4
1	0.031	0.005	0.005
2		0.025	0.019
3			0.667

Table 4. Statistical significance (p-value) of the difference in the total number of AE events between the groups with different C-factors

4.3. Influence of the filling technique

The composite placement techniques (bulk vs. incremental) have been shown to affect the magnitude of the polymerization shrinkage stresses and cuspal deformations in restored teeth [33-37]. Compared with the bulk filling method, the incremental filling method has been reported to produce lower shrinkage stresses. Therefore, the incremental filling method is expected to produce less interfacial debonding than the bulk filling method. This hypothesis will be tested in this section with the AE technique.

12 incisor bovine teeth with similar geometries were selected and randomly divided into 2 groups of 6. A cylindrical cavity, which was 2mm in depth and 4mm in diameter, was cut into the top surface of each tooth; see Figure 16(a). All samples were prepared by a single operator following typical clinical procedures with a high-speed handpiece and dental cutting burs. Each specimen was then treated with the bonding agent Adper™ Single Bond Plus (3M ESPE, USA) to the cavity surfaces and then restored with the composite resin Z100™ (3M ESPE, US) using either the bulk or incremental filling technique, as shown in Figure 16(b). The AE sensor was attached onto the opposite surface of the tooth. For the group restored with the incremental technique, recording of the AE events was interrupted in between placements of the 2 layers.

Figure 17 shows the total number of AE events for all the specimens. Note that 4 out of 6 of the specimens with the incremental filling method did not produce any AE events at all, while the remaining 2 produced only a small number of AE events. Compared with the incremental-filling group, specimens restored with the bulk-fill method produced much more AE events, indicating more interfacial debonding in those specimens.

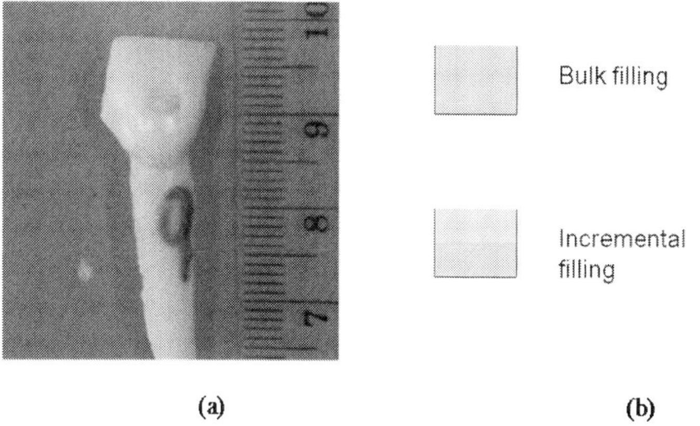

(a)   **(b)**

Figure 16. (a) Cylindrical cavity on a bovine tooth, and (b) the two different filling techniques

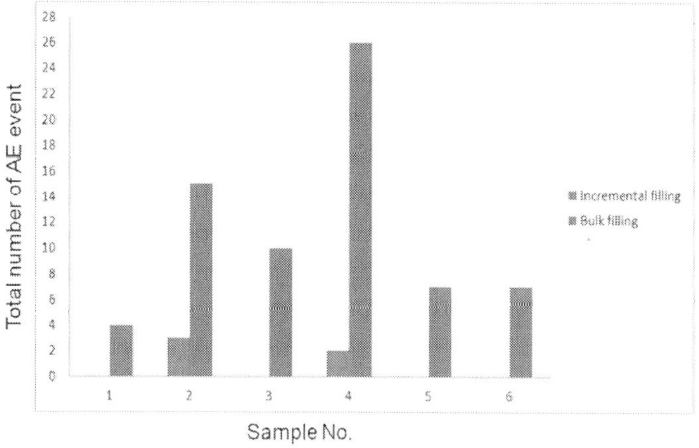

Figure 17. The total number of AE events for specimens restored with the bulk and incremental techniques

5. Discussion and concluding remarks

The AE technique, which is widely used for fracture monitoring in many areas, was for the first time introduced here to monitor *in-situ* the interfacial deboning of composite restorations. Its effectiveness was verified by several groups of experiments using specimens with different boundary constraints: freestanding blobs of composite, ring specimens cut from a tooth root and whole human molars. It was also shown from the AE results that interfacial debonding in composite restorations is greatly influenced by the restorative composite material, cavity configuration and filling technique. In general, composites with lower shrinkage, cavities with smaller C-factors and use of the incremental filling method can produce less interfacial debonding. These findings agreed with the results from other researchers using different evaluative methods.

The AE technique has several advantages over the other methods in evaluating the interfacial debonding of dental composite restorations: 1) it is non-destructive; 2) it is not affected by the light-curing process and can therefore be used to make *in-situ* measurement during the curing; and 3) it is very sensitive, being able to detect micro-cracking that cannot be seen by micro-CT. However, there are still certain limitations to the AE technique. The AE results are dependent on the operational parameters of the AE system, e.g. the signal threshold, band pass, and the gain of the preamplifier. Therefore, it is important to specify those parameters when comparing AE results from different AE measurements. Also, false signals which have similar frequencies with the interfacial debonding may not be filtered out. Efforts are therefore needed to minimize such false signals, e.g. by keeping the tooth samples wet to avoid enamel cracking caused by dehydration. Finally, unlike the optical methods, the AE technique cannot provide direct visual images of the interfacial debonding.

In the future, more efforts should be made to increase the amount of information the AE technique can provide in evaluating interfacial debonding in dental composite restorations. For example, it may be possible to locate the source of the AE events by using the times-of-arrival captured by several AE sensors placed at different positions on the tooth [38-40]. This will help to establish where the critical regions of debonding are in a restored tooth. Also, the AE technique can be used to study other factors that may influence interfacial debonding, e.g. the adhesive/bonding agent, cavity shapes (with a constant C-factor), environmental challenges such as thermal and occlusal load, etc. Overall, it is a powerful tool which can be used to evaluate the interfacial debonding of composite restorations under more clinically relevant conditions in a more systematic way. This will help to improve the quality of bonding at the tooth-restoration interface and, thus, increase the longevity of composite restorations.

Author details

Haiyan Li, Jianying Li and Alex Fok
Minnesota Dental Research Center for Biomaterials and Biomechanics, School of Dentistry, University of Minnesota, Minneapolis, MN, United States

Xiaozhou Liu
China Medical University School and Hospital of Stomatology, Shenyang, PR China

Acknowledgement

The authors would like to acknowledge 3M ESPE for providing the restorative materials and the Minnesota Dental Research Center for Biomaterials and Biomechanics (MDRCBB) for providing the test devices. They also acknowledge Mr. Xiaofei Yun and Mr. Max Hofer for their contributions to the AE tests. Xiaozhou Liu would like to thank the China Scholarship Council and the MDRCBB for financially supporting her study visit to the MDRCBB.

6. References

[1] Ferracane, J.L., *Developing a more complete understanding of stresses produced in dental composites during polymerization*. Dental Materials 2005. 11: p. 7.

[2] Forss, H. and E. Widstrom, *Reasons for restorative therapy and the longevity of restorations in adults*. Acta Odontol Scand 2004. 62: p. 5.

[3] Dauvillier, B.S., M.P. Aarnts, and A.J. Feilzer, *Developments in shrinkage control of adhesive restoratives*. Jounal of Esthetic Dentistry, 2000. 12: p. 9.

[4] Brunthaler, A., et al., *Longevity of direct resin composite restorations in posterior teeth*. Clinical Oral Investigations 2003. 7: p. 8.

[5] Yap, A.U.J., et al., *An in vitro microleakage study of three restorative techniques for class II restorations in posterior teeth*. Biomaterials, 1996. 17: p. 5.

[6] Lopes, G.C., M. Franke, and H.P. Maia, *Effect of finishing time and techniques on marginal sealing ability of two composite restorative materials*. Journal of Prosthetic Dentistry 2002. 88: p. 5.

[7] Abdalla, A.I. and C.L. Davidson, *Effect of mechanical load cycling on the marginal integrity of adhesive class I resin composite restorations*. Journal of Dentistry 1996. 24: p. 4.

[8] Ausiello, P., et al., *Debonding of adhesively restored deep class II MOD restorations after functional loading*. American Journal of Dentistry, 1999. 12: p. 5.

[9] Shono, Y., et al., *Effects of cross-sectional area on resin-enamel tensile bond strength*. Dental Materials, 1997. 13: p. 7.

[10] Goracci, C., et al., *Influence of substrate, shape, and thickness on microtensile specimens' structural integrity and their measured bond strengths*. Dental Materials 2004. 20: p. 12.

[11] Phrukkanon, S., M.F. Burrow, and M.J. Tyas, *The influence of crosssectional shape and surface area on the microtensile bond test*. Dental Materials 1998. 14: p. 10.

[12] McMurray, J., *Fundamentals of Organic Chemistry*1986, Monterey, CA: Brooks/Cole.

[13] Watts, D.C., A.S. Marouf, and A.M. Al-Hindi, *Photo-polymerization shrinkage-stress kinetics in resin-composites: methods development*. Dental materials 2003. 19: p. 11.

[14] Gonçalves, F., et al., *Contraction Stress Determinants in Dimethacrylate Composites*. Journal of Dental Reseach, 2008. 87: p. 5.

[15] Li, J., *The Biomaterials And Biomechanics Of Dental Restorations: Property Measurement And Simulation*, 2008, University of Manchester.

[16] Li, J., H. Li, and S.L. Fok, *A mathematical analysis of shrinkage stress development in dental composite restorations during resin polymerization*. Dental materials 2008. 24: p. 9.

[17] Kanemura, N., H. Sano, and J. Tagami, *Tensile bond strength to and SEM evaluation of ground and intact enamel surfaces*. Journal of Dentistry, 1999. 27: p. 8.

[18] Sano, H., et al., *Comparative SEM and TEM observations of nanoleakage within the hybrid layer*. Operative Dentistry 1995. 20: p. 8.

[19] Santis, R.D., et al., *A 3D analysis of mechanically stressed dentin–adhesive–composite interfaces using X-ray micro-CT*. Biomaterials, 2005. 26: p. 14.

[20] Nair, A. and C.S. Cai, *Acoustic emission monitoring of bridges: Review and case studies*. Engineering Structures, 2010. 32(6): p. 1704-1714.

[21] Bohse, J., *Acoustic emission characteristics of micro-failure processes in polymer blends and composites*. Composites Science and Technology, 2000. 60(8): p. 1213-1226.

[22] Skåre, T. and F. Krantz, *Wear and frictional behaviour of high strength steel in stamping monitored by acoustic emission technique*. Wear, 2003. 255(7–12): p. 1471-1479.

[23] Ereifej, N., N. Silikas, and D.C. Watts, *Initial versus final fracture of metal-free crowns, analyzed via acoustic emission*. Dental Materials, 2008. 24(9): p. 1289-1295.

[24] Vallittu, P.K., *Use of woven glass fibres to reinforce a composite veneer. A fracture resistance and acoustic emission study*. Journal of Oral Rehabilitation, 2002. 29: p. 7.

[25] Kim, K.H. and O. Okuno, *Microfracture behaviour of composite resins containing irregular-shaped fillers Journal of Oral Rehabilitation*, 2002. 29: p. 7.

[26] Li, H., et al., *Non-destructive examination of interfacial debonding using acoustic emission*. Dental Materials, 2011. 27: p. 8.

[27] Lu, H., et al., *Probing the origins and control of shrinkage stress in dental resin-composites: I. Shrinkage stress characterization technique*. Journal of Materials Science: Materials in Medicine, 2004. 15: p. 7.

[28] Liu, X., et al., *An acoustic emission study on interfacial debonding in composite restorations*. Dental Materials, 2011. 27: p. 8.

[29] Braga, R.R., et al., *Influence of cavity dimensions and their derivatives (volume and 'C' factor) on shrinkage stress development and microleakage of composite restorations*. Dental Materials, 2006. 22(9): p. 818-823.

[30] Feilzer, A.J., A.J.D. Gee, and C.L. Davidson, *Setting stress in composite resin in relation to configuration of the restoration*. J Dent Res, 1987. 66(11): p. 4.

[31] Nikolaenko, S.A., et al., *Influence of c-factor and layering technique on microtensile bond strength to dentin*. Dental Materials, 2004. 20(6): p. 579-585.

[32] Watts, D.C. and J.D. Satterthwaite, *Axial shrinkage-stress depends upon both C-factor and composite mass*. Dental Materials, 2008. 24(1): p. 1-8.

[33] Park, J., et al., *How should composite be layered to reduce shrinkage stress: Incremental or bulk filling?* Dental Materials, 2008. 24(11): p. 1501-1505.

[34] McCullock, A. and B. Smith, *In vitro studies of cuspal movement produced by adhesive restorative materials*. Br Dent J, 1986. 161: p. 5.

[35] Lee, M.-R., et al., *Influence of cavity dimension and restoration methods on the cusp deflection of premolars in composite restoration*. Dental Materials, 2007. 23(3): p. 288-295.

[36] Versluis, A. and W. Douglas, *Does an incremental filling technique reduce polymerization shrinkage stresses?* . J Dent Res, 1996. 75: p. 8.

[37] Abbas, G., et al., *Cuspal movement and microleakage in premolar teeth restored with a packable composite cured in bulk or in increments*. J Dent Res, 2003. 31: p. 8.

[38] Mavrogordato, M., et al., *Real time monitoring of progressive damage during loading of a simplified total hip stem construct using embedded acoustic emission sensors*. Medical Engineering & Physics, 2011. 33(4): p. 395-406.

[39] Lympertos, E.M. and E.S. Dermatas, *Acoustic emission source location in dispersive media*. Signal Processing, 2007. 87(12): p. 3218-3225.

[40] Salinas, V., et al., *Localization algorithm for acoustic emission*. Physics Procedia, 2010. 3(1): p. 863-871.

Natural Fiber, Mineral Filler Composite Materials

TEMPO-Mediated Oxidation of Lignocellulosic Fibers from Date Palm Leaves: Effect of the Oxidation on the Processing by RTM Process and Properties of Epoxy Based Composites

Adil Sbiai, Abderrahim Maazouz, Etienne Fleury,
Henry Sautereau and Hamid Kaddami

Additional information is available at the end of the chapter

http://dx.doi.org/10.5772/47763

1. Introduction

Lignocellulosic fibers display many well-known advantages as compared to their synthetic counterparts, including their being ecologically and toxicologically harmless, biologically degradable, and carbon dioxide (CO_2) neutral. Furthermore, natural fibers are characterized by a huge degree of variability and diversity in their properties. As they could be extracted from wood and annual plants, they are available in various forms, give a feeling of warmth to the touch, and have a pleasant appearance. None of these properties are offered by other non-wood engineering fibers.

Over the last two decades, a great deal of work has been dedicated to composites reinforced with natural fibers. Indeed the use of such natural product for the reinforcement of thermoplastic or thermosetting resins, leads to composites with lower density, higher specific stiffness and strength, together with a better biodegradability (Bledzki et al., 1999; Mishra et al., 2004; Zimmermann et al., 2004; Gandini, 2008). However, only few studies have dealt with polymers reinforced with lignocellulosic fibers obtained from palm trees (Abu-Sharkh and Hamid 2004; Wan Rosli et al. 2004; Kaddami et al. 2006; Bendahou et al. 2008 ; Sbiai et al. 2008; Bendahou et al. 2009). In the previous investigations (Kaddami et al. 2006; Bendahou et al. 2008; Sbiai et al. 2008), the reinforcing capability of palm tree fibers in thermoset or thermoplastic polymer matrices was demonstrated. In the case of epoxy-based composites, expected and strong interactions gave rise to enhanced mechanical and thermal characteristics. An increase in the glass transition temperature and an improvement of the thermo-mechanical properties, bending moduli, stress at break values, and maximum absorbed energies were

reported for composites based on fibers modified with acetic anhydride (Kaddami et al. 2006). The size of the fibers was also found to have an effect on the properties (Sbiai et al. 2008).

Reinforcement is the physical expression of the microscopic balance at the matrix/filler interface which makes up a filler network. Thus, the adhesion filler/matrix is the most important parameter governing reinforcement. It is required to render possible the transfer of mechanical constraints to the fiber when a load is applied to the composite. This adhesion can be enhanced through chemical or physical modification of the polymer and/or the filler. Such modifications depend on the physico-chemical nature of the matrix.

In most cases, the hydrophilic nature of the lignocellulosic fibers is detrimental for their interface interactions with common resins, which are mostly hydrophobic. Thus, chemical treatments have to be applied to overcome this issue (Reich et al., 2008). For this purpose, various surface modifications have been devised to selectively replace hydroxyl functions with hydrophobic groups (Duanmu et al., 2007; Biagiotti et al., 2004; Goussé et al., 2004; Gandini, 2008). Coupling agents giving rise to covalent junctions between the fibers and the matrix have also been described (Abdelmouleh et al., 2002; Paunikallio et al., 2006; Gonzalez-Sanchez et al., 2008; Bendahou et al., 2008).

Among modifications used to improve interfacial adhesion in natural fiber/polymer composites, oxidative treatments have received much attention during the seventies and eighties. Corona and plasma treatments were found to effectively enhance the interface in epoxy-based composites (Sakata et al 1993a,b), and chemical oxidative treatments have been widely reported in several studies for numerous composites of natural fiber and polymers. Many types of oxidants have been employed, e.g. dichromate/oxalic acid, ozone, potassium ferricyanide, ferric chloride, nitric acid, hydrogen peroxide, dicumyle peroxide, etc. (Sapieha et al 1989; Felix et al 1994; Cousin et al 1989; Kaliński et al. 1981; Raj et al. 1990; Felix and Gatenholm 1991; Flink et al. 1988; Young 1978; Moharana et al. 1990; Gardinera and Cabasso 1987; Zang and Sapieha 1991; Iwakura et al. 1965; Jutier et al. 1988; Michell et al. 1978; Coutts and Campbell 1979; Tzoganakis et al. 1988; Sung et al. 1982; Philippou et al. 1982; Manrich et al. 1989; Sapieha et al. 1991).

More recently, 2,2,6,6-tetramethylpiperidine-1-oxyl radical (TEMPO)-mediated oxidation of polysaccharides bearing primary alcohols has been intensively studied. This type of oxidation makes it possible to selectively oxidize, in aqueous medium, primary alcohol groups into carboxyl groups in natural polysaccharides (Isogai, and Kato 1998; Isogai and Saito 2005; Isogai et al. 2005; Davis and Flitsch 1993; De Nooy et al. 1995 Fukuzumi et al. 2009; Chang and Robyt 1996; Tahiri and Vignon 2000; Habibi et al. 2006).

Among composite processing techniques, resin transfer molding technique (RTM) has been widely used. This technique was used for high-performance air craft and automotive structures. This process can be divided into four stages: performing, mould filling, curing, and demoulding. First, the dry fiber preform is made and placed in the mould. The mould is filled with liquid resin which is then solidified in the curing stage. Note that the curing stage actually begins as soon as the resin is mixed. However, usually the process is designed such that cure proceeds slowly until the mould has been filled. Once cured, the part is removed from the mould. In some cases, the part is demoulded before complete cure and post-cured in an oven. The purpose of this process was to improve the quality (dry spots, voids) and processability and to minimize the material wastage. In conventional fibers as glass fibers and carbon fibers when these fibers are presented in the form of bundles, the flow of resin through the preform is governed by two mechanisms: bulk flow and fiber wetting. Bulk flow occurs in the space between the fiber tows, whereas fiber wetting occurs within the fiber tows (O'Flynnet al 2007, Octeau 2001, Chu 2003). The interaction between matrix and reinforced fiber was particularly important to the RTM process (Nguen-Thuc et al 2004).

Many studies have been done on the kinetics of the polymerization systems epoxy / amine (Eloundou et al. 1996a and 1996b, Halley et al. 1996, Pichaud 1996, Pascault et al 2002, Nguen-Thuc et al 2004). The chemical reaction between the epoxy prepolymer and diamine hardener can lead, depending on the stoichiometry to formation of a network. It can occur during growth of the macromolecular chains, two structural transformations: gelation and vitrification. At the beginning of the reaction, when the glass transition temperature Tg is less than the reaction temperature, the reaction is controlled by chemical kinetics. At the gel point, the formation of macromolecular chains leads to a gelling phenomenon that characterizes the transition from a liquid to a rubbery state while the vitrification phenomenon reflects a shift from the rubbery state to glassy state (Eloundou et al. 1996a, Halley et al. 1996, Pascault et al 2002). At our knowledge no studies had been dedicated to the effect of the lignocellulosic fibers on the polymerization kinetic.

2. Part 1: TEMPO-mediated oxidation

2.1. Experimental section

2.1.1. Materials

Leaflets from palm leaves of *P. dactylifera* were cut into pieces with lengths of 3 to 5 cm. After 24 h soxhlet extraction with acetone/ ethanol (75/25 by vol), the resulting product was slightly discolored. It was then ground and sieved in order to eliminate the fine powder and to keep only the particles with length ranging from 2 to 10 mm and widths from 0.2 to 0.8 mm. this material is hereafter called lignocellulosic fibers from date palm leaflets or DPLF. For XRD analysis, the samples were ground again to obtain a fine powder.

Cellulose, lignin and hemicellulose were extracted from the DPLF according to a previously described procedure (Bendahou et al., 2007). Cellulose whiskers were prepared from the rachis of *P. dactylifera* leaves (Bendahou et al., 2009).

2,2,6,6-tetramethylpiperidine-1-oxyl radical (TEMPO), sodium bromide, and 10% sodium hypochlorite solution were purchased from Sigma-Aldrich and used as received.

2.1.2. General procedure for oxidative treatment (Gomez-Bujedo et al., 2004)

2g of DPLF were stirred in 200 mL distilled water for 1 min. TEMPO (32 mg, 0.065 mmol) and NaBr (0.636 g, 1.9 mmol) were dissolved in the suspension maintained at 4 °C. TEMPO-mediated oxidation was initiated by adding dropwise the sodium hypochlorite solution (10 %, 32 ml, 43 mmol) and adjusting the pH at 10 by addition of a 0.1 M aqueous HCl. The suspension was then maintained at pH 10 ± 0.5 by continuous addition of 0.5 M NaOH and at 4 °C by means of a thermo-controlled bath. When the pH became stable, the reaction was quenched by adding methanol (5 mL). After neutralization of the reaction mixture to pH 7 by addition of 0.1M HCl, the suspension was filtrated and the solid particles were washed with distilled water and dried under vacuum at 30 °C for 18 hours. The soluble part was freeze-dried.

The same oxidation procedure was applied to the other samples, namely the cellulose, the hemicellulose and the lignin extracted from the DPLF, together with the cellulose whiskers extracted from the leave rachis.

2.1.3. Analysis techniques

The composition of the DPLF before and after oxidation *i.e.* the weight fraction of their cellulose, hemicellulose, and lignin contents was determined by selective extraction according to the NF T12-011 standard.

The carboxylate content of oxidized cellulose samples was determined by conductometric titrations (Saito & Isogai, 2004).

Infrared spectra were recorded on a FT-IR Perkin-Elmer 1720X spectrometer collecting 20 scans from 400 to 4000 cm^{-1}. Oxidized DPLF was converted to its acid form by ion exchange in order to displace the carboxyl absorption band toward higher wavelengths, thus eliminating any interference with the absorbed water band at 1640 cm^{-1}. Consequently the superposition of the sodium carboxylate peak with those of the hydrogen bonds could be avoided. The samples were dried with acetone and analyzed as KBr.

The XPS spectra were measured using an AXIS ULTRA DLD X-ray photoelectron spectrometer (KRATOS ANALYTICAL). Samples were dried at 50°C under vacuum during 24h, and set with conductive carbon adhesive tabs. The measurements were performed using a monochromatic Al Kα X-ray with a pass energy of 160 eV and a coaxial charge neutralizer. The base pressure in the analysis chamber was smaller than 5×10-8 Pa. XPS spectra of O1s and C1s levels were measured at a normal angle with respect to the plane of the surface. High resolution spectra were corrected for charging effects by assigning a value of 284.6 eV to the C1s peak (adventitious carbon). Binding energies were determined with an accuracy of ±0.2 eV. The spectral decomposition of the C1s and O1s photoelectron peaks was carried out using a Shirley background at positions and FWHMs corresponding to known components, applying a 30% Lorentzian-to-Gaussian peak shape ratio.

The distribution of sodium and calcium ions present in a cross-section of the TEMPO-oxidized fibers was evaluated by the line analysis mode using an energy dispersive X-ray (EDX) device fitted into a scanning electron microscope (SEM) (Hitachi S3500 N). The X-ray detector was a Thermo Noran Superdry II having a resolution of 143eV. The analyses were performed without prior metallization of the samples.

A Philips X'Pert diffractometer equipped with a ceramic X-ray diffraction tube operated at 40 kV and 40 mA with CuKα radiation (wavelength 0.15418 nm) was used to determine the crystalline index of the specimens. The samples were compressed into disks using a cylindrical steel mould (∅ = 15 mm) with an applied pressure of 32 MPa. The diffracted intensity was recorded for 2θ angles in the range of 10 to 40°. Data were treated according to the empirical Segal method (Segal et al., 1959).

2.2. Results and discussion

2.2.1. TEMPO-mediated oxidation

DPLF samples were oxidized with TEMPO in aqueous medium. The experiments were performed (i) at low temperature (4 °C) to avoid the formation of side products and (ii) in the presence of a large quantity of sodium hypochlorite, i.e., 21 mmol per gram of the DPLF, in order to reach a high degree of oxidation (Montanari et al., 2005; Okita et al., 2009). The selective oxidation of a hydroxymethyl group to a carboxyl via an aldehyde requires two moles of NaOCl per mole of hydroxyl according to the known TEMPO mechanism (Figure 1) (De Nooy et al., 1995). The kinetic of the reaction was directly followed by addition of an aqueous NaOH solution which neutralizes the carboxylic acid functions resulting from the oxidation. Figure 2 clearly illustrates the efficiency of the reaction with a continuous increase in NaOH consumption over a period of 720 minutes. According to FTIR analysis on the water insoluble fraction (Figure 3), a stretching vibration C=O from the free COOH band near 1737 cm^{-1} was seen to have a higher intensity in the case of oxidized fibers as opposed to their non-oxidized counterparts, thus providing evidence of the oxidation. Finally, as expected, the degree of oxidation (*Do*) was found to be 0.71 ± 0.02 (Table 1).

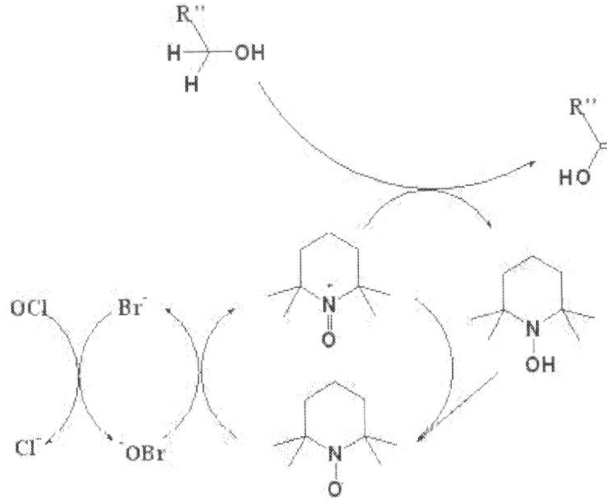

Figure 1. Scheme of the catalytic cycle for the oxidation of a primary hydroxyl.

The composition of DPLF was also investigated since a significant amount (30%) of the material became soluble in water during the oxidation. The results presented in Table 1 show that the proportions between components of DPLF have changed before and after the chemical oxidative treatment. Thus cellulose content has increased from 35 to 46 %, whereas lignin and hemicelluloses contents have decreased. Nevertheless both lignin and hemicelluloses were still present in the oxidized DPLF at high concentration respectively 12 and 34 weight percent. Okita et al., have reported on a complete elimination of lignin and hemicelluloses in the case of TEMPO-mediated oxidation of thermomechanical softwood pulp under similar processing conditions (Okita et al., 2009). Pulp used by these authors contained similar percentage of lignin and hemicelluloses but was pried open by the mechanical refining process. One can thus deduce that thermomechanical pulps are much more reactive than lignocellulosic fibers having undergone no activation. On the other hand, a weak increase of the crystalline index is observed. This could be explained by the dissolution of the lignin after oxidation. Actually, this dissolution would induce a weak decrease of the intensity of the amorphous peak I$_{am}$ at $2\theta = 18°$ (Table 1 and supporting info S1).

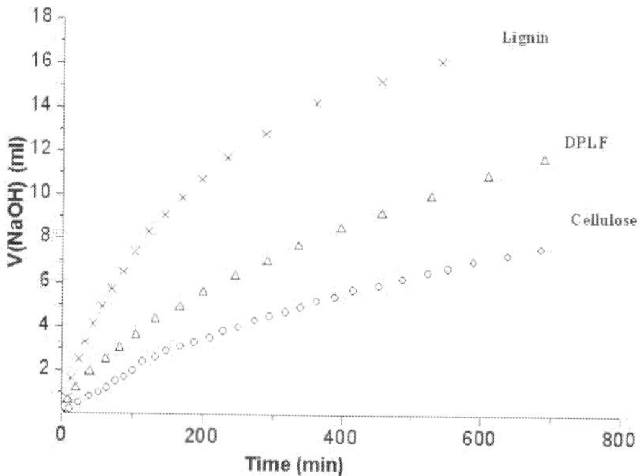

Figure 2. Kinetic TEMPO-mediated oxidation of DPLF, the extracted cellulose and lignin from DPLF: Consumption of the NaOH with the reaction time.

In order to compare the reactivity toward oxidation of each component of the DPLF, cellulose, hemicelluloses and lignin were extracted and subjected to the oxidation procedure. Kinetic data were recorded, except for hemicelluloses which presented very fast and uncontrollable kinetic. Data illustrated in Figure 2 clearly show that the extracted lignin reacted faster than the extracted cellulose, whereas the DPLF presented an intermediate rate. The structural characterization of the oxidized lignin and hemicellulose was not undertaken due to the complexity of possible structures. However, a previous study has shown that hemicelluloses extracted from the leaflets and rachis of the *P. dactylifera* palm consist of arabinoglucuronoxylans, which are either soluble or insoluble in hot water depending on the molar content of 4-O-methyl-glucuronic acid (Bendahou et al., 2007). One can thus deduce that the oxidative treatment under aqueous conditions allowed the conversion of the hydroxymethyl group of the arabinose moieties into carboxylic functions, which further facilitated the hemicellulose extraction in reaction medium maintained at 4°C. For lignin, the conversion of primary hydroxyl functions led to hydrosoluble products as already described (Okita et al., 2009).

Samples	[COOH] mmol/g	Crystalline Index %	Composition Cellulose./ Hemicellulose./Lignin %	Water insoluble fraction %
Control DPLF	0	49.5+/-0,5	35 / 28 / 27	100
Oxidized DPLF	0.71	52.5+/-0,5	46 / 34 / 12	70

Table 1. Chemical characterization of DPLF before and after TEMPO-mediated oxidation.

The reason for the moderate reactivity of the cellulose was already described. It originates from the crystalline nature of cellulose and from the difficulty to access to the hydroxyl groups at the surface of the cellulose crystals. In contrast, lignin and hemicelluloses, which are amorphous, react faster. The TEMPO-mediated oxidation of the cellulose sample was confirmed by the appearance of the characteristic carboxylic signal at 176 ppm in the ^{13}C CP-MAS NMR spectrum (Wikberg & Maunu, 2004; Martins et al., 2006) corresponding to a degree of oxidation (*Do*) equal to 0.07 (See supporting info S2 and S3).

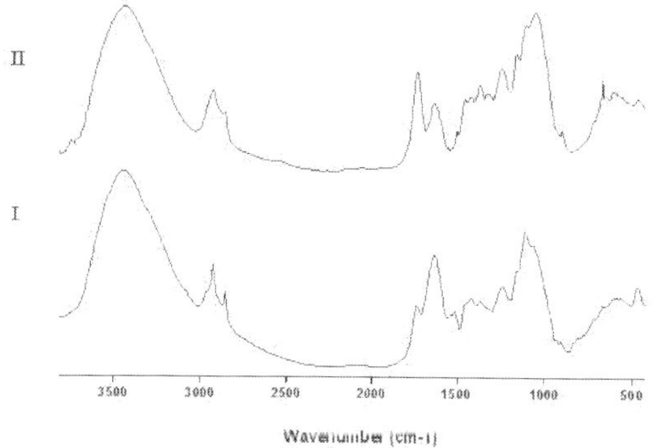

Figure 3. FT-IR spectra for DPLF before (I) and after TEMPO-mediated oxidation (II).

2.2.2. Heterogeneous kinetic modeling

Only few kinetic studies of TEMPO-mediated selective oxidation of cellulosic material have been reported in the literature (De Nooy et al., 1995; Sun et al., 2005; Mao et al., 2010). The authors have observed a first-order kinetic with respect to TEMPO and NaBr as well as a rate constant determined by their concentrations. It has also been claimed that the NaOCl concentration affects mainly the level of the conversion, leading to the possibility of reusing the reaction liquid by addition of more NaOCl. However, although TEMPO-mediated oxidation of cellulose corresponds to a reaction between a solid phase and a liquid phase, the heterogeneous aspect of the reaction was never envisaged.

In fact, cellulose fibers can be represented as accessible porous networks, where reagents can diffuse to the periphery of entities made up of bundles of cellulose microfibrils surrounded with lignin and/or hemicelluloses. This system can be described as a cylinder (Figure 4) with a diameter that becomes decreased when the Tempo-mediated oxidative process progresses.

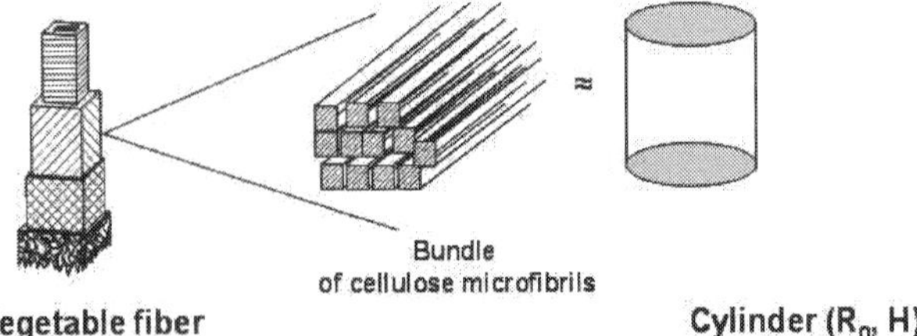

Figure 4. Schematization of the reactive substrate presented as a cylinder.

The interfacial specific rate r_{is} of an heterogeneous reaction can be expressed by equation 1 where R and H are respectively the radius and the height of the cylinder and k the rate constant for a given temperature and concentration (Levenspiel, 1972).

$$r_{is} = 2\pi\ RH\ k \tag{1}$$

Considering R_0 as the initial radius, the conversion of the reaction at time t, is given by

$$X = \frac{R_0^{\ 2} - R^2}{R_0^{\ 2}} \tag{2}$$

Since the initial molar concentration of the reactive function is n_0, its decrease with time can be described by equation 3:

$$dn = 2\pi\ R\ H\ r_{is}dt = n_0 dX = -\frac{2n_0}{R_0^{\ 2}}R\ dR \tag{3}$$

Where:

$$dX = d(\frac{R_0^{\ 2} - R^2}{R_0^{\ 2}})$$

Equation 3 can be simplified into:

$$dn = 2\pi Hr_{is}dt = -\frac{2n_0}{R_0^{\ 2}}dR$$

Which, after integration, gives the following relation:

$$R = R_0 - \frac{\pi\ H\ R_0^{\ 2}r_{is}}{n_0}t \tag{4}$$

By expressing the time required for a complete conversion to occurs as:

$$t_{Final} = \frac{n_0}{\pi\ H\ R_0\ r_{is}}$$

and rearranging the relation, one obtains:

$$(1-X)^{1/2} = 1 - \frac{t}{t_{Final}} \tag{5}$$

The conversion X was calculated from the evolution of the NaOH consumption given in Figure 2 and the evolution of $(1 - X)^{1/2}$ was plotted as a function of t. The results are presented in Figure 5, and as illustrated, the experimental values followed a linear equation as predicted by the model. The value of t_{Final} calculated from the slope of the curves was 11×10^3 minutes.

Additional experiments were attempted, using cellulose whiskers extracted from the rachis of the from *P. dactylifera* palm. Here, the advantage was to have a pure cellulose substrate with a known length and a squarish cross-section. Consequently, the shape of the cellulose whiskers could be modeled as that of a cylinder (Elazzouzi-Hafraoui et al., 2008). The results presented in Figure 5 are also compatible with a heterogeneous model and clearly highlight that cellulose whiskers behave in a manner analogous to that of DPLF. However for the whiskers, the t_{Final} was higher (1.1×10^4 minutes) than for DPLF which caused the specific rate r_{is} of the TEMPO-mediated oxidation to decrease. The difference can be ascribed to high reactivity of DPLF which contain lignin, hemicelluloses and amorphous cellulose being more reactive than cellulose whiskers.

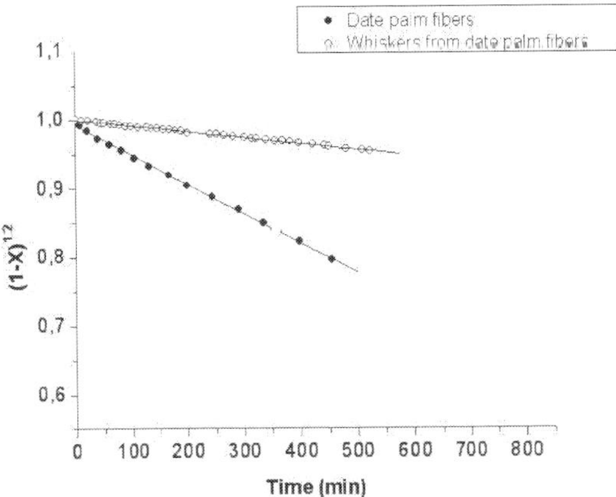

Figure 5. Modeling of the kinetics of TEMPO-mediated oxidation for (♦) DPLF: $(1-X)^{1/2} = 1 - 4{,}93 .10^{-4}t$ and (o) cellulose whiskers from data palm rachis: $(1-X)^{1/2} = 1 - 8{,}71 .10^{-5}t$.

2.2.3. Topological characterization

From the aforementioned model it can be inferred that the oxidation reaction occurs through a heterogeneous process. To characterize the topology of the chemical changes, X-rays photoelectron spectroscopy (XPS) and energy dispersive X-rays analysis were carried out on DPLF before and after oxidative treatment.

XPS revealed the modifications resulting from the oxidation (Figure 6). Indeed the C1s spectrum of the control product (spectrum a) consists of three major peaks at 284.6 (peak 1), 286.2 (peak 2) and 287.8 eV (peak 3), corresponding to C-C + C-H, C-O and O-C-O (and / or C=O) groups respectively. After oxidation (spectrum b), a new peak can be clearly distinguished at 288.8 eV and is attributed to the O=C-O function, typical of carboxylate group. This signal is overlapped with the one of the C=O group and is probably also present in the C1 spectrum of the control product but at lower level. In addition, a quantitative analysis indicates that the atomic concentration of non-oxidized carbon (peak 1) is of 71.2 % and 64.4 % before and after oxidation respectively. Assuming that XPS analysis only characterize a depth of 10 nm, these results clearly highlight that carboxylate functions are present at the top layers of DPLF.

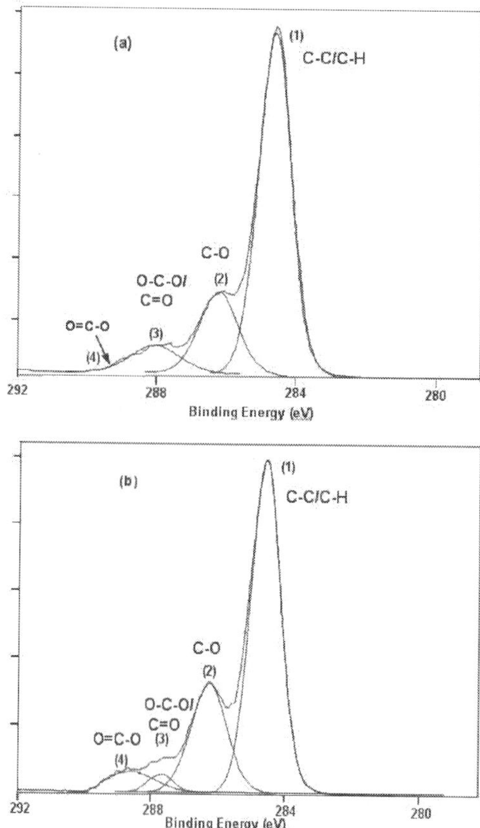

Figure 6. High resolution C1s spectra for DPLF (a) and after (b) TEMPO-mediated oxidation.

The location of carboxylate groups was clearly evidenced by the characterization with SEM/EDX analysis. In Figure 7 are represented typical cross-sections of DPLF showing also the external wall of the fibers before oxidation (Figure 7a), after oxidation (Figure 7b) and after oxidation and calcium ions exchange (Figure 7c). Zones which are colored correspond to the location of the counter-ion of carboxylate functions. The more intense the coloration, the higher is the concentration of the counter-ion and therefore the carboxylate function.

Figure 7a shows the DPLF before oxidation and the coloration represents the mapping of sodium ions due to the presence of 4-O-methyl-glucuronic acid units within the raw fibers as mentioned above. The density of initial carboxylate is low in comparison to the one observed in the case of oxidized fibers (Figure 7b) indicating the success of the oxidation treatment. TEMPO-oxidized fibers treated within an aqueous calcium chloride solution led to fibers-COOCa$^+$ structures (Saito et al., 2005). As shown in Figure 7c, the density of calcium ion is high and comparable to that obtained with sodium.

It can also be noted that carboxylate groups are found to be slightly more dense at the surface of the fibers than inside, whereas the distribution at the surface is homogeneous and the one inside inhomogeneous. Such difference between the core and the surface which is confirmed with both mapping of calcium and sodium ions was never reported before.

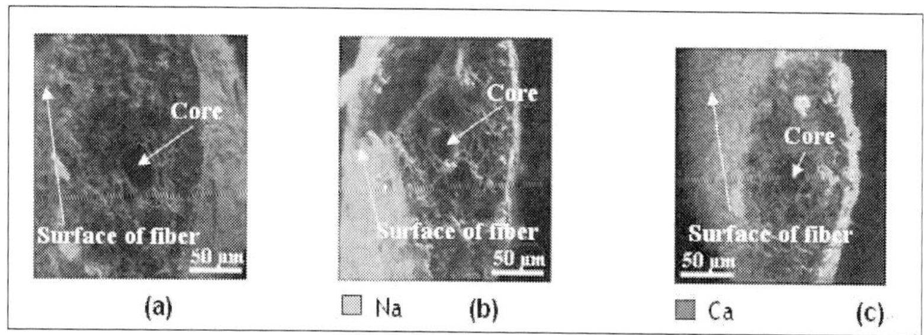

Figure 7. Energy dispersive X-ray analysis on fiber cross-section (a) before oxidation, (b) after oxidation, (c) after oxidation and ion-exchanged with CaCl₂. The colored parts correspond to the presence of the counter-ion of the carboxylate function: sodium in case of 7a and 7b and calcium in case of 7c.

This feature may be ascribed to the presence of hemicelluloses and lignin, which either limit the diffusion of the reagent within the fibers or –which is more probable - migrate to the surfaces after being oxidized. It is also important to underline that the morphology of the fibers remains identical after the TEMPO mediated-oxidation as shown in Figure 7.

3. Part 2: Effect of the oxidation on the processing by RTM process

3.1. Experimental section

3.1.1. Materials

3.1.1.1. Polymer matrix

The polyepoxy matrix was obtained through a polymerization reaction of an epoxy prepolymer with an amine curing agent. The selected epoxy resin was di-glycidyl ether of bisphenol A (DGEBA) (Ref.: LY 556) supplied by Ciba-Geigy, and the curing agent was isophorone diamine (IPD) supplied by Fluka-Chemika. The characteristics of these components are presented in Table 2. The curing was carried out according to the following setup: 2 h at 80 °C, 2 h at 120 °C, and 2 h at 160 °C.

Component	Chemical structure	Characteristics
Prepolymer DGEBA	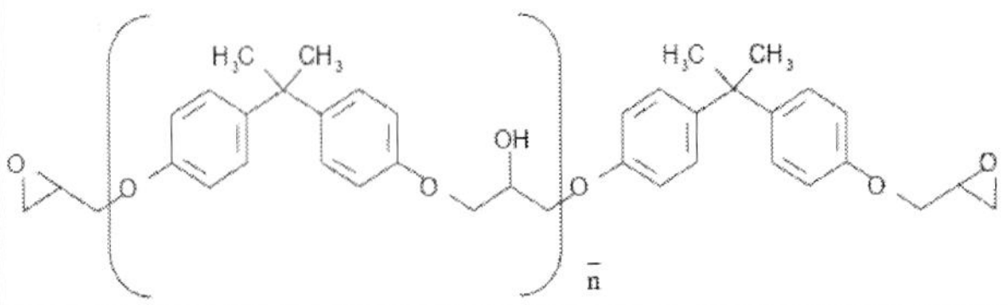	Ciba Geigy LY 556 n = 0.15 M = 380 g/mol d = 1.169 g/cm³ f = 2
isophorone diamine	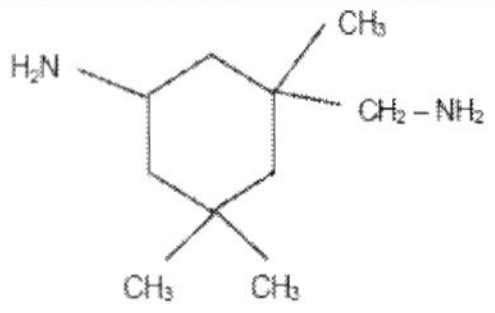	Fluka–Chemika M = 170 g/mol d = 0.92 g/cm³ g = 4

Table 2. Chemical characteristics

3.1.1.2. Reinforcement fiber preparation

The lignocellulosic fibers were obtained by cutting date palm tree leaves into small pieces of approximately 5 cm long and 10 mm wide. The fibers were then extracted for 24 h in a Soxhlet reflux of a solvent mixture composed of acetone/ethanol (75/25). Subsequently, the discolored fibers were dried at room temperature. The used fibers were denoted as unmodified. The length and width of these fibers ranged from 2 to 10 mm, and 0.2 to 0.8 mm, respectively. They were obtained by grinding and sieving the bleached fibers in a 0.1mm-hole sieve to eliminate particles designated as fines, after which they were further sieved through 0.8 mm holes to eliminate bigger fibers.

Fibers oxidation experiments were made under the following conditions. The following samples were oxidized separately. Samples (fiber, cellulose, lignin, and hemicellulose) (2 g, i.e., 12.35 mmol of anhydroglucose units) were dispersed in distilled water (200 ml) for 1 min with a mechanical agitator. TEMPO (32 mg, 0.065 mmol) and NaBr (0.636 g, 1.9 mmol) were added to the suspension, which was maintained at 4 °C. The sodium hypochlorite solution (10 %, 32.17 ml, 43.21 mmol) with pH adjusted to 10 by addition of a 0.1 M aqueous HCl was set at 4 °C by means a thermocontrolled bath. The mixture was then added to the suspension, which was stirred mechanically. The pH was maintained at 10 during the reaction by addition of a 0.5-M NaOH solution. The temperature of the suspension was maintained at 4 °C by means of a thermocontrolled bath during the oxidation reaction. When the reaction time exceeded 12 h, the kinetics became very slow and the solution turned a yellowish white. The reaction was stopped by adding 5 ml of methanol.

The reaction mixture was neutralized to pH 7 with 0.1 M HCl. The oxidized sample was washed with distilled water, after which it was filtered and dried at room temperature. The fiber oxidation was characterized by various methods (IR, conductimetry, solid-state NMR, XPS, MEB, EDX, X-ray diffraction) (Sbiai and al 2010). In the following, the oxidized fibers are referred to as modified fibers.

3.1.1.3. Composite processing

Composites were processed using the resin transfer molding (RTM) method. This process can be divided into four stages: performing, mould filling, curing, and demoulding. The epoxy resin was stored in container A while container B contained the curing agent IPD. The resin mixtures were preheated at approximately 60 °C to reduce the viscosity. The resin was degassed for 20 min to prevent voids from forming during pumping. In container B, IPD was kept at room temperature under an argon atmosphere in order to avoid evaporation and carboxylation. A good circulation of the resin throughout the pump and pipes was necessary. A mold ($100 \times 60 \times 6$ mm^3) made of a composite material, was preheated at 80 °C for 2 h before injection. A continuous mat of date palm tree fibers (either unmodified or modified (oxidized)) used as reinforcement was placed in the mold cavity under isothermal conditions. To observe the flow of the resin during the injection process, a transparent mold made of glass was used under equivalent conditions. A camera was employed to observe the process, which was deemed to have come to an end when the resin was seen to exit from the vent at the other side of the mold. Upon completion of the cure cycle, the solid composite parts were ejected and post-cured under the same conditions as the pure matrices.

3.1.2. Analysis techniques

3.1.2.1. Chemical composition of the fiber

The chemical compositions of the dried date palm tree fibers were determined according to French Standards (NFT12-011). It was thus possible to assess the weight fraction of cellulose, hemicelluloses, and lignin.

3.1.2.2. Scanning Electron Microscopy (SEM)

SEM was used to investigate the morphology of the different types of materials, as well as the filler/matrix interface. The microscope was an ABT-55. The specimens were frozen in liquid nitrogen, fractured, mounted, coated with gold/palladium, and observed using an applied voltage of 10 kV.

3.1.2.3. Differential Scanning Calorimetry (DSC)

A Mettler TA3000 calorimeter was used to measure the glass transition temperature, Tg, which was taken as the onset temperature of the specific heat increment. The heating rate was fixed at 10 °C min^{-1}, and scans were recorded under an argon atmosphere (flowrate10 mL min^{-1}) in a temperature range between -100 and +200 °C.

3.1.2.4. Dynamic Mechanical Analysis (DMA)

DMA experiments were performed with a Rheometrics RDAII, equipped for rectangular samples and working in shear mode. Values of the shear storage, G′, and shear loss, G″, moduli as well as the tangent of the loss angle, $\tan\delta = G''/G'$, were determined. This apparatus was especially dedicated to the study of films and composite materials. The average typical dimensions of the composite samples were 20x4x1 mm³. The tests were performed under isochronal conditions at 1 Hz, and each sample was heated from -120 to +200 °C at a heating rate of 2K/min. The maximum shear strain was equal to 0.2%.

3.1.2.5. Non-linear mechanical properties

Three-point bending tests were performed according to the international ISO178 standard to determine the flexural strength (MPa), the flexural modulus (GPa), and the total absorbed energy (J) of the composites. The testing machine was a 2/M type supplied by MTS (load cell: 10kN). The samples were parallelepiped bars with dimensions close to 60x10x5 mm³ and the distance between the supports was fixed at 50 mm. Tests were carried out at room temperature, and the data collected on five samples was averaged.

3.2. Results and discussion

3.2.1. Chemical analysis of the fibers

Results of the chemical composition of the different fibers are presented in Table 3. It can be clearly seen that the chemical oxidation induced a significant decrease of the lignin content and an increase of that of the cellulose and hemicelluloses. This was explained by the oxidation followed by the dissolution of lignin during the TEMPO-mediated oxidation.

Constituent	Raw dried palm tree fibers (wt %)	Oxidized fibers (wt %)
Cellulose	35 %	46 %
Hemicelluloses	28 %	34 %
Lignin	27 %	12 %

Table 3. Chemical Composition of Date Palm Tree Fibers Before (raw dried palm tree fibers) and After (modified (oxidized) fibers) the Chemical Oxidative Treatment (Sbiai et al. 2010).

3.2.2. Observation of resin flow during composite preparation

With regard to the observation of front displacement in the mat during the RTM experiments, there was a large difference between the two kinds of fibers. In fact, in the case of unmodified fibers, the front displacement was slow and heterogeneous, whereas in the case of the oxidized fibers, it was faster and homogeneous. This difference is portrayed in Fig. 8, presenting the photos of the fronts taken after 15 seconds of resin injection. In fact, the distance covered by the resin front was higher in the case of the oxidized fibers. On the other hand, the mat of oxidized fibers was homogeneously traveled by the resin, as compared to the mat of unmodified fibers.

These differences could be explained by variations in compatibility between the resin and the filler in the two systems, giving rise to a difference of interaction at the resin/fiber interface. In fact, the carboxylic groups at the fiber surface, in addition to the low amount of lignin in the case of the modified fibers, helped increase the affinity of the epoxy resin with the oxidized fibers. These observations were very important for the control of the process. On the other hand, one can predict some effects of the fiber oxidation on the morphologies and the properties.

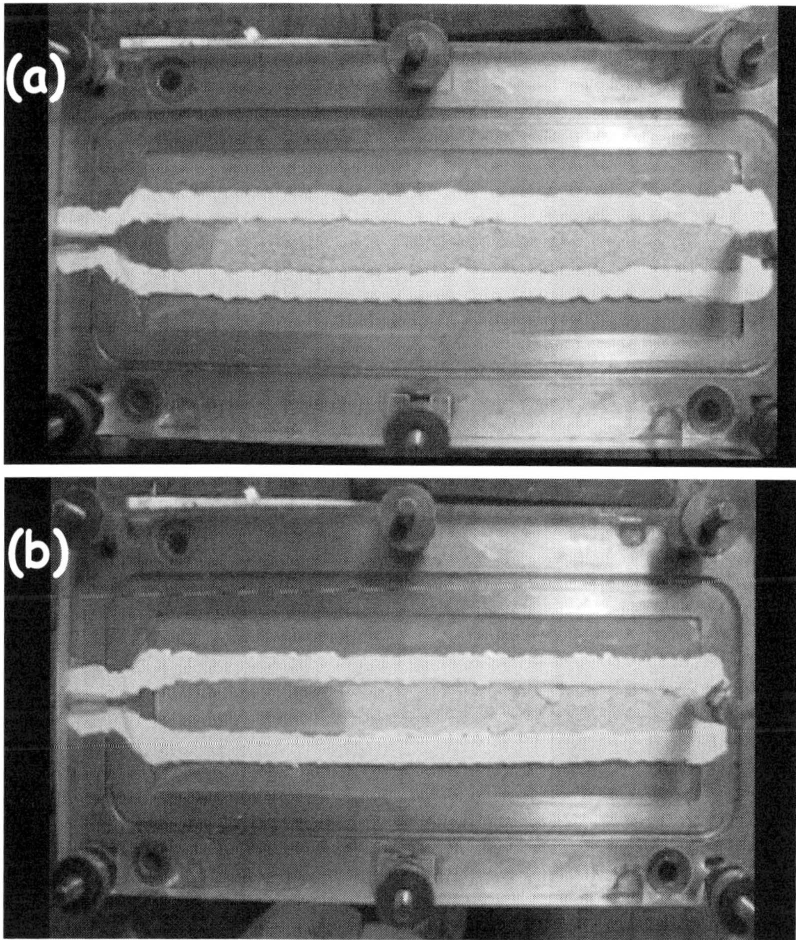

Figure 8. RTM experiment: the resin front after 15 second of injection (at T° = 25°C and P = 1.5 bar) on a mat of **(a)** unmodified and **(b)** TEMPO-mediated oxidized date palm tree fibers

3.2.3. Morphological investigation of the interfaces

Figures 9 and 10 show SEM micrographs of freshly fractured surfaces of composite materials based on the polyepoxy matrix filled with unmodified and modified fibers, respectively. Reinforced materials were investigated. For each composite material, at least tree magnifications were used to reveal the effect of the fiber treatment on the interfacial adhesion. For the unfilled material, i.e. the thermoset matrix (Fig 9-a), the fracture surface

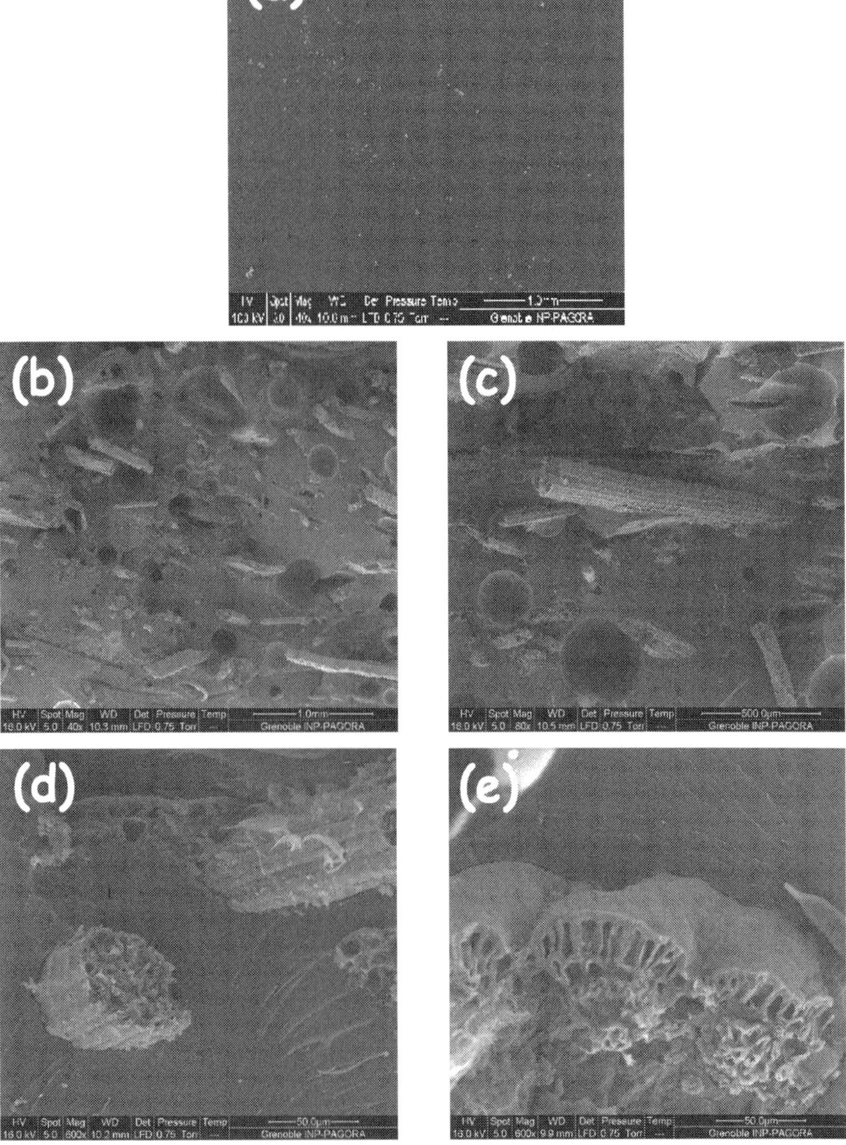

Figure 9. Scanning electron micrographs of freshly fractured surfaces of polyepoxy /unmodified fiber composites with (a) 0 wt%, and (b, c, d, e) 10 wt % of unmodified date palm tree fibers at various magnifications

was rather smooth, as could be expected for brittle polymers. By comparing these micrographs with those of the composite materials (Fig 9 b, c, d and e), the fibers could be clearly identified. The SEM micrographs in Fig. 9 indicated that the interfacial adhesion between the filler and the matrix was not very strong in the case of composites based on unmodified fibers. In fact, the fibers were pulled out from the matrix and their surface remained practically clean (see Figs 9-b and 9-c). On the other hand, fracturing the samples did not lead to the palm tree fibers breakage (Figs 9-d and 9-e). However, it is worth noting that the interaction between the unmodified fibers and the matrix was superior to that of the composite constituted of a hydrophobic matrix filled with unmodified fibers, such as unsaturated polyesters, polypropylenes or polyethylenes.

In contrast, for the composites containing modified fibers, the micrographs in Fig. 10 are evidence of a better adhesion between the matrix and the filler. One can observe the absence of holes around the fillers on the fractured surface, i.e. no debonding occurred. Nor was there any breakage of fibers during fracture (See Fig. 10-c and 10-d). On the other hand, the area surrounding the cellulosic filler seemed to be continuous with the matrix phase, and the epoxy resin appeared to be polymerized within the fiber lumens (see Figs. 10c and 10-d). This variation in interfacial adhesion between the composites based on unmodified and modified fibers is attributed to difference in the nature of physico-chemical interactions that can be created at the interface. This difference can be explained from the stronger interaction developed by the carboxylic groups created on the modified fibers. On the other hand, the dissolution of lignin after fiber oxidation gave rise to an increase of the hydrophilic character of the fibers. As a consequence, the wettability of the fiber surface with regard to the epoxy resin - a necessary condition for good interfacial adhesion - was superior in the case of the modified fibers. The introduction of the epoxy resin within the lumen was evidence of this higher thermodynamic affinity between the fibers and the polyepoxy matrix.

3.2.4. Thermal behavior of palm tree fiber-based composite materials

As mentioned above Section 2, the thermal behavior of date palm tree fiber-based composites was investigated by DSC. The glass transition temperatures, Tg's, of these materials are listed in Table 4. The Tg of the unfilled epoxy matrix was around 155°C. Table 3 clearly shows that the introduction of the lignocellulosic fibers led to a decrease in Tg. This decrease was more pronounced in the case of composites based on unmodified fibers.

The decrease in Tg could be explained by an unbalance of the stoichiometric ratio in the matrix as well as in the vicinity of the fibers after mixing with fibers. The fibers could have more affinity with one component as opposed to with another. This resulted in a hindering of the cross-linking process of the polyepoxy resin.

These results were completely opposite to those obtained in the case of composites based on an industrial epoxy matrix (DGEAB (AW106)/Jeffamin (HV953U)) supplied by CIBA – GEIGY. For the latter composites, an increase in Tg was observed after the introduction of the lignocellulosic filler (Kaddami et al 2006; Sbiai et al 2008). This difference could be attributed to the difference of resin and polymerization kinetics.

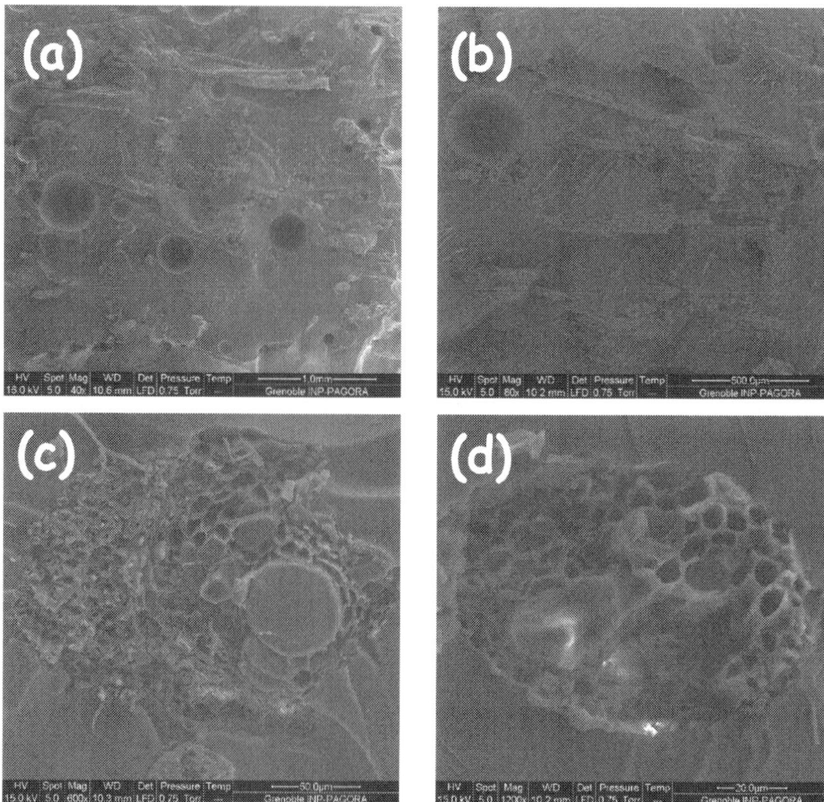

Figure 10. Scanning electron micrographs of freshly fractured surfaces of polyepoxy /modified fiber composites with 10 wt % of modified date palm tree fibers at various magnifications.

Sample		Tg	$T\alpha$	G'_c	G'_c/G'_m
		°C	°C	Mpa	
Neat epoxy		155	150	9.69 (G'e)	1
Composites based on unmodified fibers	5 wt%	147	148	14.4	1.49
	10 wt%	148	148	22.2	2.29
	15 wt%	145	149	25.9	2.67
Composites based on oxidized fibers	5 wt%	147	146	11.5	1.19
	10 wt%	137	145	17.8	1.84
	15 wt%	137	148	16.4	1.69

Table 4. The Glass Transition Temperature, Tg, Determined from DSC Measurements, the Main Relaxation Temperature, Tα, the Rubbery Storage Shear Modulus at Tg + 50 °C, the G'c, of the Composite Materials and the Relative Shear Modulus, G'c/G'm (where G'm refers to the rubbery shear storage modulus of the neat epoxy) Determined from DMA Experiments

3.2.5. Mechanical behavior

The mechanical behavior of all specimens was investigated under both linear (DMA measurements), non-linear conditions (three-point bending experiments), and Charpy impact tests.

3.2.5.1. Dynamical mechanical analysis

The dependence of log G′, i.e. the logarithm of the shear storage modulus, and the loss factor tanδ, vs. the temperature at 1Hz are displayed in Figs. 11 and 12, for composite materials based on unmodified and modified fibers, respectively.

All materials exhibited a relaxation process that was associated with the glass-rubber transition of the matrix, displayed as a sharp decrease in modulus and a concomitant maximum of the loss factor. This relaxation process, denoted α, involved the release of cooperative motions of the chains between crosslinks. The relaxation temperature, $T\alpha$, corresponding to the maximum of the loss factor is listed in Table 3, and was found to be approximately 150°C for all materials. A slight decrease in $T\alpha$ was observed for the composites based on modified fibers; however it was less significant than the one observed for the Tg, as obtained by DSC.

From the dependence of log G′ vs. temperature, it was clear, for both kinds of fibers (oxidized and not oxidized), that the modulus at the rubbery state increased with the fibers content. However, it was difficult to observe any significant effect of the filler at low temperature, i.e. in the glassy state. A simple mixing rule rendered it possible to account for this fact. As is well known, the exact determination of a sample's glassy modulus depends on the precise knowledge of the sample dimensions. On the other hand, the water absorption could affect the exact determination of the glassy modulus. Therefore, the reinforcing effect of the filler was estimated in the rubbery region of the polymer matrix. The values of the rubbery shear modulus are reported in Table 4, as are the relative rubbery modulus values corresponding to the ratio of the rubbery modulus of the composites, G'_c, divided by that of the neat matrix, G'_m . Since the modulus was not perfectly constant as a function of the temperature, the G′ values reported in Table 3 correspond to averages.

For all the composites, the reinforcement effect of the lignocellulosic filler (modified or unmodified) was observed in the rubbery sate. It could be quantified through the values of the relative rubbery modulus, which increased up to 1.84 and 2.67, respectively, for the composites based on the modified and unmodified fibers. The increase in modulus upon filler addition was ascribed to the difference between the modulus of the neat matrix (polyepoxy) and that of the lignocellulosic fibers, as well as to the decent interactions at the interfaces of these composites. No significant effect of the fiber modification was observed on the rubbery modulus despite the fact that TEM microscopy demonstrated the presence of better interactions at the interface in the case of the composites based on modified fibers.

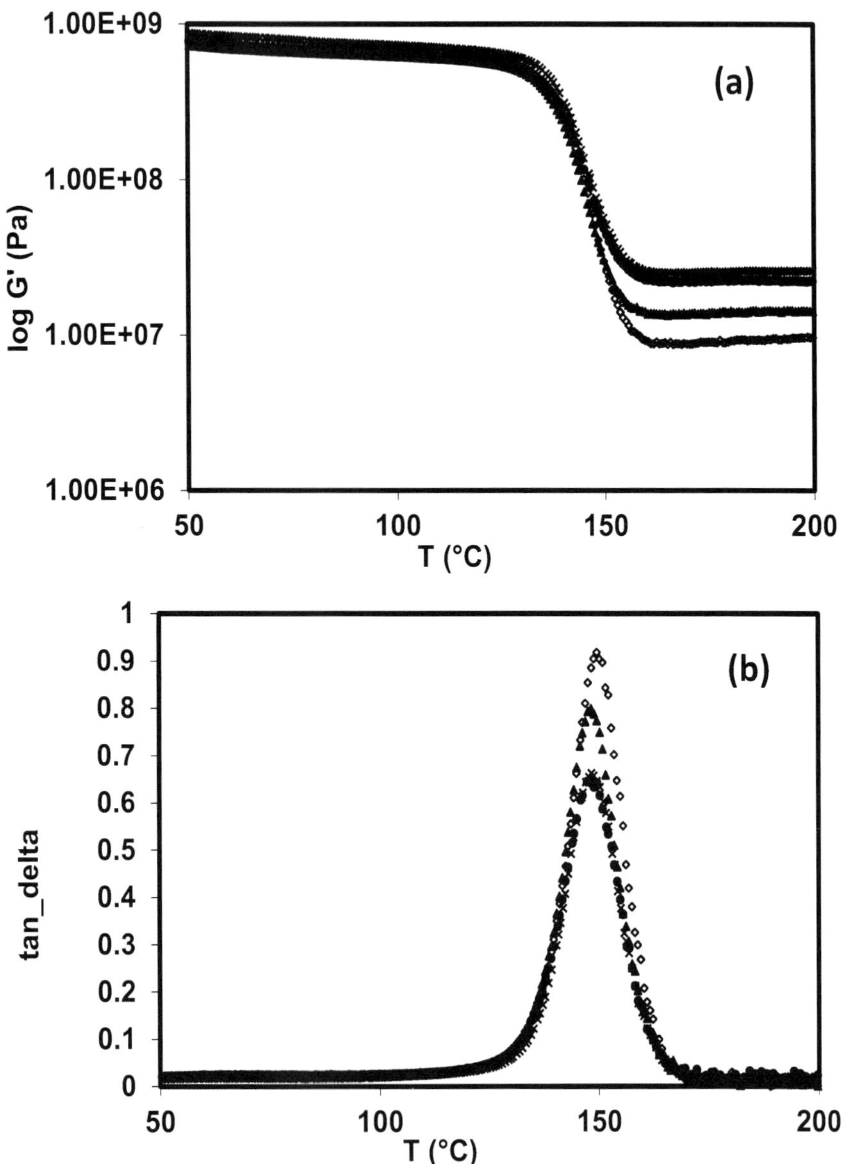

Figure 11. (a) The shear storage modulus G_o, and (b) the loss factor tanδ vs. temperature at 1 Hz for composites based on unmodified date palm tree fibers with (◊) 0, (▲) 5, (●)10 and (X) 15 wt.-% of filler

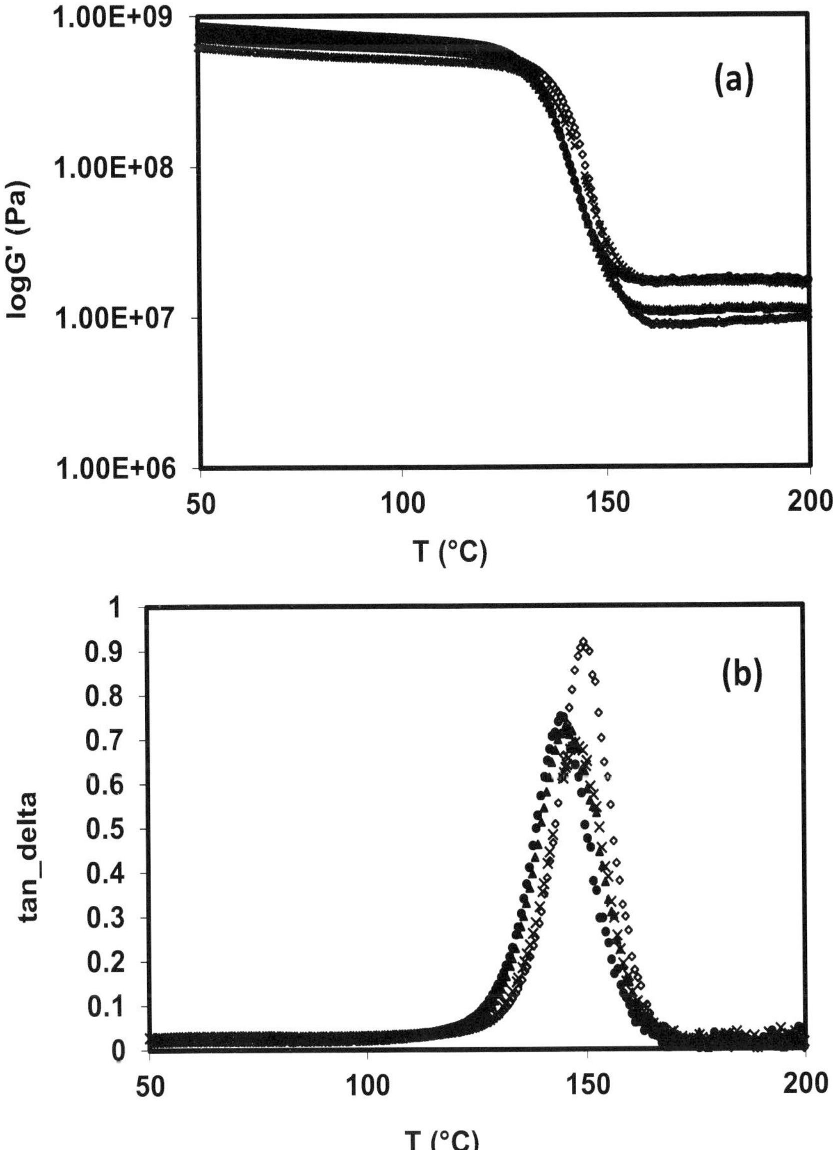

Figure 12. (a) The shear storage modulus G_o, and (b) the loss factor tanδ vs. temperature at 1 Hz for composites based on modified date palm tree fibers with (◊)0, (Δ) 5 , (O) 10, and (X) 15 wt.-% of filler.

3.2.5.2. High strain behavior (three-point bending test)

Storage shear modulus values measured through DMA experiments were determined at room temperature. High strain experiments should provide information on the mechanical properties at the glassy state. Figure 13 gives typical load vs. displacement curves obtained from the three-point bending experiments for the neat polyepoxy matrix and composites filled with 15 wt.-% of modified and unmodified date palm tree fibers. These curves were obtained in the glassy state of the matrix, and the tests were conducted for all materials filled with 5, 10, and 15 wt.-% of modified and unmodified date palm tree fibers. The mechanical properties derived from these experiments are presented in Fig. 14.

Panels a and b of Fig. 14 show the evolution of the shear modulus and the upper yield stress as a function of the filler content. The data were obtained from the three point bending tests. As expected, the composites were more brittle than the neat matrix. The composite material reinforced with the modified filler displayed higher mechanical properties as compared to the composite filled with the unmodified filler. In fact, the composites with modified fibers showed a higher modulus and a higher upper yield stress.

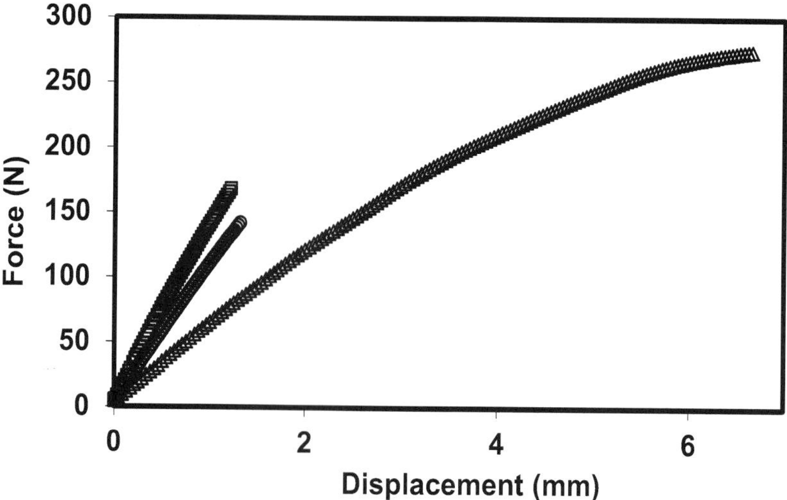

Figure 13. Load versus displacement curves obtained from three-point bending tests performed at room temperature (25 °C) for epoxy-based composites filled with : (Δ) 0, (O)15wt% of non-modified fibers and (◊)15 wt % of oxidized palm tree fibers.

3.2.5.3. Charpy impact tests

Figure 15 shows the results of Charpy impact tests. The absorbed energy at break is presented as a function of the filler content. These tests confirmed that the composites were brittle. In fact, a lower energy was required for breaking the composite materials as compared to the neat matrix. On the other hand, and within the error margins, no significant difference was observed between the composites based on modified and unmodified fibers.

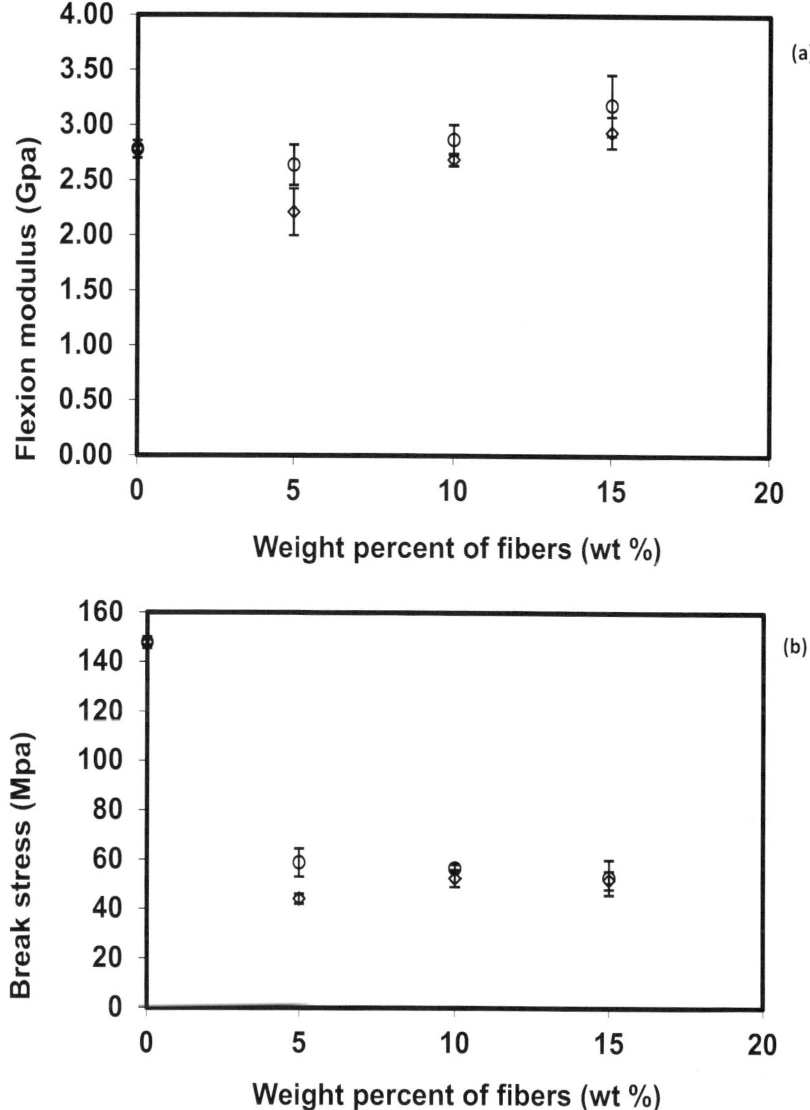

Figure 14. Mechanical properties as functions of the filler content, obtained from three-point bending tests of epoxy based composites filled with (O) modified and (◊) unmodified. a) Shear modulus; b) upper yield stress

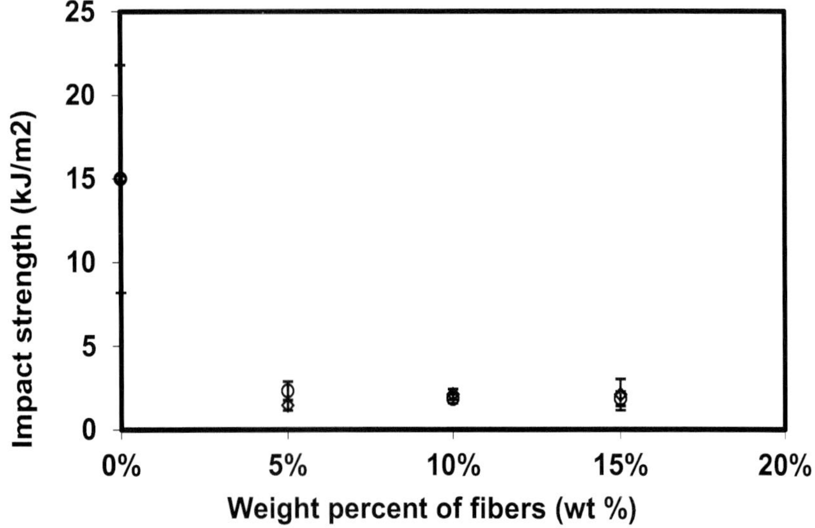

Figure 15. The absorbed energy as a function of the filler content, obtained from Charpy impact tests of epoxy based composites filled with (O) modified and (◊) unmodified fibers fibers

4. Part 3: Rheokinetic study of the reactive composite materials

4.1. Experimental section - Technical Analysis

4.1.1. Differential Scanning Calorimetry (DSC)

A Mettler Thermoanalyser TA3000 operating in the -100°C to 300°C temperature range and equipped with a liquid nitrogen cooling system, was used to determine the enthalpy of reaction, the glass transition temperature T_g, which was taken as the onset temperature of the deflection heat capacity change, and variation of the heat capacity ΔCp through the Tg of the samples. The heating rate was 10 °C min-1 in a nitrogen atmosphere (flowrate10 mL min-1). The T_{g0} and the heat capacity of the initial unreacted mixture ΔCp_0 were determined during a first scan, the fully cured material glass transition temperature $T_{g\infty}$ and its corresponding heat capacity change ΔCp_∞ were obtained after curing at 160 °C with vaccum.

Isothermal cures were examined at various reaction times. The Conversion, x, can be deduced from DSC Scans and calculated from the following equation:

$$X(t) = 1 - \frac{\Delta H_r(t)}{\Delta H_0}$$

Where $\Delta Hr(t)$ residual heat of reaction and ΔH_0 total enthalpy of the reaction.

4.1.2. Rheokinetic study

The gel time was studied by a Couette Rheometer called Rheomat 115 (Cf. Table 5) with a shear rate of 1 s^{-1} and a period of 30 to 60 minutes. The principle of the apparatus is to measure the viscosity of the mixture as a function of time. The sample in liquid form is placed in a cylindrical tube and the bucket used is the one called D.

Mode	Temperature	Sensors	Frequency	Tools
- Viscosity versus shear rate - Time study	$T_{max} = 100°C$	- Viscosity from 1 mPa.s to 5. 10^5 Pa.s - Shear stress from 2 Pa to 35.10^3 Pa	- N : from 5 to 780 min^{-1} - N/100 : from 0,05 to 780 min^{-1}	2 bucket used C & D.

Table 5. Rheometer characteristics

4.2. Results and discussions

4.2.1. Thermal characteristics

4.2.1.1. Epoxy filled by unmodified fibers

To cure thermoset polymers it's important to have information about the thermal properties of the starting monomers but also of the fully cured polymer. On the other hand, in order to understand the final morphology and properties, the study of the evolution of the glass transition during the network formation is of great interest. Thus, the glass transition temperature (Tg) of the epoxy networks based on DPLF fibers (from 5 % wt to 15 % wt) epoxy blends was investigated using DSC analysis. Table 6 lists the experimentally obtained values of Tg_0, Tg_∞, ΔCp_0, ΔH_0 (relative to the mass of the reactive system), ΔCp_∞ and λ (defined as ratio of ΔCp_0 and ΔCp_∞) for each DGEBA/IPD /fibers DPLF blend studied. Not that the ratio is still stoichimetric.

Material	Tg_0 (°C)	ΔCp_0 (J.g^{-1}.K^{-1})	ΔH_0 (J/g)	Tg_∞ (°C)	ΔCp_∞ (J.g^{-1}.K^{-1})	$\lambda = \dfrac{\Delta Cp_\infty}{\Delta Cp_0}$
Neat epoxy	-30	0,57	463	152	0,28	0,49
Epoxy with 5 % wt of unmodified fibers	-29	0,47	420	144	0,28	0,60
Epoxy with 10 % wt of unmodified fibers	-30	0,52	401	135	0,32	0,62
Epoxy with 10 % wt of TEMPO oxidized fibers	-30	0,54	413	119	0,29	0,54

Table 6. Influence of fibers on thermal properties (DSC analysis)

The thermal characteristics found for the neat epoxy are the same order of magnitude as reported by literature (Pichaud, 1997, Pichaud et al. 1999, Nguyen-Thuc 2004). The initial Tg (Tg_0) and final Tg (Tg_∞) of the neat epoxy/amine system were −30 °C and 152 °C, respectively. The total exothermic energy of the polymerization reaction ΔH_0 was 463 J g−1 K−1.

The enthalpies of reaction $\Delta H0$ decrease by introducing an increasing amount of fiber. This decrease can be explained by the change in amine/epoxy stochiometry resulting from etherification reactions. Indeed, the fibers are rich in hydroxyl function and furthermore have an amount of trapped water. Thus it is then possible to envisage a competition between the following reactions:

- Reaction between the oxirane of DGEBA and amine functions of the IPD.
- Reaction between the oxirane of DGEBA and water,
- Etherification reaction of DGEBA oxirane and the fibers hydroxyl functions.

It was not possible to highlight the presence of ether bonds, by Infrared, because the absorption bands are overlapped by those of the cellulose. But the significant decrease of glass transition temperature Tg_∞ observed when DPLF fibers are added could be attributed to the the etherification reactions mentioned above and which have caused the change in local stoichiometry (Garcia-Loera 2002, Pascault et al. 2002). The evolution of the glass transition temperature was also observed in the case of epoxy network modified with thermoplastic (Fernandez et al. 2001). On the other hand, the ratio λ which characterizes the variation of the chain mobility between the crosslinked polymer and initial monomer, varies from 0.49 to 0.62 when the rate of virgin fiber is from 0 to 10 wt% respectively. This is another indication of the evolution of the polymer network stiffness, caused by the introduction of the fibers and indicates that the fibers induce a variation of the stoichiometric ration (oxirane/amine). These evolutions were not observed when epoxy amine system (DGEBA/IPD) was filled by core-shell (probably little or no functionalized) where no/or few reactions were possible between the cross linking polymer and the core-shell (Nguyen-Thuc et al. 2002, 2003 and 2004).

4.2.1.2. Epoxy filled by TEMPO oxidized fibers

In the case of a DGEBA / IPD with modified fibers, we observe similar evolutions of the thermal properties were observed (Tg_∞ decreases and the ratio λ increases). The same discussion could be done for these materials, however the decrease of Tg_∞ is much more pronounced when compared to unmodified fibers. In this case, the system is more complex compared to the composites of unmodified fibers because the additional reaction between carboxylic acid and oxirane functions. However it's worthy noticing that the oxidized fibers are more hygroscopic than the unmodified one. Actually, the remaining water fraction after drying (at 105°C overnight) is about 8wt% for the oxidized fibers and 6wt% for the unmodified fibers. This result agrees with this obtained by Trindad et al who compared the moisture in the fibers of sugarcane bagasse before (9.5%) and after modification (11%) by oxidation with sodium periodate (Trindad et al. 2004). This remaining amount of water will induce a variation of the stoichiometry and catalyses the etherification reactions (Sherman et al. 2008). This higher amount of remaining water could explain the more pronounced decrease of Tg_∞ in the case of the composites with oxidized fibers.

4.2.2. Kinetic behaviour

The reaction kinetics of the DGEBA / IPD reactive system (r = 1) filled with 0 and 10 wt% unmodified fibers and with oxidized fibers, was studied by DSC. The increase in conversion (the extent of reaction) was monitored as a function of time and curves are plotted in fig. 16.

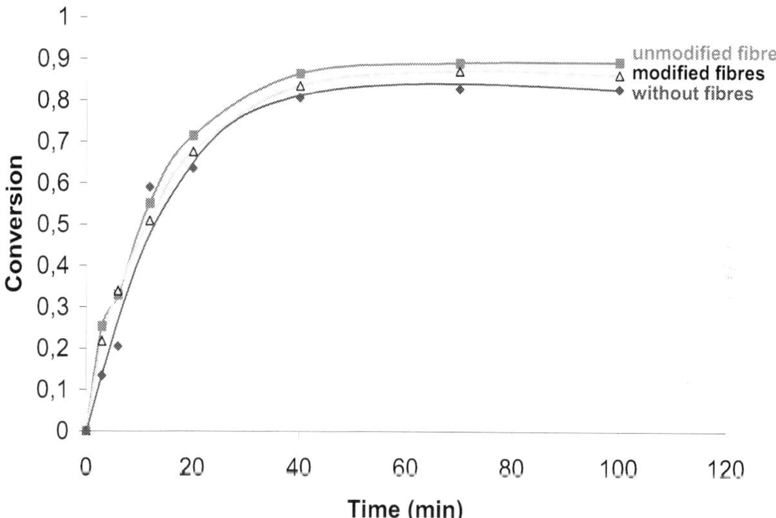

Figure 16. Evolution of Conversion x versus time of reaction at 80 °C of system DGEBA / IPD (♦) without fiber (■) with 10% wt of unmodified fibers and (Δ) with 10% wt of TEMPO oxidized fibers

Conversion values are higher for systems with PLD fibers. The polymerization kinetics is enhanced through the presence of these fibers. Indeed the introduction of hydroxyl groups (OH, COOH, H2O ...) in the reaction medium promotes interactions between the epoxy-amine and other nucleophilic molecules leading to the formation of an intermediate complex (Fig 17.) (Rozenberg 1986, Garcia-Lorea 2002).

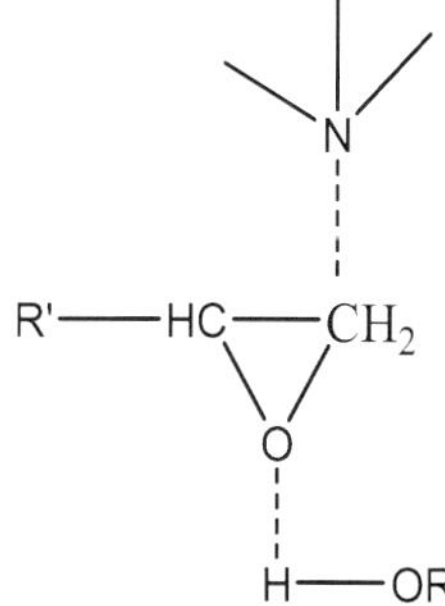

Figure 17. Trimolecular complex catalyst formed by hydroxyl functional groups

The formation of this complex makes nucleophilic attack of the primary amine easier and therefore the reaction is accelerated. Indeed, DPLF fibers added to the system contain hydroxyl groups in the case of unmodified fibers, and more acid groups in the case of oxidized fibers. Furthermore a significant amount of water is present (humidity between 6 and 8%wt) which has an additional catalytic effect in the reaction medium. The presence of water seems to be the dominant factor more than acid groups which are presented in small amounts in the oxidized fibers. Garcia-Lorea and al. (Garcia-Lorea 2002) studied the catalytic effect of water on the reaction kinetics of the system DER332 DGEBA / Jeffamine D400. They showed that the kinetics accelerates with the rate of water incorporated into the system epoxy / amine. If we compare the evolution of conversion in our case with the results, we can decide on the strong effect of humidity on the kinetics of fiber. Furthermore the rate of water incorporated studied by Garcia-Lorea (6-10%) is similar to the humidity in DPLF fibers (6-8%) (Garcia-Loera 2002).

The isothermal cure reaction of DGEBA–IPD networks at various cure temperatures was studied. The built-up in Tg and the extent of conversion x, during cure were monitored as the crosslinking reaction progressed under isothermal conditions (Pascault et al. 1990). The Tg –x relationship could be expressed based on Dibenedetto's formula. Fig 18 presents the Tg vs conversion for the different DGEBA/IPD /DPLF fiber systems. This evolution is predicted by the model based on Dibenedetto's approach modified by Pascault and William (Pascault et al. 1990) from an extension of the Couchman equation (Couchman 1987).

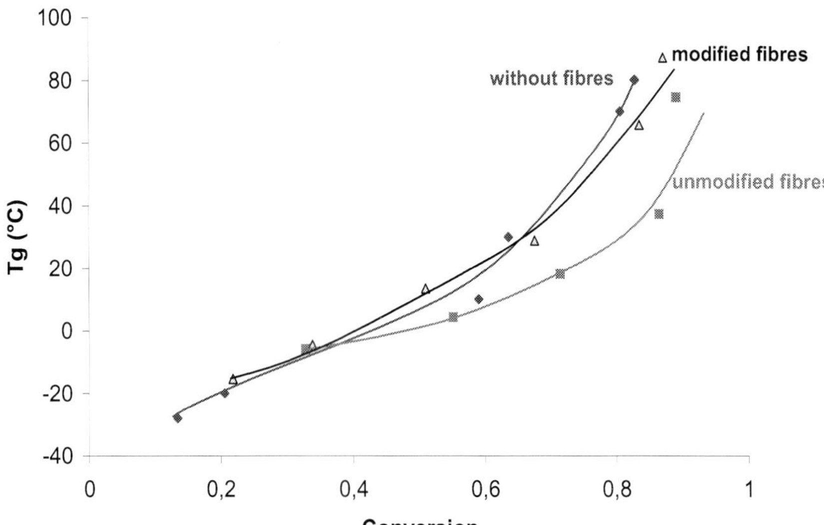

Figure 18. Dependence of Tg on conversion x of DGEBA/IPD system at 80 °C of system DGEBA / IPD (♦) without fiber (■) with 10% wt of unmodified fibers and (Δ) with 10% wt of TEMPO oxidized fibers.

$$\frac{Tg - Tg_0}{Tg_\infty - Tg_0} = \frac{\lambda x}{1 - (1 - \lambda)x}$$

They showed that the adjustable parameter λ is equal to the ratio $\Delta Cp_\infty / \Delta Cp_0$, where ΔCp_0 and ΔCp_∞ are respectively the heat capacities of the initial mixture and of the fully cured network.

This curve is very important to understand the kinetics of such systems. The correlation between Tg and X characterizes the structure of the reactive system. Three curves are obtained which correspond to the system DGEBA / IPD without fiber, DGEBA / IPD with 10% DPLF fiber and DGEBA / IPD with 10% TEMPO oxidized fiber. At low conversions the Tg values for the system without fiber are lower than those charged by fiber (modified or unmodified). This is due to the nature of chemical reactions that occur. The kinetics are different and the etherification reaction seems to be favoured by the abundance of reactive species (OH fiber, water, and COOH in the case of modified fibers) that catalyzes the crosslinking reaction. From 50% of conversion Tg values of the system without fibers are superior to those systems with fibers (modified or unmodified). In this interval of Conversion, the decrease of Tg was due to the etherification reactions that change the value of stoichiometric ratio (not equal to unity) (Garcia-Loera 2002).

Comparing the two systems, with 10% DPLF fibers and that with 10% TEMPO oxidized fibers, we note that the Tg values are higher for the system based oxidized fibers. This can be explained by the humidity slightly higher in the oxidized fiber and by the existence of carboxylic acid. These conditions make the interpretation of the kinetics very difficult because of competition between many reactive species. It's worthy noticing that in composite materials where the filler doesn't react with the croslinking polymer, Tg-X curves are coincident which indicate that the reaction involved in the croslinking process are the same (Nguyen-Thuc et al. 2003). Thus in our case the fact of the Tg-X curves are not coincident is a proof that the reactions involved in the croslinking process are not the same in the three studied systems and the functional groups of the fibers and water molecules are involved in the croslinking process.

4.2.3. Rheokinetic study

We were interested to determine the gelation time of the crosslinking material and to evaluate the effect of fiber on the gel time. The gelation time of the systems DGEBA / IPD with and without DPLF fibers were studied at different temperatures: 60, 70, 80 and 90 °C (Fig.19). These temperatures were chosen above the Tg_{gel} to avoid vitrification (Glass transition) before gelation. According to the work of Pichaud and al (Pichaud et al. 1999) the Tg_{gel} is 32°C for the system DGEBA / IPD.

The gel times determined for the system DGEBA / IPD without fiber are respectively about 30, 14, 9, 5 and 7 minutes at 60, 70, 80 and 90 °C. This result is in parfait agreement with the results found by Pichaud (Pichaud et al. 1999) with the same system. The fibers induce a reduction of the gelation time and this effect is more exacerbated at mow temperature (60°C). The fibers promote the reaction by the presence of various reactive species, including the catalytic effect which promotes the etherification reactions. This effect is less detectable at high temperatures where the reaction kinetics is much faster.

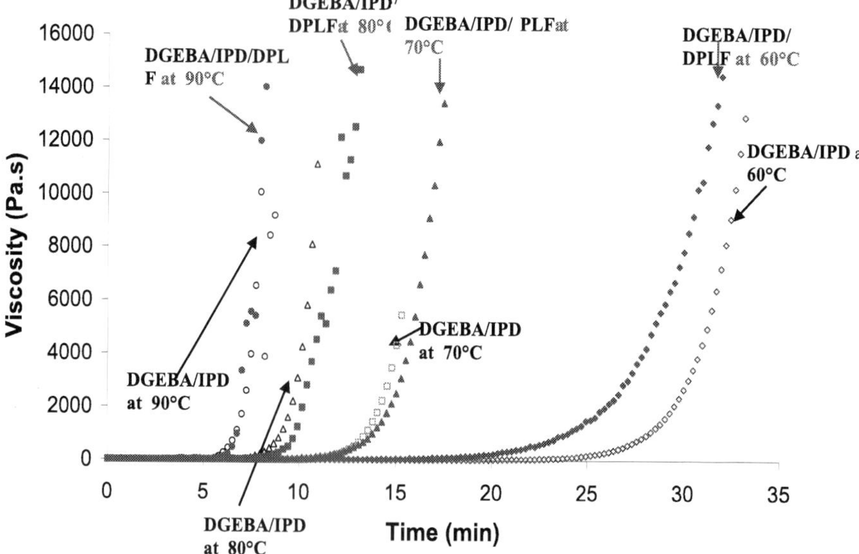

Figure 19. Viscosity vs reaction time for the two systems based DGEBA / IPD without fiber and with unmodified fiber to 5% by weight at 60 °, 70 °, 80 ° and 90 °C

The gelation phenomenon was found to obey an Arrhenius law. The value of the activation energy for the system DGEBA / IPD is about 59 kJ / mol. This value is comparable to that obtained in the literature (Ea = 61kJ/mol by Pichaud and al. (Pichaud et al. 1999)). For the system DGEBA / IPD with DPLF fiber, the activation energy is much lower and equal to 47 kJ / mol. This shows that the mechanisms of crosslinking reactions for the system epoxy / amine are not identical because others reactions are induced by the fibers.

4.2.3.1. Effect of TEMPO oxidized fibers on the gel time

The fig. 20 below shows the evolution of viscosity versus time for the three systems DGEBA / IPD without fibers (EP), DGEBA / IPD with 5 wt% DPLF fiber and DGEBA / IPD with 5 wt % TEMPO oxidized fibers.

DPLF Fibers into system DGEBA / IPD promotes the reaction kinetics. Nevertheless, viscosity versus time curve of epoxy DGEBA/IPD with oxidized fibers exhibit similar behavior as that with unmodified fibers and the gelation time is the same for both systems (27 min).

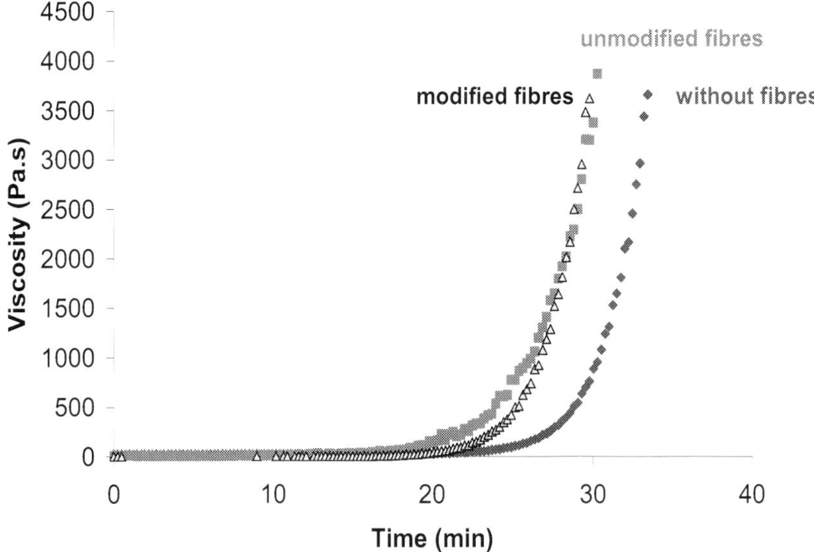

Figure 20. Evolution of the viscosity vs reaction time of DGEBA/IPD system at 60 °C (♦) without fiber (■) with 10% wt of unmodified fibers and (Δ) with 10% wt of TEMPO oxidized fibers.

5. Conclusions

The results presented in this study show that date palm lignocellulosic fibers with high lignin and hemicelluloses contents have been successively modified by a classical TEMPO-mediated oxidation process. Despite the composite character of the fibers substrate, kinetic results have proved that oxidation has occurred in a heterogeneous manner. In contrast to previous studies, it was possible to perform this efficient and aqueous chemoselective reaction without destruction of the fibers structure, while keeping a large amount of residual lignin and hemicelluloses. It was further demonstrated that the distribution of the carboxylate group was unconventional, with an inhomogeneity between the core and the surface of the fibers probably due to a migration of the oxidized and partially hydrosoluble product.

The kinetic study by DSC enables to determine the evolution of conversion degree as function of the reaction time. It was also shown that the polymerization kinetics were accelerated in the presence of the fibers. This catalytic effect is more important in the case of the oxidized fibers. This was explained by the presence of water adsorbed by the fibers but also the catalytic effect of the carboxyl groups in the case of the oxidized fibers. However, even the presence of the fibers acceleration the gelation of the system, no effect of the oxidation of the fibers on the gelation time was detected.

When studying the effect of the oxidation on the processing of the composites, it was shown that the preparation of composites using RTM process was facilitated in the case of composites based on oxidized fibers. During the process, the front displacement of injected resin was regular, homogeneous and faster in the case of oxidized fibers. The morphology, thermal and mechanical properties of polyepoxy reinforced with lignocellulosic fibers extracted from date palm trees were also investigated. Thermal properties from DSC measurements showed that the glass transition temperature of the composites, mainly those based on oxidized fibers, was lower than that of the neat matrix. Dynamic mechanical analysis showed a significant increase of the rubbery modulus when lignocellulosic, unmodified and oxidized fibers were introduced into the polymer. No significant difference of the rubbery modulus between the two families of composites was observed. Analysis of the high strain mechanical proprieties (three-point bending tests) demonstrated some reinforcement of the oxidized fibers as compared to their unmodified counterparts. This confirmed the microscopic analysis which pointed at a better adhesion at the fiber/matrix interface in the case of the composites comprising the oxidized fibers.

Author details

Adil Sbiai[1,2], Abderrahim Maazouz[2], Etienne Fleury[2], Henry Sautereau[2] and Hamid Kaddami[1]

[1]*Université Cadi Ayyad, Laboratoire de Chimie Organométallique et Macromoléculaire – Matériaux Composites, Faculté des Sciences et Techniques-Guéliz, Morocco*

[2]*INSA, Laboratoire des Matériaux Macromoléculaires / IMP, UMR CNRS n 8 5223, France*

Acknowledgement

The authors thank for their financial support the Hassan II Academy of Sciences and Technologies - Morocco and and the French Ministry ofForeign Affairs (Corus program 6046).

6. References

Abdelmouleh M, Boufi S, Salah BA, Belgacem M, Naceur MN, Gandini A (2002) Interaction of silane coupling agents with cellulose. Langmuir 18:3203-3208.

Abu-Sharkh, B. F., and Hamid, H. (2004). "Degradation study of date palm fibre/polypropylene composites in natural and artificial weathering" *Polym Degrad Stab* 85, 967-973.

Bendahou A, Dufresne A, Kaddami H, Habibi Y (2007) Isolation and structural characterization of hemicelluloses from palm of Phoenix dactylifera L. Carbohydr Polym 68: 601–608.

Bendahou A, Kaddami H, Sautereau H, Raihane M, Erchiqui F, Dufresne A (2008) Short palm tree fibers poly olefin composites: Effect of filler content and coupling agent on physical properties. Macromol Mater Eng 293:140-148.

Bendahou A, Habibi Y, Kaddami H, Dufresne A (2009) Physico-chemical characterization of palm from Phoenix Dactylifera-L, preparation of cellulose whiskers and natural rubber-based nanocomposites. J Biobased Mater Bioenergy 3:81-90.

Biagiotti J, Puglia D, Torre L, Kenny JM, Arbelaiz A, Marieta C, Llano-Ponte R, Mondragon I (2004) A systematic investigation on the influence of the chemical treatment of natural fibers on the properties of their polymer matrix composites. Poly Composite 25:470-479.

Bidstrup, S. A., Macosko, C.W. (1986) Structure-Rheology relations for model epoxy networks. 31st International SAMPE Symposium and Exhibition 551-562.

Bin S, Chunju G, Jinhong M, Borun L (2005) Kinetic study on TEMPO-mediated selective oxidation of regenerated cellulose. Cellulose 12: 59-66.

Bledzki AK, Gassan J.(1999)Composites reinforced with cellulose based. Prog Polym Sci 24:221-274.

Chang, P. S., and Robyt, J. F. (1996). "Oxidation of primary alcohol groups of naturally occurring polysaccharides with TEMPO ion," *Carbohydrate Chemistry* 15, 819-830.

Chu, J. (2003) "Development of an intelligent injection system and its application to resin transfer molding," Ph. D. Thesis, McGill University, Montreal, Canada.

Couchman, P. R. (1987) Thermodynamics and the compositional vibration of glass transition temperature, Macromol. 20 : 1712-1717.

Cousin, P., Bataille, P., Schreiber, H. P., and Sapieha, S. (1989). "Cellulose-induced crosslinking of polyethylene," *Journal of Applied Polymer Science* 37(10), 3057-3060.

Coutts, R. S. P., and Campbell, M. D. (1979). "Coupling agents in wood fibre-reinforced cement composites," *Composites* 10(4), 228-232.

Dang Z, Zhang J, Ragauskas AJ (2007) Characterizing TEMPO-mediated oxidation of ECF bleached softwood kraft pulps. Carbohydr Polym 70:310–317.

Da Silva Perez D, Montanari S, Vignon MR (2003) TEMPO-Mediated Oxidation of Cellulose III. Biomacromolecules 4:1417-1425.

Davis, N. J., and Flitsch, S. L. (1993). "Selective oxidation of monosaccharide derivatives to uronic acids," *Tetrahedron Letters* 34, 1181-1184.

De Nooy AEJ, Besemer AC, van Bekkum H (1995) Selective oxidation of primary alcohols mediated by nitroxyl radical in aqueous solution. Kinetics and mechanism. Tetrahedron 51: 8023–8032.

Duanmu J, Gamstedt EK, Rosling A (2007) Hygromechanical properties of composites of crosslinked allylglycidyl-ether modified starch reinforced by wood fibres. Compos Sci Technol 67:3090-3097.

Elazzouzi-Hafraoui S, Nishiyama Y, Putaux JL, Heux L, Dubreuil F, Rochas C (2008) The shape and size distribution of crystalline nanoparticles prepared by acid hydrolysis of native cellulose. Biomacromolecules 9:57-65.

Eloundou, J. P., Feve, M., Gerard, J.F., Harran, D., Pascault, J.P. (1996) Temperature dependence of the behavior of an epoxy-amine system near the gel point through viscoelastic study. 1. Low-T_g epoxy-amine system. Macromol. 29(21) :6907-6917.

Eloundou, J. P., Gerard, J.F., Harran, D., Pascault, J.P. (1996) Temperature dependence of the behavior of a reactive epoxy-amine system by means of rheology. 2. High-T_g epoxy-amine system. Macromol. 29(21): 6917-6927.

Felix, J. M., and Gatenholm, P. (1991). "The nature of adhesion in composites of modified cellulose fibers and polypropylene," *Journal of Applied Polymer Science* 42(3), 609-620.

Felix, F., Gatenholm, P., and Schreiber, H. P. (1994). "Plasma modification of cellulose fibers: Effects on some polymer composite properties," *Journal of Applied Polymer Science* 51(2), 285-295.

Fernandez, B., Corcuera, M.A., Marieta, C., Mondragon, I. (2001) Rheokinetic variations during curing of a tetrafunctional epoxy resin modified with two thermoplastics. Eur.Polym. J. 37:1863-1869.

Flink, P., Westerlind, B., Rigdahl, M., and Stenberg, B. (1988). "Bonding of untreated cellulose fibers to natural rubber," *Journal of Applied Polymer Science* 35(8), 2155-2164.

Fukuzumi, H., Saito, T., Iwata, T., Kumamoto, Y., and Isogai, A. (2009). "Transparent and high gas barrier films of cellulose nanofibers prepared by tempo oxidation," *Biomacromol.* 10, 162-165.

Gandini A (2008) Polymers from renewable resources: A challenge for the future of macromolecular materials. Macromolecules 41:9491-9504.

Garcia-Loera (2002) A. Mélanges réactifs thermodurcissable / Additifs extractibles : Phénomènes de séparation de phase et morphologies. Application aux matériaux poreux, Thèse INSA - Lyon, 161 p.

Gardinera, E. and Cabasso, I. (1987). "On the compatibility and thermally induced blending of poly(styrene phosphonate diethyl ester) with cellulose acetate," *Polymer* 28(12), 2052-357.

Gomez-Bujedo S, Fleury E, Vignon MR (2004) Preparation of cellouronic acids and partially acetylated cellouronic acids by TEMPO/NaClO oxidation of cellulose acetate. Biomacromolecules 5:565-571.

González-Sánchez C, González-Quesada M, De La Orden MU, Urreaga JM (2008) Comparison of the effects of polyethylenimine and maleated polypropylene coupling agents on the properties of cellulose-reinforced polypropylene composites. J Appl Polym Sci 110:2555-2562.

Goussé C, Chanzy H, Cerrada ML, Fleury E (2004) Surface silylation of cellulose microfibrils: Preparation and rheological properties. Polymer 45:1569-1575.

Habibi Y, Chanzy H, Vignon MR (2006) TEMPO-mediated surface oxidation of cellulose whiskers. Cellulose 13:679-687.

Halley, P. J., Mackay, M.E. (1996) Thermorheology of thermosetting-An overview. Polym. Eng. Sci. 36(5) : 593-609.

Han Y, Law K-N, Daneault, C, Lanouette R(2008) Chemical and mechanical techniques for improving the papermaking properties of jack pine TMP fibers. Tappi J 7: 13-18.

Isogai, A., and Kato, Y. (1998). "Preparation of polyuronic acid from cellulose by TEMPO-mediated oxidation," Cellulose 5, 153-164.

Isogai, A., and Saito, T. (2005). "Ion-exchange behavior of carboxylate groups in fibrous cellulose oxidized by the TEMPO-mediated system," Carbohydrate Polymers 61(2), 183-190.

Isogai, T., Yanagisawa, M., and Isogai, A. (2005). "Degrees of polymerization (DP) and DP distribution of cellouronic acids prepared from alkali-treated celluloses and ball-milled native celluloses by TEMPO-mediatedoxidation," Cellulose 16, 117-127.

Iwakura, Y., Kurosaki, T., and Imai, Y. (1965). "Graft copolymerization onto cellulose by the ceric ion method," Journal of Polymer Science Part A: General Papers 3(3), 1185-1193.

Jutier, J.-J., Lemieux, E., and Prud'homme, R. E. (1988). "Miscibility of polyester/ nitrocellulose blends: A DSC and FTIR study," Journal of Polymer Science - Part B: Polym. Phys. 26(6), 1313-1329.

Kaddami H, Dufresne A, Khelifi B, Bendahou A, Taourirte M, Raihane M, Issartel N, Sautereau H, Gérard JF, Sami N (2006) Short palm tree fibers - Thermoset matrices composites. Compos Part A-Appl S 37:1413 – 1422.

Kali ński, R., Galeski, A., and Kryszewski, M. (1981). "Low-density polyethylene filled with chalk and liquid modifier," Journal of Applied Polymer Science 26(12), 4047-3058.

Le Guen MJ, Newman RH (2007) Pulped Phormium tenax leaf fibres as reinforcement for epoxy composites. Compos Part A-Appl S 38:2109-2115.

Levenspiel O (1972) Fluid-particle Reactions in Chemical Reaction Engineering Oregon State University 357-373.

Li Z, Renneckar S, Barone JR (2010) Nanocomposites prepared by in situ enzymatic polymerization of phenol with TEMPO-oxidized nanocellulose. Cellulose 17:57-68.

Manrich, S., and Agnelli, J. A. M. (1989). "The effect of chemical treatment of wood and polymer characteristics on the properties of wood-polymer composites," Journal of Applied Polymer Science 37(7), 1777-1790.

Mao L, Law K-N, Daneault C, Brouillette F (2008) Effects of carboxyl content on the characteristics of TMP fibers. Ind Eng Chem Res 47: 3809-3812.

Mao L, Ma P, Law K, Daneault C, Brouillette F (2010) Studies on kinetics and reuse of spent liquor in the TEMPO-mediated selective oxidation of mechanical pulp. Ind Eng Chem Res 49:113–116.

Martins MA, Forato LA, Mattoso LHC, Colnago LA (2006) A solid state 13C high resolution NMR study of raw and chemically treated sisal fibers. Carbohydr Polym 64:127-133.

Michell, A. J., Vaughan, J. E., and Willis, D. (1978). "Wood fiber-synthetic polymer composites. II. Laminates of treated fibers and polyolefins," Journal of Applied Polymer Science 22(7), 2047-2061.

Mishra S, Mohanty AK, Drzal LT, Misra M, Hinrichsen G (2004) A review on pineapple leaf fibers, sisal fibers and their biocomposites. Macromol Mater Eng 289:955-974.

Moharana, S., Mishra, S. B., and Tripathy, S. S. (1990). "Chemical modification of jute fibers. I. Permanganate-initiated graft copolymerization methyl methacrylate onto jute fibers," Journal of Applied Polymer Science 40(3-4), 345-357.

Montanari S, Roumani M, Heux L, Vignon MR (2005) Topochemistry of carboxylated cellulose nanocrystals resulting from TEMPO-mediated oxidation. Macromolecules 38:1665-1671.

Nguyen-Thuc, B. H., Maazouz, A.,(2002) Morphology and rheology relationships of epoxy/core-shell particles blends. Polym. Eng. Sci. 42:120-133

Nguyen-Thuc, B. H. (2003) Etude rhéocinétique et mécanique des réseaux époxydes modifiés par des élastomères. Mise en forme par le procédé RTM. Thèse INSA-Lyon, 181 p.

Nguyen-Thuc, B. H., and Maazouz, A. (2004). "Elastomer- modified epoxy/amine systems in a resin transfer moulding process," Polymer International 53, 591-602.

Octeau, M.-A. (2001). "Composite bicycle fork design for vacuum assisted resin transfer moulding," Ph. D. Thesis, McGill University, Montreal, Canada.

O'Flynn,J. (2007) "Design for manufacturability of a composite helicopter structure made by resin transfer moulding," Ph. D. Thesis, McGill University, Montreal, Canada.

Okita Y, Saito T, Isogai A (2009) TEMPO-mediated oxidation of softwood thermomechanical pulp. Holzforschung 63: 529-535.

Okita Y, Saito T, Isogai A (2010) Entire surface oxidation of various cellulose microfibrils by TEMPO-mediated oxidation. Biomacromolecules 11:1696-1700.

Pascault, J. P., Williams, R. J. J. (1990) Glass transition temperature versus conversion relashionship for the thermosetting polymers. J. Polym. Sci. Part B : Polymer Physics 28 : 85-95.

Pascault, J. P., Sautereau, H., Verdu, J., Williams, R. J. J. (2002) Thermosetting polymers. New York & Basel, Ed. Marcel DEKKER. 2002.

Paunikallio T, Suvanto M, Pakkanen TT (2006) Viscose fiber/polyamide 12 composites: Novel gas-phase method for the modification of cellulose fibers with an aminosilane coupling agent. J Appl Polym Sci 102:4478-4483.

Philippou, J. L., Johns, W. E., and Tinh, N. (1982). "Bonding wood by graft polymerization. The effect of hydrogen peroxide concentration on the bonding and properties of particleboard," Holzforschung Internat. J. of the Biology, Chemistry, Physics and Technology of Wood 36(1), 37-42.

Pichaud, S. (1997) Etude d'un système réactif époxy-amine en vue du contrôle du procédé d'injection RTM à l'aide de la microdiélectrométrie. Thèse C.N.A.M, 174 p.

Pichaud., S., Duteurtre, X., Fit, A., Stephan, F., Maazouz, A., Pascault, J. P. (1999) Chemorheological and dielectric study of epoxy-amine for processing control. Polym. Inter 48 : 1205-1218.

Raj, R. G., Kokta, B. V., and Daneault, C. (1990). "A comparative study on the effect of aging on mechanical properties of LLDPE-glass fiber, mica, and wood fiber composites," Journal of Applied Polymer Science 40(5), 645-655.

Reich S, El Sabbagh A, Steuernagel L (2008) Improvement of fibre-matrix-adhesion of natural fibres by chemical treatment. Macromol Symp 262:170-181.

Rozenberg, B. A. (1986) Kinetics, thermodynamics and mechanism of reaction of epoxy oligomer with diamines. In: K. Dusek (Ed.). Epoxy resins and composites I. berlin: springer Verlag, Adv. Polym. Sci., Vol. 72, 1986. Kinetics, thermodynamics and mechanism of reaction of epoxy oligomer with diamines 113-165.

Saito T, Isogai A (2004) TEMPO-mediated oxidation of native cellulose. The effect of oxidation conditions on chemical and crystal structures of the water-insoluble fractions. Biomacromolecules 5:1983-1989.

Saito T, Shibata I, Isogai A, Suguri N, Sumikawa N (2005) Distribution of carboxylate groups introduced into cotton linters by the TEMPO-mediated oxidation. Carbohyd Polym 61: 414-419.

Saito T, Hirota M, Tamura N, Kimura S, Fukuzumi H, Heux L, Isogai A (2009) Individualization of nano-sized plant cellulose fibrils by direct surface carboxylation using TEMPO catalyst under neutral conditions. Biomacromolecules 10:1992–1996.

Sakata, I., Morita, M., and Pandey, S. N. (1993a). "Decrystallization of jute by cyanoethylation," *J. Appl. Polym. Sci.* 47(1), 73-83.

Sakata, I., Morita, M., Tsuruta, N., and Morita, K. (1993b). "Activation of wood surface by corona treatment to improve adhesive bonding," *Journal of Applied Polymer Science* 49(7), 1251-1258.

Sapieha, S., Allard, P., and Zang, Y. H. (1991). "Dicumyl peroxide-modified cellulose/LLDPE composites," *Journal of Applied Polymer Science* 41(9-10), 2039-2048.

Sapieha, S., Pupo, J. F., and Schreiber, H. P. (1989). "Thermal degradation of cellulose-containing composites during processing," *Journal of Applied Polymer Science* 37(1), 233-240.

Sbiai, A., Kaddami, H., Fleury, E., Maazouz, A., Erchiqui, F., Koubaa, A., Soucy, J. , and Dufresne, A. (2008). "Effect of the fibers size on the physico-chemical properties of composites based on palm tree fibers and epoxy thermoset polymer," *Macromol. Materials & Engineering* 293(8),684-691.

Sbiai A, Maazouz A, Fleury E, Sautereau H, Kaddami H (2010) Short date palm tree fibers / polyepoxy composites prepared using RTM process: effect of Tempo-mediated oxidation of the fibers. BioResources 5:672-689.

Sbiai, A., Kaddami, H., Sautereau, H., Maazouz, A., and Fleury, E. (2011). "TEMPO mediated oxidation of date palm tree fibers Characterization of modification," Carbohy. poly.86 (4), 1445-1450.

Segal L, Creely JJ, Martin AE, Conrad CM (1959) An empirical method for estimating the degree of crystallinity of native cellulose using the X-ray diffractometer. Text Res J 29:786-794.

Sherman, C. L., Zeigler, R.C., Verghese, N.E., Marks, M J (2008) Structure-property relationships of controlled epoxy networks with quantified levels of excess epoxy etherification. Poly. 49:1164-72.

Shibata I, Isogai A (2003) Depolymerization of cellouronic acid during TEMPO-mediated oxidation. Cellulose 10:151-158.

Sung, N. H., Kaul A., Chin I., and Sung, C. S. P. (1982). "Mechanistic studies of adhesion promotion by γ-aminopropyl triethoxy silane in α-Al$_2$O$_3$/polyethylene joint," *Poly. Eng.and Sci.* 22(10), 637-644.

Tahiri, C., and Vignon, M. R. (2000). "TEMPO-oxidation of cellulose: Synthesis and characterisation of polyglucuronans," *Cellulose* 7, 177-188.

Trindade, W. G., Hoareau, W., Razera, I.A.T., Ruggiero, R., Frollini, E., Castellan, A. (2004) Phenolic Thermoset Matrix Reinforced with sugar Cane Bagasse fibers: Attempt to develop a new fiber surface chemical modification involving formation of quinones followed by reaction with furfuryl alcohol. Macromol. Mater. Eng. 289:728-736.

Tzoganakis, C., Vlachopoulos, J., and Hamielec, A. E. (1988). "Production of controlled-rheology polypropylene resins by peroxide promoted degradation during extrusion," *Polymer Engineering and Science* 28(3), 170-180.

Wan Rosli, W. D., Law, K. N., Zainuddin, Z., and Asro, R. (2004). "Effect of pulping variables on the characteristics of oil-palm frond-fiber," *Biores. Technol.* 93, 233-240.

Wikberg H, Maunu SL (2004) Characterisation of thermally modified hard and softwoods by ^{13}C CPMAS NMR. Carbohyd Polym 58:461-466.

Young, R. A. (1978). "Bonding of oxidized cellulose fibers and interaction with wet strength agents," *Wood and Fiber Science* 10(2), 123-139 .

Zang, Y. H., and Sapieha, S. (1991). "A differential scanning calorimetric characterization of the sorption and desorption of water in cellulose/linear low-density polyethylene composites," *Polymer* 32(12), 489-492.

Zimmermann T, Pöhler E, Geiger T (2004) Cellulose fibrils for polymer reinforcement. Adv Eng Mat 6:754-761.

Oil Palm Biomass Fibres and Recent Advancement in Oil Palm Biomass Fibres Based Hybrid Biocomposites

H.P.S. Abdul Khalil, M. Jawaid, A. Hassan, M.T. Paridah and A. Zaidon

Additional information is available at the end of the chapter

http://dx.doi.org/10.5772/48235

1. Introduction

Worldwide 42 countries cultivate *Elaeis guineensis* (oil palm tree) on about 27 million acres. Oil palm is one of the most valuable plants in Malaysia, Indonesia and Thailand. Oil palm tree (**Figure 1**) generally has an economic life span of about 25 years, and it contributes to a high amount of agricultural waste in Malaysia. The oil palm tree is ≈ 7–13m in height and 45–65 cm in diameter, measuring 1.5m above the ground level (Abdul Khalil et al. 2010d) and one of the commercial crop in Malaysia. Malaysia is the world's largest producer and exporter of the oil palm, accounting for approximately 60% of the world's oil and fat production. The oil palm industry in Malaysia, with its 6 million hectares of plantation, produced over 11.9 million tons of oil and 100 million tons of biomass (Abdul Khalil et al. 2010b). The amount of biomass produced by an oil palm tree, inclusive of the oil and lignocellulosic materials, is on the average of 231.5 kg dry weight/year (Abdul Khalil et al. 2010c). An estimation based on a planted area of 4.69 million ha (MPOB 2009) and a production rate of dry oil palm biomass of 20.34 tonnes per ha per year (Lim 1998) show that the Malaysian palm oil industry produced approximately 95.3 million tonnes of dry lignocellulosic biomass in 2009. This figure expected to increase substantially when the total planted hectarage of oil palm in Malaysia could reach 4.74 million ha in 2015 (Basiron and Simeh 2005), while the projected hectarage in Indonesia is 4.5 million ha.Oil palm production has nearly doubled in the last decade, and oil palm has been the world's foremost fruit crop, in terms of production, for almost 20 years (Abdul Khalil et al. 2010c). Oil palm industries generate abundant amount of biomass say in million of tons per year (Rozman et al. 2005) which when properly used will not only be able to solve the disposal problem but also can create value added products from this biomass.

OPB is an agricultural by-product periodically left in the field during the replanting, pruning, and milling processes of oil palm. Oil palm biomass (OPB) is classified as lignocellulosic residues that typically contain 50% cellulose, 25% hemicellulose, and 25% lignin in their cell wall(Alam et al. 2009).The biomass from oil palm residue include the oil palm trunk (OPT), oil palm frond (OPF), kernel shell, empty fruit bunch (EFB), presses fruit fibre (PFF), and palm oil mill effluent (POME). Oil palm fronds accounts for 70% of the total oil palm biomass produced, while the EFB accounts for 10% and OPT accounts for only about 5% of the total biomass produced (Ratnasingam 2011). They also stated that 89% of the total oil palm biomass produced annually used as fuel, mulch and fertilizer. In 2006, Malaysia alone produced about 70 million tonnes of oil palm biomass, including trunks, fronds, and empty fruit bunches (Yacob 2007). Despite this enormous production, oil comprises only a small fraction of the total biomass produced by the plantation. The remaining biomass is an immense amount of lignocellulosic materials in the form of fronds, trunks and empty fruit bunch. As such, the oil palm industry must be prepared to take advantage of the situation and utilize the available biomass in the best possible manner (Basiron 2007). Oil palm biomass waste can create substantial environmental problems when simply left on the plantation fields. Presently, EFB mainly used as mulch, but the economic are marginal due to the high transport cost. It is seldom burnt as fuel, as the shell and fruit fibres are sufficient for oil palm mills (Abdul Khalil 2004). It reported that oil palm biomass burnt as fuel in the boiler to produce steam for electricity generation in the processing of oil palm (Nasrin et al. 2008). Researchers stated that a large amount of oil palm residues resulting from the harvest can be utilized as by–products, and it can also help to reduce environmental hazards (Sulaiman et al. 2011). Researchers carried out an extensive study on utilization of OPB as a source of renewable materials (Sumathi et al. 2008). Oil palm biomass fibres offer excellent specific properties and have potential as outstanding reinforcing fillers in the matrix and can be used as an alternative material for bio-composites, hybrid composites, pulp, and paper industries (Abdul Khalil et al. 2010d; Abdul Khalil et al. 2009).

Natural fibres such as hemp, kenaf, jute, sisal, banana, flax, oil palm etc. have been in considerable demand in recent years due to their eco-friendly and renewable nature. Natural fibres received considerable attention as potential reinforcements in polymer composites (Wong et al. 2010; Wan Nadirah et al. 2011; Bledzki and Gassan 1999). The attraction towards utilization of natural fibres as a reinforcement of polymer-based composites is mainly due to their various advantages over synthetic fibres such as are low density, lower cost, light weight, high strength to weight ratio, biodegradability, acceptable specific properties, better thermal and insulating properties (Rout et al. 2001; Rana et al. 2003; Joshi et al. 2004; Nayak et al. 2009). Natural fibre are also less wear and tear in processing, lower energy requirements for processing, wide availability and relative non abrasiveness over traditional reinforcing fibres such as glass and carbon. Natural fibre based polymer composites made of jute, oil palm, flex, hemp, kenaf have a low market cost, attractive with respect to global sustainability and find increasing commercial use in different applications (Jawaid et al. 2011b). Despite the advantages, use of natural fibre reinforced composites has been restricted due to its high moisture absorption tendency, poor wettability, and low thermal stability during processing and poor adhesion with the synthetic counterparts (Demir et al. 2006; Son et al. 2001). Natural fibres are not suitable for high performance military and aerospace applications due to its low strength, environmental sensitivity, and poor moisture resistance which results in degradation in strength and stiffness of natural fibre reinforced composites. Most of the drawbacks that have been identified can be overcome by effective hybridization of natural fibre with synthetic fibre or natural fibre.

Figure 1. Oil palm Tree

Hybrid composites are these systems in which one kind of reinforcing material incorporated in a mixture of different matrices (blends) (Thwe and Liao 2003), or two or more reinforcing and filling materials are present in a single matrix (Karger-Kocsis 2000; Fu et al. 2002) or both approaches are combined. Hybrid composite which contain two or more types of fibre in a single matrix, the advantages of one type of fibre could complement with what are lacking in the other. Hybrid composites fabrication by proper material design could help in achieve balance in cost and performance (John and Thomas 2008). Various researchers have tried blending of two fibres in order to achieve the best utilization of the positive attributes of one fibre and to reduce its negative attributes as far as practicable(Abdul Khalil et al. 2009; Abu Bakar et al. 2005; Jawaid et al. 2012; Jacob et al. 2004a; Akil et al. 2009). One another reasons for blending of one fibre with other natural fibres are to impart fancy effect, reduce cost of the end product, and find out suitable admixture of natural origin to mitigate the gap between demand and supply

(Basu and Roy 2007). It is possible to combine two or more existing materials and allow a superposition of their properties—in short, to create a *hybrid* (Figure 2)(Ashby and Brechet 2003). Hybrid composites reinforced with natural fibres, well often combined with synthetic fibres such as glass/Carbon fibres, can demonstrate exemplary mechanical performance (Abu Bakar et al. 2005; Wan Busu et al. 2010; Noorunnisa Khanam et al. 2010). Sisal/oil palm fibres and jute/oil palm fibres appear to be promising materials because of the high tensile strength of sisal and jute fibres and the toughness of oil palm fibre (Jacob et al. 2004b; Jawaid et al. 2010). Therefore, any composite comprised of these two fibres will exhibit the desirable properties of the individual constituents. The primary advantages of using oil palm fibres in hybrid composites are its low densities, non abrasiveness and biodegradability. Mixing natural fibres like hemp and kenaf with thermoplastics put Flex Form Technologies (Jon Fox-Rubin 2010) on the map and in the door panels of Chrysler's Sebring convertible. However, the combination of rising oil prices and exterior applications could drive its utilization even higher. Flex Form is also looking to produce vehicle load floors, headliners, seatbacks, instrument panel top covers, knee bolsters, and trunk liners.

2. Oil palm fibres

Oil palm industries generates massive quantities of oil palm biomass such as oil palm trunk (OPT), oil palm frond (OPF) and oil palm empty fruit bunch (EFB) as shown in **Figure 3**. The OPF and OPT generated from oil palm plantation while the oil palm EFB from oil palm processing. In Malaysia, oil palm EFB is one of the biomass materials, which is a by-product from the palm oil industry. EFB are left behind after the fruit of the oil palm harvested for the oil refining process. EFB amounting to 12.4 million tonnes/year (fresh weight) are regularly discharged from palm oil refineries (Abdul Khalil et al. 2010c). This oil palm EFB has high cellulose content and has potential as natural fiber resources, but their applications account for a small % of the total biomass productions. Several studies showed that oil palm fibres have the potential to be an effective reinforcement in thermoplastics and thermosetting materials (Khalil et al. 2008; Hassan et al. 2010; Shinoj et al. 2011). In order to develop other applications for oil palm fibres they need to be extracted from the waste using a retting process (Shuit2009). Oil palm frond (OPF) is one of the most abundant by-products of oil palm plantation in Malaysia. Oil palm fronds are available daily throughout the year when the palms are pruned during the harvesting of fresh fruit bunches for the production of oil. OPF contains carbohydrates as well as lignocellulose and it amounting to 24 million tons/year discharged from oil palm mills. Oil palm frond, consisting of leaflets and petioles, is a by-product of the oil palm industry in Malaysia and their abundance has resulted in major interest in their potential use for livestock feed (Dahlan 2000). OPF are left rotting between the rows of palm trees, mainly for soil conservation, erosion control and ultimately the long-term benefit of nutrient recycling (Abu Hassan 1994). The large quantity of fronds produced by a plantation each year makes these a very promising source of roughage feed for ruminants.

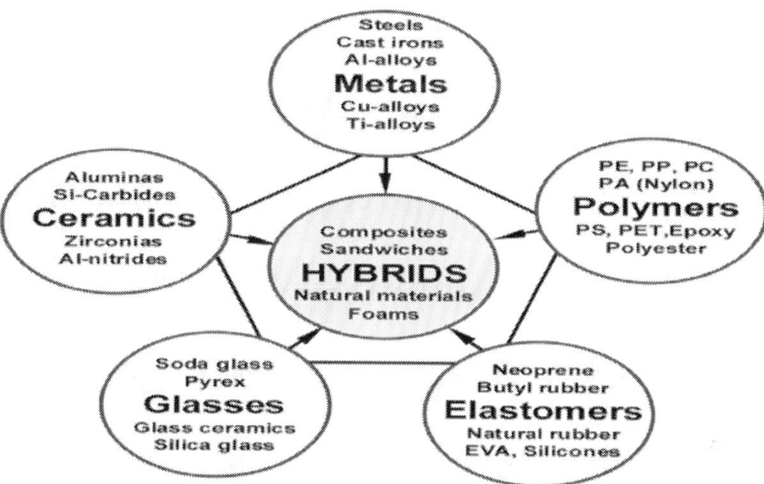

Figure 2. Hybrid materials combine the properties of two (or more) monolithic materials, or of one material and space. They include fibrous and particulate composites, foams and lattices, sandwiches and almost all natural materials. One might imagine two further dimension: those of shape and scale (Ashby and Brechet 2003 with permission).

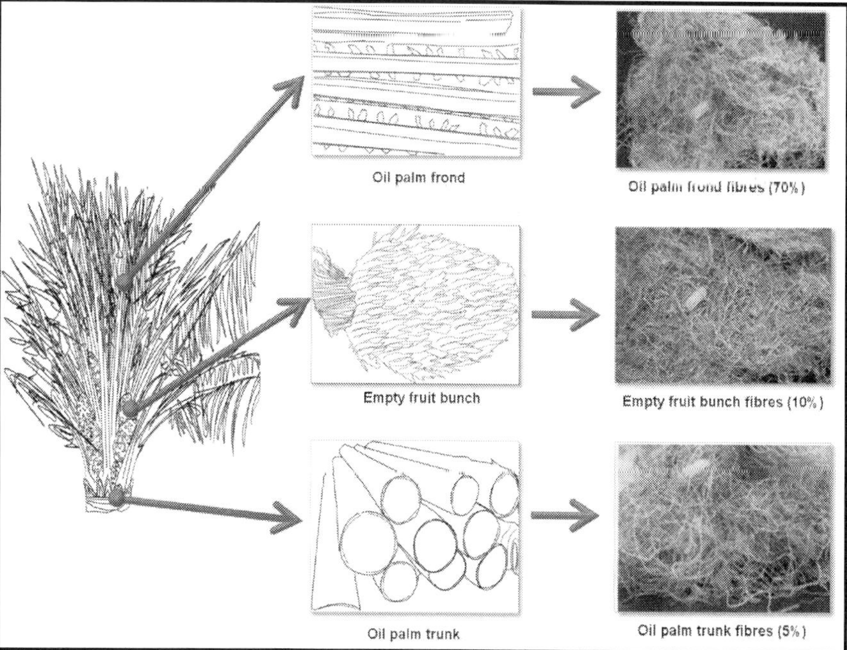

Figure 3. Oil palm biomass and oil palm biomass fibres from oil palm tree

Oil palm tree discarded for replantation after 25~30 years of oil production. Related to the large production of main products from oil palm in Malaysia, there is abundance of oil palm trunk. A large quantity of cellulosic raw material generated in the form of felled trunks during replanting can be utilized. Oil palm trunk obtained from oil palm tree and it consists of vascular bundles and parenchayma. Oil palm trunk amounted to approx 3 million tonnes/year. Up to now, there is no economical value of oil palm trunk from the structural point of view and ultimately it becomes a hazardous material to farmers. To increase the added value of these residues, several investigations have been carried out to produce hybrid plywood, MDF, polymer composites, particle boards, paper, pulp, furniture, bio fuels etc. from oil palm biomass. Many studies have been carried out on the utilization of the oil palm EFB fibres such as in particleboard, pulp, medium density fibreboard, and composites (Rozman et al. 2005). For non-structural applications, works are continuing to look into the possibilities of using OPT as furniture and particleboard raw material (Chew 1985). At present, most of the oil palm biomass are disposed off at the oil palm plantation or burned at the mills to produce oil palm ash. Thus, finding useful utilization of the oil palm biomass in fabrication of natural fibre based composites/hybrid composites will surely alleviate environmental problems related to the disposal of oil palm wastes. Application of oil palm biomass fibre in different sector is shown in **Table 1**.

Oil palm biomass	Products	References
Oil palm EFB fibres	Plywood	Abdul Khalil et al. 2010c
	MDF	Abdul Khalil et al. 2008; Abdul et al. 2010b
	Polymer biocomposite	Chai et al. 2009
	Hybrid composite	Jawaid et al. 2010
	Particle boards	Zaidon et al. 2007
	Biofuel	Shuit et al. 2009
Oil palm frond fibres	Pulp, Paper	Wan Rosli et al. 2004; Abu Hassan 1994
	Nutrient recycling	Abu Hassan 1994
	Fibreboard	Abu Hassan 1994
	Biodegradable film	Noor Haliza 2006
	Animal feed	Hassim et al. 2010; Dahlan 2000
	Downdraft gasifier	Sulaiman 2011
Oil palm trunk fibres	Lignin	Xiao et al. 2001
	Plywood, Furniture	Abdul Khalil et al. 2010c; Abdul Khalil et al. 2010a; Chew 1985

Table 1. Application of oil palm fibres

2.1. Chemical composition of oil palm fibres

It is well know that chemical constituents of oil palm biomass significantly vary due to their diverse origins and types (Chew and Bhatia 2008). Oil palm biomass is lignocellulosic residues composed of cellulose, hemicellulose, lignin and ash (Raveendran et al.1995). **Table 2** shows the chemical composition of different oil palm biomass. All the oil palm biomass are rich in lignin, cellulose and hemicellulose (Meier and Faix 1999; Demirbaş 2000). Oil palm EFB fibres are lignocellulosic fibres where the cellulose and hemicellulose are reinforced in a lignin matrix similar to that of other natural fibres. These oil palm empty fruit bunch consist of high cellulose content and is a potential natural fibre resources, but its applications account for a small percentage of the total biomass productions. High cellulose content and high toughness value of oil palm EFB fibres make it suitable for application in polymer composites (Sreekala et al., 2004; John et al., 2008). The cell wall of OPF also consists of cellulose, hemicellulose and lignin. In addition to these main components, ash, glucose and xylose are also present in cell wall of oil palm fibres. It was revealed that oil palm fibre from oil palm frond contain highest composition of hemicellulose compared to coir, pineapple, banana, and even soft and hardwood fibres (Abdul Khalil et al. 2006). Cellulose, hemicellulose, and lignin that form major constituents of the oil palm EFB fibre might differ depending on plant age, environment, soil condition, weather effect, and testing methods used (**Table 3**). Researchers reported that chemical composition of EFB and OPF are quite comparable with coir but lower in cellulose content as compared to jute and flax fibres (Khalil et al. 2000). Oil palm trunk fibre is strong and high content of lignin (23%) as lignified cellulose fibres retain their strength better than delignified fibres. The chemical compositions of a lignocellulosic fibre vary according to the species, growing conditions, method of fibre preparations and many other factors (Bledzki and Gassan 1999).

Composition	Oil Palm biomass chemical composition (wt%)		
	Oil Palm EFB	Oil palm Frond	Oil Palm Trunk
Cellulose	43-65	40-50	29-37
Hemicellulose	17-33	34-38	12-17
Holocellulose	68-86	80-83	42-45
Lignin	13-37	20-21	18-23
Xylose	29-33	26-29	15-18
Glucose	60-66	62-67	30-32
Ash	1-6	2-3	2-3

Sources: Law et al., 2007; Abdul Khalil 2006;Punsuvon 2005); Shinoj et al. 2011; Chew and Bhatia 2008.Mohamad and Abdul Halim 1985;Abdul Khalil 2004);Law and Jiang 2001); Sreekala et al. 2001

Table 2. Chemical composition of oil palm biomass

Hemicellulose (%)	Cellulose (%)	Lignin (%)	Location	References
22	48	25	Malaysia	Hill and Abdul Khalil 2000
30	36	22	Indonesia	Minowa et al.1998
30	50	18	Malaysia	Khalil et al. 2008
28	65	19	India	Sreekala et al.1997; Law et al. 2007

Table 3. Chemical composition of oil palm EFB fibres from different researchers

2.2. Physical properties of oil palm fibres

Table 4 provides data on physical properties of different oil palm biomass fibres. Fibre strength is a crucial factor to choose fibre that is specific for certain usage. Fibre length is an vital factor in determining bonding and stress distribution (Khalil et al. 2008). Oil palm EFB fibre length values between hardwood and softwood fibre length (Hassan et al. 2010). Researchers also reported that aspect ratio (l/d) of fibre has a significant effect on the properties of final composite materials. Researchers also reported that OPF fibres are shorter and thicker as compared with EFB and OPT fibres. Fibre with thicker cell wall resists collapse and do not contribute to interfibre bonding to the same extent (Reddy and Yang 2005). The microfibril angle, cell dimensions, and the chemical composition of fibres are the essential variables that determine the over all properties of the fibres (John and Thomas 2008). OPT fibres show high density which also indicates that fibre is strong. Researchers reported that fibers with higher lignin content, lower l/d ratio and higher microfibrillar angle show lower strength and modulus but have higher extensibility (Reddy and Yang 2005). The shape and size of cell lumen depends on the cell wall thickness, source of the fibres and it affect the bulk density of fibres (Reddy and Yang 2005). The physical properties such as length, diameter, lumen width, density and microfibril angle of oil palm fibres have revealed momentous changes in the physical and mechanical properties of the composite materials (Hassan et al. 2010; Shinoj et al. 2011; Jawaid and Abdul Khalil 2011a). Owing to their low specific gravity, which is about 1.25–1.50 g/cm^3 as compared to glass fibers which is about 2.6 g/cm^3, the lignocellulosic fibers are able to provide a high strength-to-weight ratio in plastic materials (Abu Bakar et al. 2005).

Fibre	Fibre length (mm)	Fibre Dia (μm)	Lumen width (μm)	Density (g/cm³)	Fibril Angle (⁰)	Ref
Oil Palm EFB	0.89-142	8 –300	8	0.7-1.55	46	Mohamad and Abdul Halim 1985; Law and Jiang 2001; Bismarck 2005; Amar 2005),Zulkifli et al. 2009),Khalil et al. 2008,Hassan et al. 2010
Oil Palm Frond	0.59 -1.59	11-19.7	8.20-11.66	0.6-1.2	40	Mohamad and Abdul Halim 1985; Law and Jiang 2001; Amar 2005, Khalil et al. 2008, Law and Jiang 2001
Oil Palm Trunk	0.60 -1.22	29.6 –35.3	17.60	0.5-1.1	42	Mohamad and Abdul Halim 1985; Khoo 1985; Amar 2005;Khalil et al. 2008;Ahmad et al.2010

Table 4. Physical properties of oil palm biomass fibres

2.3. Mechanical properties of oil palm biomass fibres

Table 5 gives the data for mechanical properties of oil palm biomass fibres. Mechanical properties such as tensile strength and modulus related to the composition and internal structure of the fibers. It reported that generally the tensile strength and young's modulus of plant fibre increases with increasing cellulose content of the fibres (Aji et al. 2009). Oil palm trunk fibre found to be suitable as reinforcement because it possesses high tensile strength (300-600 MPa) which is considered high when compared with other natural fibre. The properties of cellulosic fibers are strongly influenced by chemical composition, fibre structure, microfibril angle, cell dimensions and defects, it differs from different parts of a plant as well as from different plants (Dufresne 2008). The thicker walled fibre tend to produce an open and bulky sheet with low burst/tensile strength and high tearing resistant (Mishra et al. 2004). The mechanical properties of natural fibers also depend on their own cellulose and its crystalline organization, which can determine the mechanical properties (Bledzki and Gassan 1999). Oil palm fibres are hard and tough and found to be a potential reinforcement in polymer composites (Jawaid and Abdul Khalil 2011a).

Fibres	Tensile strength (MPa)	Young's modulus (GPa)	Elongation at break (%)	Ref
Oil Palm EFB	50-400	0.57-9	2.5-18	(Sreekala et al. 2004; Bismarck 2005; Kalam et al. 2005; Bakar et al. 2006)
Oil Palm Frond	20-200	2-8	3-16	Lab sources
Oil Palm Trunk	300-600	8-45	5-25	Ahmad et al. 2010; Killman and Hong 1989; Lim and Khoo 1986

Table 5. Mechanical properties of oil palm fibre

3. Oil Palm Fibre based hybrid composites

The oil palm fibres have been focus of study in Malaysia and around the world but still oil palm fibres have not found a solid economic value. Oil palm fibres are abundant in nature and due to its low density, non abrasiveness and biodegradability can be used in hybrid composites. In order to take full advantage of the oil palm fibres, it can be combined with other high strength fibres in the same matrix to produce hybrid composites, and thereby an economically viable composite can be obtained. Researchers reported that fibre content, fibre length, orientation, extent of intermingling of fibres, fibre/matrix interface and arrangement of both the fibres mainly affect the over all properties of a hybrid composite (Munikenche Gowda et al. 1999; Sreekala et al. 2002). Hybrid effect is defined as a positive or negative deviation of a certain mechanical property from the rule of mixture behaviour (Kickelbick 2007). Researchers explained that the rule of mixture defines a composite property as weighted average of the properties of its constituents. For example, if two fibres of different properties such as oil palm fibres and jute fibres are incorporated in the polymer, the resulting hybrid composite would most probably exhibit properties which are some sort of an average between those individual fibre components. Reported work on oil palm fibre based hybrid composites are shown in **Table 6**.

3.1. Physical and mechanical properties of oil palm fibre based hybrid composites

The physical properties are used to calculate the product weight, which relates to manufacturing costs, injection molding machine melt capacity (machine size), and the product dimensional control (mold shrinkage). The mechanical properties of natural fibre composites depends on fibre properties, surface character of material, types of matrix, fibre-matrix bonding, volume fraction of fibre and alignment of fibers etc. Mechanical properties of natural fibre composite also depend on composite processing method and effectiveness of the coupling between the fibre and matrix phases. Most of the studies on natural fibre hybrid composite involve study of mechanical properties as a function of fibre length, fibre loading, extent of intermingling of fibres, fibre to matrix bonding and arrangement of both the fibres, effect of various chemical treatments of fibers, and use of coupling agents.

Hybrid	Matrix	References
Oil Palm/Glass	Polyester	Kumar et al. 1997;Agrawal et al. 2000;Abdul Khalil et al. 2007; Karina et al. 2008; Wong et al. 2010)
	Polypropylene(PP)	Rozman et al.2001a,b
	Phenol	Sreekala et al. 2002a,b; Sreekala et al. 2004;
	Formaldehyde(PF)	Sreekala et al. 2005)
	Epoxy	Hariharan et al. 2004; Abu Bakar et al. 2005; Mridha et al. 2007
	Vinyl Ester	Abdul Khalil et al. 2009
	Natural rubber	Anuar et al. 2006
Oil palm/Sisal	Natural rubber	Jacob et al. 2004a,b; Jacob et al., 2005;Jacob et al. 2006a,b,c,d;Jacob et al. 2007; John et al. 2008
Oil Palm/Jute	Epoxy	Jawaid et al., 2010;2011a-g; Khalil et al., 2011
Oil Palm/Kalonite	Polyurethane	Anuar and Badri, 2007

Table 6. Reported works on Oil Palm Fibre based Polymer Hybrid Composites

3.1.1. Oil palm/Glass fibres reinforced hybrid composites

Hybridization of oil palm fibres with glass fibres carried out by several researchers in Malaysia, India and Indonesia. It's clear from the literature review that first time oil palm/glass hybrid composites fabricated and researchers developed fire resistant sheet moulding hybrid composites by measuring fire retardancy through limiting oxygen index (Kumar et al. 1997). After initiative from Kumar et al. (1997), another researchers carried out non-isothermal crystallization kinetics of oil palm/glass hybrid composites and analyzed it in the light of Ozawa's Theory (Agrawal et al. 2000). Preliminary study on mechanical properties of oil palm/glass fibres as reinforcing agents in polypropylene (PP) matrix was carried out (Rozman et al. 2001a,b). Results indicated that the incorporation of oil palm and glass fibres into PP matrix resulted in the reduction of tensile and flexural strength (Rozman et al. 2001a). They also studies the effect of coupling agents on tensile and flexural properties of hybrid composites and concluded that only slight improvements in some cases were shown for those composites treated with polymethylenepolyphenyl isocyanate. In another interesting research, they reported that effect of extraction of the oil palm EFB fibres show significant improvement in flexural and tensile strength and toughness, with a slight increase in the flexural and tensile modulus and elongation at break of oil palm/glass hybrid composites (Rozman et al. 2001b).

Researchers reported extensive study on effect of glass fibre loading on tensile, flexural and impact response of oil palm EFB fibre/phenol formaldehyde(PF) composite (Sreekala et al. 2002a). The over all mechanical performance of the hybrid composite was improved except impact strength and density of the hybrid composite decreases as volume fraction of oil palm EFB fibre increases while hardness of the hybrid composite also showed a slight decrease on an increased volume fraction of oil palm EFB fibre in hybrid composites. Similar study carried out on the influence of glass fibre loading, relative volume fraction of fibres in hybrid composites, and surface modification of fibres on water sorption kinetics in oil palm/glass fibres reinforced PF hybrid composites (Sreekala et al. 2002b). Results out put indicated that Hybridization of oil palm fibre with glass fibre considerably decreased the water sorption of oil palm composite. Accelerated weathering studies of untreated and treated oil palm/glass reinforced PF composites conducted and observed biodegradation and irradiation effects in light of variations in tensile and impact properties (Sreekala et al. 2004). Results indicated that mechanical performance of hybrid composites decreased with thermal and radiation ageing. Researchers studied physical and mechanical properties of the oil palm/glass fibres reinforced epoxy resin bi-layer and tri-layer hybrid composites (Hariharan et al. 2004; Abu Bakar et al. 2005). They observed that hybridization of oil palm with glass fibres increased impact strength of the hybrid composites but tensile strength and young's modulus show negative hybrid effect while positive hybrid effect was observed for the elongation at break of the hybrid composites.

In 2006, one researchers studied tensile and impact properties of thermoplastic natural rubber hybrid composite with short glass fibre and oil palm EFB fibres for the first time (Anuar et al. 2006). Researchers focused on the effect of treatment of glass and oil palm fibres with coupling agent such as silane, and maleic anhydride grafted polypropylene (MAgPP). Obtained results show that hybrid composites containing 10% EFB and 10% glass fibre gave an optimum tensile and impact strength for treated and untreated hybrid composites and tensile properties increased with addition of coupling agent. In another study on physical (density and water absorption), and mechanical (tensile, flexural and impact) properties of oil palm EFB/glass fibres reinforced polyester hybrid composites investigated with increasing loading of both oil palm EFB and glass fibres (Abdul Khalil et al. 2007). Hybridization of glass fibres with oil palm EFB fibres improved mechanical and physical properties of hybrid polyester composite as compared to oil palm EFB/Polyester composites due to the high strength and modulus value of glass fibre than the inferior EFB fibre. In 2007, first type researchers hybridized oil palm wood flour particle (Size < 250μm) with woven glass fibre reinforced epoxy composite and fabricate hybrid composites by hand lay-up method (Mridha et al. 2007). They reported that impact strength reduced with increasing the filler content and also several damages found in specimens at higher filler content resulting higher energy absorption during impact. Similar work reported by Indonesian researchers who studied physical and mechanical properties of oil palm/glass fibre reinforced polyester hybrid composites related to EFB fibre specimen length and fibre loading (Karina et al. 2008). Results show that oil palm fibre length did not show any significance effect on the flexural strength and density of hybrid composites but shorter EFB fibre show low dimensional stability as compared to longer

Until now only one research paper reported on mechanical and physical properties of the oil palm EFB/glass fibre reinforced vinyl ester hybrid composites at a different layer arrangement (Abdul Khalil et al. 2009). Mechanical and physical properties of hybrid composites were found higher than that of mechanical and chemical treated oil palm fibres reinforced composites. It also clear from the results that layering pattern of oil palm EFB and glass fibres within hybrid composites affect dimensional stability and decrease by the incorporation of glass fibre as compared to mechanical and chemical treated composites. Recently conducted a study on impact behaviour of E-glass/oil palm fibres reinforced polyester hybrid composites showed that impact strength improved with increasing number of glass fibre layer and increment in fibre length (Wong et al. 2010).

3.1.2. Oil palm/Sisal fibres reinforced hybrid composites

In 2004, Ist time any researchers try to fabricate oil palm based hybrid biocomposites by hybridization of oil palm fibres with other natural fibres by the unique combination of sisal and oil palm fibres reinforced rubber composites (Jacob et al. 2004a,b). Researchers studied the effect of fibre loading, fibre ratio, and treatment of fibres on mechanical properties of sisal/oil palm fibre reinforced hybrid composites (Jacob et al. 2004a). Results indicated that increasing the concentration of fibres reduced tensile and tear strength, but enhanced tensile modulus of the hybrid composites. They also reported that 21g sisal and 9 g oil palm based hybrid composite show maximum tensile strength and concluded that tensile strength of hybrid composites depend on weight of sisal fibres rather than oil palm fibres due to high tensile strength of sisal fibres. It also seen that treatments of both sisal and oil palm fibres causes better fibre/matrix interfacial adhesion and resulted in enhanced mechanical properties. Similar studies carried out on the influence of fibre length on the mechanical properties of untreated sisal/oil palm fibre based hybrid composites and reported that increase in fibre length decreases the mechanical properties of hybrid composites due to fibre entanglements (Jacob et al. 2004b).

Researchers also studied water sorption characteristics of the oil palm/sisal hybrid composites with reference to fibre loading, chemical modification and influence of temperature (Jacob et al. 2005). Mercerization and silanation treatment of sisal and oil palm fibres decreased the water uptake in the hybrid composites and moisture uptake found to be dependent on the properties of the biofibres. In an interesting study, researchers reported durability and ageing characteristics of oil palm/sisal hybrid composites (Jacob et al. 2007). Researchers again fabricated oil palm /sisal hybrid composites by using surface treated oil palm and sisal fibre with varying concentration of NaOH solution and different coupling agents (John et al. 2008). They compared fibre reinforcing efficiency of the chemically treated biocomposites with that of untreated composites. Results demonstrated that chemical treatment of sisal and oil palm fibres enhanced mechanical properties of hybrid composites and 4% NaOH show superior tensile properties due to better fibre/matrix interaction. Hybrid composites prepared from fluorosilane treated fibres exhibited better mechanical properties as compared to other silane treated hybrid composites. They also investigated anisotropic swelling studies of hybrid composites and revealed that fluorosilane treated fibres based hybrid composites has the highest degree of fibre alignment.

3.1.3. Oil palm/Jute fibre reinforced hybrid composites

Recently we carried out an extensive study on mechanical and physical properties of oil palm/jute fibres reinforced epoxy hybrid composites in our laboratory. Research output show that dimensional stability, density, void content, and chemical resistance of oil palm EFB fibres reinforced epoxy composite improved with hybridization of oil palm EFB fibres with non woven/woven jute fibres (Jawaid et al. 2011d,f,g; Abdul Khalil et al. 2011). Hybridization of oil palm EFB composites with jute fibres improved physical properties of oil palm/jute hybrid composites due to packed and hybrid arrangement of fibre, less hydrophilic nature of jute fibres, better jute/epoxy interaction. It also observed that woven hybrid composites display poor dimensional stability, density and chemical resistant due to high void content of woven jute fabrics as compared to non woven fibres.

Mechanical performance of oil palm EFB fibres with non woven/woven jute fibres reinforced hybrid composites were also carried out (Jawaid et al. 2010;2011b,c,e). Results indicated that oil palm/non woven jute fibre based hybrid composites has higher tensile and flexural properties due to high tensile strength of jute fibres and better jute/epoxy matrix interaction. Hybrid composite show less impact strength as compared to oil palm EFB composite because oil palm EFB fibre has high fracture toughness as compared to jute fibres (Jawaid et al. 2010; 2011c,g). It observed that mechanical properties of woven hybrid composites significant improved work as compared to non woven hybrid composites due to woven fibres are tightly bonded, low deformation at break, stretching nature of fabrics or fabrics mats have higher fibre count than chopped strand (non woven) mat (Jawaid et al. 2011b,e).

3.1.4. Oil palm/Kaolinite hybrid composites

Mechanical and water sorption properties of oil palm empty fruit bunch fibres and kaolinite reinforced polyurethane hybrid bio-composites were studied (Amin and Badri 2007). Hybrid bio-composites showed maximum flexural and impact strengths were at 15% kaolinite loading. It observed that oil palm fibre hybridization with kaolinite improved stiffness, strength, and display better water resistance to extent. Mechanical properties of oil palm/Kaolinite hybrid enhanced due to interaction between kaolinite with PU matrix and EFB fibres.

3.2. Electrical properties of oil palm fibre based hybrid composites

In the field of natural fibre reinforced polymer composites, extremely limited research has been reported on the electrical properties. In 1981, Ist time researchers reported electrical resistivity and dielectric constant of some natural fibre (Coir, banana, sisal pineapple leaf, and palmyra fibres) and results indicated that natural fibre has high electric resistivity and dielectric strength (Kulkarni et al. 1981). They also suggested that natural fibre can be used as a replacement for wood in insulating applications. Further study reported by another researcher on electrical resistivity of various part of the coconut tree and confirm that it can used as good insulators (Satyanarayana et al. 1982). Recently electrical properties of oil palm fibres reported and concluded that dielectric constant value of oil palm fibre is in agreement with the values reported earlier (Chand 1992) for other natural fibres (Shinoj et al. 2010). Results obtained by researchers also clear that fibre size reduction caused an increase in dielectric constant, while alkali treatment on fibres caused a decrease in dielectric constant. Several researchers worked on electrical properties of natural fibre reinforced polymer composites (Reid et al.1986; Paul and Thomas 1997; Datta et al.1984; Jayamol et al.1998; Chand and Jain 2005). Researchers studied electrical properties of untreated and treated oil-palm fibre reinforced polymer composites by using transient plane source technique at room temperature. Results indicated that all the silane and alkali treated fibres shows enhancement in thermal conductivity and diffusivity of the composites as compared to acetylated composite (Agrawal et al. 2000). In this work, the dielectric properties of oil palm/natural rubber composite materials of untreated and treated oil palm fibres were characterized. It noticeable that oil palm fibre reinforced in rubber matrix affects the dielectric properties of composites and treatment of fibres increase the loss factor. They also reported that low fibre content (20%) show less significant effect on the dielectric properties of composites (Marzinotto et al. 2007). Recently, another researchers investigate the influence of the fibre orientation in the dielectric properties of the oil palm tree fibre reinforced polyester composite by using a dielectric spectroscopy (Ben Amor et al. 2010). Results demonstrated that orientation of the fibre can strongly influence the dielectrical properties and interfacial polarization processes in composites and this technique can be used to probe the composite interphase and investigate the effect of fibre orientation on the evolution of composite interfacial properties.

In an attractive research work carried out on thermal conductivity and diffusion of banana/glass fibres reinforced rubber hybrid composites in relation to fibre loading, fibre ratio, frequency, chemical modification of fibres and the presence of a bonding agent (Agarwal et al. 2003). Electrical strength and volume resistivity of jute/wheat and jute/jamun fibres reinforced bispheniol-c-formaldehyde has been evaluated (Mehta and Parsania 2006). Researchers reported that there is no much difference in dielectric breakdown strength between hybrid composites but volume resistivity of hybrid composites increased by 197-437% as compared to jute fibres reinforced bispheniol-c-formaldehyde composite. Researchers investigated dielectric behaviours of glass and jute fibres reinforced polyester hybrid composites and later on electrical properties of banana/glass hybrid fibre reinforced composites with varying hybrid ratios and layering patterns were analyzed (Fraga et al. 2006; Joseph and Thomas 2008). In 2006, Ist time dielectric characteristics and volume resistivity of sisal-oil palm hybrid biocomposites investigated (Jacob et al. 2006). Results obtained show that dielectric constant increases with fibre loading at all frequencies, and volume resistivity decreases with frequency, and fibre loading, this implies that the conductivity increases upon addition of lignocellulosic fibres. Researchers also reported that chemical modification of fibres resulted in a decrease in dielectric constant and increase in volume resistivity due to increase in hydrophobicity of fibres. It also observed that addition of a two-component dry bonding agent consisting of hexamethylene tetramine and resorcinol, used for the improvement of interfacial adhesion between the matrix and fibres reduced the dielectric constant of the hybrid composites. They also reported that dissipation factor was seen to increase with fibre loading which indicates that the electrical charges can be retained over a longer period of time. Researchers studied dielectric constant and volume resistivity values of sisal/coir hybrid composites and reported similar results to oil palm/sisal hybrid composites (Haseena et al. 2007).

3.3. Thermal and dynamic mechanical properties of oil palm fibre based hybrid composites

Thermal stability of natural fibres and natural fibre based composites can be analyzed by two techniques viz., thermogravimetric analysis (TGA) and differential scanning calorimeter (DSC). Weight and energy loss with heating is common phenomena for natural fibre reinforced polymer composites due to degradation and loss of residual solvents and monomers. Weight loss on heating is studied by TGA and measurement of relative changes in temperature and heat or energy either under isothermal or adiabatic conditions studied by DSC. Dynamic mechanical analysis (DMA) is a technique for measuring the modulus and damping factor of a sample. The modulus is a measure of how stiff or flimsy sample is, and amount of damping a material can provide is related to energy it can absorb (Duncan 2008). DMA is generally used for thermoplastics, thermosets, composites and biomaterials. DMA is one of the most powerful tools to study the behaviour of polymer composite materials and it allows for a quick and easy measurement of material properties (Swaminathan and Shivakumar 2009). Previously thermal properties of hybrid composites such as banana/glass, sisal/glass, bamboo/glass, hemp/glass, etc. were studied by researchers and observed that addition of glass fibre improved thermal properties of natural fibre based composites. Researchers reported that hybridization of banana fibre reinforced polymer composites with glass fibres enhanced melting point, crystallization temperature and onset thermal degradation temperature of maleic anhydride grafted polypropylene (MAPP) treated banana/glass hybrid composites (Samal et al. 2009a). It is due to SiO groups in glass fibre which interlinks with anhydride group of MAPP, providing synergism between glass and banana fibres. Similar study on thermal stability of banana/glass hybrid composites carried out by TGA and DSC and revealed that MAPP treated banana and glass fibres enhanced thermal stability of polypropylene (Nayak et al. 2010a). Dynamic mechanical properties of banana/glass hybrid composites have been analyzed to investigate the interfacial properties (Samal et al. 2009a; Nayak et al. 2010a). Results show that the storage modulus and loss modulus of MAPP treated hybrid composites improved over the whole temperature range, indicating better adhesion between fibre/matrix. Dynamic mechanical properties of banana/glass hybrid composites also studied over a range of temperature and three different frequencies (Pothan et al. 2010). Storage modulus values of hybrid composite decreases above the glass transition temperature (Tg) where glass is the core material.

Researchers studied thermal properties of sisal/glass hybrid composites and confirm that addition of glass fibre improved thermal properties of sisal/PP composites (Jarukumjorn and Suppakarn 2009; Nayak and Mohanty 2010b). In an interesting study on thermal properties such as crystallization, melting behaviour and thermal stability of bamboo/glass hybrid composites and obtained results indicate an increase in thermal stability of bamboo composites due to the higher thermal stability of glass fibre than bamboo fibre (Samal et al. 2009b; Nayak and Mohanty 2010b; Lee and Wang 2006). Hybridization of hemp fibre composites with glass fibres shifts the temperature of degradation to a higher value indicating an increased thermal stability of the hybrid composites (Panthapulakkal and Sain 2007). Effect of glass fibre loading and MAPP treatment on dynamic mechanical properties of sisal/glass hybrid composites carried out and reported improvement in storage modulus and loss modulus of hybrid composites with hybridization of sisal/pp composite with glass fibres and treatment with MAPP (Ornaghi Jr et al. 2010; Nayak and Mohanty 2010b). Dynamic mechanical properties of bamboo/glass fibre hybrid composites were also studied and observed results indicate an increase in storage modulus and decrease in damping properties due to hybridization and treatment of fibres with MAPP as compared to untreated composites and pure matrix (Nayak et al. 2009,2010c; Samal et al. 2009b). Until now, there is no any researcher work carried out on thermal stability of banana/sisal and pineapple leaf/glass reinforced hybrid composites but few work reported on dynamic mechanical properties of these hybrid composites. They concluded that storage modulus and damping factor of banana/sisal and pineapple leaf/glass hybrid composites also enhanced due to hybridization with natural and synthetic fibres (Idicula et al. 2005a,b; Uma Devi et al. 2010).

Literature review indicated that particularly limiting work reported on thermal and dynamic mechanical properties of oil palm based hybrid composites. In 2006, Ist time any researchers studied thermal properties of sisal/oil palm fibres reinforced natural rubber hybrid composites and observed that sisal fibre loading and chemical modification of fibres enhanced thermal stability of the oil palm composites (Jacob et al. 2006b). In case of oil palm/sisal hybrid composite, the peak temperatures have decreased to 356.3^{0}C, and a new peak has come at 489.9^{0}C due to hemicellulose and alpha-cellulose degradation and addition of fibers results in an increase of thermal stability as indicated by the higher peak temperatures. Chemical modification of fibres results in a further increase of thermal stability as evident from the peak temperatures of 524^{0}C and 518.3^{0}C of 4%NaOH and aminosilane treated composites.

They also carried out an extensive study on dynamic mechanical properties of oil palm/sisal fibre reinforced natural rubber hybrid composites as a function of temperature (Jacob et al. 2006c). The storage modulus of hybrid composites increased with sisal fibre loading while the damping factor found to decrease due to the increased stiffness imparted by oil palm and sisal fibres. Treatment of fibres with NaOH also enhanced the storage modulus value of hybrid composites while chemically treated composites show decreased in damping factor value as compared to untreated composites. Similar study on dynamic mechanical properties of oil palm/sisal hybrid composites with reference to the role of coupling agents were carried out (Jacob et al. 2006d). It observed that storage and loss modulus of treated hybrid composites increased while damping properties decreased due to better interfacial interface between fibre and matrix. The dynamic mechanical properties of oil palm fibre/glass fibre reinforced phenol formaldehyde hybrid composites as a function of fibre content, and hybrid fibre ratio was investigated (Sreekala et al. 2005). The glass transition temperature of the hybrid composites found to be lower than that of the unhybridized composites. Storage modulus of the hybrid composites was also found to be lower than that of unhybridized oil palm fibre/PF composite. Loss modulus of hybrid composites increased with increase in glass fibres, and gradual decrease in loss modulus is observed with the increase in frequency. It also noted that the activation energy decreases with incorporation of glass fibres in hybrid composites.

In our laboratory, we carried out research work on thermal and dynamic mechanical properties of oil palm/jute fibre reinforced epoxy hybrid composites (Jawaid and Abdul Khalil 2011e; Jawaid et al. 2012). The thermal behaviour of the oil palm/jute based hybrid composites were determined (**Figure 4**), and it noticed that initial degradation temperature of hybrid composites shifted to a higher temperature well over 268-271°C as compared with the EFB composite (260°C), which indicate higher thermal stabilities of the hybrid composites (Jawaid and Abdul Khalil 2011e). Final degradation temperature of hybrid composites also shifted between 441 to 443°C due to complex reaction. Hybridization of oil palm EFB fibres with jute fibres, the final decomposition temperature and ash content of the hybrid composite shifted slightly towards higher temperature as a result of the high thermal stability of jute fibre which acts as barriers to prevent the degradation of oil palm EFB fibres. Similar study carried out on thermal stability of oil palm/woven jute fibre reinforced epoxy hybrid composites and results indicated that woven hybrid composites initial and final decomposition temperature higher than oil palm/jute hybrid composites (Jawaid et al. 2012). It clear from **Figure 4** that oil palm EFB/woven jute hybrid composites are more thermal stable as compared to oil palm EFB composite which is possibly due to the higher thermal stability of woven jute fibre than oil palm EFB fibre.

Dynamic mechanical properties of oil palm/jute and woven hybrid composites also carried out in our laboratory (Jawaid and Abdul Khalil 2011e; Jawaid et al. 2012). Figure 5 shows storage modulus values of hybrid composites at a frequency of 1 Hz. On investigating the variation of storage modulus with temperature, storage modulus was found to be decreased with the increase in temperature in all cases at low temperature because fibre do not contribute much to imparting stiffness to the material. As temperature increases, the components become more mobile and lose their mobility and lose their close packing arrangement. The high stiffness of hybrid composites is in agreement with their tensile property reported in our previous research (Jawaid et al. 2011g). In particular the composite stiffness is substantially increased with woven jute and EFB fibre incorporation, and it causes a drop in the modulus above T_g and comparatively less than epoxy matrix (Jawaid et al. 2012). The stiffness of the oil palm EFB composite increases with hybridization with woven jute fibres, resulting in high storage modulus (**Figure 5**).Moreover, the addition of woven jute fibres allows greater stress transfer at the interface and ultimately increases the storage modulus.

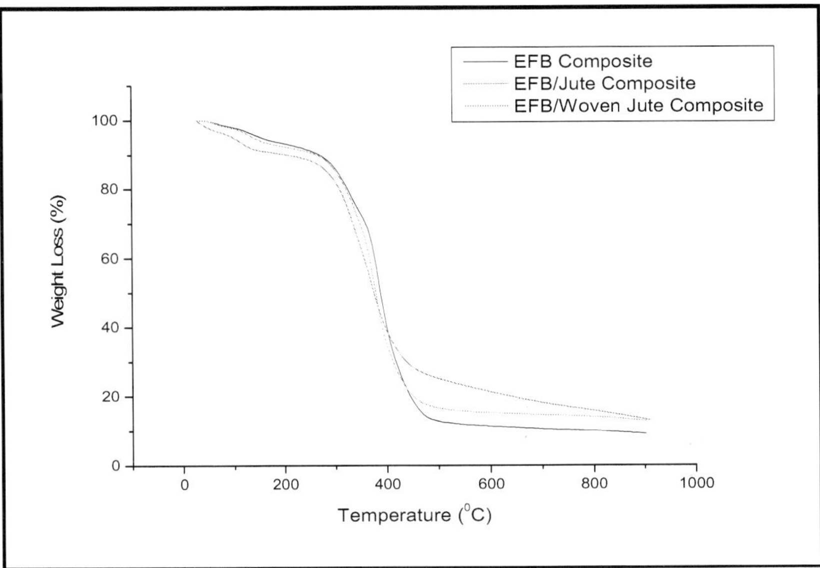

Figure 4. Thermogravimetric analysis curve of Oil Palm EFB, oil palm EFB/jute, and Oil Palm EFB/Woven Jute Hybrid Composites.

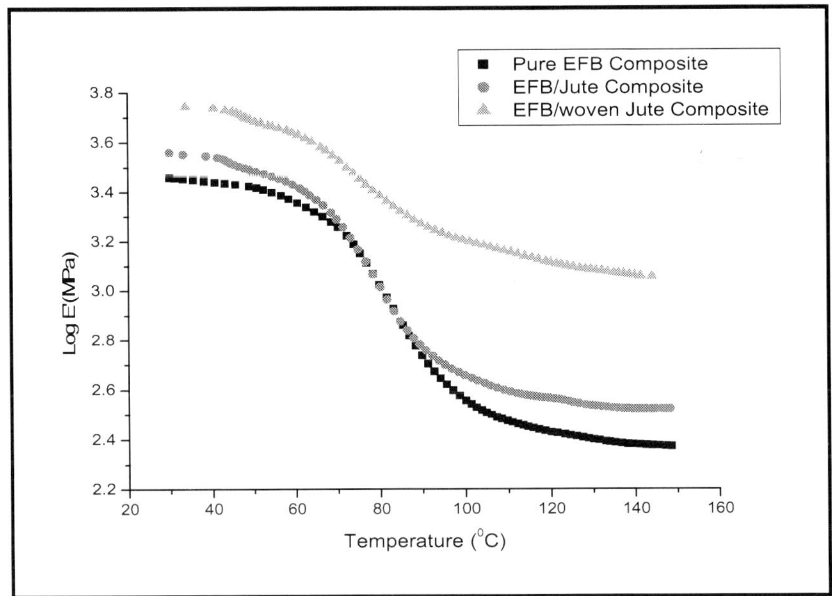

Figure 5. Storage modulus of oil palm EFB, oil palm EFB/jute, and oil palm EFB/woven jute hybrid compossites

Trends of change in the damping factor of the hybrid composites with temperature are shown in Figure 6. Damping factor of oil palm/non woven jute hybrid composite indicate that incorporation of the small amount of jute fibre to oil palm EFB/epoxy composite enhances the damping characteristics of the hybrid composites (Jawaid and Abdul khalil 2011e). With incorporation of oil palm EFB and woven jute fibres, the Tan δ peak was lowered as expected (Jawaid et al. 2012). Reinforcement of fabrics results in the formation of barriers that restrict the mobility of polymer chain, leading to lower flexibility, lower degree of molecular motion and hence lower damping characteristics (Jacob et al. 2006).

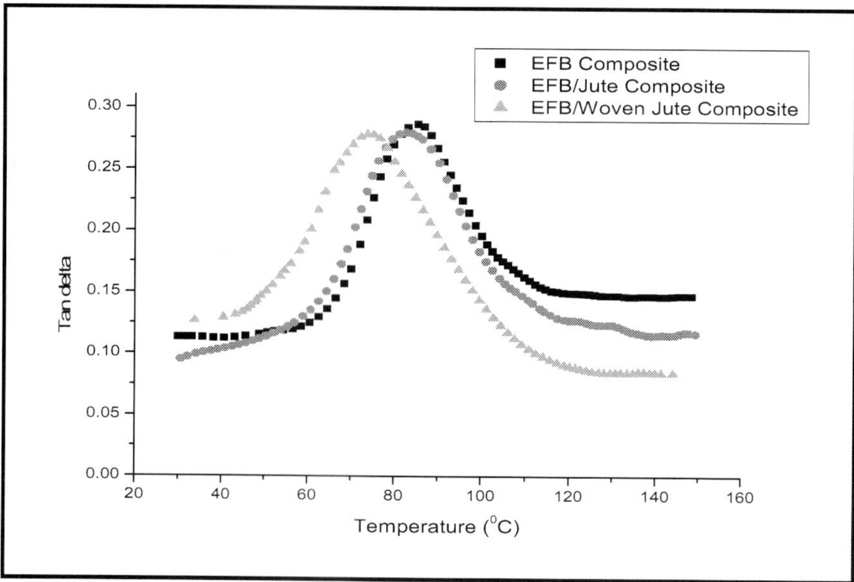

Figure 6. Damping factor of oil palm EFB, oil palm EFB/jute, and oil palm EFB/woven jute hybrid composites

4. Application of oil palm fibres based composites

The use of bio-fibres for composite applications is being investigated throughout the world (Bledzi et al, 2006). Biocomposites have received increasing attention from both of the academic world and industries as building, automotive, packaging, and so on. Green composites or biocomposites based on natural fibres and resins are increasingly used for various applications as replacements for non-degradable materials. Green composites made entirely from renewable agricultural resources could offer a unique alternative for building, construction, furniture and automotive application (Gupta 2009). Biocomposites find applications in a number of fields viz., automotive industry and construction industry such as for panels, frames, ceilings, and partition boards (Abu Bakar et al. 2005), structural applications, aerospace, sports, recreation equipment, boats, office products, machinery, etc.

(Sreekala et al. 2002a), and structural and non structural components such as window, door, siding, fencing, roofing, decking, etc. (Yatim et al. 2011). Bledzki et al. (2006) point out the importance of natural fibres and natural fibres composites in the automotive industry. Researchers already developed many automotive components (interior and exterior) from bio-fibre reinforced composite materials and designed Eco car which is built from bio fibre composite panels that incorporate biodegradable resins as the matrix material. (Bledzi et al, 2006). A modern automobile requires recyclable or biodegradable parts because of the increasing demand for a clean environment (Gupta 2009). Composites are also finding application in electrical and electronic industries.

Recently trends developed in the automotive industry to use natural and recycled materials to fulfil "The End of Life Vehicle (ELV) Directive", which came into force in October 2000, requires all member states to de-pollute all scrapped vehicles, avoid hazardous waste and reduce the amount of waste placed in landfill sites to a maximum of five per cent per car by 2015. Ford automotive stated to develop one million Lincoln and Mercury vehicles that feature recycled plastic bottles and even worn-out jeans in their seat cushions and backs by the end of the year, while French carmaker Citroën has turned to natural materials derived from hemp, bamboo and kenaf for the door linings of its 'premium' models. A team at Baylor University in the USA has made trunk liners, floorboards and car-door interior covers using fibres from the outer husks of coconuts, replacing the synthetic polyester fibres typically used. So, the main driving force to use natural fibres or recycled materials to develop cost effective, low weight and sustainable car to fulfil ELV directive and reduce hazardous waste.

Oil palm fibres can be used as fillers in thermoplastics and thermoset composites. These composites have wide applications in furniture and automobile components. Malaysian Palm Oil Board (MPOB) has developed a series of technologies on manufacturing thermo-formable plastic composites through compression or extrusion process for car components such as bumbers, trimmings, rear parcel shell, spare wheel cover and splash shield and also plastic pellets. Blending of oil palm fibres with polyols can produce cost effective lighter products such as packaging materials and roof insulators, which only require low compressive hardness. Oil palm fibre-filled automotive upholstery parts, dampening sheets for automotive industry using oil palm fibres, moulded particleboard, pulp and paper from oil palm fibres also developed. Researchers developed composite plastic fiber for automotive application by using oil palm biomass (Viva 2011). Progress in utilization of oil palm fibre finally reached to commercialization stage when PROTON (Malaysian National Carmaker) entered into an agreement with PORIM (Palm Oil Research Institute of Malaysia) to develop the thermoplastic and thermoset composites and used it in PROTON car (Panigrahi 2010). Sheffield University (UK) research team started to work with a producer of palm oil fibre and show interest in sustainable car production by replacing traditional materials and say that once the research will be in a later stage they will try to motive car manufacturers to use new materials and they hope that once the regulations become more stringent, this will force to do the change. In an interesting research, oil palm EFB fibre reinforced concrete roof slates produced, and preliminary results suggest that oil palm EFBs have a potential to be a material component of reinforced mortar roofing slates with the appropriate water cement ratio and concrete mix (Kaliwon 2010).

Researcher have developed, manufactured and assembled a small prototype car with all body panels made from jute fibre reinforced composite and hybrid composite. It also reported that door panels manufactured from flax/sisal hybrid epoxy composite shown remarkable weight reduction of about 20% (Schuh 2004). In 2000, Audi launched the A2 midrange car in which door trim panels were made of polyurethane reinforced with mixed flax/sisal mat. Researcher developed bamboo/glass fibre biocomposites for applications in wall panels, roofing panels, floors (Van Rijswijk et al. 2001). From literature view, we unable to find any application of oil palm fibres based hybrid composites which is required to explore the application in automobile, construction, packaging, etc sectors.

5. Conclusions

Oil palm biomass is an agricultural by-product periodically left in the field after oil extraction and considers as hazardous material creating environmental problems. The oil palm biomass residue include the oil palm trunks, oil palm fronds, kernel shell, empty fruit bunch, presses fruit fibre, and palm oil mill effluent. Oil palm fronds accounts for 70%, EFB accounts for 10%, and OPT accounts for only about 5% of the total biomass produced. Oil palm biomass residues can be utilized as by–products, and it can also help to reduce environmental hazards. Oil palm biomass fibres consist of cellulose, hemicelluose, and lignin while oil palm empty fruit bunch consist of high cellulose content and is a potential natural fibre resource, but its applications account for a small percentage of the total biomass productions. The properties of oil palm fibres are strongly influenced by chemical composition, fibre structure, microfibril angle, cell dimensions and defects, it differs from different parts of a plant as well as from different areas. Oil palm biomass fibres offer excellent specific properties and have potential as outstanding reinforcing fillers in the matrix and can be used as an alternative material for bio-composites, hybrid composites, pulp, and paper industries. Natural fibres such as oil palm biomass fibres are not suitable for high performance applications due to its low strength, environmental sensitivity, and poor moisture resistance which results in degradation in strength and stiffness of natural fibre reinforced composites. Most of the drawbacks that have been identified can be overcome by effective hybridization of oil palm fibres fibre with glass fibre or natural fibres (jute/sisal).

The primary advantages of using oil palm fibres in hybrid composites are its low densities, non abrasiveness and biodegradability. Oil palm fibres with jute/sisal fibres appear to be promising materials because of the high tensile strength of jute and sisal fibres and the toughness of oil palm fibre. Now most of automobile manufacturer try to replace synthetic fibres with natural fibres, but it's not comparable in properties while fabricating hybrid composites by the combination of two natural fibres give them advantage to replace synthetic fibres. Hybridization of oil palm EFB fibres with glass/jute/sisal fibres enhanced physical and mechanical properties of oil palm EFB composites. It would appear that a wide variety of work carried out worldwide on physical and mechanical properties of oil palm/glass, oil palm /sisal, and oil palm/jute hybrid composites but still not fully explore electrical, thermal, and dynamic mechanical properties of oil palm based hybrid composites. Over all conclusions is that hybrid composites fabrication by using oil palm fibres will help in development of unique cost effective advanced composites possessing superior mechanical properties, dimensional stability, appropriate stiffness, damping behaviour and thermal stability.

Extensive research still required to do on oil palm based hybrid composite while exploring compatible of oil palm fibres with other natural fibres by compounded with several other polymers in the line of previous work. Microstructure of the interface between the oil palm fibre with other natural fibres, and matrix needs to be investigated and the interfacial properties should be studied with single fibre pull out, micro-bond test and single fibre fragmentation test. Complementary techniques such as X-ray photoelectron spectroscopy, time-of-flight secondary ion mass spectrometry may be studied to get more information about the chemical composition of the surface, its wetting behaviour, etc., that, coupled with the mechanical assessment of the interface, can shed more light on the structural characteristics of the interface. Future research on oil palm fibre based hybrid composites not only limited to its automotive applications but it also required to explore its application in aircraft components, construction industry, rural housing and biomedical applications.

Author details

H.P.S. Abdul Khalil*
School of Industrial Technology, Universiti Sains Malaysia, Penang, Malaysia

M. Jawaid and A. Hassan
Department of Polymer Engineering, Faculty of Chemical Engineering,
Universiti Teknologi Malaysia, Skudai, Johor, Malaysia

M.T. Paridah
Institute of Tropical Forestry and Forest Products,
Universiti Putra Malaysia, Serdang, Selangor Darul Ehsan, Malaysia

A. Zaidon
Faculty of Forestry, Universiti Putra Malaysia, Serdang, Selangor Darul Ehsan, Malaysia

Acknowledgement

The author, Mohammad Jawaid would like to thank Universiti Teknologi Malaysia for the Visiting Lecturer Position that has made this research work possible.

---

* Corresponding Author

6. References

Abdul Khalil, H. P. S., A. H. Bhat, M. Jawaid, P. Amouzgar, R. Ridzuan, and M. R. Said. 2010a. Agro-Wastes: Mechanical and physical properties of resin impregnated oil palm trunk core lumber. *Polymer Composites* 31 (4):638-644.

Abdul Khalil, H. P. S., M. Y. Nur Firdaus, M. Jawaid, M. Anis, R. Ridzuan, and A. R. Mohamed. 2010b. Development and material properties of new hybrid medium density fibreboard from empty fruit bunch and rubberwood. *Materials & Design* 31 (9):4229-4236.

Abdul Khalil, H. P. S., S. Hanida, C. W. Kang, and N. A. Nik Fuaad. 2007. Agro-hybrid composite: The effects on mechanical and physical properties of oil palm fiber (EFB)/glass hybrid reinforced polyester composites. *Journal of Reinforced Plastics and Composites* 26 (2):203-218.

Abdul Khalil, H. P. S., M. Jawaid, and A. Abu Bakar. 2011. Woven hybrid composites: Water absorption and thickness swelling behaviours. *BioResources* 6 (2):1043-1052.

Abdul Khalil, H. P. S., C. W. Kang, A. Khairul, R. Ridzuan, and T. O. Adawi. 2009. The effect of different laminations on mechanical and physical properties of hybrid composites. *Journal of Reinforced Plastics and Composites* 28 (9):1123-1137.

Abdul Khalil, H. P. S., M. Y. Nur Firdaus, M. Anis, and R. Ridzuan. 2008. The Effect of Storage Time and Humidity on Mechanical and Physical Properties of Medium Density Fiberboard (MDF) from Oil Palm Empty Fruit Bunch and Rubberwood. *Polymer-Plastics Technology and Engineering* 47 (10):1046 - 1053.

Abdul Khalil, H. P. S., M. R. Nurul Fazita, A. H. Bhat, M. Jawaid, and N. A. Nik Fuad. 2010c. Development and material properties of new hybrid plywood from oil palm biomass. *Materials & Design* 31 (1):417-424.

Abdul Khalil, H. P. S., B. T. Poh, A. M. Issam, M. Jawaid, and R. Ridzuan. 2010d. Recycled polypropylene-oil palm biomass: The effect on mechanical and physical properties. *Journal of Reinforced Plastics and Composites* 29 (8):1117-1130.

Abdul Khalil, H. P. S., Rozman, H.D. 2004. *Gentian dan lignoselulosik*: Universiti Sains Malaysia: Pulau Pinang.

Abdul Khalil, H.P.S., Siti Alwani, M., Mohd. Omar, A.K.,. 2006. Chemical composition, anatomy, lignin distribution, and cell wall structure of Malaysian plant fibers. *BioResources* 1 (2):220-232.

Abu Bakar, A., Hariharan,, and H. P. S. Abdul Khalil. 2005. Lignocellulose-based hybrid bilayer laminate composite: Part I - Studies on tensile and impact behavior of oil palm fiber-glass fiber-reinforced epoxy resin. *Journal of Composite Materials* 39 (8):663-684.

Abu Hassan, O., Ishida, M., Mohd. Shukri, I., Ahmad Tajuddin, Z.,. 1994. Oil-Palm Fronds as a Roughage Feed Source for Ruminants in Malaysia. Malaysia Agriculture Research and Development Institute (MARDI), Kuala Lumpur, Malaysia.

Agarwal, R., N. S. Saxena, K. B. Sharma, S. Thomas, and L. A. Pothan. 2003. Thermal conduction and diffusion through glass-banana fiber polyester composites. *Indian Journal of Pure and Applied Physics* 41 (6 SPEC.):448-452.

Agrawal, R., N. S. Saxena, M. S. Sreekala, and S. Thomas. 2000. Non-isothermal crystallization kinetics of glass oil palm fibre reinforced phenolformaldehyde composites. *Journal of Scientific and Industrial Research* 59 (2):136-139.

Agrawal, Richa, N. S. Saxena, M. S. Sreekala, and S. Thomas. 2000. Effect of treatment on the thermal conductivity and thermal diffusivity of oil-palm-fiber-reinforced phenolformaldehyde composites. *Journal of Polymer Science Part B: Polymer Physics* 38 (7):916-921.

Ahmad, Z.,H.M.Saman, and P.M.Tahir. 2010. Oil palm trunk fibre as a bio-waste resource for concrete reinforcement. *International Journal of Mechanical and Materials Engineering* 5 (2):199-207.

Aji, I. S., S. M. Sapuan, E. S. Zainudin, and K. Abdan. 2009. Kenaf fibres as reinforcement for polymeric composites: A review. *International Journal of Mechanical and Materials Engineering* 4 (3):239-248.

Akil, H. M., L. W. Cheng, Z. A. Mohd Ishak, A. Abu Bakar, and M. A. Abd Rahman. 2009. Water absorption study on pultruded jute fibre reinforced unsaturated polyester composites. *Composites Science and Technology* 69 (11-12):1942-1948.

Al-Qureshi, H.A. 2001. The application of jute fibre reinforced composites for the development of a car body. Paper read at UMIST Conference, at UK.

Alam, Md Z., A. A. Mamun, I. Y. Qudsieh, S. A. Muyibi, H. M. Salleh, and N. M. Omar. 2009. Solid state bioconversion of oil palm empty fruit bunches for cellulase enzyme production using a rotary drum bioreactor. *Biochemical Engineering Journal* 46 (1):61-64.

Amar, K. M., Manjusri, M. & Lawrence, T. D., ed. 2005. *Natural Fibers, Biopolymers, and Biocomposites*. USA: CRC Press, Tayler and Francis Group.

Anuar, H., S. H. Ahmad, R. Rasid, and N. S. Nik Daud. 2006. Tensile and impact properties of thermoplastic natural rubber reinforced short glass fiber and empty fruit bunch hybrid composites. *Polymer - Plastics Technology and Engineering* 45 (9):1059-1063.

Ashby, M. F., and Y. J. M. Brechet. 2003. Designing hybrid materials. *Acta Materialia* 51 (19):5801-5821.

Bakar, A. A., A. Hassan, and A. F. Mohd Yusof. 2006. The effect of oil extraction of the oil palm empty fruit bunch on the processability, impact, and flexural properties of PVC-U composites. *International Journal of Polymeric Materials* 55 (9):627-641.

Basiron, Y. 2007. Palm oil production through sustainable plantations. *European Journal of Lipid Science and Technology* 109 (4):289-295.

Basiron, Y., and M.A. Simeh. 2005. Vision 2020 – The oil palm phenonmenon. *Palm Industry Economic Journal*. 5 (2):1-10.

Basu, G., and A. N. Roy. 2007. Blending of jute with different natural fibres. *Journal of Natural Fibers* 4 (4):13-29.

Ben Amor, I., H. Rekik, H. Kaddami, M. Raihane, M. Arous, and A. Kallel. 2010. Effect of palm tree fiber orientation on electrical properties of palm tree fiber-reinforced polyester composites. *Journal of Composite Materials* 44 (13):1553-1568.

Bismarck, A., Mishra, S., Lampke, T.,. 2005. Plant fibres as reinforcement for green composites. In *Natural fibres biopolymers and biocomposites*, edited by A. K. Mohanthy, Misra, M., Drzal, L.T. USA: CRC Press, Taylor and Francis Group.

Bledzki, A. K., Faruk, O., Sperber, V. E. 2006. Cars from bio fibres. *Macromolecular Materials and Engineering* 291:449–457.

Bledzki, A. K., and J. Gassan. 1999. Composites reinforced with cellulose based fibres. *Progress in Polymer Science (Oxford)* 24 (2):221-274.

Chai, L. L., S. Zakaria, C. H. Chia, S. Nabihah, and R. Rasid. 2009. Physico-mechanical properties of PF composite board from EFB fibres using liquefaction technique. *Iranian Polymer Journal (English Edition)* 18 (11):917-923.

Chand, N. 1992. Electrical characteristics of sunhemp fibre. *Journal of Materials Science Letters* 11 (3):138-139.

Chand, N., and D. Jain. 2005. Effect of sisal fibre orientation on electrical properties of sisal fibre reinforced epoxy composites. *Composites Part A: Applied Science and Manufacturing* 36 (5):594-602.

Chew, L.T., Ong,C.L. 1985. Particleboard from oil palm trunk. Paper read at Proceedings of the National Symposium on Oil Palm By-products for Agro-based Industries, at Kaula Lumpur.

Chew, T. L., and S. Bhatia. 2008. Catalytic processes towards the production of biofuels in a palm oil and oil palm biomass-based biorefinery. *Bioresource Technology* 99 (17):7911-7922.

Dahlan, I. 2000. Oil palm frond, a feed for herbivores. *Asian-Australasian Journal of Animal Sciences* 13 (SUPPL. C):300-303.

Datta, A. K., B. K. Samantaray, and S. Bhattacherjee. 1984. Mechanical and dielectric properties of pineapple fibres. *Journal of Materials Science Letters* 3 (8):667-670.

Demir, H., U. Atikler, D. Balköse, and F. TihmInlIoglu. 2006. The effect of fiber surface treatments on the tensile and water sorption properties of polypropylene-luffa fiber composites. *Composites Part A: Applied Science and Manufacturing* 37 (3):447-456.

Demirbaş, A. 2000. Mechanisms of liquefaction and pyrolysis reactions of biomass. *Energy Conversion and Management* 41 (6):633-646.

Dufresne, A. 2008. Cellulose-based composites and nanocomposites. In *Monomers, Polymers and Composites from Renewable Resources*, edited by A. Gandini, Belgacem, M.N. Oxford, UK: Elsevier.

Duncan, John. 2008. Principle and applications of mechanical thermal analysis. In *Principles and Applications of Thermal Analysis* edited by P. Gabbott. Oxford: Blackwell Publishing Ltd, UK.

Fraga, A. N., E. Frullloni, O. De La Osa, J. M. Kenny, and A. Vaİ zquez. 2006. Relationship between water absorption and dielectric behaviour of natural fibre composite materials. *Polymer Testing* 25 (2):181-187.

Fu, S. Y., G. Xu, and Y. W. Mai. 2002. On the elastic modulus of hybrid particle/short-fiber/polymer composites. *Composites Part B:Engineering* 33 (4):291-299.

Gupta, A. K. 2009. Natural plant fibre based green composite for automobile application.

Hariharan, A., Abu Bakar, Abdul Khalil,H.P.S. 2004. Influence of Oil Palm fibre loading on the mechanical and physical properties of glass fibre reinforced epoxy bi-layer hybrid laminated composite. Paper read at Proceeding of 3rd USM-JIRCAS Joint International Symposium, 9-11 March, 2004, at Penang, Malaysia.

Haseena, A. P., G. Unnikrishnan, and G. Kalaprasad. 2007. Dielectric properties of short sisal/coir hybrid fibre reinforced natural rubber composites. *Composite Interfaces* 14 (7-9):763-786.

Hassan, Azman, Arshad Adam Salema, Farid Nasir Ani, and Aznizam Abu Bakar. 2010. A review on oil palm empty fruit bunch fiber-reinforced polymer composite materials. *Polymer Composites* 31 (12):2079-2101.

Hassim, H. A., M. Lourenço, G. Goel, B. Vlaeminck, Y. M. Goh, and V. Fievez. 2010. Effect of different inclusion levels of oil palm fronds on in vitro rumen fermentation pattern, fatty acid metabolism and apparent biohydrogenation of linoleic and linolenic acid. *Animal Feed Science and Technology* 162 (3-4):155-158.

Hill, C. A. S., and H. P. S. Abdul Khalil. 2000. Effect of fiber treatments on mechanical properties of coir or oil palm fiber reinforced polyester composites. *Journal of Applied Polymer Science* 78 (9):1685-1697.

Idicula, M., S. K. Malhotra, K. Joseph, and S. Thomas. 2005a. Dynamic mechanical analysis of randomly oriented intimately mixed short banana/sisal hybrid fibre reinforced polyester composites. *Composites Science and Technology* 65 (7-8):1077-1087.

Idicula, M., S. K. Malhotra, K. Joseph, and S. Thomas. 2005b. Effect of layering pattern on dynamic mechanical properties of randomly oriented short banana/sisal hybrid fiber-reinforced polyester composites. *Journal of Applied Polymer Science* 97 (5):2168-2174.

Jacob, M., B. Francis, S. Thomas, and K. T. Varughese. 2006c. Dynamical mechanical analysis of sisal/oil palm hybrid fiber-reinforced natural rubber composites. *Polymer Composites* 27 (6):671-680.

Jacob, M., B. Francis, K. T. Varughese, and S. Thomas. 2006d. The effect of silane coupling agents on the viscoelastic properties of rubber biocomposites. *Macromolecular Materials and Engineering* 291 (9):1119-1126.

Jacob, M., S. Jose, S. Thomas, and K. T. Varughese. 2006b. Stress relaxation and thermal analysis of hybrid biofiber reinforced rubber biocomposites. *Journal of Reinforced Plastics and Composites* 25 (18):1903-1917.

Jacob, M., S. Thomas, and K. T. Varughese. 2004a. Natural rubber composites reinforced with sisal/oil palm hybrid fibers: Tensile and cure characteristics. *Journal of Applied Polymer Science* 93 (5):2305-2312.

Jacob, M., S. Thomas, and K. T. Varughese. 2004b. Mechanical properties of sisal/oil palm hybrid fiber reinforced natural rubber composites. *Composites Science and Technology* 64 (7-8):955-965.

Jacob, M., K. T. Varughese, and S. Thomas. 2005. Water sorption studies of hybrid biofiber-reinforced natural rubber biocomposites. *Biomacromolecules* 6 (6):2969-2979.

Jacob, M., K. T. Varughese, and S. Thomas. 2006a. Dielectric characteristics of sisal-oil palm hybrid biofibre reinforced natural rubber biocomposites. *Journal of Materials Science* 41 (17):5538-5547.

Jacob, Maya; , Sabu; Thomas, and K.T. Varughese. 2007. Biodegradability and Aging Studies of Hybrid Biofiber Reinforced Natural Rubber Biocomposites. *Journal of Biobased Materials and Bioenergy* 1:118-126.

Jarukumjorn, K., and N. Suppakarn. 2009. Effect of glass fiber hybridization on properties of sisal fiber-polypropylene composites. *Composites Part B: Engineering* 40 (7):623-627.

Jawaid, M., and H. P. S. Abdul Khalil. 2011a. Cellulosic/synthetic fibre reinforced polymer hybrid composites: A review. *Carbohydrate Polymers* 86 (1):1-18.

Jawaid, M., H. P. S. Abdul Khalil, and A. Abu Bakar. 2010. Mechanical performance of oil palm empty fruit bunches/jute fibres reinforced epoxy hybrid composites. *Materials Science and Engineering A* 527 (29-30):7944-7949.

Jawaid, M., H. P. S. Abdul Khalil, and A. Abu Bakar. 2011b. Woven hybrid composites: Tensile and flexural properties of oil palm-woven jute fibres based epoxy composites. *Materials Science and Engineering A* 528 (15):5190-5195.

Jawaid, M., H. P. S. Abdul Khalil, and O. S. Alattas. 2012. Woven hybrid biocomposites: Dynamic mechanical and thermal properties. *Composites Part A: Applied Science and Manufacturing* 43 (2):288-293.

Jawaid, M., H. P. S. Abdul Khalil, A. H. Bhat, and A. Abu Baker. 2011c. Impact properties of natural fiber hybrid reinforced epoxy composites. *Advanced Materials Research* 264-265:688-693.

Jawaid, M., H. P. S. Abdul Khalil, P. Noorunnisa Khanam, and A. Abu Bakar. 2011d. Hybrid Composites Made from Oil Palm Empty Fruit Bunches/Jute Fibres: Water Absorption, Thickness Swelling and Density Behaviours. *Journal of Polymers and the Environment* 19 (1):106-109.

Jawaid, M., and Abdul Khalil, H.P.S. 2011e. Effect of layering pattern on the dynamic mechanical properties and thermal degradation of oil palm-jute fibers reinforced epoxy hybrid composite. *BioResources* 6 (3):2309-2322.

Jawaid, M., H. P. S. A. Khalil, and A. A. Bakar. 2011f. Hybrid composites of oil palm empty fruit bunches/woven jute fiber: Chemical resistance, physical, and impact properties. *Journal of Composite Materials* 45 (24):2515-2522.

Jawaid, M., H. P. S. A. Khalil, A. A. Bakar, and P. N. Khanam. 2011g. Chemical resistance, void content and tensile properties of oil palm/jute fibre reinforced polymer hybrid composites. *Materials and Design* 32 (2):1014-1019.

Jayamol, G., S. S. Bhagawan, and S. Thomas. 1998. Electrical properties of pineapple fibre reinforced polyethylene composites. *Journal of Polymer Engineering* 17 (5):383-404.

John, M. J., B. Francis, K. T. Varughese, and S. Thomas. 2008. Effect of chemical modification on properties of hybrid fiber biocomposites. *Composites Part A: Applied Science and Manufacturing* 39 (2):352-363.

John, M. J., and S. Thomas. 2008. Biofibres and biocomposites. *Carbohydrate Polymers* 71 (3):343-364.

Jon Fox-Rubin, David Cramer. 2010. *Thermoplastic Composites*. Flex Technologies 2010 [cited 29Th October 2010]. Available from http://www.fiberforge.com/thermoplastic-composites/thermoplastic-composites.php.

Joseph, S., and S. Thomas. 2008. Electrical properties of banana fiber-reinforced phenol formaldehyde composites. *Journal of Applied Polymer Science* 109 (1):256-263.

Joshi, S. V., L. T. Drzal, A. K. Mohanty, and S. Arora. 2004. Are natural fiber composites environmentally superior to glass fiber reinforced composites? *Composites Part A: Applied Science and Manufacturing* 35 (3):371-376.

Kalam, A., B. B. Sahari, Y. A. Khalid, and S. V. Wong. 2005. Fatigue behaviour of oil palm fruit bunch fibre/epoxy and carbon fibre/ epoxy composites. *Composite Structures* 71 (1):34-44.

Kaliwon, J., Sh Ahmad, S., Abdul Aziz, A. 2010. International Conference on Science and Social Research (CSSR). 5-7 Dec. 2010. 528 – 531. Kuala Lumpur, Malaysia

Karger-Kocsis, J. 2000. Reinforced polymer blends. In *Polymer Blends*, edited by D. R. Paul, Bucknall, C.B.,. New York: John Wiley & Sons.

Karina, M., H. Onggo, A. H. Dawam Abdullah, and A. Syampurwadi. 2008. Effect of oil palm empty fruit bunch fiber on the physical and mechanical properties of fiber glass reinforced polyester resin. *Journal of Biological Sciences* 8 (1):101-106.

Khalil, H. P. S. A., H. D. Rozman, M. N. Ahmad, and H. Ismail. 2000. Acetylated plant-fiber-reinforced polyester composites: A study of mechanical, hygrothermal, and aging characteristics. *Polymer-Plastics Technology and Engineering* 39 (4):757-781.

Khalil, H. P. S. Abdul, M. Siti Alwani, R. Ridzuan, H. Kamarudin, and A. Khairul. 2008. Chemical Composition, Morphological Characteristics, and Cell Wall Structure of Malaysian Oil Palm Fibers. *Polymer-Plastics Technology and Engineering* 47 (3):273 - 280.

Khoo, K. C., Lee, T. W. 1985. Sulphate pulping of the oil palm trunk. Paper read at National Symposium on Oil Palm By-Products for Agro-Based Industries, at Kaulalumpur, Malaysia.

Kickelbick, G., ed. 2007. *Introduction to Hybrid Materials*. Edited by G. Kickelbick, *Hybrid Materials:Synthesis, Characterization, and Applications* Weinheim: WILEY-VCH Verlag GmbH & Co. KGaA.

Killman, W. and Hong, L.T. 1989. Anatomy and properties of oil palm stem.Proceeding of the national symposium of oil palm by-products for agro-based industries, Kaula Lumpur, pp. 18-42.

Kulkarni, A.G., K.G. Satyanarayana, and P.K. Rohatgi. 1981. Electrical resistivity of some plant fibers. *J. Mater. Sci.* 16 1719-1726.

Kumar, R. N., L. M. Wei, H. D. Rozman, and A. Abusamah. 1997. Fire resistant sheet moulding composites from hybrid reinforcements of oil palm-fibres and glass fibre. *International Journal of Polymeric Materials* 37 (1-2):43-52.

Law, K. N., W. R. W. Daud, and A. Ghazali. 2007. Morphological and chemical nature of fiber strands of oil palm empty-fruit-bunch (OPEFB). *BioResources* 2 (3):351-362.

Law, Kwei-Nam, and Xingfen Jiang. 2001. Comparative papermaking properties of oil-palm empty fruit bunch. *TAPPI Journal* 84 (1):95.

Lee, S. H., and S. Wang. 2006. Biodegradable polymers/bamboo fiber biocomposite with bio-based coupling agent. *Composites Part A: Applied Science and Manufacturing* 37 (1):80-91.

Lim, K. O. 1998. Oil palm plantations – A plausible renewable source of energy. *International Energy Journal* 20 (2):107-116.

Lim, S.C. and Khoo, K.C.1986.Characterization of oil palm trunk and its potential utilization.*Malaysian Forester* 49:3-22.

Marzinotto, M., C. Santulli, and C. Mazzetti. 2007. Dielectric properties of oil palm-natural rubber biocomposites.Conference on Electrical insulation and Dielectric Phenomena, CEIDP 2007.

Mat Amin, Khairul Anuar, and Khairiah Haji Badri. 2007. Palm-based bio-composites hybridized with kaolinite. *Journal of Applied Polymer Science* 105 (5):2488-2496.

Mehta, N. M., and P. H. Parsania. 2006. Fabrication and evaluation of some mechanical and electrical properties of jute-biomass based hybrid composites. *Journal of Applied Polymer Science* 100 (3):1754-1758.

Meier, D., and O. Faix. 1999. State of the art of applied fast pyrolysis of lignocellulosic materials - A review. *Bioresource Technology* 68 (1):71-77.

Minowa, Tomoaki, Teruo Kondo, and Soetrisno T. Sudirjo. 1998. Thermochemical liquefaction of indonesian biomass residues. *Biomass and Bioenergy* 14 (5–6):517-524.

Mishra, Supriya, Amar K. Mohanty, Lawrence T. Drzal, Manjusri Misra, and Georg Hinrichsen. 2004. A Review on Pineapple Leaf Fibers, Sisal Fibers and Their Biocomposites. *Macromolecular Materials and Engineering* 289 (11):955-974.

Mohamad, H., Zin Zawawi, Z. , and H. Abdul Halim. 1985. Potentials of oil palm by-products as raw materials for agro-based industries. Paper read at National Symposium on Oil Palm By-Products for Agro-Based Industries., at Kaulalumpur, Malaysia.

MPOB. *Overview of the Malaysian Oil Palm Industry 2009.* MPOB 2009.

Mridha, S., S. B. Keng, and Z. Ahmad. 2007. The effect of OPWF filler on impact strength of glass-fiber reinforced epoxy composite. *Journal of Mechanical Science and Technology* 21 (10):1663-1670.

Munikenche Gowda, T., A. C. B. Naidu, and R. Chhaya. 1999. Some mechanical properties of untreated jute fabric-reinforced polyester composites. *Composites Part A: Applied Science and Manufacturing* 30 (3):277-284.

Nasrin, A. B., A. N. Ma, Y. M. Choo, S. Mohamad, M. H. Rohaya, A. Azali, and Z. Zainal. 2008. Oil palm biomass as potential substitution raw materials for commercial biomass Briquettes production. *American Journal of Applied Sciences* 5 (3):179-183.

Nayak, S. K., S. Mohanty, and S. K. Samal. 2010a. Influence of interfacial adhesion on the structural and mechanical behavior of PP-banana/glass hybrid composites. *Polymer Composites* 31 (7):1247-1257.

Nayak, S. K., and S. Mohanty. 2010b. Sisal glass fiber reinforced PP hybrid composites: Effect of MAPP on the dynamic mechanical and thermal properties. *Journal of Reinforced Plastics and Composites* 29 (10):1551-1568.

Nayak, S. K., S. Mohanty, and S. K. Samal. 2009. Influence of short bamboo/glass fiber on the thermal, dynamic mechanical and rheological properties of polypropylene hybrid composites. *Materials Science and Engineering A* 523 (1-2):32-38.

Nayak, S. K., S. Mohanty, and S. K. Samal. 2010c. Hybridization effect of glass fibre on mechanical, morphological and thermal properties of polypropylene-bamboo/glass fibre hybrid composites. *Polymers and Polymer Composites* 18 (4):205-218.

Noor Haliza , A.H., Fazilah , A., and Azemi, M.N.,. 2006. Development of hemicelluloses biodegradable films from oil palm from (Elais Guinnesis). In *International Conference on*

Green and Sustainable Innovation. Chang Mai, Thailand: Energy Management and Conservation Centre, Chang Mai university.

Noorunnisa Khanam, P., H. P. S. Abdul Khalil, M. Jawaid, G. Ramachandra Reddy, C. Surya Narayana, and S. Venkata Naidu. 2010. Sisal/Carbon Fibre Reinforced Hybrid Composites: Tensile, Flexural and Chemical Resistance Properties. *Journal of Polymers and the Environment*:1-7.

Oksman, K. 2000. Mechanical properties of natural fibre mat reinforced thermoplastic. *Applied Composite Materials* 7 (5-6):403-414.

Ornaghi Jr, H. L., A. S. Bolner, R. Fiorio, A. J. Zattera, and S. C. Amico. 2010. Mechanical and dynamic mechanical analysis of hybrid composites molded by resin transfer molding. *Journal of Applied Polymer Science* 118 (2):887-896.

Panigrahi, S. , R. L. Kushwaha, Sujata Panigrahi.2010. Characterization of Palm Fiber for Development of Biocomposites Material for Automotive Industries.*SAE International Journal of Commercial Vehicles* 3 (1):304-312.

Panthapulakkal, S., and M. Sain. 2007. Injection-molded short hemp fiber/glass fiber-reinforced polypropylene hybrid composites -mechanical, water absorption and thermal properties. *Journal of Applied Polymer Science* 103 (4):2432-2441.

Paul, A., and S. Thomas. 1997. Electrical properties of natural-fiber-reinforced low density polyethylene composites: A comparison with carbon black and glass-fiber-filled low density polyethylene composites. *Journal of Applied Polymer Science* 63 (2):247-266.

Pothan, L. A., C. N. George, M. J. John, and S. Thomas. 2010. Dynamic mechanical and dielectric behavior of banana-glass hybrid fiber reinforced polyester composites. *Journal of Reinforced Plastics and Composites* 29 (8):1131-1145.

Punsuvon, P., Anpanurak, W., Vaithanomsat, P., Tungkananuruk, N.,. 2005. Fractionation of chemical components of oil palm trunk by steam explosion. Paper read at 31st Congress on Science and Technology of Thailand, at Suranaree University of Technology, Thailand.

Rana, A. K., A. Mandal, and S. Bandyopadhyay. 2003. Short jute fiber reinforced polypropylene composites: effect of compatibiliser, impact modifier and fiber loading. *Composites Science and Technology* 63 (6):801-806.

Ratnasingam, Jegatheswaran. *Oil Palm Biomass Utilization – Counting the Successes in Malaysia*. 2011. Available from http://www.woodmagmagazine.com/node/417.

Raveendran, K., A. Ganesh, and K. C. Khilar. 1995. Influence of mineral matter on biomass pyrolysis characteristics. *Fuel* 74 (12):1812-1822.

Reddy, Narendra, and Yiqi Yang. 2005. Biofibers from agricultural byproducts for industrial applications. *Trends in Biotechnology* 23 (1):22-27.

Rcid, Jonathan D., William H. Lawrence, and Richard P. Buck. 1986. Dielectric Properties Of An Epoxy Resin And Its Composite. I. Moisture Effects On Dipole Relaxation. *Journal of Applied Polymer Science* 31 (6):1771-1784.

Rout, J., M. Misra, S. S. Tripathy, S. K. Nayak, and A. K. Mohanty. 2001. The influence of fibre treatment of the performance of coir-polyester composites. *Composites Science and Technology* 61 (9):1303-1310.

Rozman, H. D., L. Musa, and A. Abubakar. 2005. Rice husk-polyester composites: The effect of chemical modification of rice husk on the mechanical and dimensional stability properties. *Journal of Applied Polymer Science* 97 (3):1237-1247.

Rozman, H. D., G. S. Tay, R. N. Kumar, A. Abusamah, H. Ismail, and Z. A. Mohd. 2001b. The effect of oil extraction of the oil palm empty fruit bunch on the mechanical properties of polypropylene-oil palm empty fruit bunch-glass fibre hybrid composites. *Polymer-Plastics Technology and Engineering* 40 (2):103-115.

Rozman, H. D., G. S. Tay, R. N. Kumar, A. Abusamah, H. Ismail, and Z. A. Mohd. Ishak. 2001a. Polypropylene-oil palm empty fruit bunch-glass fibre hybrid composites: A preliminary study on the flexural and tensile properties. *European Polymer Journal* 37 (6):1283-1291.

S.Yacob. 2007. Progress and challenges in utilization of oil palm biomass. In *Asian Science and Technology Seminar*. Jakarta, Indonesia.

Samal, S. K., S. Mohanty, and S. K. Nayak. 2009a. Banana/glass fiber-reinforced polypropylene hybrid composites: Fabrication and performance evaluation. *Polymer - Plastics Technology and Engineering* 48 (4):397-414.

Samal, S. K., S. Mohanty, and S. K. Nayak. 2009b. Polypropylene-bamboo/glass fiber hybrid composites: Fabrication and analysis of mechanical, morphological, thermal, and dynamic mechanical behavior. *Journal of Reinforced Plastics and Composites* 28 (22):2729-2747.

Satyanarayana, K. G., C. K. S. Pillai, K. Sukumaran, S. G. K. Pillai, P. K. Rohatgi, and K. Vijayan. 1982. Structure property studies of fibres from various parts of the coconut tree. *Journal of Materials Science* 17 (8):2453-2462.

Schuh, T. G. 2010. *Renewable materials for automotive applications* 2004 [cited 29.09.2010]. Available from www.ienica.net/fibresseminar/schuh.pdf.

Sheffield University. 2011. Biodegradable cars in the future? http://www.cars-tips.com/best-car-tips/biodegradable-cars-in-the-future/

Shinoj, S., R. Visvanathan, and S. Panigrahi. 2010. Towards industrial utilization of oil palm fibre: Physical and dielectric characterization of linear low density polyethylene composites and comparison with other fibre sources. *Biosystems Engineering* 106 (4):378-388.

Shinoj, S., R. Visvanathan, S. Panigrahi, and M. Kochubabu. 2011. Oil palm fiber (OPF) and its composites: A review. *Industrial Crops and Products* 33 (1):7-22.

Shuit, S. H., K. T. Tan, K. T. Lee, and A. H. Kamaruddin. 2009. Oil palm biomass as a sustainable energy source: A Malaysian case study. *Energy* 34 (9):1225-1235.

Son, Jungil, Hyun-Joong Kim, and Phil-Woo Lee. 2001. Role of paper sludge particle size and extrusion temperature on performance of paper sludge–thermoplastic polymer composites. *Journal of Applied Polymer Science* 82 (11):2709-2718.

Sreekala, M. S., J. George, M. G. Kumaran, and S. Thomas. 2001. Water-sorption kinetics in oil palm fibers. *Journal of Polymer Science, Part B: Polymer Physics* 39 (11):1215-1223.

Sreekala, M. S., J. George, M. G. Kumaran, and S. Thomas. 2002a. The mechanical performance of hybrid phenol-formaldehyde-based composites reinforced with glass and oil palm fibres. *Composites Science and Technology* 62 (3):339-353.

Sreekala, M. S., M. G. Kumaran, M. L. Geethakumariamma, and S. Thomas. 2004. Environmental effects in oil palm fiber reinforced phenol formaldehyde composites: Studies on thermal, biological, moisture and high energy radiation effects. *Advanced Composite Materials: The Official Journal of the Japan Society of Composite Materials* 13 (3-4):171-197.

Sreekala, M. S., M. G. Kumaran, and S. Thomas. 1997. Oil palm fibers: Morphology, chemical composition, surface modification, and mechanical properties. *Journal of Applied Polymer Science* 66 (5):821-835.

Sreekala, M. S., M. G. Kumaran, and S. Thomas. 2002b. Water sorption in oil palm fiber reinforced phenol formaldehyde composites. *Composites - Part A: Applied Science and Manufacturing* 33 (6):763-777.

Sreekala, M. S., S. Thomas, and G. Groeninckx. 2005. Dynamic mechanical properties of oil palm fiber/phenol formaldehyde and oil palm fiber/glass hybrid phenol formaldehyde composites. *Polymer Composites* 26 (3):388-400.

Sulaiman, F., N. Abdullah, H. Gerhauser, and A. Shariff. 2011. An outlook of Malaysian energy, oil palm industry and its utilization of wastes as useful resources. *Biomass and Bioenergy* 35 (9):3775-3786.

Sulaiman, S.A., Ahmad, M. R.T., Atnaw, S.M.,. 2011. Prediction of Biomass Conversion Process for Oil Palm Fronds in a Downdraft Gasifier. Paper read at The 4th International Meeting of Advances in Thermofluids, 3-4th October, 2011, at Melaka, Malaysia.

Sumathi, S., S. P. Chai, and A. R. Mohamed. 2008. Utilization of oil palm as a source of renewable energy in Malaysia. *Renewable and Sustainable Energy Reviews* 12 (9):2404-2421.

Swaminathan, G., and K. Shivakumar. 2009. A Re-examination of DMA testing of polymer matrix composites. *Journal of Reinforced Plastics and Composites* 28 (8):979-994.

Thwe, M. M., and K. Liao. 2003. Durability of bamboo-glass fiber reinforced polymer matrix hybrid composites. *Composites Science and Technology* 63 (3-4):375-387.

Uma Devi, L., S. S. Bhagawan, and S. Thomas. 2010. Dynamic mechanical analysis of pineapple leaf/glass hybrid fiber reinforced polyester composites. *Polymer Composites* 31 (6):956-965.

Van Rijswijk, K., W.D. Brouwer, and A. Beukers. 2001. Application of Natural Fibre Composites in the Development of Rural Societies. edited by A. Engineering: FAO.

Viva, Edmund. 2011. Palm oil waste to be processed into composite board. http://www.varieart.com/article137-Palm-Oil-Waste-to-be-Processed-into-Composite-Board [Accessed on 13th April 2012]

Wan Busu, W. N., H. Anuar, S. H. Ahmad, R. Rasid, and N. A. Jamal. 2010. The mechanical and physical properties of thermoplastic natural rubber hybrid composites reinforced with Hibiscus cannabinus, L and short glass fiber. *Polymer - Plastics Technology and Engineering* 49 (13):1315-1322.

Wan Nadirah, W. O., M. Jawaid, A. A. Al Masri, H. P. S. Abdul Khalil, S. S. Suhaily, and A. R. Mohamed. 2011. Cell Wall Morphology, Chemical and Thermal Analysis of

Cultivated Pineapple Leaf Fibres for Industrial Applications. *Journal of Polymers and the Environment*:1-8.

Wan Rosli, W. D., K. N. Law, Z. Zainuddin, and R. Asro. 2004. Effect of pulping variables on the characteristics of oil-palm frond-fiber. *Bioresource Technology* 93 (3):233-240.

Wong, K. J., U. Nirmal, and B. K. Lim. 2010. Impact behavior of short and continuous fiber-reinforced polyester composites. *Journal of Reinforced Plastics and Composites* 29 (23):3463-3474.

Xiao, B., X. F. Sun, and R. Sun. 2001. Chemical modification of lignins with succinic anhydride in aqueous systems. *Polymer Degradation and Stability* 71 (2):223-231.

Yatim, J.M., Abd Khalid, N.H, Reza. 2011. Seminar Embracing Green Technology In Construction – Way Forward., 26th April, Grand Margherita Hotel, Kuching Sarawak, Malaysia

Zaidon, A., A. M. Norhairul Nizam, M. Y. Mohd Nor, F. Abood, M. T. Paridah, M. Y. Nor Yuziah, and H. Jalaluddin. 2007. Properties of particleboard made from pretreated particles of rubberwood, EFB and rubberwood-EFB blend. *Journal of Applied Sciences* 7 (8):1145-1151.

Zulkifli, R., M. J. M. Nor, A. R. Ismail, M. Z. Nuawi, S. Abdullah, M. F. M. Tahir, and M. N. A. Rahman. 2009. Comparison of acoustic properties between coir fibre and oil palm fibre. *European Journal of Scientific Research* 33 (1):144-152.

Properties of Basalt Plastics and of Composites Reinforced by Hybrid Fibers in Operating Conditions

N.M. Chikhradze, L.A. Japaridze and G.S. Abashidze

Additional information is available at the end of the chapter

http://dx.doi.org/10.5772/48289

1. Introduction

Polymeric reinforced materials are characterized by a number of advantages over traditional structural materials since they offer such unique properties as high specific strength in some cases, in combination with light transmission, radio transparency, high electrical insulating characteristics, non-magnetic properties, corrosion resistance.

The possibility of the preparation of new materials with predetermined characteristics is one of the main advantages of reinforced plastics. In Table 1 the standard mechanical characteristics of the most abundant structural materials as well as averaged standard characteristics, for example, of oriented and randomly reinforced glass-plastics are given.

In recent years an information on new type of polymeric composite-basalt plastic (BP), in which the basalt fiber is used instead of glass reinforcing one [1,2], is of frequent occurrence. Basalt fibers are practically highly competitive with glass ones by main mechanical characteristics and surpasses them by some of them, in particular, by water-resistance and chemical stability (is shown below). But in the form of twisted and non-twisted threads, rovings, roving cloth and discrete fibers, basalt ones represent an alternative and promising reinforcing element for composites. In addition, at solving of a series of specific problems, for example, for preparation of materials with predetermined strength and deformation characteristics in different directions of load application, the combination of glass, high-strength basalt, high-strength and high-modulus carbon fibers were used, that is to say, the production of composites, reinforced by hybrid fibers (HFRC) was organized [3-8].

Here we don't detail the properties of these and other types of reinforcing fibers. We have restricted ourselves to the comparison of glass, basalt and carbon fibers (Table 2).

Material	Density Kg.m^{-3}	Tensile strength, MPa	Tensile modulus, GPa	Specific strength km	Specific rigidity km
Metals					
Steel	7800	400	200	5.1	2560
Aluminum alloy	2800	300	72	10.7	2580
Titanium	4500	350	115	17.8	2560
Wood					
Oak	720	130	15	15.2	1750
Plastics					
Polyethylene	960	20	0.5	2.1	52
Vinyl plastic	1400	60	3	4.3	210
Glass reinforced plastics					
Unidirectional	2000	1600	56	80.0	2800
Glass-cloth-base laminate	1900	500	30	26.2	1570
Randomly oriented	1400	100	8	6.7	530

Table 1. Mechanical properties of structural materials

Fiber	Density gr. m^{-3}	Failure stress, GPa	Extension at failure, %	Elasticity modulus GPa
Glass	2.4-2.5	2.8-3.0	4.7-5.6	74-95
Basalt	2.6-2.8	1.9-2.6	3.5-4.5	70-90
Carbon	1.9-2.1	2.2-7.2	0.5-2.4	200-785

Remark: Indexes are given for the production of the firms of various countries.

Table 2. Properties of reinforcing fibers

In parallel with the advantages, BP and HFRC, undoubtedly, are characterized by some disadvantages, which must be taken into account at the preparation and operation of structures and items with the use of BP and HFRC. These disadvantages involve:

- Structural non-uniformity and inadequate stability of the technology of preparation leads to considerable dissipation of mechanical and other indexes, which may attain to 15-20% in relation to average values even at standard short-term testing. At long-term testing the dissipation increased.
- Polymeric nature of a matrix determines an enhanced sensitivity of materials to the prehistory of preparation and to temperature-time regime of further operation, which is responsible for determines strength and deformation properties of BP and HFRC. At moderate temperatures for traditional structural materials a temperature-time dependence of mechanical and other properties appears only slightly, whereas the

presence of polymeric matrix in considered materials predetermines an impossibility of the evaluation of strength or deformability at room temperature without specifying of time in the course of which the materials are in stressed state.

- Directional locating of reinforcing fibers in the plane of reinforcement as well as a lamination of the structure in the direction perpendicular to mentioned plane, causes an anisotropy of mechanical and other properties. As a rule, a number of characteristics, necessary for determination of one or another properties of reinforced plastics, is considerably more than for isotropic materials. Moreover, the regularities of the behavior of reinforced plastics at mechanical testing depend on the direction of load application. For example, for oriented composites a tension diagram in the direction of reinforcement is governed by Hooke's law. At loading at an angle to the direction of reinforcement, this diagram becomes essentially non-linear. A lamination of the structure of polymeric composites predetermines their low resistance to interlayer shear and to transverse breaking off. Therefore, at bending, these materials may be destroyed because of the fact that tangential stresses will be higher than material's resistance to interlayer shear instead of the fact that normal stresses (extending or compressing) may attain the limiting values.

- Deformations, generated perpendicularly to reinforcing fibers, are mainly realized in matrix interlayers because of low rigidity of the latter in comparison with glass, basalt or carbon fibers; this fact leads to the formation of the cracks in the interlayers of a binder between the fibers or at phase boundaries. Low crack resistance is particularly characteristic of oriented plastics. The cracks have little or no effect on the values of characteristics, obtained as a result of short-term testing. However, such characteristics of a material as hermeticity, resistance to corrosive media, mechanical and electrotechnical properties in the conditions of long-term operation at the appearance and intergrowth of the track are significantly impaired.

- Relatively low value of elasticity modulus of reinforced plastics and composites leads to the fact that load-carrying ability of thin-wall structures is limited by deformability and stability instead of the strength. For complete use of high strength characteristics of the composites it is profitable to design the item and structure as three-layered or to provide the stiffening ribs. Designing must be carried out in such a way as to the material will operate on tension instead of compression, whenever possible. However, it should be noted that in some cases the low elasticity modulus is a definite advantage of reinforced polymers (for example, pipe-lines from mentioned materials without the compensators of temperature deformation and etc.)

Considered peculiarities of reinforced polymers, generally, and of BP and HFRC, in particular, must be taken into account at designing and at the use of structures and items from mentioned materials.

Appearance of new generation of reinforced polymers – BP and HFRC is due to the quest for preparation of the materials, characterized by higher initial mechanical and other indexes and by higher stability of these indexes at the action of various operating factors.

At the present time the volumes of the production and of the use of the composites, reinforced by high-strength and high-modulus fibers, are insignificant. The main consumers of mentioned materials are aviation and rocket-space engineering. The main barrier for widening of the fields of the use of such materials (for example, in wind power engineering, chemical production and etc.) is their high cost.

In regard to the cost of BP and HFRC, it should be noted that the ways for their cost reduction, probably, are associated with a cheapening of initial materials as well as of mechanization of the production instead of the increase of the output, concentration and specialization of the production. In this connection, the problem of the cost reduction for composites by, if only for, partial replacement of expense and scarce carbon fiber by considerably cheaper (by an order) basalt one without significant impairment of main operating properties of the material is highly topical. Moreover, the share of reinforcing fibers as well as of a binder in the expenses of raw materials is distinct for the composites of various types. For the composites, in which the nonwoven reinforcing elements are used in the form of threads and mats, the expenses of reinforcing materials attain to 30-35% of all material expenses. At the same time the expenses of reinforcing materials in the form of the cloths may attain to 50-70% at the preparation of basalt plastics. Therefore an essential effect in the reduction of composite's cost may be attained by replacing of the cloths from twisted threads by nonwoven reinforcing materials and roving cloths.

Price cost reduction for BP and HFRC is also possible at the expense of introducing of efficient fillers-reinforcers into the matrix composition. This method allows a considerable decrease of fiber content without an essential reduction of characteristics of the resulting material.

One more way for enhancement of the efficiency of the use of BP and HFRC in action is a rational design and the use of the items from them with regard to the effect of real environment on the material. Below the primary attention is given to more or less detailed consideration of these problems.

2. General methodology of investigations

At designing of structures and items from composite materials, primarily the values of their calculated resistances are necessary. By long-term calculated resistance (R_{cl}) of the material in normal conditions the product of normative resistance of the material by coefficient of long-term resistance and by coefficient of the uniformity of its mechanical characteristics is meant:

$$R_{cl} = R_{nor} K_{\ell\text{-}t} K_u$$

Normative resistance (R_{nor}) was determined a strength limit of the materials under study by the results of short-term testing of small samples, carried out in accordance with acting standards. Coefficient of long-term resistance ($K_{\ell\text{-}t}$) was determined by testing to failure of the series of the samples of the materials at long-term loading at the stresses comprising a definite part from a strength limit of the material. Uniformity coefficient (K_u) was determined by well-known three sigma rule by calculation of arithmetic mean and by root-mean-square deviation of the strength, which are defined on the basis of statistical analysis of the results of mass testing of strength properties of BP and HFRC.

Calculated resistances of the materials, operating at the joint action of static load and regimes, different from normal ones (elevated temperature, high humidity, corrosive medium and etc.) were determined by multiplying the long-term calculated resistances into corresponding coefficients of operating conditions:

$$R_{c\ell}^{T} = R_{c\ell} \cdot K_{T}, \qquad R_{c\ell}^{w} = R_{c\ell} \cdot K_{w}, \qquad R_{c\ell}^{cor} = R_{c\ell} \cdot K_{cor}, \qquad R_{c\ell}^{atm} = R_{c\ell} \cdot K_{atm}$$

where K_{T}, K_{w}, K_{cor}, K_{atm} - coefficients of operating conditions of composites, service of which is provided, respectively, at elevated temperature, in water or at high humidity at the action of corrosive media, in atmospheric conditions, as well as at synchronous long-term action of load as well as of external factors. In some cases the coefficients of operating conditions were determined at the joint action of various factors, for example, of temperature, water /humidity ($K_{T,w}$).

The objects of investigations were:

2.1. Basalt reinforced plastics

BP-1. Sheet basalt plastic. Thickness (δ)-1.5-2.5mm; density (ρ)-1360-1380 kg.m^{-3}; Matrix – unsaturated polyester resin of Turkish production (65 mass%). Reinforcing element – chopped fiber, obtained by cutting of basalt roving of Georgian production with following characteristics: rectilinear density 600-4800 tex; elemental fiber diameter 10-16 µ; specific tenacity 350-450 mN/tex.

The mode of preparation: contact moulding (without pressure and temperature).

Expected field of application: light-transparent guarding building structure.

BP-2. Basalt cloth – based laminate. δ = 0.7-5.0 mm; ϱ = 1530-1560 kg.m^{-3} . Matrix-phenol-formaldehyde resin of Ukrainian production (25-35 mass%). Reinforcing element – cloth from twisted threads of Georgian production with following characteristics: thickness 0.25-0.35 mm; surface density 150-450 g/m^{2}; density in warp 4-8 th/cm; density in weft 6-12 th/cm, or cloth from basalt roving with the indexes: thickness 0.4-0.9 mm; surface density 300-700 g/m^{2}; density in warp – 1.7-3.5 th/cm; density in weft 2.9-4.0 th/cm;

The mode of preparation: direct pressing, pressure 45-55 kgf/cm^{2}, pressing temperature 413-443K, holding time on 1 mm-5-12 min.

Expected field of application: shells of three–layered building panels (among them for corrosive media).

BP-3. Oriented basalt plastic. δ = 1.0-7.0 mm; ϱ = 1520-1540 kg.m^{-3}. Matrix-epoxy-phenol resin of Ukrainian production (25-32 mass%). Reinforcing element-permanently oriented basalt fiber in the form of roving (data see in BP-1).

The mode of preparation: production of veneer, its impregnation by a binder, direct pressing of semifinished item.

Expected field of application- auxiliary structural elements and details.

BP-4. Pressed basalt plastic. δ = 2.0-8.0 mm; ϱ = 1850-1950 kg.m^{-3}. Matrix – modified phenol-formaldehyde resin of Ukrainian production (25-35 mass%). Reinforcing element-chopped or permanently oriented basalt fiber (data see in BP-1, BP-2).

The mode of preparation: preliminary impregnation of reinforcing element, direct pressing at the temperatures of 140-160^{0}C; pressure 250-350 kgf/cm^{2}, holding time 2-5 min on 1 mm of material.

Expected field of application: auxiliary structural details for corrosive media.

2.2. Composites on the basis of hybrid fibers

HFRC-1 . Oriented bi-directional composite. δ = 1.5-2.5 mm; ϱ = 1450-1550 kg.m^{-3}. Matrix – epoxy resin of Ukrainian production (70-75 mass%). Reinforcing elements – glass and carbon fibers (GF, CF), located in the composite by the scheme, presented in Fig.1. Glass fibers of alkalineless composition are presented in the form of roving of Ukrainian production. Polyacrylonitrile carbon rovings of Russian production offer the strength 2.3 GPa and elasticity modulus 220 GPa. Ratio GF : CF = 0.3÷0.7 (by mass).

The mode of preparation: production of prepreg, its direct pressing.

Expected field of application: shell of wind turbine blade.

HFRC-2. The same, but 20% of carbon fiber is replaced by basalt one in the form of roving.

HFRC-3. Oriented composite. δ = 2-3 mm, ϱ = 1450-1550 kg.m^{-3}. Matrix – the same as in the case of HFRC-1 and HFRC-2; Reinforcing elements – the same as in the case of HFRC-1 and HFRC-2 . They are located by the scheme, shown in fig. 2.

The mode of preparation: the same as in the case of HFRC-1 and HFRC-2.

Expected field of application: the spar of wind turbine blade.

HFRC-4. The material similar to the composite HFRC–3, but 20% of carbon fiber is replaced by basalt one in the form of roving.

The results of determination of normative resistances at various types of stressed state (tension, bending, compressing, shear - R_{nor}^{t}, R_{nor}^{b}, R_{nor}^{c}, R_{nor}^{sh} as well as of short-term elasticity modulus at tension, bending and compressing (E_{s-t}^{t}, E_{s-t}^{b}, E_{s-t}^{c}) and coefficients of uniformity of strength properties of the materials under study are given in Table 3.

Returning to the problem on uniformity coefficient of material, it should be noted that testing, carried out for its determination were performed at room temperature – humid conditions. Incidentally, in the course of operating of the structures by the use of plastic materials, they may undergo to various temperature- humid effects and it may be suggested

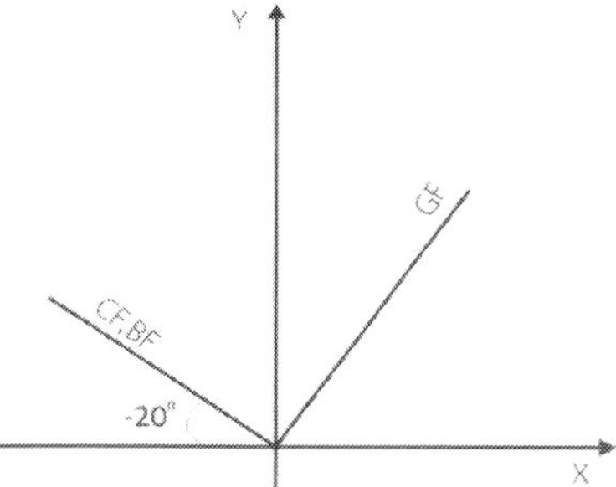

Figure 1. Structure of composite intended for shell

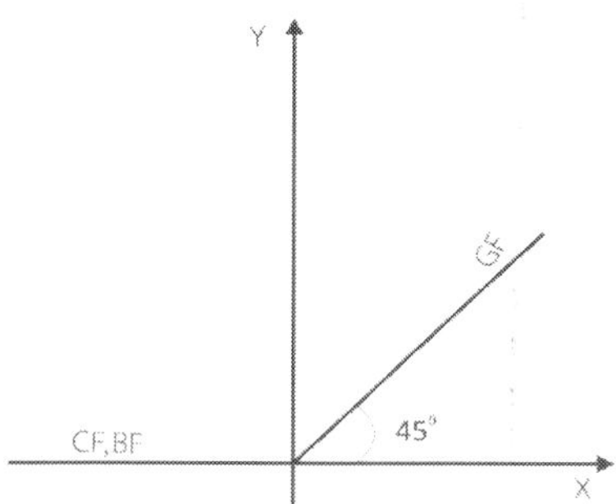

Figure 2. Structure of composite intended for spar

that these effects may exert some influence not only to the variation of absolute values of mechanical properties of BP and HFRC, but, to some extent, they may reflect on the indexes of uniformity of strength properties of the materials. To check this suggestion the investigations were carried out to reveal the influence of preliminary action on the indexes of uniformity of strength properties of BP-1, BP-2, HFRC-2 and HFRC-4 at tension. In Table 4 the results of these testing are given. They involve the data necessary to calculate the uniformity coefficient (number of testing – n, arithmetic – mean value for strength - σ_{av}, mean square deviation - σ' as well as variation coefficient – V). The regimes of preliminary action on the samples of the material were: №1 – holding in laboratory room; №2 – heating at 353K over 10 days; №3 – steeping over 1 day; №4 - steeping over 10 days; №5 – steeping over 10 days by further drying over 10 days; №6 - steeping over 10 days at 353K.

As is seen from Table 4, the variation of K_u is relatively small and maximum reduction of K_u comprises 9 % for BP and 6 % for HFRC. Hence, the value of K_u, obtained by testing in normal temperature- humid conditions, may be used by confidence at practical calculations.

Under prolonged (long-term) strength of the solid the dependence of time duration up to its failure on the stress and temperature is meant. The coefficient of long-term resistance is a value, determined by testing of a series of the materials samples under prolonged loading to failure at the stresses, constituent a definite part from material strength limit. Thus, in the terms "long-term resistance" and "durability" an equal meaning is assigned.

Material	R_{nor}^{t}, MPa	R_{nor}^{b}, MPa	R_{nor}^{c}, MPa	R_{nor}^{sh}, MPa	E_{s-t}^{t}, GPa	E_{s-t}^{b}, GPa	E_{s-t}^{c}, GPa	K_u
BP - 1	69.0	145.0	105.6	55.0	6.0	-	-	0.65
BP - 2	250.8	130.0	105.5	75.0	26.0	-	-	0.78
BP - 3	480.6	750.8	410.2	190.0	31.5	-	-	0.75
BP - 4	85.0 / 560.0	130.0 / 260.0	110.0 / 210.5	-	19.3 / 19.5	-	-	0.75
HFRC-1	195.6 / 163.1	480.2 / 270.3	261.1 / 219.1	10.2 / 8.1	9.4 / 3.9	14.5 / 3.4	10.6 / 4.2	0.72
HFRC-2	292.5 / 228.2	567.2 / 351.4	410.2 / 319.9	12.2 / 10.2	15.0 / 5.4	23.4 / 5.4	14.6 / 5.8	0.68
HFRC-3	455.4 / 6.9	718.1 / 19.9	420.8 / 8.0	24.8 / 1.2	96.9 / 5.8	78.6 / 5.2	78.1 / 4.1	0.74
HFRC-4	132.2 / 85.6	415.7 / 95.9	160.2 / 107.7	8.8 / 2.4	49.7 / 19.8	58.8 / 16.4	51.0 / 16.7	0.70

Remark: 1 BP-1. Resistances at shear are given in the direction, perpendicular to sheet plane.

2. BP-2. For efforts acting in the direction of the base of basalt cloth (δ=7mm).

3. BP-3. At the ratio between longitudinal and transverse fibers, equal to 1:1 for efforts, acting in the direction of fibers.

4. BP-4. In numerator and denominator at reinforcing by chopped and oriented fibers, respectively.

5. HFRC-1, HFRC-2, HFRC-3, HFRC-4. In numerator and denominator the values along and transversely to X axis, respectively (Fig. 1,2).

Table 3. Normative resistances, short-term elasticity modulus and uniformity coefficients for BP and HFRC

Material	Act. reg.	n	σ_{av}, MPa	σ', MPa	V, %	K_u	Material	Act. reg.	n	σ_{av}, MPa	σ', MPa	V, %	K_u
BP-1 (δ=1.5mm)	№1	95	72	8	11.1	0.67	HFRC-2 (δ=1.5mm)	№1	92	292	29	9.9	0.70
	№2	100	75	9	12.0	0.64		№2	95	262	25	9.5	0.71
	№3	96	65	8	12.3	0.63		№3	99	277	29	10.5	0.69
	№4	98	62	7	11.3	0.66		№4	100	295	30	10.2	0.69
	№5	92	66	7	10.6	0.68		№5	93	290	31	10.7	0.68
	№6	97	69	9	13.0	0.61		№6	95	277	30	10.8	0.68
BP-2 (δ=0.8mm)	№1	98	359	28	7.8	0.77	HFRC-4 (δ=0.8mm)	№1	100	155	10	6.5	0.80
	№2	95	349	30	8.6	0.74		№2	100	166	18	10.8	0.67
	№3	90	367	30	8.2	0.72		№3	95	125	12	9.6	0.71
	№4	92	228	18	7.9	0.80		№4	96	115	12	10.4	0.69
	№5	93	341	27	7.9	0.76		№5	99	140	14	10.0	0.70
	№6	95	361	30	8.3	0.74		№6	91	111	12	10.8	0.68

Table 4. Statistical processing of the result of testing to reveal the influence of preliminary action of various factors on uniformity indexes.

The equation of temperature-time dependence of the solids, as it well known, relates durability (τ), stress (σ) and temperature (T) to each other [9,10]:

$$\tau = \tau_0 \exp\left(\frac{U_0 - \gamma\sigma}{kT}\right) \qquad (1)$$

where τ_0 - is a constant, approximately equal to 10^{-13} sec, which in order of a value is near to the period of thermal oscillations of atoms; U_0 – initial activation energy of the process of material destruction; k – Boltzmann's constant; γ - average coefficient of overstresses.

Physical meaning of the formula (1) may be explained by means of thermal fluctuation theory of strength. According to this theory, destruction is kinetic, thermally fluctuating process of permanent accumulation of damages, developed in the body since the load application to its destruction. Breaking of interatomic bonds, activated by applied stress, is an elementary act of destruction process.

In the course of experiments it has been established that BP and HFRC under our study are, mainly, obey the temperature-time dependence. Along with it, it should be noted, that in relation to BP and HFRC, which are bi or more component composites, physical meaning of the values τ_0, U_0 and γ is not reasonably evident. But it should be taken into account that the main goal of our investigations is to obtain the empirical relationships between long-term resistance (durability) of new types of structural materials and the conditions of their operation. In this case the question about physical meaning of above-mentioned values does not need to be posed.

From (1) the following is obtained:

$$\sigma = \frac{U_0 - \dfrac{kT}{\lg e}\lg\dfrac{\tau}{\tau_0}}{\gamma} \qquad (2)$$

If in formula (2) it is granted that $\tau = \tau_i = const$, we obtain

$$\sigma = \frac{U_0}{\gamma} - BT \qquad (3)$$

where $B = \dfrac{K\lg\dfrac{\tau_i}{\tau_0}}{\gamma\lg e}$

Thus, the linear dependence between material strength and temperature at $\tau = const$ is in existence. At elevating of testing temperature, primarily, the adhesion bonds are broken at the boundaries of basalt (glass) – binder and in the matrix the cracks, parallel to the fiber, are formed, since U_0 values for silicate fibers are equal to 350-385 kJ/mole and the values of activation energy of destruction of polymeric matrix comprise 125-190 kJ/mole. Intensity of breaking of adhesion bond: carbon-binder is of lesser importance, since U_0 for polyacrylonitrile is near to U_0 for binder (200 kJ/mole). In spite of this fact, in materials under study, breaking of adhesion bond doesn't lead to their destruction since reinforcing element continues an operation as the bundle of non-bound fibers. In this case minimal breaking stress for uni-directional basalt plastics is estimated as a half of breaking stress, obtained by standard testing of the material at normal temperature.

3. Mechanical properties of BP and HFRC

3.1. Effect of duration of static loading and temperature on strength

Dependence of breaking stress of BP and HFRC on temperature was estimated by the coefficients of operating conditions $K_T = \sigma_T / \sigma_{s\text{-}t}$, where σ_T and $\sigma_{s\text{-}t}$ are breaking stresses for the samples after temperature action and at short-term testing, respectively. The values of K_T are presented in Table 5.

If in formula (1) it is granted that

$$T = T_i = const, \text{ then } \sigma = \frac{U_0}{\gamma} - A \lg \frac{\tau}{\tau_0} \tag{4}$$

where $A_i = \dfrac{kT_i}{\gamma \lg e}$.

Material	Temperature, K					
	313			333		
	Tension, Compression	Bending	Shearing	Tension, Compression	Bending	Shearing
BP-1	0.65	0.85	0.67	0.60	0.79	0.63
BP-2, BP-3	0.88	0.79	0.78	0.77	0.72	0.72
BP-4	0.70	0.88	0.72	0.63	0.78	0.66
HFRC-1	0.72	0.88	0.77	0.70	0.85	0.72
HFRC-2	0.72	0.82	0.82	0.70	0.79	0.77
HFRC-3	0.74	0.86	0.88	0.71	0.81	0.80
HFRC-4	0.75	0.85	0.79	0.74	0.78	0.75

Remark: 1. Coefficients of operating conditions of materials in the structures at temperature 273 K are taken to be unity; 2. At intermediate temperatures K_T may be determined by interpolation.

Table 5. Coefficients of operating condition K_T in the structures, operating at elevated temperatures

In regard to material under study, the linear relationship between strength and durability logarithm is mainly obeyed at normal as well as at the temperatures of 313K and 333K. Mentioned temperatures are considerably less than temperature of forced elasticity of BP and HFRC and because of this fact the dependence $\sigma - \lg \tau$ on tension hasn't a kink, characteristic of the case of the action of high (>350K) temperatures. Below these dependences are presented in the coordinates $\sigma - \lg \tau$ in contrast to the coordinates $\lg \tau - \sigma$, used in the theory of the strength of the solids (Fig. 3, 4).

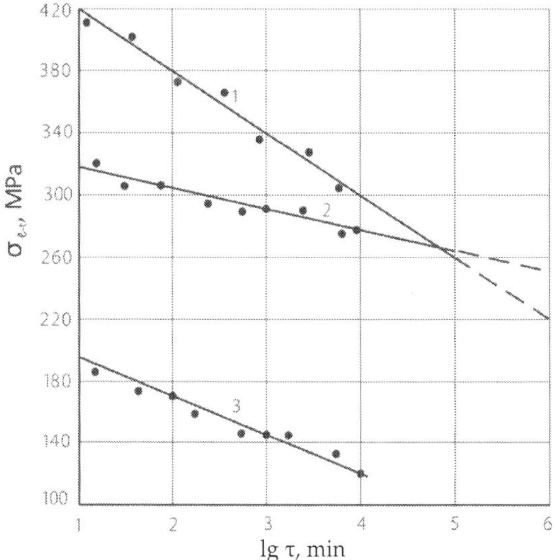

Figure 3. Curves of long-term strength of BP-3, BP-4 (tension): 1-BP-3; 2-BP-4, uni-directional;   3 – BP-4, full strength.

To estimate a time dependence of the strength, the coefficients of operating conditions of the material were used: $K_\tau - \sigma_\tau / \sigma_{st}$, where σ_τ breaking stress after the time interval, corresponding to service life of the structure or item. The values K_τ for BP and HFRC are given in Table 6.

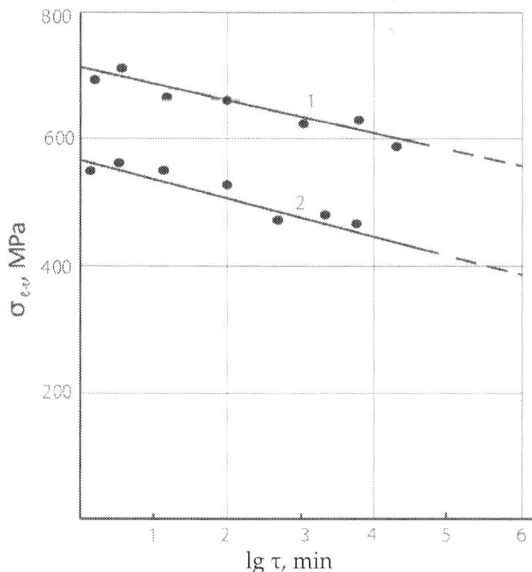

Figure 4. Curves of long-term strength of BP-1 at various temperatures (bending). 1 – 313K; 2 – 333K

Material	Tension			Compression			Bending		
	10^3h	10^4h	10^5h	10^3h	10^4h	10^5h	10^3h	$10\ ^4$h	10^5h
BP-1	0.71	0.65	0.60	0.71	0.68	0.58	0.68	0.56	0.45
BP-2	0.86	0.75	0.68	0.81	0.67	0.59	0.83	0.72	0.55
BP-3	-	0.68	0.56	0.88	0.77	0.61	0.87	0.74	0.59
BP-4	0.81	0.77	0.70	0.89	0.78	0.51	0.78	0.72	0.63
HFRC - 1	0.91	0.79	0.69	0.85	0.76	0.62	0.78	0.69	0.55
HFRC - 2	0.88	0.73	0.66	0.82	0.71	0.59	0.80	0.63	0.51
HFRC - 3	0.90	0.76	0.71	0.88	0.77	0.68	0.82	0.66	0.59
HFRC - 4	0.81	0.72	0.62	0.79	0.70	0.62	0.80	0.63	0.49

Table 6. The values of the coefficient K_τ at various types of stressed state

Direct experimental determination of σ_τ is fraught with great difficulties: the maintenance of constant external conditions and predetermined stress over a long period of time is necessary. Therefore the values σ_τ were determined for three values: 1, 10^2, 10^3 hours by extrapolating on the basis of equation (1), obtained curve and by assuming that external factors don't distort a linear character of temporal dependence of the strength.

Hence, we have, separately, the coefficients of operating conditions providing the temperature influence as well as considering the loading duration. Over many years the method of multiplying of these coefficients has been used to account the joint effect of these factors on long-term resistance. But as it was shown in [11], this method leads to considerable overstating of calculated resistances of glass plastics, especially at the temperatures close to glass transition temperature of a binder. To check this fact, a materials under study were subjected to the joint action of loading and temperature (313 K, 333 K), correlating the data, obtained in this case with the values of the coefficients of operating conditions K_τ and K_T (Table 7).

Coefficient	Temperature, K	BP - 1	BP - 2	BP - 3	BP - 4
K_T	313	0.84	0.88	0.90	0.85
	333	0.65	0.72	0.85	0.79
K_τ (τ = 5 years)		0.52	0.82	0.89	0.83
$K_T \cdot K_\tau$	313	0.44	0.72	0.80	0.71
	333	0.34	0.59	0.76	0.66
K_τ^T (τ = 5 years)	313	0.41	0.69	0.75	0.65
	333	0.28	0.49	0.69	0.59
$\dfrac{K_\tau^T}{K_T \cdot K_\tau}$	313	0.93	0.95	0.94	0.91
	333	0.82	0.83	0.91	0.89

Table 7. Values of coefficients of operating conditions of BP at bending

The analysis of the data of Table 7 confirms the fact that the method of coefficients multiplying really leads to enhanced values of the coefficients of operating conditions and consequently to overstating of calculated resistances of BP. Thus, it was decided to determine the coefficients of operating conditions for the joint action of external factors and loading.

3.2. Effect of the time of loading action and temperature on deformation characteristics

Deformability of the materials under study, caused by force action, was estimated by short-term and long-term elasticity and shear modulus ($E_{s\text{-}t}$, $E_{\ell\text{-}t}$, $G_{s\text{-}t}$, $G_{\ell\text{-}t}$) were determined by short-term static testing of small standard samples as a ratio between the increment of stress and the increment of relative deformation of a sample. $E_{\ell\text{-}t}$, $G_{\ell\text{-}t}$ were obtained by long-term static testing of the samples at stresses equal to calculated long-term resistance of materials as a ratio between the stress and maximum relative deformation of the sample at damping of creeping. It should be noted that the term "long-term elasticity modulus" is conventional in this case, since deformations of polymeric composites at long-term loading, in reality, aren't elastic.

At the temperatures, no greater than the temperature of beginning of binder destruction, reinforcing fibers act as linear-elastic materials. Binders are characterized by visco-elastic properties. Therefore, deformations of BP and HFRC, generally, depend significantly on duration and temperature of operation.

As might be expected, in uni–directional or orthogonally-reinforced BP and HFRC, the creeping is formed at the action of constant loading applied to the direction of reinforcement. But after a time, this process is practically terminated. This fact is quite clear since at first an effort is distributed between fibers and binder, but stresses in binder relax and all stresses are progressively imparted to fibers.

At loading of BP and HFRC, randomly reinforced and oriented at angle to loading direction, creeping isn't damped and is continued up to material destruction. Creeping anisotropy of these materials is expressed to a considerable more extent than an anisotropy of elastic properties and sharply enhances with temperature elevation.

The goal of experiments for determination of creeping of BP and HFRC was an establishment of functional relationship between stress, deformation and duration of loading action. Some results of the study of BP and HFRC creeping are given in Fig. 5.

As is seen, experimental points of all curves, plotted for various levels of stresses in the coordinates $\varepsilon - \tau$, fall on one curve, constructed in the coordinates $\varepsilon / \varepsilon_0 - \tau$, which allows to consider the material under study as linearly visco-elastic. Isochrones, constructed for this reason, are straight lines, reflecting a linear relationship between deformations and stresses for each fixed instant of time τ_i.

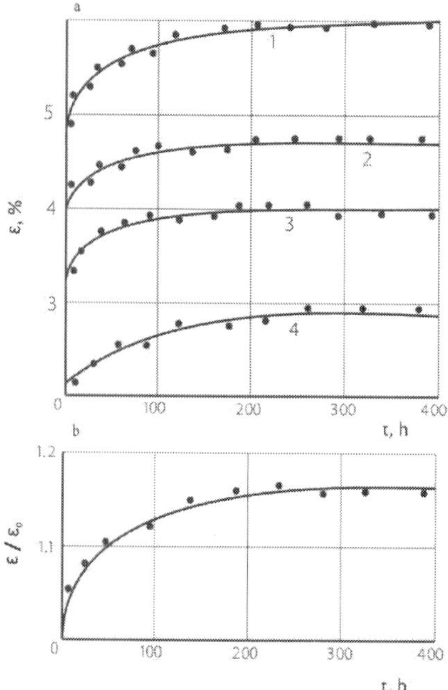

Figure 5. Curves of creeping (a) and generalized curve of creeping (b) for basalt cloth-base laminate at compression on cloth weft. Stresses: 1-90 MPa, 2 – 85MPa; 3 – 65MPa; 4 – 50 MPa.

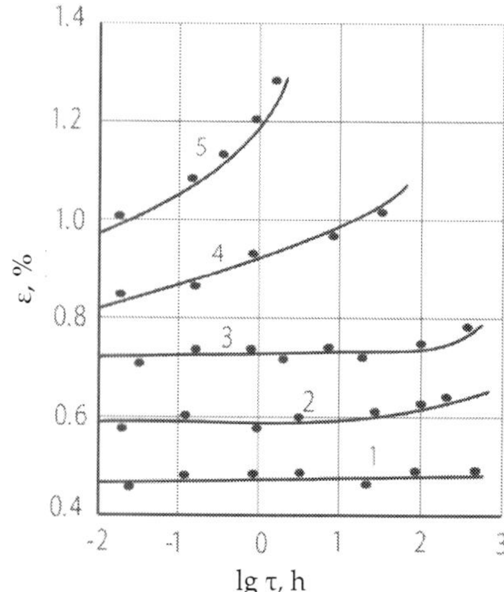

Figure 6. Creeping of the composite HFRC–4. 1,2,3,4,5, respectively, are the levels of relative stress: 0.4; 0.5; 0.6; 0.7; 0.8.

Creeping of HFRC has been investigated. The levels of relative stress, that is to say, ratio between acting stress and breaking one comprised 0.4; 0.5; 0.6; 0.7; 0.8; Experimental curves of creeping of the composite HFRC are presented in Fig. 6. By creeping curves the families of isochronic curves $\sigma - \varepsilon$ were constructed, by which the dependence of elasticity modulus on time was found. After 10^4 hours the reduction of elasticity modulus comprised 10-12%.

Creeping of BP-1 exhibits non-damped character. Deformation of creeping at tension over $5 \cdot 10^3 - 1 \cdot 10^4$ hours appears to be several times more than instantly-elastic ones (Fig.7). Similar picture is observed for other types of stressed state (Fig.8).

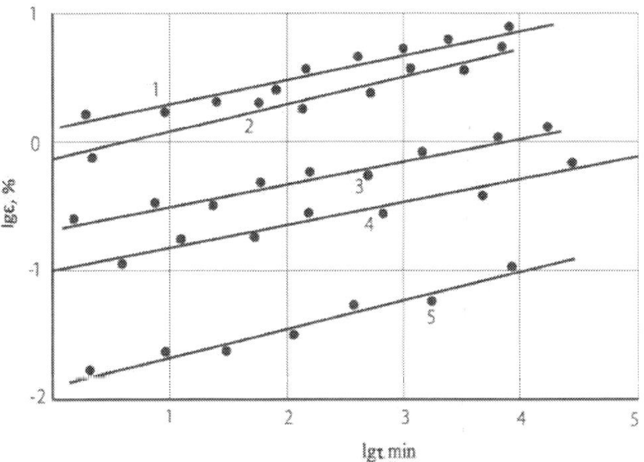

Figure 7. Curves of BP-1 creeping at tension 1-70MPa; 2-65 MPa; 3-40 MPa; 4-20 MPa; 5-10MPa.

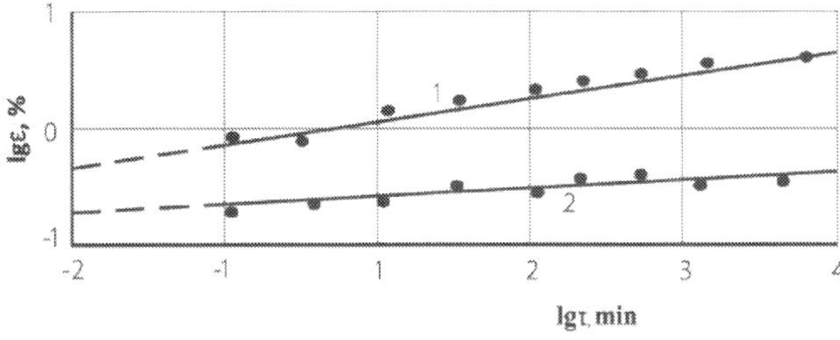

Figure 8. Curves of BP-1 creeping at bending for stress levels: 1-50MPa; 2-20 MPa.

Creeping plots at the coordinates $\lg\varepsilon$ - $\lg\tau$ present a series of parallel straight lines, which is indicative of the presence of power dependence of deformation creeping on time.

In Table 8, 9 the values of the coefficients of operating conditions are given, considering the variation of deformation characteristics of BP and HFRC depending on duration of loading action and on temperature variation.

Material	Index	K_τ, after hours		
		10^3	10^4	10^5
BP-1	E^t	0.72	0.61	0.52
	G	0.69	0.59	0.50
BP-2	E^t	0.91	0.82	0.78
	G	0.89	0.85	0.79
BP-3	E^t	0.95	0.89	0.82
	G	0.72	0.68	0.62
BP-4	E^t	0.91	0.86	0.82
	G	0.85	0.79	0.71
HFRC - 1	E^t	0.92	0.87	0.85
	G	0.85	0.76	0.69
HFRC - 2	E^t	0.88	0.82	0.79
	G	0.79	0.71	0.62
HFRC - 3	E^t	0.95	0.89	0.85
	G	0.91	0.86	0.78
HFRC - 4	E^t	0.89	0.82	0.75
	G	0.81	0.72	0.69

Table 8. The values of the coefficients of operating conditions K_τ by deformation properties

Material	Index	K_T at temperatures		Material	Index	K_T at temperatures	
		313K	333K			313K	333K
BP-1	E^t	0.53	0.42	HFRC-1	E^t	0.85	0.79
	G	0.51	0.39		G	0.82	0.75
BP-2	E^t	0.71	0.64	HFRC-2	E^t	0.82	0.72
	G	0.81	0.74		G	0.79	0.71
BP-3	E^t	0.88	0.76	HFRC-3	E^t	0.89	0.82
	G	0.83	0.74		G	0.85	0.79
BP-4	E^t	0.78	0.61	HFRC-4	E^t	0.82	0.85
	G	0.77	0.85		G	0.80	0.72

Table 9. The values of the coefficients of operating conditions K_T by deformation properties

4. Resistance of BP and HFRC to environment

4.1. Mechanical characteristics at atmospheric action

It is well known that by selecting of corresponding regimes of accelerated testing on atmospheric resistance the reduction of mechanical characteristics of materials may be relatively readily attained. But the determination of reasonably accurate correlation between the results of natural and accelerated testing on ageing of polymeric composites was unsuccessful. So far as we know, this problem wasn't solved up till now in relation to inorganic as well as to organic materials. In this connection it was decided to perform the bench testing on natural ageing of BP and HFRC in environmental conditions of South Caucasus.

Ageing of BP and HFRC, intended for operation in atmospheric conditions, is a result of complex action of such factors as chain reaction of oxidation, temperature-humid deformation of a binder, penetration of moisture into material with further leaching of fiber, abrasive action of dust.

Exposure of BP and HERC in unloaded state and under stresses, close to calculated resistances of materials, revealed a considerable difference in the character of development of ageing processes in loaded and unloaded composites, as well as a difference between the values of long-term resistance and creeping of the samples, tested in atmospheric and laboratory conditions. For example, in Fig 9 the behavior of the samples of BP-2, unloaded and loaded by various intensity are shown.

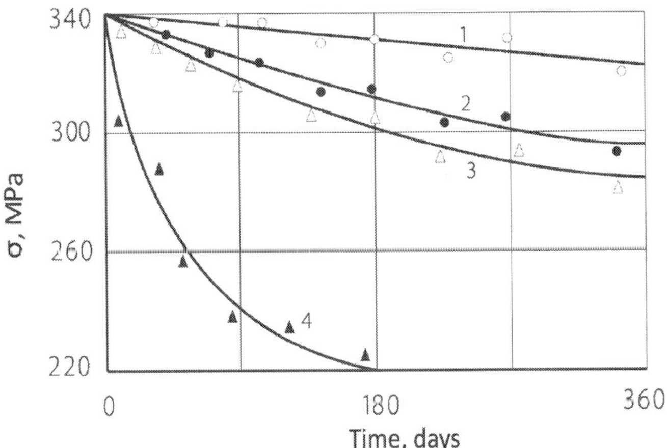

Figure 9. Dependence of strength at BP-2 tension on ageing time at open air and on the level of action stress: 1 - zero stress, 2 – 0.2 σ_{s-t}, 3 – 0.45 σ_{s-t}, 4 – 0.70 σ_{s-t}

Binder nature has a pronounced effect on atmospheric resistance of BP and HFRC. It is difficult to judge about atmospheric ageing of BP and HFRC on the basis of polyester matrix, since the process of binder hardening isn't finished; this leads to enhancement of elasticity modulus by 5-15%. True enough, the strength reduction by 8-17% still takes place.

As a result of atmospheric ageing, mechanical characteristics of BP and HFRC on the basis of epoxy and phenol-formaldehyde matrixes are gradually decreased depending on material thickness and applied stress. Difference in the variation of mechanical indexes of loaded and unloaded samples is revealed to a greater extent than in BP and HFRC on the basis of polyester matrix. Ageing process of materials are developed, mainly, on the surface; in this connection their atmospheric resistance significantly depends on material thickness. Existence of stressed state has a pronounced effect on intensification of ageing of BP and

HFRC on the basis of epoxy and phenol-formaldehyde binders in atmospheric conditions; in this case the effect of materials thickness is reflected to the greatest extent. Thus, a great number of BP-2 samples of 1.0 mm thickness, exposed at stress of 0.75 $\sigma_{s\text{-}t}$, were destructed immediately at the stand before expected exposure time, whereas the samples of 4.5 mm thickness weren't destroyed. The results of mechanical testing have shown a slight reduction of the strength of these samples (stresses in both cases were equal).

As a result of performed investigations, the values of the coefficients of operating conditions of BP and HFRC in service, have been obtained (Table 10).

Material	For calculated resistances	For long-term elasticity modulus	Remark
BP-1	0.69	0.80	Presented coefficients
BP-2	0.75	0.82	are given for BP-1 of
BP-3	0.79	0.83	1.5-3.0 mm thickness,
BP-4	0.72	0.75	for BP-2, BP-3 and BP-4 of 2.0-7.0 mm
HFRC-1	0.82	0.85	thickness and for all
HFRC-2	0.79	0.81	types of HFRC – 5.0-
HFRC-3	0.85	0.88	8.0 mm of thickness.
HFRC-4	0.79	0.82	

Table 10. Coefficients of operating conditions – K_{atm} of BP and HFRC

4.2. Mechanical characteristics at the action of water and some corrosive liquid media

Stability of mechanical properties of BP and HFRC is determined by resistance of reinforcing component of material, matrix and adhesion bond between them to aqueous and chemical media. An advantage of basalt fiber over other ones, in addition to higher thermal stability, must consist in water resistance and chemical endurance. In order to prove this fact, the action of alkali and mineral acids of various concentration (up to 60%) was studied on threads strength from alkalineless, alkaline and basalt glass, used in composites. Chemical composition of these glasses is the following (in mass%): aluminumborosilicate (alkalineless) – SiO_2-54; Al_2O_3 – 14; B_2O_3 – 10; CaO-16; MgO-4; Na_2O – 2; sodiumcalciumsilicate (alkaline) - SiO_2-71; Al_2O_3 – 3; CaO-8; MgO-3; Na_2O – 15; basalt one: SiO_2-49; Al_2O_3 – 16; Fe_2O_3 – 10; CaO – 9; MgO-7; Na_2O-4; MnO<1; TiO_2<1. It was studied the behavior of roving, offering the non-twisted strand with a diameter of elementary fiber of the order of 10-16μ.

Testing was carried out in water and in the media, most characteristic for chemical production: in caustic soda, sulphuric and nitric acids. Solution temperature comprised 291-295K. Before testing the samples were preliminary conditioned. Duration of static action of corrosive medium on threads was taken to be 240 hours, since the most intensive reduction of strength of fibers and of materials on their base at normal temperature is observed within first 200-240 hours, after which some stabilization of their strength indexes takes place. After maintenance in corrosive media the samples of threads were tested on the machine at relative humidity of air – 75-78% by determination of breaking load – P (P_0 – breaking load of dry threads). Because of some reversibility of strength reduction in fibers at the action of corrosive media, the samples of threads didn't dried before testing on the machine.

As is seen from the data of Table 11, 5-10% solutions of caustic soda have the most destructive effect on alkalineless threads. With increasing of the concentration of NaOH, the stability of strength characteristics of threads enhanced.

With increasing of the temperature of alkaline solution the total solubility of glass fiber enhances. In this case maximum solubility is slightly smoothed, in doing so its displacement is observed to the realm of higher concentrations. For basalt threads, in principle, the same character of the strength dependence on temperature and solution concentration is observed, but the values of mass loss is significantly lower (Table 12).

The most intensive strength reduction for glass threads is also observed at the action of 5-10% NaOH. Residual strength of basalt threads is considerably higher (Fig10).

Corrosive media[*]	Residual strength $\dfrac{P}{P_0}$ 100%				
	1%	5%	10%	25%	≥ 45%[**]
Caustic soda	$\dfrac{35}{45}$	$\dfrac{32}{41}$	$\dfrac{33}{45}$	$\dfrac{56}{67}$	$\dfrac{59}{72}$
Sulphuric acid	$\dfrac{22}{27}$	$\dfrac{4}{18}$	$\dfrac{5}{21}$	$\dfrac{16}{31}$	$\dfrac{18}{42}$
Nitric acid	$\dfrac{6}{21}$	$\dfrac{4}{19}$	$\dfrac{5}{24}$	$\dfrac{17}{38}$	$\dfrac{19}{42}$

Remark: in numerator – indexes of threads from alkalineless glass; In denominator - indexes of basalt threads.
[*] reduction of glass threads in water comprised 40%, of basalt threads 15%.
[**] maximum concentration of H_2SO_4 – 45%, of HNO_3-60%.

Table 11. Reduction of breaking strength of threads in corrosive media of various concentration

Temperature, K	313	323	333	343	363
Mass loss, mg.cm^{-2}, hour^{-1}	$\dfrac{0.03}{0.01}$	$\dfrac{0.06}{0.03}$	$\dfrac{0.03}{0.01}$	$\dfrac{0.03}{0.01}$	$\dfrac{2.33}{1.49}$

Remark: in numerator – the values of glass threads of alkalineless composition; In denominator-the values for basalt threads.

Table 12. Effect of temperature and caustic soda concentration on mass variation of fibers from alkalineless glass and from basalt.

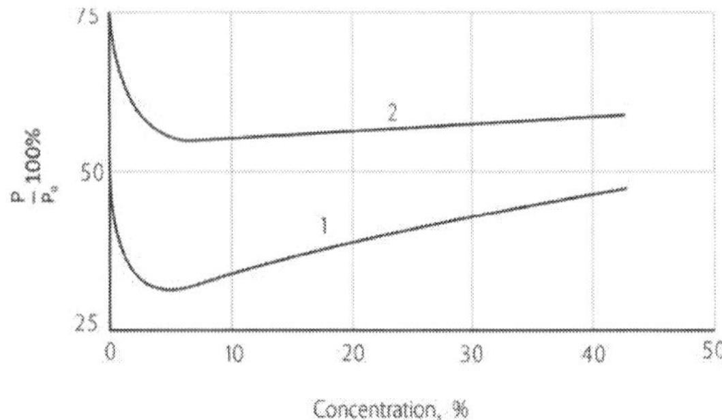

Figure 10. Strength reduction of threads from alkaline glass (1) and from basalt (2) in caustic soda.

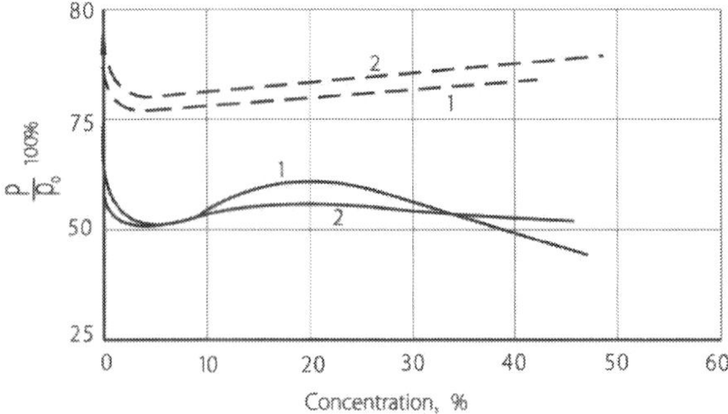

Figure 11. Variation of strength of alkaline glass threads (——) and basalt ones (---) in sulphuric (1) and nitric (2) acids.

Testing of glass and basalt threads in sulphuric and nitric acids has shown that a factor of concentration exerts a lesser influence on the variation of strength characteristics. By the example of behavior of alkaline glass threads it may be noted a some increased influence of sulphuric and nitric acids on them in concentration range of 5-10%. But effect of this influence is very limited. Especially it may be said in relation to basalt threads (Fig 11).

Action of acids in the range of higher (>20%) concentrations is different. Concentration factor of nitric acid doesn't effect on strength of glass threads as well as of basalt ones. At the action of sulphuric acid (40%) a sharp decrease of the strength of glass threads is observed and an effect on the strength of basalt threads is not so noticeable (Fig 11).

Data on strength reduction in threads from alkalineless glass and basalt threads in sulphuric and nitric acids are presented in Table 11. Sulphuric acid acts considerably more aggressively in concentration range of 5-10%. Concentration factor of nitric acid doesn't effect significantly on the value of strength reduction of glass threads. The character of acid action on basalt threads remains identical but a level of residual strength of these threads after exposure in the media remains higher in comparison with a level of the strength of alkalineless glass threads.

Similar character in discordance of the values of the coefficients of operating conditions was also found at accounting of one more factor – temperature. Diagrams of long-term resistance of BP-1 at air and in water at the temperatures of 293K, 313K and 333K, presented in Fig. 12, as an example, permit to determine $K_{T,\tau}^{W}$, the comparison of which with $K_T \cdot K_\tau \cdot K_W$ once again convinces in incorrectness of the method of coefficients multiplying (Table 14).

As indicated earlier, the joint effect of environment and time of its action on long-term resistance of materials to destruction is every so often estimated by multiplying of corresponding coefficients of operating conditions. But as with separate accounting of the effect of duration of the action stresses and temperature, this method causes the considerable errors in direction of reduction of safety factor. This fact is evident by the data of Table 13, where K_W and K_τ^w are coefficients of operating conditions in aqueous medium, and at the joint effect of water and duration of its action, respectively.

Material	K_W	K_τ	$K_\tau \cdot K_W$	K_τ^w	$\dfrac{K_\tau^w}{K_\tau \cdot K_w}$
BP-1	0.89	0.52	0.46	0.40	0.87
BP-2	0.79	0.82	0.65	0.59	0.91
BP-3	0.82	0.89	0.72	0.60	0.83
BP-4	0.85	0.83	0.71	0.66	0.93
HFRC-1	0.90	0.88	0.79	0.70	0.89
HFRC-2	0.83	0.83	0.69	0.62	0.90
HFRC-3	0.89	0.82	0.73	0.68	0.93
HFRC-4	0.83	0.79	0.66	0.61	0.92

Table 13. Comparison of the coefficients of operating conditions obtained by accounting of separate and joint action of water and of duration of its action on materials (at $\tau = 10^4$ hours).

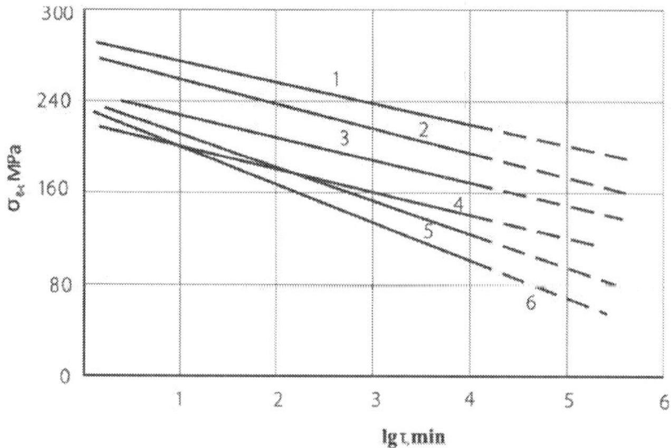

Figure 12. Long-term resistance of HFRC-2 at air and in water at the temperatures 293K(1,4), 313K (2,5) and 333K (3,6). 1, 2, 3 - at air; 4, 5, 6 - in water

Material	Temperature, K	K_T	K_w	K_τ	$K_T \cdot K_w \cdot K_\tau$	$K_{T,\tau}^w$	$\dfrac{K_{T,\tau}^w}{K_T \cdot K_w \cdot K_\tau}$
BP-1	293	0.95	0.89		0.44	0.42	0.95
	313	0.85	0.86	0.52	0.38	0.30	0.79
	333	0.72	0.86		0.32	0.20	0.62
BP-2	293	0.96	0.79		0.62	0.61	0.98
	313	0.87	0.78	0.82	0.56	0.35	0.62
	333	0.73	0.76		0.45	0.27	0.60
BP-3	293	0.95	0.82		0.69	0.67	0.97
	313	0.83	0.82	0.89	0.61	0.43	0.70
	333	0.70	0.69		0.43	0.28	0.65
BP-4	293	0.97	0.85		0.68	0.65	0.96
	313	0.82	0.85	0.83	0.58	0.57	0.98
	333	0.74	0.80		0.49	0.30	0.61
HFRC-1	293	0.98	0.90		0.78	0.77	0.99
	313	0.85	0.85	0.88	0.64	0.54	0.84
	333	0.78	0.81		0.56	0.33	0.59
HFRC-2	293	1.00	0.83		0.69	0.66	0.96
	313	0.90	0.81	0.83	0.61	0.54	0.89
	333	0.81	0.79		0.53	0.32	0.60
HFRC-3	293	1.00	0.89		0.73	0.70	0.96
	313	0.88	0.87	0.82	0.63	0.57	0.90
	333	0.80	0.82		0.54	0.33	0.61
HFRC-4	293	1.00	0.84		0.66	0.61	0.92
	313	0.84	0.83	0.79	0.55	0.43	0.78
	333	0.72	0.80		0.46	0.27	0.59

Table 14. Comparison of the coefficients of operating conditions, calculated by accounting of separate and joint effects of water, temperature and duration of their action on materials (at $\tau=10^4$ hours)

Analysis of obtained data permits to conclude that if at room temperature the physical character of water action on BP and HFRC is dominated, then by temperature elevation the chemical activity of aqueous medium becomes predominant. It may be also concluded that a long-term operation of materials under study don't cause a sharp decrease of their load-carrying capacity and destruction, since the medium temperature (313K, 333K) is significantly lower then the glass transition temperature of the binders and reinforcing elements (basalt, glass, carbon) are sufficiently stable in these conditions.

Going to the problem of chemical resistance of BP and HFRC, first and foremost it should be noted that the character of variation of their strength in alkaline media to a large extent depends on the composition of reinforcing component of materials, whereas an acid effect depends, mainly, on acid resistance of a matrix. Correlation of strength indexes of materials with epoxy, phenol-formaldehyde and polyester binders depending on the concentration of caustic soda after 240-hour maintenance shows that general character of strength variation is identical in all cases (Fig. 13). The greatest reduction of strength is observed in 5-10% solutions of caustic soda. In concentrated solutions the strength, practically, remains at initial level. Such character of strength variation is retained for all structural directions of BP and HFRC. In Fig.14 strength variation for main structural directions is presented at bending of HFRC-4 after 240 hour maintenance in the solution of caustic soda. It is evident that maximum reduction of strength limit, in all cases, takes place in the range of concentration of caustic soda of 5-10%.

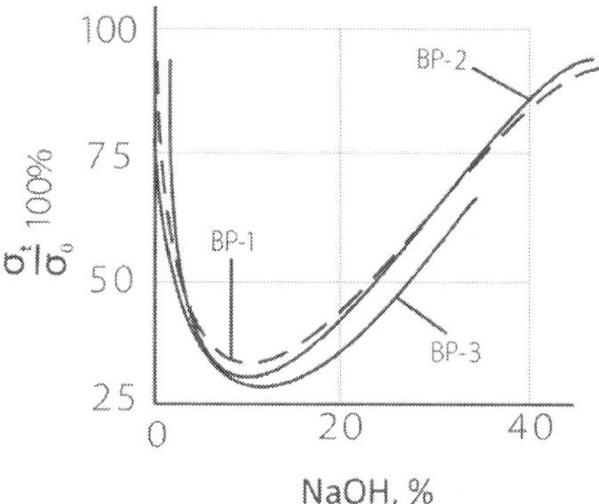

Figure 13. Strength reduction at tension of BP in caustic soda

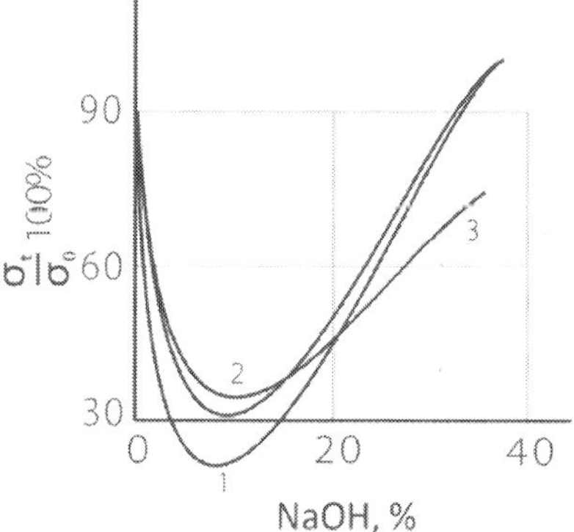

Figure 14. Strength reduction at various structural directions at bending of BP-2 in caustic soda. 1-on warp; 2-on weft; 3-at an angle of 45°

The character of strength variation for BP and HFRC, depending on acid concentration, is determined, mainly, by inertia of a binder and by its adhesion to reinforcing fiber. Polyester BP, in comparison with phenol-formaldehyde ones, exhibit better resistance to the action of mineral acids, their strength properties are more stable. In 5-10% solutions of sulphuric acid the penetration of liquid into phenol-formaldehyde BP takes place; this fact reduces an adhesion of a binder to basalt fiber and causes the swelling of the samples of BP-4. Its strength limit at tension is gradually reduced and in the range of medium concentrations is practically unchanged (Fig. 15).

Nitric acid is a powerful oxidizer and even at low temperatures causes the breakage of the surface of phenol-formaldehyde BP, with leads to definite losses in its mass. By increasing of acid concentration the swelling of the samples of BP-2 takes place but its strength in the range of low concentrations is varied moderately. By increasing of acid concentration strength drop is continued (Fig.15).

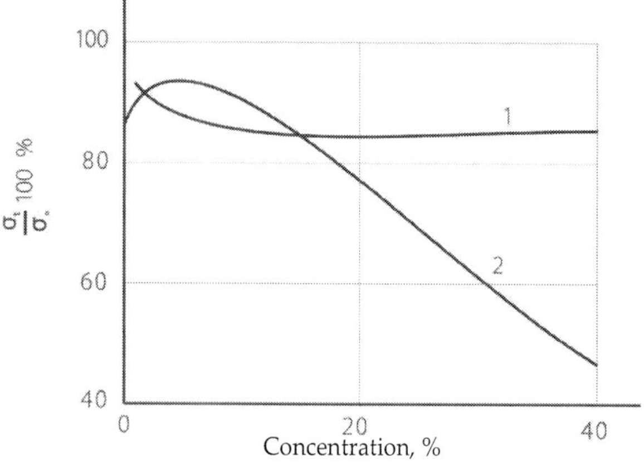

Figure 15. Strength variation at tension of BP-2 in sulphuric (1) and nitric (2) acids

At long-term action of sulphuric acid on polyester BP the breaking of the surface of the samples of BP-1 takes place accompanying by the loss of its mass and strength reduction. True enough, this process occurs only at low concentrations (up to 5%) and in the solutions of higher concentrations the strength remains at original level. Concentrated nitric acid depending on the time of action on polyester BP, causes an intensive swelling or washing-out of the samples. Strength of BP-1 is gradually decreased by increasing of the concentration of nitric acid.

The further stage of the work was the determination of long-term resistance of BP and HFRC in water and in 1 % solutions of caustic soda, sulphuric and nitric acids. Long-term resistance of BP and HFRC on bending was studied by the procedure, described in [12]. Long-term resistance of the materials, operating in water and corrosive liquid media, is nothing more nor less than the coefficient of their operating conditions, accounting the joint action of temporal factor and of water or anyone corrosive medium on the materials. For example, in fig.16, the dependence of $K_{\ell-t}^{w(cor)}$ on logarithm of durability for BP-1 is presented. At predetermined operating time for this material – 10^5 hours (11.4 years) conventional $K_{\ell-t}^{w(cor)}$ obtained by extrapolating of experimental data, depending on testing medium, comprises from 0.28 to 0.11. Results of testing presented in Fig. 16 , permit to propose the following coefficients of long-term resistance of BP-1 on bending at its assumed operation in stressed state in water and in 1% solutions of caustic soda, sulphuric and nitric acids (Table15).

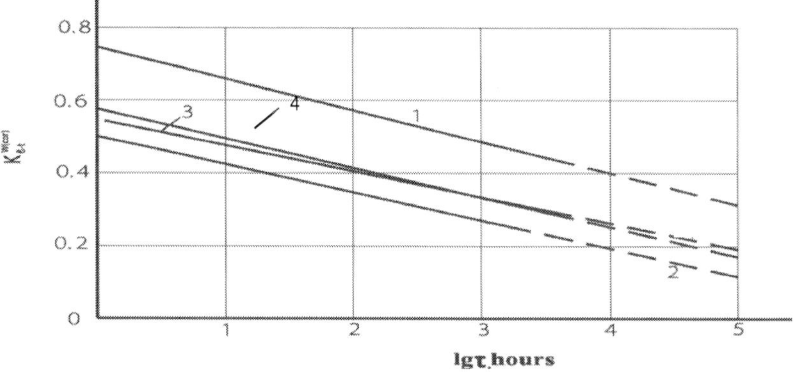

1 – water; 2 – H_2SO_4; 3 – HNO_3; 4 – NaOH

Figure 16. Dependence of $K_{\ell-t}^{w(cor)}$ on a logarithm of durability at the bending of BP-1.

Operating conditions	Proposed time of exploitation			
	Up to 1 year	To 3 years	To 5 years	To 10years
Dry state	0.67	0.64	0.62	0.58
Water	0.42	0.36	0.33	0.29
1% NaOH	0.22	0.20	0.18	0.15
1% H_2SO_4	0.20	0.18	0.15	0.11
1 % HNO_3	0.25	0.21	0.19	0.16

Table 15. Coefficients of long-term resistances of BP-1 in water and in the solutions of caustic soda and acids

Processing of these and other above presented results of investigations, for all materials, considered in this work, permitted to obtain the reference data on long-term calculated resistances of BP and HFRC. These data, by our opinion, may be beneficial at designing of structures and items by the use of BP and HFRC.

5. Conclusion

At present a sufficient experience is accumulated in world practice in the field of the technology of preparation of basalt plastics and composite materials with reinforcing structures from hybrid fibers. Development of the works along this line is determined by possibility of preparation of new generation of materials with a wide spectrum of properties. Along with it, a diversity of the requirements, imposed to structural materials, tended to the fact that none of newly elaborated materials can occupy the dominant place at current stage of technology development, at least, in the immediate future. Each type of materials may be optimal in certain specific cases, as shown in presented work. A wide spectrum of needed materials, apart from considered here, may be prepared by the use of a number of reinforcing fibers with various elastic and strength characteristics and combined with carbides and oxides of the binders, as well as of reinforcing schemes which permit a purposeful control of strength, rigidity and other properties of materials.

In parallel with it, it should be noted that at present the data for physical-mechanical properties of basalt plastics and composites on the basis of hybrid fibers as well as for variation of these properties in expected operating conditions are extremely limited, which retards their use as structural materials.

We hope that the engineers of various specialties may find in this work the practical recommendations on the approach of estimation of serviceability of materials in operating media, generally, as well as the data on calculated resistances of basalt plastics and composites on the basis of hybrid reinforcing fibers, in particular.

Author details

N.M. Chikhradze, L.A. Japaridze and G.S. Abashidze
G. Tsulukidze Mining Institute, Georgia

6. References

[1] J. M. Park, W. G. Shin, D.J. Yoon (1999) Composites Science and Technology, v. 59, 1.3: 355-370.

[2] K.E. Perepelkin (2006) Polymer fibrous composites, their main types, production principles and properties. Chemical fibers. №1: 41-50 (in Rus.).

[3] P. Bronds, H. Lilhopt, A. Lystrup (2005) Composite Materials for wind power turbine blades. Annual Review of Materials research, vol. 35: 505-538.

[4] D.A. Griffin (2002) SAND 2002-1879, vol. I, Albuquerquer, NM: Sandia National Laboratories.

[5] D.A. Griffin, T.D. Aswill (2003) Proceedings of the 48 International SAMPE Symposium and Exhibition. Long Beach, CA.

[6] D.A. Griffin (2004) SAND 2004-0073, vol. II, Sandia National Laboratories.

[7] E.S. Zelenski et al (2001) Reinforced plastics-modern structural materials. Russian Chemical journal. v.XIV, №2: 56-74 (in Rus.)

[8] Carbon/glass hybrids used in composite wind turbine rotor blade design. By Karen Fisher Mason, Contributing Writer (2004) Composites Technology.

[9] S.N. Zhurkov, E.E. Tomashevski (1959) In: Some problems of the strength of the solid. M., Publishing Hous of Academy of Sciences of USSR, p.61-66 (in Rus.)

[10] V.R. Regel et al (1974). Kinetic nature of the strength of the solid. M. , "Nauka" (in Rus.).

[11] V.I. Alperin et al (1975) – In: Hand book on plastic masses. v. II, M. "Khimia", pp. 442-512 (in Rus.).

[12] G.S. Abashidze, F.D.S. Marquis, N.M. Chikhradze (2007). Basalt reinforced plastics: Some operating properties. Materials Science Forum Vols. 561-565, pp. 671-674. 2007 Traus Tech Publications, Switzerland.

Catalysts and Environmental Pollution Processing Composites

Heterogeneous Composites on the Basis of Microbial Cells and Nanostructured Carbonized Sorbents

Zulkhair Mansurov, Ilya Digel, Makhmut Biisenbaev, Irina Savitskaya, Aida Kistaubaeva, Nuraly Akimbekov and Azhar Zhubanova

Additional information is available at the end of the chapter

http://dx.doi.org/10.5772/47796

1. Introduction

The fact that microorganisms prefer to grow on liquid/solid phase surfaces rather than in the surrounding aqueous phase was noticed long time ago [1]. Virtually any surface – animal, mineral, or vegetable – is a subject for microbial colonization and subsequent biofilm formation. It would be adequate to name just a few notorious examples on microbial colonization of contact lenses, ship hulls, petroleum pipelines, rocks in streams and all kinds of biomedical implants. The propensity of microorganisms to become surface-bound is so profound and ubiquitous that it vindicates the advantages for attached forms over their free-ranging counterparts [2]. Indeed, from ecological and evolutionary standpoints, for many microorganisms the surface-bound state means dwelling in nutritionally favorable, non-hostile environments [3]. Therefore, in most of natural and artificial ecosystems surface-associated microorganisms vastly outnumber organisms in suspension and often organize into complex communities with features that differ dramatically from those of free cells [4].

Initially introduced as just an imitation of Mother Nature, artificial immobilization of cells and enzymes has now transformed itself into a valuable biotechnological instrument. Its growing practical application and development over years led to appearance of fascinating novel microbial and enzymatic technologies [5-7]. Research on the immobilized biocatalysts is currently conducted in many laboratories around the world. In Japan, USA and other countries immobilized microbial cells have been successfully applied for adsorption of heavy metals from dilute solutions [8, 9], for purification of sewage [10] as well as for intensification of microbiological technologies (production of antibiotics, organic acids, sugar syrups, fermented drinks, etc.) [11]. It was shown that immobilized cells allow

conducting biotechnological process over extended periods of time, under strict control of the process kinetics, product quality and microbial activity [12].

Immobilization of cells can be carried out mainly by two methods: by entrapment of the microorganisms into porous polymers or microcapsules or by binding to an organic or inorganic support matrix (adsorption methods). The latter is considered to be more suitable for retaining cell viability [13]. Adsorption is also one of the easiest methods of immobilization of microbial cells, especially those that adhere naturally to the surfaces of materials [14]. It should be noted here that rapid development of technology of receipt of the immobilized biocatalysts resulted in contradictory results. So, the first attempts of immobilization were related with adsorption of enzymes and cells on arboreal sawdust and coal. In these experiments, adsorption was accompanied by a considerable desorption. In this connection, regarding the simplicity and availability of adsorption immobilization it has been having a reputation like "easy come easy go". Though never forgotten, in the last decade adsorption methods of immobilization gained increasingly more interest caused by considerable expansion in assortment of carriers with outstanding absorption properties, by better understanding of mechanisms and approaches aimed on firm attachment of biocatalyst to a carrier and by development of new methods of surface conditioning [12].

The adhesion of microbial cells to surfaces is rendered mainly by Van der Waals forces, ionic and covalent interactions, with considerable contribution of various microbial exopolymers [13]. Traditionally, adsorption immobilization is regarded as consisting of several relatively distinct stages, including a) adsorption of dissolved macromolecules on the surface; b) diffusion and concentration of cells from the bulk phase to the surface; c) reversible attachment of cells; d) biosynthesis of anchoring polymers by the cells which leads to an irreversible attachment stabilized by covalent bonds and entropy-driven interactions.

Selection of an appropriate adsorbent, especially for industrial process is based on several criteria. Most important among them are: a) material's costs and availability in large amounts; b) simplicity and efficacy of the immobilization process; c) preservation of cell viability; d) adsorbent's specific surface (capacity). There is no an ideal material so far but many these requirements are met by inorganic (sand particles, ceramics, metallic hydroxides and porous glass) and organic (charcoal, wood shavings and cellulose, polyurethanes) carriers. For example, porous glass-based fixed-bed reactors are successfully used for of the aerobic [15] and anaerobic [16] biotechnological transformations.

The immobilization process can be characterized by several parameters: initial biomass loading, retainment of biomass, strength of the adhesion, retainment of the activity of the biocatalyst, effectiveness of mass transfer, engineering realization and general operational stability. When microorganisms are immobilized by adsorption the initial cell loading of the immobilization matrix is one of the limiting factors [17]. The cell loading on the adsorbent is influenced by the physical and chemical properties of the adsorption material, of the microorganism to be immobilized and by the composition and parameters of the surrounding medium. Another critical point for a system with the cells immobilized by adsorption is the retainment of the biomass on the surface. The retainment is generally ruled by the adhesion strength, which can be described in kinetic and in thermodynamic terms.

Concerning the biocatalyst viability/activity retaining, the immobilization by adsorption is probably the gentlest existing method [14]. Because the adsorptive fixation occurs under "standard" conditions, no changes of the cultivation parameters are necessary to produce the immobilized biocatalysts. Compared to cell entrapment in organic polymers it can generally be assumed that during adsorption also the enzymatic activity can be preserved at a high level. Very often the activity of only one enzyme is responsible of the catalytic process of interest. In such a process the stability is characterized by the half-life of the enzyme. Enzymatic "half-lives" up to two years have been reported [18].

Though adsorbed biocatalyst systems are easy to run and used for many years, there is still enough space for optimization [13]. Development and probation of new types of heterogeneous composite materials, possessing advanced properties for biological catalysts, as carrier systems, as filters etc. on the basis of attached enzymes or whole microbial cells is of great importance for biotechnological processes. These and other tasks are addressed by *engineering enzymology* – a scientific and technical discipline combining principles, theoretical approaches and practical methods of chemical and enzymatic catalysis, microbiology, chemical technology and biochemistry. Recent efforts in engineering enzymology are focused (among others) on the following directions:

- development and optimization of immobilization methods leading to novel biotechnological and biomedical applications;
- search of materials satisfying strict requirements of biotechnological processes (such as non-toxicity, mechanical stability, etc.);
- construction of bio-composite materials based on individual enzymes, multi-enzyme complexes and whole cells, targeted on realization of specific industrial processes;
- development of methods for modification of surface properties aimed on fine tuning and better control of the "biocatalyst-carrier" interface

In the light of these challenges, nanostructured carbonized materials appear as an attractive substrate for designing and production of cost-effective high-performance bio-composite materials.

2. Synthesis of nanostructured carbonized materials

Adsorption properties of carbonaceous adsorbents are used in purification and recovery of valuable substances for very long time. Active carbons are used in oil processing, petroleum chemistry, wine making, butter production, etc. [19-21]. They are increasingly applied in medicine, for example, to remove toxins from physiological liquids [22]. The last years are characterized by the intensive studies on carbon nanotubes and nanostructured carbon sorbents (NCS).

There are many methods suitable for synthesis of NCS, such as electric arc discharge, laser vaporization and chemical vapor deposition techniques [23-25]. In the Institute of Combustion Problems (Almaty, Kazakhstan) following methods are used for obtaining of NCS: flame carbonization, catalytic carbonization and synthesis of carbon nanotubes by microwave plasma enhanced chemical vapor deposition (MPECVD). It was found that the transition metals like Fe, Ni, Co, their oxides and alloys are very effective catalysts for carbon nano-structuring. Another interesting approach used was the carbonization of walnut shells, grape seeds, apricot stones, wheat bran, rice husk, etc. in presence of activating agents. The samples were carbonized according to the procedure developed in the R.M. Mansurova Laboratory of Carbon Nanomaterials at the Institute of Combustion Problems, using a gas-flow setup **(Figure 1)** within temperature range of 250–900 °C in argon flow (50–90 cm³/min).

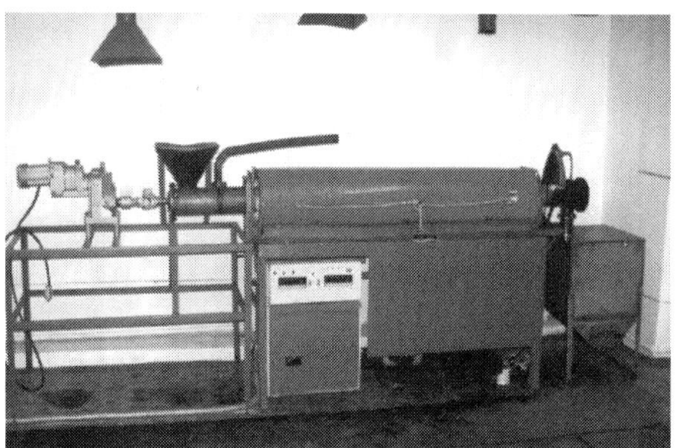

Figure 1. Pilot setup for flame carbonization of diverse raw plant materials.

During carbonization the major mass loss occurred within temperature range 150-500°C, where large amount of volatile and liquid products (65-75 % of total mass) were released. In the case of rice husk, the reduction of mass was found to be around 50% which is related to high content of silicon in the samples.

3. Surface structure and composition of the plant-derived carbonized sorbents

Carbon surface has a unique character. It has a porous structure which determines its high adsorption capacity; it has a chemical composition which enables numerous interactions with both polar and nonpolar molecules. Besides, it has active sites in the form of edges, dislocations and discontinuities which facilitate its chemical reactions with many compounds and functional groups.

Carbonized sorbents obtained by us on the basis of plant materials possess extended macro- and mesoporous structure, favorable for the adsorption of large molecules and cells [20, 22]. One can see in the **Table 1** that the specific surface (S_{sp}) and size of pores increased proportionally to the carbonization temperature up to 700 °C. However, further increase of temperature caused decrease of these parameters due to the increase of the density of the samples as reported also by Banerjee and coworkers [26].

Raw Material	T, °C	Size, µm			S_{sp}, m²/g
		Macropores	Mesopores	Micropores	
Walnut Shells	300	25	12	1.8	250
	500	30	13	2.3	770
	600	30	16	2.4	780
	700	30	16	2.3	800
	800	28	14	1.7	830
	850	29	15	2.4	800
Grape Stones	300	18	12	3	200
	600	22	14	6	500
	700	27	15	7	530
	800	25	13	5	540
	850	26	14	6	500

Table 1. Specific surface and pore size of the samples carbonized at different temperatures

Electron microscopy images (**Figure 2**) show the meso- and macro-porous structure of the materials appeared as a result of flame carbonization. A drastic contrast is visible between the structures of the raw material and the material after temperature treatment. Interestingly, flame carbonization of the raw plant materials often led to formation of complex carbon nanostructures (**Figure 3**) of various size and morphology. Treatment at 500°C resulted in appearance of transparent thin membrane sheets of 20-40 µm size. Prolonged heating (>30 min) at 600°C cased the formed translucent films to roll into 1400 nm long tubular structures of a diameter 400-500 nm. Further increase in the carbonization temperature and duration initiated the appearance of variety of nanostructures of diverse morphologies.

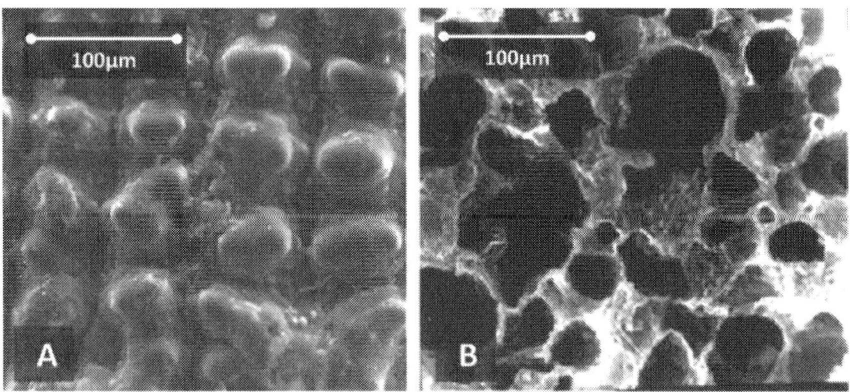

Figure 2. Electron microscopic images of rice shells in native state (A) and after carbonization at 650°C (B)

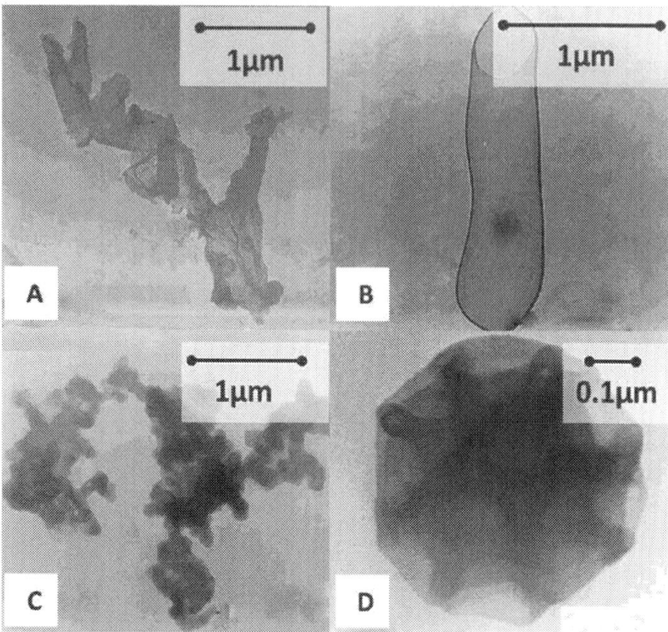

Figure 3. Diverse micro- and nanoscale structures observed with electron microscopy in apricot stones carbonized at different temperatures: (A) at 500°C; (B) at 600°C for 30 min; (C) at 700°C; (D) at 750°C.

Together with flame carbonization, microwave plasma-enhanced chemical vapor deposition (MPECVD) is considered to be a very promising method for the carbon nanotubes synthesis due to the lower growth temperature, uniform heat distribution and the ability to control different growth parameters [27]. Carbon nanostructures having different shapes have been synthesized using this method: aligned and curly filaments, flat and coiled carbon nanosheets. We also investigated the impact of different growth parameters such as temperature, pressure and hydrogen/methane exchange rate on the morphology of the carbon nanotubes. The results showed that there is a strong dependence of the morphology of the carbon nanotubes on the experimental conditions. For example, the quality of the carbon nanotubes was greatly affected by nitrogen influx during the growth process. Moreover, the diameter of the carbon nanotubes became smaller as nitrogen concentration in the gas mixture dropped. This implies the potential way to control the diameter of the carbon nanotubes precisely. It was also found that the threshold field required for the field emission can be reduced if nitrogen gas was introduced.

Adsorption behavior of the NCSs cannot be interpreted on the basis of surface area, pore size and nanostructural features alone. Specificity, affinity and capacity of such materials are strongly determined by chemical groups on their surface. These groups mostly appear due to controlled oxidation of carbon material's surface and can be roughly classified as phenolic (hydroxyl), carbonyl, carboxyl, ester, lactone and other groups (**Figure 4**).

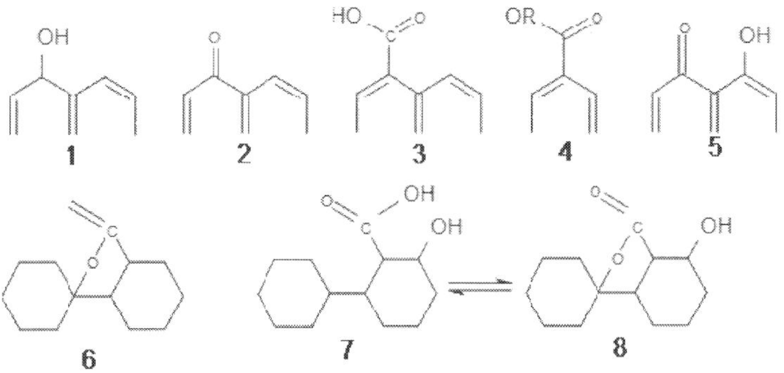

Figure 4. Major functional groups on the surface of carbonized adsorbents: 1) phenol (hydroxyl); 2) quinone (carbonyl); 3) carboxyl; 4) ester; 5) enol; 6)-8) different kinds of lactone groups.

Functional chemical groups on the NCS were analyzed by infrared spectroscopy using IR-spectrophotometer UR-20 (Zeiss Co., Germany). The IR-spectra of the native (raw) plant materials were mainly composed of characteristic absorption bands of NH_2 (3431.92 cm^{-1}), OH (3009.97 cm^{-1}), C=O (1643.25 cm^{-1}), C–O (1241.55 cm^{-1}), C–OH (1055.64-1157.28 cm^{-1}), C=C, C=N (1662.55 cm^{-1}) groups (**Figure 5a**). After carbonization at temperatures of 600–850 °C, a sharp (10-fold) drop of the intensity of characteristic bands of OH and NH groups was observed. In turn, the intensity of characteristic C–O–C bands increased. Also, bands related to CO_3^{2-} groups appeared and their intensity increased substantially with temperature rise. We observed also the bands related to CO_2 group in the region 2486.19 cm^{-1} (**Figure 5b**). Carbonization of apricot stones and rice husks proceeded similarly, but the latter displayed lower intensity of the corresponding bands.

In general, the higher was the temperature of the carbonization process the more intense were the characteristic absorption bands of the groups of NH_2, COH, C=O, OH as well as valence vibrations of CH-group in the aromatic rings. Furthermore, the IR-spectra obtained after carbonization exhibited characteristic absorption bands at 883 and 1050 cm^{-1} corresponding to C=C deformation vibrations of the aromatic ring, also those related to valence vibrations of aromatic ring 1600, 1578 and 1510 cm^{-1} as well as –C–H-vibrations at 3053 and 3030 cm^{-1}. In the sorbents carbonized at 300-850°C, the following polyaromatic hydrocarbons were identified: pyrene, (due to the presence of its characteristic absorption bands at 720 and 850 cm^{-1}); coronene: 547 and 1320 cm^{-1}; fluoranthene: 600, 740 and 820 cm^{-1}. Thus, we established the appearance of multiple polyaromatic hydrocarbons structures after carbonization. Further increase of carbonization temperature led to increase in intensity of absorption bands related to aromatic condensed systems.

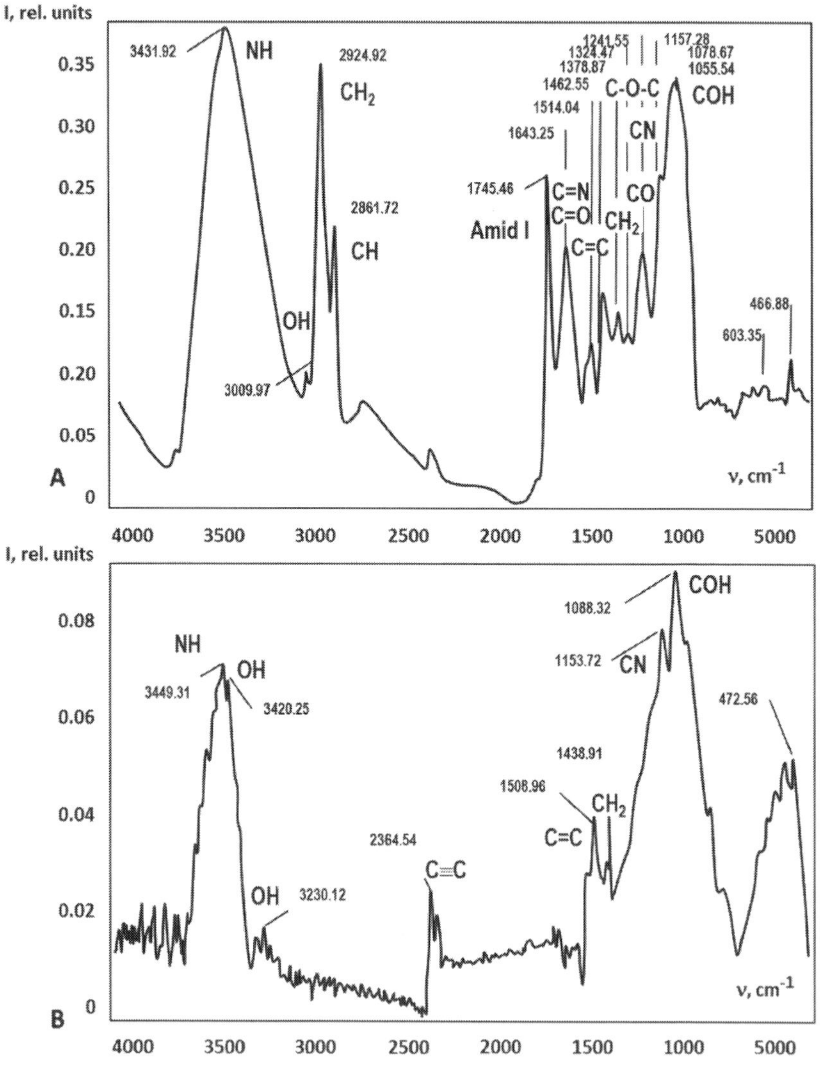

Figure 5. IR-spectra of native (non-carbonized) (A) and carbonized at 800°C (B) apricot stone surfaces.

4. Adsorption characteristics of the nanostructured carbonized materials in respect of microbial cells

Activated carbons are known as excellent adsorbents. Their applications include the adsorptive removal of color, odor, taste, undesirable organic and inorganic pollutants from drinking and waste water; air purification in inhabited spaces; purification of many chemical, and pharmaceutical products, etc. [19-21, 28]. Their use in medicine and health applications to combat certain types of bacterial ailments and for the adsorptive removal of certain toxins and poisons, and for purification of blood, becomes increasingly popular [29, 30]. Studies in last years have brought data on high adsorption ability of carbonized materials in respect of mammalian cells [31], microbial cells [32] and enzymes [33, 34].

The samples we used for microbial adsorption had been carbonized according to the procedure developed at the Laboratory of Hybrid Technologies in the Institute of Combustion Problems, Almaty, Kazakhstan. A flow set-up was used with following parameters: temperature range 650-800°C in argon flow (50-90 cm³/min). Different temperatures and flow regimes caused alterations in the pore structure and therefore resulted in different properties of the activated carbon [35, 36]. The adsorbents obtained from plant material showed themselves as very versatile and efficient because of their extremely high surface area, multiple functional groups and the macro-pore structure which is highly suitable for bacterial adhesion.

The interaction between the cell and the adsorbing surface is dictated by multiple physicochemical variables, reviewed in many brilliant works [37-39]. Obviously, an effective attachment depends on chemical and physical properties of both adsorbent and cells. The chemical groups on the surface of the carbonized materials were mentioned in the previous section. In this respect, microbial cells demonstrate even larger versatility. Their surfaces can be hydrophilic or hydrophobic, carry positively or negatively charges; expose various specialized chemical groups and even release polymers (adhesive glycoproteins, polysaccharides, proteins, teichoic acids, etc. (**Figure 6.**).

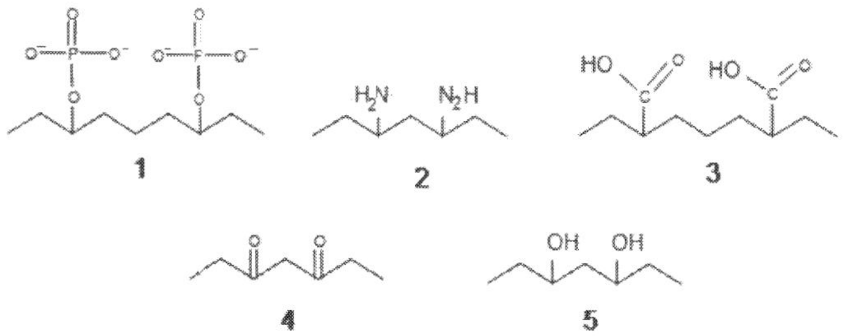

Figure 6. Principal functional groups on the surface of microbial cells: 1) phosphate; 2) amine; 3) carboxyl; 4) carbonyl; 5) hydroxyl.

Molecular biological and biochemical studies on cell adhesion focus predominantly on identification, isolation and structural analysis of attachment-responsible biological molecules and their genetic determinants. Physiological aspects of cellular adsorption concern mainly the influence of cultivation parameters (temperature, nutrition compounds, oxygen concentration, presence of antibiotics and vitamins) on bacterial adherence-related phenotype, adhesion molecules metabolism and surface structural organization [40]. Once in initial contact with a surface, microbes develop different types of attachment behaviors. Motile attachment behavior of *P. fluorescens* allows the flagellated cells to move along surfaces in a semi-attached condition within the hydrodynamic boundary layer, independent of the flow direction [41]. Reversible adhesion of *E. coli* cells with residence times of over several minutes on a surface has been described as "near-surface swimming"[42]. In the case that microbes can no longer move perpendicularly away from the surface the term "irreversible attachment" is used [14].

A net electrostatic charge on the NCS and the cell surfaces affects the distribution of ions in the surrounding interfacial region, resulting in an increased concentration of counter ions (ions of opposite charge to that of the particle) close to the surface that results in the formation of an electric double layer. This layer consists of two parts: an inner region (Stern layer) where the ions are strongly bound and an outer (diffuse) region where they are less firmly associated.

Thermodynamically, spontaneous cell adsorption onto a surface results in decrease of Gibbs free energy but sometimes there is a significant energy barrier due to electrostatic repulsion. Existing theoretical models predict that there are two regions where the strongest attraction forces between two surfaces occur (the "primary" and "secondary" minima, at distances of ~0.5 nm and ~5 nm, correspondingly). Generally it is assumed that microbes adhere reversibly to the "secondary minimum" and irreversibly to the "primary minimum" with the aid of cell surface appendages that can pierce the repulsive energy barrier [12, 14].

Our previous experiments have shown that, together with electrostatic properties, the hydrophobicity of both cells and NCSs plays a crucial role in both adsorption capacity and biomass retainment on the surface. Hydrophobicity of the carbonized materials can be easily controlled by the activation of the surfaces by water steam. The nature of the exposed chemical groups enables formation of multiple covalent bonds between the surfaces (**Figure 7**). Large number of different interactions involved in the cellular attachment to the carbonized surfaces makes possible fine tuning of the immobilization process in order to achieve versatility and adaptability of the bio-composite materials for different applications.

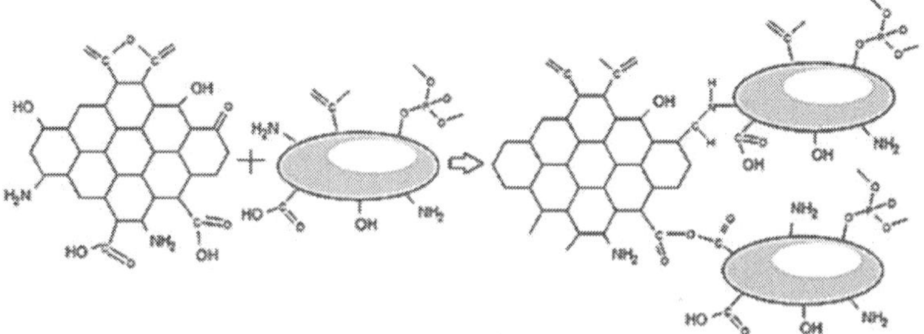

Figure 7. Formation of covalent bonds between the surfaces of microbial cells and the carbonized materials considerably contributes to stability of the bio-composite materials.

Electron microscopy examinations suggested that there is strong bonding interaction between microbial cells and the NCSs. In case of optimal incubation parameters, the cell load reaches ~62 %, corresponding to ~10^8 colony-forming units (~viable cells) per gram of NCS. The microbial cells were distributed on the surfaces not homogenously but rather formed clusters (micro-colonies). Taking into consideration potential intestinal and biomedical applications of the bio-composites, this fact is of particular importance because inter-cellular interactions and aggregation processes in the micro-colonies point out initial stages of biofilm formation, which in turn is an essential factor for bacterial survival and adaptability.

Figure 8 shows subsequent stages of rice husk colonization by Lactobacilli. It is clearly visible that the number of cells in a micro-colony varies between around 20 and 200 corresponding to the natural micro-colony structure in the epithelial layer of the intestine. The appeared bacterial colonies demonstrated almost irreversible adhesion in the absence of a competitive substrate (intestinal surface).

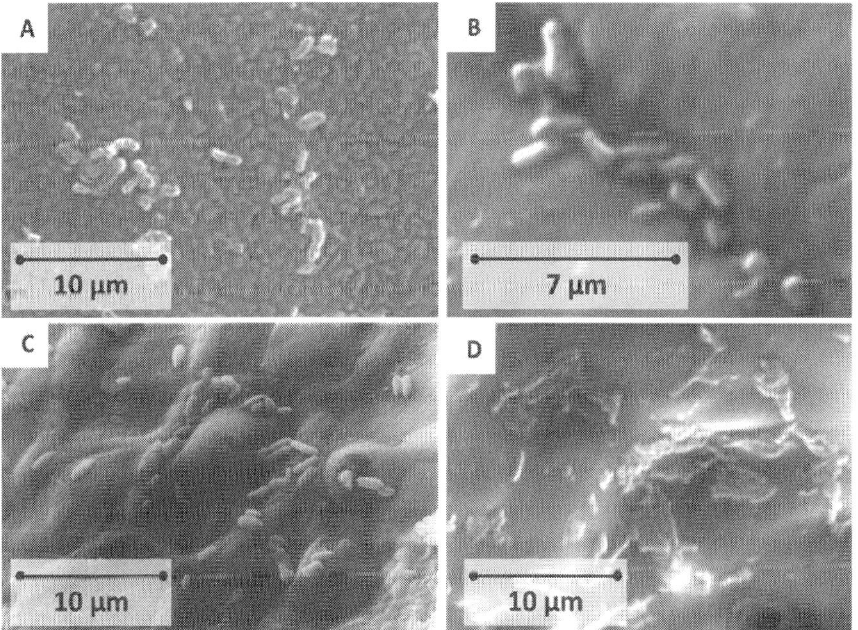

Figure 8. Subsequent stages of colonization of NCSs by Lactobacilli. A: carbonized rice husk, initial adsorption; B: carbonized grape stones, initial adsorption, C and D – carbonized grape stones, micro-colony formation

5. Performance of bio-composite carbonized materials in probiotic applications

In our model experiments in vitro, NCSs showed outstanding compatibility with many bacterial strains, indicating their high potential in miscellaneous branches of biotechnology and medicine. One of such applications of great interest is design and approbation of new generation of probiotic preparations for preventions and correction of micro-ecological disorders in gastrointestinal tract of the humans and animals. Environmentally, nutritionally and infection-induced pathologic shifts of gastrointestinal tracts' micro-ecology often lead to the increase in amount of gram negative bacteria, particularly of *Enterobacteria*. It leads to the translocation of bacterial toxic products from bowels to other organs causing development of endotoxinemia and other pathologies.

Probiotic is a viable mono- or mixed culture of beneficial microorganisms applied to animals or humans that sustainably improves properties of the indigenous microflora. The term "probiotics" was first coined by Lilley and Stillwell in 1965 [43]. R. Fuller later defined probiotics as "A live microbial feed supplement which beneficially affects the host animal by improving its intestinal microbial balance." [44]. Over years, the term "probiotics" has undergone several more definitions arriving at the final one, officially adopted by the International Scientific Association for Probiotics and Prebiotics, outlining the breadth and scope of probiotics as they are known today: "Live microorganisms, which when administered in adequate amounts, confer a health benefit on the host" [45]. Mechanisms of probiotic action are numerous and include:

- Prevention of adhesion of pathogen to host tissues;
- Stimulation and modulation of the mucosal immune system by reducing the production of pro-inflammatory cytokines through action on NF-kB pathways;
- Improvement of intestinal barrier integrity and up-regulation of mucin production;
- Killing or inhibiting the growth of pathogens through the production of bactriocins or other products such as acids or peroxides, which are antagonistic toward pathogenic bacteria.

Over past decades, probiotics have been extensively studied for their health-promoting effects and have been successfully used to control gastro-intestinal diseases. As shown above, the mechanisms of probiotic action appear to link with colonization resistance and immune modulation. Since *Bifidobacterium* and *Lactobacillus* species belong to normal intestinal microflora of humans, majority of probiotics was created on the basis of these bacteria. Lactic acid bacteria can produce numerous antimicrobial components such as organic acids, hydrogen peroxide, carbon peroxide, diacetyl, bacteriocins, as well as adhesion inhibitors, which strongly affect microflora.

Many probiotic preparations serve for an improvement of micro-ecological situations in bowels. Therefore, to reach the destination place, probiotic preparations have to pass through the stomach and the small intestine, which is unavoidably connected with significant reduction probiotic bacteria viability. To reduce this undesirable effect, several approaches have been suggested so far. Montalto et al. administered probiotic mix both in capsules and in liquid form without observing statistically significant difference in bacterial survival [46]. A specially designed tube with a reservoir containing probiotics has been suggested by Çaglar et al [47] with some encouraging results. However, the search for most suitable means of delivery and dosages of probiotics continues. One of our aims in this respect was to investigate the capacity of the carbonized materials as protective media for probiotic bacteria immobilized in their pores.

6. Biological objects

The type strain of lactic acid bacteria, *Lactobacillus fermentum AK-2* was used in our probiotic studies. It possesses excellent probiotic potencies due to its high antagonistic and adhesive activities. Rice husk and grapes stones carbonized were produced in the Institute of Combustion Problems at the al-Farabi Kazakh National University in Almaty as described above. *Lactobacillus* cells were adsorbed onto the carrier for 24 hours. Unattached cells were rinsed away by the isotonic NaCl solution and the firmly attached bacteria were incubated for several more days for micro-colony formation. After that the prepared bio-composite material was examined microscopically to ensure successful settlement of bacteria.

In bacteria survival experiments, gastric conditions were modeled in vitro by using gastric juice received from clinical gastroscopy. Different preparations of *L. fermentum* in MRS-1 medium were incubated in the gastric juice for 1 hour. After that the number of viable cells was quantified.

In vivo experiments were conducted on 6-8 week old wild rats, previously subjected to an experimental dysbacteriosis induced by the antibiotic ciprofloxacin. The animals were divided into several experimental groups. The control group received only the antibiotic in therapeutic dose of 5 mg/ kg body mass; the first group, in addition, was fed with liquid suspension of *L. fermentum* AK-2; the second and the third groups received, after the induced dysbacteriosis, the same amounts of *L. fermentum* but the bacteria were immobilized on grape stones and rice husk, correspondingly.

As an indicator of the probiotic activity, the number of viable *Enterobacteria* in different parts of the rat intestine was measured. Changes in detected amounts of gram-negative *Enterobacteria* such as *Escherichia coli, Klebsiella pneumoniae, Proteus vulgaris, Proteus mirabilis, Enterobacter aerogenes, Salmonella typhimurium, Shigella zonnei* and *Shigella fleexneri* were considered as a measure of the antagonistic action strength of the preparations. For *Enterobacteria* quantification, suspended gut content was incubated on Petri dishes with Endo agar. The analyses were conducted for 15 days, every day starting from a day of antibiotic treatment. Finally, the amount of *Lactobacillus* cells attached to the rat intestinal epithelium was directly counted as described elsewhere [48].

7. Results

Twenty four hours after immobilization *Lactobacillus* displayed very good growth rate and began forming micro-colonies on the NCS. The data on gastric juice resistance of suspended and immobilized preparations of *Lactobacillus* are shown in **Table 2.**

Experimental group	Concentration of viable cells, ml^{-1}	
	Before treatment	After treatment
Suspended culture	3.7×10^9	5.2×10^5
Grape stone-based bio-composite	1.6×10^9	3.1×10^6
Rice husk-based bio-composite	1.1×10^9	8.2×10^7

Table 2. Influence of in-vitro gastric juice treatment on viability of suspended and immobilized cells of Lactobacillus fermentum AK-2

In the suspended *Lactobacillus* culture after the gastric juice treatment the concentration of living cells decreased more than 7000 times. In contrast to that, the cells being a part of the bio-composite materials showed significantly (~500 times) better survival rate. The obtained data strongly suggested the protective action of NCSs on the immobilized *Lactobacillus* cells. These results look very encouraging in respect of construction of highly efficient bio-composite materials having extended probiotic activities.

The next series of experiments was devoted to comparative analysis of the antagonistic activity of suspended and immobilized probiotic preparations. After induced dysbacteriosis, intestinal microflora of rats was observed for the period of 15 days. The data are presented in the **Table 3**.

Experimental group	Number of bacteria in 1g			
	Large intestine		Small intestine	
	wall	lumen	wall	lumen
Before treatment with ciprofloxacin				
The control	$(3.0\pm0.3) \times 10^4$	$(6.9\pm0.7) \times 10^6$	$(1.1\pm0.2)\times10^2$	$(1.5\pm0.4)\times10^3$
1 day after treatment with ciprofloxacin				
Without probiotics	$(2.9\pm0.5) \times 10^5$	$(7.9\pm0.6) \times10^7$	$(2.9\pm0.3) \times10^4$	$(1.2\pm0.4) \times10^5$
Probiotics on GS*	$(9.7\pm0.6) \times10^4$	$(1.2\pm0.3) \times10^7$	$(7.8\pm0.8) \times10^2$	$(8.9\pm0.2) \times10^3$
Probiotics on RH**	$(6.9\pm0.4) \times10^4$	$(9.5\pm0.2) \times10^6$	$(0.9\pm0.3) \times10^2$	$(5.3\pm0.8) \times10^3$
15 days after treatment with ciprofloxacin				
Without probiotics	$(2.4\pm0.4) \times10^5$	$(2.1\pm0.3) \times10^8$	$(8.4\pm0.2) \times10^4$	$(9.5\pm0.8) \times10^5$
Probiotics on GS*	$(1.2\pm0.7) \times10^5$	$(9.6\pm0.6) \times10^7$	$(2.9\pm0.4) \times10^3$	$(2.8\pm0.4) \times10^4$
Probiotics on RH**	$(3.4\pm0.4) \times10^4$	$(8.5\pm0.7) \times10^6$	$(1.2\pm0.3) \times10^3$	$(1.2\pm0.7) \times10^3$

*GS: grape stone based carbonized adsorbent, **RH: rice husk-based carbonized adsorbent.

Table 3. Influence of the probiotic bio-composites containing L. fermentum AK-2 on the quantity of Enterobacteria in the intestine of rats after ciprofloxacin-induced dysbacteriosis.

The data in the table show that after ciprofloxacin-induced dysbacteriosis, significant increase (~2 orders) of the undesirable *Enterobacteria*-group microflora was observed, manifesting even more during the following 15 days after the antibiotic administration. This occurred both in the gut lumen and in its walls. Application of probiotics in the bio-composite forms, using carbonized rice husk and carbonized grape stones led to significant suppression in *Enterobacteria* proliferation and spread. Being immobilized on NCS, probiotic bacteria effectively inhibited growth of unhealthy bacterial forms, this counteracting development of dysbacteriosis. The measured inhibitory effects were much higher than those shown by suspended probiotic preparation. This effect can have been brought by different mechanisms, including better survival of probiotic bacteria (as was demonstrated above), by their increased antagonistic metabolic activity, and possibly also by exchange of the bacteria adsorbed on NCS and the bacteria attached to the intestinal cell walls.

The possibility that some exchange between the cells adsorbed in different locations indeed could take place has been demonstrated by counting of lactic acid bacteria cells attached on the surface of intestinal epithelium after NCS-adsorbed *Lactobacillus* cells were brought in contact with cultured intestinal cells (**Figure 9**).

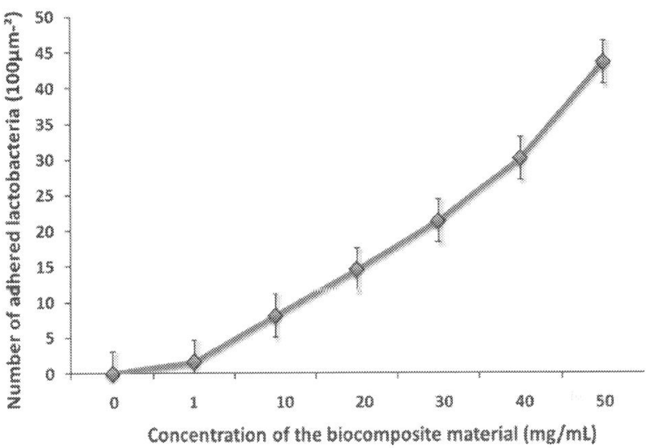

Figure 9. Colonization of cultured intestinal epithelial cells by Lactobacillus initially immobilized on the carbonized rice husk as a function of the applied bio-composite material.

As shown in the Figure 9, increase in concentration of the applied bio-composite material resulted in corresponding rise in the number of *Lactobacillus* cells firmly attached to the intestinal epithelium. Most active adhesion of probiotic cells occurs within the first two hours of incubation. Our calculations also showed that around 50% of the cells detached from the carbonized adsorbent later strongly attach to the surface of the epithelial cells. Since attachment to the recipient's cells is one of the most important indicators of a probiotic preparation activity, the obtained data suggest that the created bio-composite probiotic preparations successfully performed their functions.

We suppose that all the above-mentioned mechanisms (better survival, shifts in physiological activity, etc.) could contribute to the increased activity of immobilized probiotic strains. Our previous data suggest that the immobilization of *Lactobacillus* on carbonized materials (rice husk, grape stones) increased their physiological activity and the quantity of the antibacterial metabolites by 25-60%, which consequently would lead to increase of the antagonistic activity of *Lactobacillus*. High in-vitro and in-vivo efficiency of the immobilized probiotics can be also ascribed by the specific micro-environmental physicochemical conditions on the interface "sorbent/microbe" [14]. Moreover, besides delivery of bacteria in intestine the NCSs can possibly contribute to detoxification by absorption of intestinal toxins by the active sites on the surface not occupied by microbial cells. All these considerations suggest synergistic summation of multiple beneficial effects.

The collected evidences clearly demonstrate that the use of the nano-structured carbonized sorbents as delivery vehicles for the oral administration of probiotic microorganisms has a very big potential for improving functionality, safety and stability of probiotic preparations. A novel probiotic preparation named "Riso-Lact" has been recently developed at the al-Farabi Kazakh National University. The preparation consists of the *Lactobacillus fermentum AK-2* cells immobilized on rice husk under well-defined optimized laboratory conditions and will possibly find its application in treatment of dysbacteriosis in humans and animals. The great binding strength and capacity of the material in respect of cells and dissolved compounds are mainly conditioned by the extended network of nanotubes but appears also due to high hydrophobicity of the surface. Although many more studies and tests are necessary, and a lot of work needs yet to be done, we can now envision the creation of a new generation of NCS-based probiotic preparations, effectively normalizing the intestinal microflora, bringing relief to millions of patients around the world.

8. Future prospects for biomedical and environmental engineering applications

In the finalizing part of this chapter we would like to give a short overview of possible further applications of the bio-composite materials based of nanostructured carbonized adsorbents. Without any doubt, the field of future applications of such materials is vast and hardly foreseeable. The topics presented bellow will underline mostly biomedical and environmental applications where certain preliminary experimental material has been collected by our and other working groups.

8.1. Bio-composite material on the basis of carbonized rice husk and micro-algae *Spirulina*

As we have shown above, NCSs with immobilized *Lactobacillus* cells have a good potential as future probiotic preparations. One of key elements of their probiotic action is the involvement of the immobilized cells into the physiological processes on the surface of the intestinal epithelium. The NCS themselves also are able to adsorb significant amounts of gram-negative cells and their toxins, thus contributing to the beneficial effects of the preparations.

Together with known probiotic strains, we have recently attempted to apply the approaches previously had been probed on *Lactobacillus*, on micro-algae *Spirulina* in order to check its behavior as a component of bio-composite materials. *Spirulina platensis* is a blue-green alga (photosynthesizing cyanobacterium) having diverse biological activities. Due to high content of highly valuable proteins, indispensable amino acids, vitamins, beta-carotene and other pigments, mineral substances, indispensable fatty acids and polysaccharides, *Spirulina* has been found suitable for use as bioactive additive [49]. *Spirulina* produces an immune-stimulating effect by enhancing the resistance of humans, mammals, chickens and fish to infections, has capacity of influencing hemopoiesis, stimulating the production of antibodies and cytokines [50]. Moreover, *Spirulina* preparations are regarded as functional products contributing to the preservation of the resident intestinal microflora, especially lactic acid bacilli and bifidobacteria, and to a decrease in the level of undesirable microorganisms like *Candida albicans* [51].

In our experiments, both NCSs and *Spirulina* cells (the strain CALU-532m) applied alone to cultured rat epithelial cells, (IEC-6) at concentrations 5-50 µg/mL stimulated their proliferation and regeneration (**Figure 10**). Remarkably, being combined together in a bio-composite material, they showed a distinct synergy in their action, seemingly enhancing each other's activities.

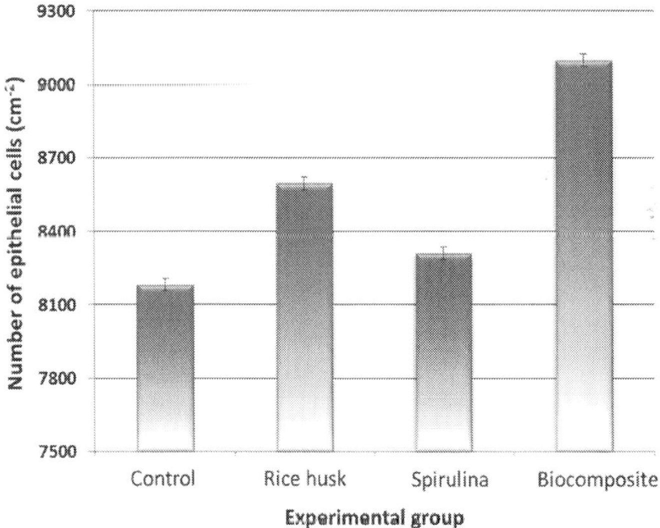

Figure 10. Stimulation of growth of cultured intestinal epithelial rat cells by carbonized materials (rice husk) and Spirulina platensis cells applied alone and in combination.

The epithelial cells treated by the NCS-Spirulina bio-composite reached confluence after 30 hours of incubation, whereas the control group required 48 h. The number of viable cells per square centimeter was found to be 9100±4.6 (SD) and 8180±3.5(SD), respectively.

8.2. Nanostructured carbonized materials for treatment of chronic wounds and sores

Problem of intractable wounds is one of persisting challenges in current clinical practice. In spite of modern antibiotics, hormonal and anti-inflammatory drugs, some chronic wounds and sores resist any treatment for weeks and months. The problem is especially serious in case of diabetic patients. The contributing factors are well-known and include high acidic wound environment, high concentration of bacteria (typically over ~10^5 cells per gram tissue) and their toxins, continuous production of inflammatory cytokines, products of tissue necrosis and high osmotic pressure.

All chronic wounds are colonized by bacteria. The delayed closure for many chronic and acute wounds is associated with high levels of bacteria in the wounded tissues. Even at lower levels bacteria hinder wound healing due to toxin secretion either directly from viable cells (exotoxins) or as a result of cell lysis (endotoxins). These toxins tend to cause local necrosis and disrupt the delicate balance of critical mediators such as cytokines and proteases necessary for healing progression. The swelling of the surrounding tissues (edema) often causes disturbances in oxygen delivery, which in turn leads to gangrenous and secondary necrotic processes. Therefore, control and removal of toxins and bacteria should be considered as key elements of a successful wound-healing therapy.

Taking these circumstances into account, it becomes obvious that chemical (pharmacological) treatment alone is rather inefficient in treatment of such wounds. A common topical agent used to combat bacterial burden in chronic wounds is disperse silver [52]. Recent comparative evaluation of various therapeutic methods has shown that the application of carbonized adsorbing materials to the necrosis zones is even more efficient: it lowers intoxication, stabilizes the blood level of glucose, improves indices of immunological reactivity, promotes a more dynamic course of a wound process, and reduces the time of treatment [53]. Some possible factors contributing to the efficiency of the NCS in wound healing are presented in **Figure 11**.

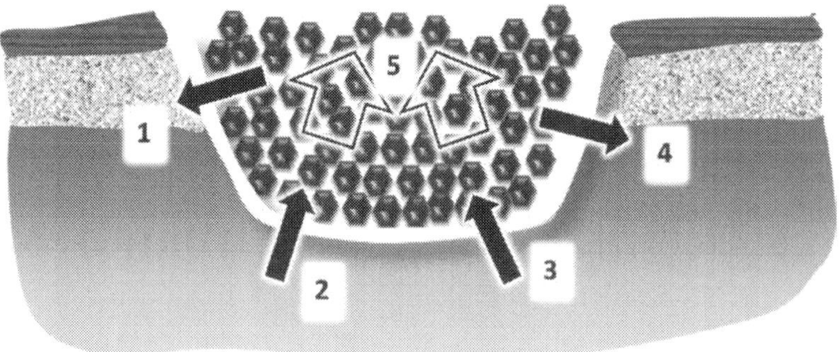

Figure 11. Hypothetical mechanisms of beneficial effects of NCS in wound healing processes: 1) stimulation of tissue regeneration; 2) binding of microorganisms and their toxins; 3) adsorption of inflammation factors and products of necrosis; 4) direct antimicrobial action; 5) capillary drainage.

Our studies have demonstrated that the use of NCS indeed can develop into an outstanding method for stimulation of wound healing. We produced in rats infected injuries, which healing typically occurs in 10-12 days. In the case the NCS were applied directly after the injury, an improvement and acceleration in wound healing was systematically observed (**Figure 12**).

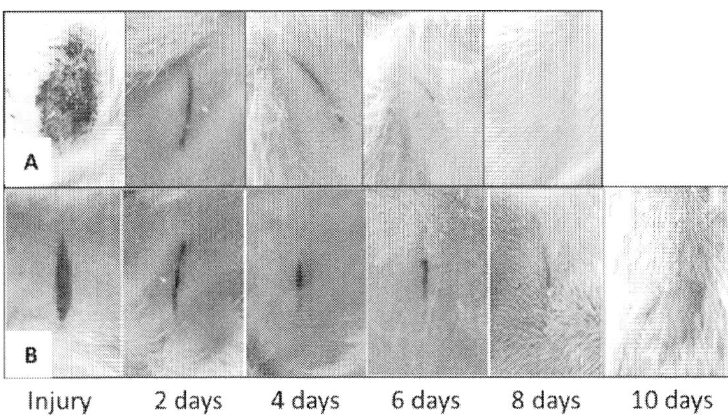

Figure 12. Dynamics of infected wound healing in rats. A: Nanostructured carbonized sorbents were applied after injury. B: Wound healing in the control group.

Multiple data obtained on rats with different levels of bacterial contamination suggest that NCSs may offer multiple specific advantages in topical wound management through their high adsorption ability in respect of both gram-positive and gram-negative bacteria as well as bacterial toxins. High adsorbing activity in respect of bacterial lipopolysaccharides has been measured by our group in a model studies. The other possible mechanism of beneficial action of the NCS, such as stimulation of tissue regeneration and binding of inflammation mediators, are yet need to be studied.

8.3. NCS in bioremediation

Bioremediation is the use of microorganisms, their structures and their metabolic pathways to remove pollutants. Bioremediation is the most promising and cost effective technology widely used nowadays to clean up both soils and wastewaters containing organic or inorganic contaminants [9]. Discharge of pollutant-containing wastes has led to destruction of many agricultural lands and water bodies. Utilization of various microbes and their products to adsorb, transform and inactivate pollutants enhances the efficiency of the environment decontamination significantly. For bioremediation purposes, microbial cells (bacteria, fungi, algae, etc.) can be applied alone or in combination with some adsorbent, which can greatly enhance the viability and activity of the biological component. Such bio-composite sorbent, unlike mono-functional ion exchange resins, contains variety of functional sites including carboxyl, imidazole, sulphydryl, amino, phosphate, sulfate, thioether, phenol, carbonyl, amide and hydroxyl moieties.

Compared to "classical" sorbents, the bio-sorbents are cheaper, more effective alternatives for the removal of metallic elements, especially heavy metals from aqueous solution. Therefore, the bio-composite sorbents are increasingly widely used for heavy pollutants removal. This is now a field of intensive investigations focusing on microbial cellular structure, biosorption performance, material pretreatment, modification, regeneration/reuse, modeling of biosorption (isotherm and kinetic models), the development of novel bio-sorbents, their evaluation, potential application and future. A potent supportive discipline in bioremediation studies is molecular biotechnology, capable to elucidate the mechanisms at molecular level and to construct engineered organisms with higher biosorption capacity and selectivity.

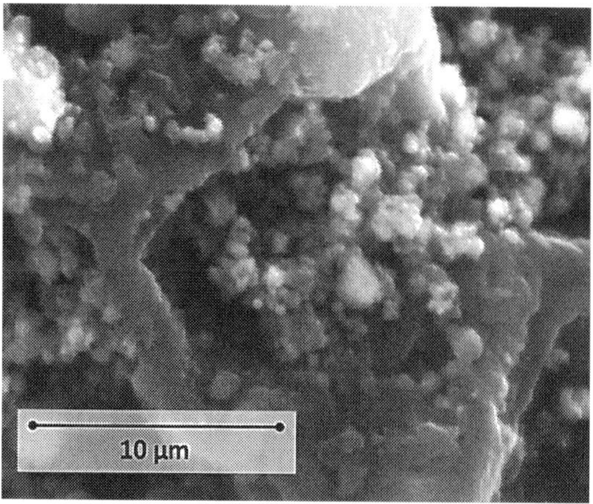

Figure 13. Electron microphotograph of the heterogeneous bio-composite bioremediation material created on the basis of NCS and bacterial cells Pseudomonas aeruginosa.

Due to their remarkable properties, nanostructured carbon materials such as carbonized grape stones and rice husk can be used as sorbents for extraction of toxic and radioactive elements. Current joint research conducted at the al-Farabi Kazakh National University (Microbiology Dept. of the Biology Faculty together with the Institute of Combustion problems) is aimed to create cost-effective and sustainable bio-composite materials on the basis of microbial cells adsorbed NCS of plant origin. Electron microscopy observations confirmed that multiple bioremediation-valuable cells can successfully attach, survive and proliferate inside the porous network of the NCS (**Figure 13**). The resulting heterogeneous biological composite materials possess outstanding pollutant-binding and transforming properties accompanied by high specificity, depending on the particular microbial strain used. In our model experiments, the obtained materials specifically adsorbed up to 95% metals from solutions. For cleaning of oil-polluted soils, we are currently developing a heterogeneous composite on the basis of carbonized sorbent and an immobilized microbial consortium consisting of bacterial strains with high oil-oxidizing activity. First encouraging results were obtained in the field experiments on oil-polluted soils.

Summarizing, we would like to emphasize that further studies and better understanding of the interactions between CNS and microbial cells are necessary. The future use of living cells as biocatalysts, especially in the environmental field, needs more systematic investigations of the microbial adsorption phenomenon. For this purpose it is necessary to develop and expand interdisciplinary collaboration networks connecting biologists, chemists, physicists and biochemical engineers.

This newly gained interdisciplinary knowledge could significantly stimulate development of novel immobilized bio-catalysts possessing high activity, selectivity and stability. Taking into account the wide spectrum of abilities of microorganisms and carbonized surfaces carriers, this ambitious mission does not look like an impossible one. Undoubtedly, in the coming years we will see the expanding of the application spheres of CNS-based heterogeneous composite biomaterials.

Author details

Zulkhair Mansurov, Makhmut Biisenbaev, Irina Savitskaya, Aida Kistaubaeva, Nuraly Akimbekov and Azhar Zhubanova
Al-Farabi Kazakh National University, Microbiology Department, Almaty, Kazakhstan

Ilya Digel
Aachen University of Applied Sciences, Institute of Bioengineering, Jülich, Germany

Acknowledgement

The authors thank Prof. Dr. Gerhard Artmann and Prof. Dr. Dr. Aysegül Temiz Artmann (Institute of Bioengineering, Aachen University of Applied Sciences) for their all-round support in performing these studies.

9. References

[1] Zobell, C.E., *The Effect of Solid Surfaces upon Bacterial Activity.* J Bacteriol, 1943. 46(1): p. 39-56.

[2] John, D.E. and J.B. Rose, *Review of factors affecting microbial survival in groundwater.* Environ Sci Technol, 2005. 39(19): p. 7345-56.

[3] Costerton, J.W., et al., *Microbial biofilms.* Annu Rev Microbiol, 1995. 49: p. 711-45.

[4] Stoodley, P., et al., *Biofilms as complex differentiated communities.* Annu Rev Microbiol, 2002. 56. p. 187-209.

[5] Kannan, A.M., et al., *Bio-batteries and bio-fuel cells: leveraging on electronic charge transfer proteins.* J Nanosci Nanotechnol, 2009. 9(3): p. 1665-78.

[6] Shibasaki, S., H. Maeda, and M. Ueda, *Molecular display technology using yeast--arming technology.* Anal Sci, 2009. 25(1): p. 41-9.

[7] Willner, I., B. Willner, and E. Katz, *Biomolecule-nanoparticle hybrid systems for bioelectronic applications.* Bioelectrochemistry, 2007. 70(1): p. 2-11.

[8] Gupta, R. and H. Mohapatra, *Microbial biomass: an economical alternative for removal of heavy metals from waste water.* Indian J Exp Biol, 2003. 41(9): p. 945-66.

[9] Kamaludeen, S.P., et al., *Bioremediation of chromium contaminated environments.* Indian J Exp Biol, 2003. 41(9): p. 972-85.

[10] Hunt, P.G., et al., *Denitrification of agricultural drainage line water via immobilized denitrification sludge.* J Environ Sci Health A Tox Hazard Subst Environ Eng, 2008. 43(9): p. 1077-84.

[11] Akin, C., *Biocatalysis with immobilized cells.* Biotechnol Genet Eng Rev, 1987. 5: p. 319-67.

[12] Digel, I., *Effect of transition metal ions and water soluble polymers on microbial cells adhesion to solid surfaces,* in *Biology Faculty.* 1998, Kazakh National University: Almaty.

[13] Klein, J. and H. Ziehr, *Immobilization of microbial cells by adsorption.* J Biotechnol, 1990. 16(1-2): p. 1-15.

[14] Digel, I., *Controlling microbial adhesion: a surface engineering approach,* in *Bioengineering in Cell and Tissue Research,* S.C. G.M. Artmann, Editor. 2008, Springer: Berlin. p. 601-625.

[15] Paca, J., et al., *Factors influencing the aerobic biodegradation of 2,4-dinitrotoluene in continuous packed bed reactors.* J Environ Sci Health A Tox Hazard Subst Environ Eng, 2011. 46(12): p. 1328-37.

[16] Park, C.H., M.R. Okos, and P.C. Wankat, *Characterization of an immobilized cell, trickle bed reactor during long term butanol (ABE) fermentation.* Biotechnol Bioeng, 1990. 36(2): p. 207-17.

[17] Cassidy, M.B., H. Lee, and J.T. Trevors, *Environmental applications of immobilized microbial cells: A review.* Journal of Industrial Microbiology & Biotechnology, 1996. 16(2): p. 79-101.

[18] Chibata, I., *Application of immobilized enzymes and immobilized microbial cells for productions of L-amino acids and organic acids.* Hindustan Antibiot Bull, 1978. 20(3-4): p. 58-67.

[19] Smíšek, M. and S. *Cerný, *Active carbon: manufacture, properties and applications.* Topics in inorganic and general chemistry,. 1970, Amsterdam, New York,: Elsevier Pub. Co. xii, 479 p.

[20] Jankowska, H., et al., *Active carbon.* Ellis Horwood series in physical chemistry. 1991, New York: E. Horwood. 280 p.

[21] Burchell, T.D., *Carbon materials for advanced technologies.* 1999, Amsterdam ; New York: Pergamon. xvii, 540 p.

[22] Gorchakov, V.D., et al., *[Immunosorption on active carbon].* Vopr Med Khim, 1981. 27(4): p. 544-7.

[23] Scott, L.T. and M.A. Petrukhina, *Fragments of fullerenes and carbon nanotubes : designed synthesis, unusual reactions, and coordination chemistry.* 2011, Hoboken, N.J.: Wiley. xviii, 413 p.

[24] Jorio, A., G. Dresselhaus, and M.S. Dresselhaus, *Carbon nanotubes : advanced topics in the synthesis, structure, properties, and applications.* Topics in applied physics,. 2008, Berlin ; New York: Springer. xxiv, 720 p.

[25] Blank, V. and B. Kulnitskiy, *Carbon nanotubes and related structures 2008.* 2008, Kerala, India: Research Signpost. 197 p.

[26] Banerjee, S., S. Naha, and I.K. Puri, *Molecular simulation of the carbon nanotube growth mode during catalytic synthesis.* Applied Physics Letters, 2008. 92(23): p. 233121.

[27] Zheng, J., et al., *Plasma-assisted approaches in inorganic nanostructure fabrication.* Adv Mater, 2010. 22(13): p. 1451-73.

[28] Son, H.J., Y.H. Park, and J.H. Lee, *Development of supporting materials for microbial immobilization and iron oxidation.* Appl Biochem Biotechnol, 2004. 112(1): p. 1-12.

[29] Mansurov, Z.A. and M.K. Gilmanov, *Nanostructural Carbon Sorbents for Different Functional Application*, in *Sorbents: Properties, Materials and Applications*, T.P. Willis, Editor. 2009, Nova Science.

[30] Kerimkulova, A.R., et al., *Nanoporous carbon sorbent for Molecular – sieve Chromatography of Lipoprotein Complex.* Russian J. of Physical Chemistry A, 2012. 86(6): p. 1004-1007.

[31] Beg, S., et al., *Advancement in carbon nanotubes: basics, biomedical applications and toxicity.* J Pharm Pharmacol, 2011. 63(2): p. 141-63.

[32] Grigor'ev, A.V., et al., *[The adhesion of pathogenic microflora on carbon sorbents].* Zh Mikrobiol Epidemiol Immunobiol, 1991(7): p. 11-4.

[33] Feng, W. and P. Ji, *Enzymes immobilized on carbon nanotubes.* Biotechnol Adv, 2011. 29(6): p. 889-95.

[34] Lenihan, J.S., et al., *Protein immobilization on carbon nanotubes through a molecular adapter.* J Nanosci Nanotechnol, 2004. 4(6): p. 600-4.

[35] Mansurov, Z.A., *Some Applications of Nanocarbon Materials for Novel Devices Nanoscale Devices - Fundamentals and Applications*, R. Gross, A. Sidorenko, and L. Tagirov, Editors. 2006, Springer Netherlands. p. 355-368.

[36] Mansurov, Z., *Flame synthesis of carbon nanomaterials: An overview.* International Journal of Self-Propagating High-Temperature Synthesis, 2011. 20(4): p. 266-268.

[37] Korber, D.R., et al., *Reporter systems for microscopic analysis of microbial biofilms.* Methods Enzymol, 1999. 310: p. 3-20.

[38] Verran, J. and K. Whitehead, *Factors affecting microbial adhesion to stainless steel and other materials used in medical devices.* Int J Artif Organs, 2005. 28(11): p. 1138-45.

[39] Geoghegan, M., et al., *The polymer physics and chemistry of microbial cell attachment and adhesion.* Faraday Discuss, 2008. 139: p. 85-103; discussion 105-28, 419-20.

[40] Junter, G.A. and T. Jouenne, *Immobilized viable microbial cells: from the process to the proteome em leader or the cart before the horse.* Biotechnol Adv, 2004. 22(8): p. 633-58.

[41] Korber, D.R., J.R. Lawrence, and D.E. Caldwell, *Effect of Motility on Surface Colonization and Reproductive Success of Pseudomonas fluorescens in Dual-Dilution Continuous Culture and Batch Culture Systems.* Appl Environ Microbiol, 1994. 60(5): p. 1421-9.

[42] Vigeant, M.A. and R.M. Ford, *Interactions between motile Escherichia coli and glass in media with various ionic strengths, as observed with a three-dimensional-tracking microscope.* Appl Environ Microbiol, 1997. 63(9): p. 3474-9.

[43] Lilly, D.M. and R.H. Stillwell, *Probiotics: Growth-Promoting Factors Produced by Microorganisms.* Science, 1965. 147(3659): p. 747-8.

[44] Fuller, R., *Probiotics : prospects of use in opportunistic infections.* 1995, Herborn-Dill: Institute for Microbiology and Biochemistry.

[45] Guarner, F., et al., *Should yoghurt cultures be considered probiotic?* Br J Nutr, 2005. 93(6): p. 783-6.

[46] Montalto, M., et al., *Probiotic treatment increases salivary counts of lactobacilli: a double-blind, randomized, controlled study.* Digestion, 2004. 69(1): p. 53-6.

[47] Caglar, E., et al., *Salivary mutans streptococci and lactobacilli levels after ingestion of the probiotic bacterium Lactobacillus reuteri ATCC 55730 by straws or tablets.* Acta Odontol Scand, 2006. 64(5): p. 314-8.

[48] Sadykov, R., et al., *Oral lead exposure induces dysbacteriosis in rats.* J Occup Health, 2009. 51(1): p. 64-73.

[49] Gershwin, M.E. and A. Belay, *Spirulina in human nutrition and health.* 2008, Boca Raton: CRC Press. xiii, 312 p.

[50] Deng, R. and T.J. Chow, *Hypolipidemic, antioxidant, and antiinflammatory activities of microalgae Spirulina.* Cardiovasc Ther, 2010. 28(4): p. e33-45.

[51] Vonshak, A., *Spirulina platensis (arthrospira) : physiology, cell-biology and biotechnology.* 1997, London: Taylor & Francis.

[52] Elliott, C., *The effects of silver dressings on chronic and burns wound healing.* Br J Nurs, 2010. 19(15): p. S32-6.

[53] Kuliev, R.A. and R.F. Babaev, [Physical factors in the comprehensive therapy of purulent wounds in diabetes mellitus]. Probl Endokrinol (Mosk), 1991. 37(5): p. 24-6.

New Composite Materials in the Technology for Drinking Water Purification from Ionic and Colloidal Pollutants

Marjan S. Ranđelović, Aleksandra R. Zarubica and Milovan M. Purenović

Additional information is available at the end of the chapter

http://dx.doi.org/10.5772/48390

1. Introduction

Composite materials (composites) are inherently heterogeneous and represent a defined combination of chemically and structurally different constituent materials, ensuring the required properties such as mechanical strength, stiffness, low density, or other specific characteristics depending on their purpose. Therefore, composite material is a system composed of two or more physically distinct phases whose combination produces a synergistic effect and aggregate properties that are different from those of its constituents. Favorable characteristics of composite materials were known to the people even in the period BC (before Christ-Century) and were used in order to improve the quality of human daily life. For example, it is known that in the ancient period, people made bricks that were reinforced with straw, and thus secured greater longevity and durability of their buildings. The incorporation of the straw improves the strength, toughness and thermal insulation properties of these composites. In principle, the degree of reinforcement (volume fraction of straw) and the level of alignment of the straw stalks (and their lengths) may be adjusted so that not only the properties but their anisotropy may be optimised differently in various parts of the structure [1]. Significant development and application of composites began in the second half of the 20th century, wherein their diversity and areas of application are constantly increasing. Development of composite materials is resulted mainly from the increasing need for materials with better mechanical characteristics that would be used as components in various constructions. For this purpose, such composites should have an adequate strength, stiffness, good oxidation resistance and low weight. Intensive study of composite materials and their processing methods has caused that these materials replace metals and alloys and become indispensable in the manufacture of parts for automobiles, spacecrafts, sports equipment etc. In terms of exploiting modern engineering composites

this remains a central principle. Modern composites can be said to have "designed micro- and nanostructures" which means that the constituents of composites have much more finely divided structures and tend to have sizes in the micrometre or nanometre range. Basic factors affecting properties of composites are as follows:

- Properties of phases;
- Amount of phases;
- Bonding and the interface between the phases;
- Size, distribution and shape (particles, flakes, fibers, laminates) of the dispersed phase - reinforcement;
- Orientation of the dispersed phase - reinforcement (random or preferred).

Good bonding (adhesion) between matrix and dispersed phase provides a high level of mechanical properties of the composite via the interface. In addition, interfaces are responsible for numerous processes of electron transfers and play crucial role in redox processes, heterogeneous catalysis, adsorption etc. Usually, there are three forms of interface between the two phases within the composite:

1. Direct bonding with no intermediate layer. In this case adhesion ("wetting") is provided by either covalent bonding or van der Waals force;
2. Intermediate layer in form of solid solution of the matrix and dispersed phases constituents;
3. Intermediate layer (interphase) in form of a third bonding phase (adhesive).

Current challenges in the field of composite materials are associated with the extension of their application area from structural composites to functional and multifunctional composites. In this respect, a great improvement of composite materials through processing has been made enabling the development of composite materials for electrical, thermal and other functional applications that are relevant to current technological needs. Examples of functions are joining, repair, sensing, actuation, deicing (as needed for aircraft and bridges), energy conversion (as needed to generate clean energy), electrochemical electrodes, electrical connection, thermal contact improvement and heat dissipation (*i.e.*, cooling, as needed for microelectronics and aircrafts) [2]. Modern processing includes the use of additives (which may be introduced as liquids or solids), the combined use of fillers at the micrometer and nanometer scales, the formation of hybrids, the modification of the interfaces in a composite and control over the microstructure. Therefore, it can be said that the development of composite materials for current technological needs must be application driven and process oriented. The conventional composites engineering approach, which is focused on mechanics and purely structural applications, is in contrast to mentioned modern practice.

On the contemporary level of science development it is known that materials of certain characteristics can be obtained only by strictly defined procedures of processing and depend on their chemical composition and structure. Since composites are heterogeneous systems, as already has been noted, the matrix is of great importance whose structure and chemical composition determine the most dominant features of the composite as a unit. However, it should be noted here that the composite does not possess properties of a single component but exhibits qualitatively new features, because of which it is considered as a new material. In addition to the dominant use of composites as structural elements, important application of composite materials is in the water purification technologies. In this field of application, composites usually have the role of adsorbent, electrochemically active materials, catalysts, photocatalysts etc. Bearing in mind that the material efficiency in the removal of harmful substances from water is higher if greater is its surface area, there are tends of scientists to develop these materials with required and defined nanostructures. In addition to the specific surface area increasing, nanostructured materials exhibit a qualitatively new properties compared to the related structure at the micro or macro scale. In this manner, it is developed specific procedure for certain metal hydroxides and natural organic matter layering onto alumosilicate matrix as well as procedures of microalloying which both lead to significant changes of the surface acido-basic and electrical properties of the alumosilicate matrix. The nano-scale composites provide an opportunity to study the phase boundaries and phenomena occurring at the surface, interface boundaries and within intergranular area during composites synthesis or during their interaction with aqueous solutions.

2. An overview and trends in use of composites in industrial plants

Nanocomposites based on polymers represent an area of significant scientific interest and developing industrial practice. Despite the proven benefits of polymer based nano-composites in the scope of their mechanical properties, and some distinctive combination/synergism of improved structural features, the real application remains still relatively isolated and not well discussed.

An insight in the historical (re)view on polymer nano-composites showed on the first type used based on the combination of natural fillers and polymers in the 90s [3-6] up to estimated 145 million USD spent at huge market of polymer based nano-composites in 2013 [7].

3. The concepts of interphase boundaries modification, microalloying and coating/layering in the composite synthesis

Methods and techniques for managing properties of composite materials include the selection and modification of constituent materials as well as changing the interface boundaries within the composite. Some composites are most commonly fabricated by impregnation (infiltration) of the matrix or matrix precursor in the liquid state into the appropriate filler preform. The connection between the constituents depends on the microstructure and chemistry of the interface boundary. The matrix and filler are connected by chemical bonds, interdiffusion, van der Waals forces and mechanical interlocking [2]. The first three interactions require very close filler-matrix contact that can be achieved if the matrix or matrix precursor wetting the surface of filler during the infiltration of matrix or matrix precursors in the filler preform. Effective wetting means that the liquid is evenly distributed over the surface of filler, while a poor wetting means that the liquid drops formed on the surface. Wettability can be increased by applying the coatings, adding wetting agents or by chemical surface functionalization (the introduction of functional groups on the surface that increase wettability) thereby changing the surface energy. If the filler is carbon fiber, surface treatments involve oxidation treatments and the use of coupling agents, wetting agents, and/or coatings. Often, metals or ceramics are used as coatings for carbon fillers. Metallic coatings are usually formed by coating carbon fiber reinforcements with metals *i.e.* Ni, Cu and Ag. Examples of ceramic coatings are TiC, SiC, B_4C, TiB_2, TiN which are distributed by using Chemical Vapor Deposition (CVD) technique or by solution coating methods starting from organometalic compounds. Therefore, these are examples of application of coatings on carbon materials to illustrate the method of modification of surface properties.

In the case of metal-ceramic composites, certain liquid metals react with ceramic preform during infiltration. For instances, composites based on the $Al–Al_2O_3$ system can be obtained by Reactive Metal Penetration (RMP) method which is based on infiltration of ceramic preforms by a liquid metal, generally aluminium or aluminium alloys [8,9]. During the process, a liquid metal simultaneously reacts and penetrates the ceramic preform, usually silica or a silicate, resulting in a metal-ceramic composite characterized by two phases that are interpenetrated. Another example is the reaction between SiC and Al during the infiltration of molten aluminum in a preheated preform:

$$4Al + 3SiC \rightarrow Al_4C_3 + 3Si \tag{1}$$

From the equation it can be seen that Si is generated during the reaction which is then dissolved in molten aluminum, while Al_4C_3 occurs at the SiC-Al interfacial boundary. The degree of reaction increases with increasing temperature. On the contrary, there are metals that in liquid state difficult wet the surface of the ceramic resulting in metal infiltration hindering. The difficulty of wetting and bonding of liquid metals to ceramic surfaces is related to atomic bonding in the ceramic lattice and can be improved by application of coatings. Coated particles (composite particles) are composed of solid phase covered with thinner or thicker layer of another material [10.11]. These coatings - layers on the surface are important for several reasons. In such way, the surface characteristics of the initial solid phase are modified and sintering conditions as well as molten metal infiltration can be better controlled.

As can be seen from examples, the processing of composite materials often involves high temperature and pressure to cause the joining of constituent materials forming a cohesive material. Generally, the matrix dictates the required temperature, pressure and processing time during composite synthesis. Sintering is an important factor in achieving the desired microstructure of ceramic based composites and includes very complex processes. In addition to surface coatings, an important influence on sintering has been exhibited by an addition of microalloying components, which significantly determine a microstructure and properties of ceramics [12]. The presence of small amounts of impurities in the starting material can vastly influence their mechanical, optical, electrical, color, diffusivity, electrical conductivity, and dielectric properties of matrix. Microalloying, as a known modern procedure for changing the intrinsic semiconductor properties, by authors' original works (Purenovic et al.), get more and more important role in the control of some structurally sensitive properties of metals, alloys, ceramics, composites and other materials. It is known that the nature of matter is determined by its composition and structure. There are many structurally sensitive properties of materials, but among the most sensitive are the conductivity, electrode potential, magnetic, catalytic and mechanical properties. Microalloying means adding certain elements in small (ppm) quantities, thereby modified structure results in a significant change in the value of conductivity and the electrode potential. Conducted own investigation and the results obtained showed an excellent rational electrochemical behavior of composites such as microalloyed aluminum, microalloyed magnesium, as well as composite ceramics and quartz sand microalloyed with aluminum and magnesium, in contact with aqueous solutions of electrolytes or water which contain harmful ingredients in ionic, molecular and colloidal state. Microalloyed and structurally modified composite ceramics have high porosity (30%), with the macro-, meso-, micro- and submicropores. There is direct relationship between porosity and structure of these composite materials, especially when it comes to nanostructured fragmented crystals. It is worth to emphasize the domination of amorphous phases with crystalline substructure, which is impossible to be removed, and it would be inappropriate to be removed, because the contact of crystals with amorphous layer is responsible for numerous processes of electrons exchange. By certain processes and reactions in the solid phase, the amorphous microalloyed aluminum, microalloyed amorphous magnesium, amorphous-crystalline structure of composite microalloyed ceramics and amorphous-crystalline structure of microalloyed quartz sand could be obtained. Many metals, alloys and composite electrode materials manifested significant differences in the reversible thermodynamic potential and the steady corrosion potential.

The manufacturing processes used to make composite ceramics can cause the development of liquid phases during sintering, and their retention as remnant glass at triple junctions and along grain boundaries and interphase boundaries after cooling to room temperature. Formed thin intergranular films are relevant to creep behavior at high temperatures, and also responsible for the strength of the bonding at interfaces. However, the heat treatment at elevated temperatures which is used for joining constituent materials and establishing the cohesive forces shows a disadvantage because cooling can lead to disturbance of established bonds between phases. Namely, during the cooling, differences in coefficients of thermal expansion could result in unequal contraction by which established bonds are broken. This problem is particularly evident in metal-ceramic composites, where high temperatures are usually applied during synthesis.

4. Preparation of modern nano-composites

Processing of nanocomposites based on layered silicates is rather challenging activity to achieve the full technical and engineering potential, which is the field with the largest growth forecast [13-16]. The modification of silicates by use of organic components is needed to allow intercalation, and also in order to improve compatibility/nano-distribution some additional ingredients have to be applied. The thermal treatment as step in processing sequence helps proper stabilisation of nanocomposites that has to take into consideration the oxidative stability of the polymer substrate, the influence of the nano-filler and the impact of modifiers and compatibilisers.

Montmorillonite of natural origin is among the most used nano-fillers. Traditional nano-fillers contain metal ions and other contaminants that may influence the thermooxidative stability and features of the nanocomposites. Organic modification of the (natural/traditional) clay is usually realized by cation exchange with a long-chain amines or quaternary ammonium salts. Content of such involved organic material content within the clay may be up to 40 mas.%. Therefore, the total thermal resistance of the composite material highly depends on the thermal stability of the organic ingredient. The thermal stability of the ammonium salts is limited at the processing temperatures applied (ex. extrusion, injection molding, etc.). Namely, thermal degradation of ammonium salts starts at 180°C and may be even tentatively reduced by catalytically active sites on the alumosilicate layer [17].

The compatibiliser applied as organically modified filler is often polypropylene-g-maleic anhydride in amount from 5 to 25% in the final composite formulation. The inferior stability of such low molecular weight filler comparing to the parent polymer affects the total stability of the final polymer based nanocomposites.

5. An improvement of composites stability

Nanocomposites may show higher stability due to increased barrier to oxygen, or lower stability because of undergone to hydrolysis through entrapped water [18,19]. In conventional practice stabilizer systems based on phenolic antioxidants and phosphites are applied, and in recent investigations new found components of filler degradation deactivators has been tested [20].

A traditional state-of-art polypropylene (PP) nanocomposite consisting of maleated PP and nano-clay is traditionally stabilized by a proven combination of phenolic antioxidant and phosphites. The polymer degradation may be completely prevented even after 5 extrusion cycles by using the patented stabilizer system AO-2 (based on oxazoline, oxazolone, oxirane, oxazine and isocyanate groups) [20], additionally improving mechanical properties of the resulting nano-composites and discoloration during processing and application.

The underlined thermal instability of the usual ammonium organic modifiers can be diminished by using the phosphonium, imidazolium, pyridinium, tropylium ions [21]. An alternative way to produce thermally stable nano-composites is the use of unmodified clays in combination with selected copolymers playing role of dispersants, intercalants, exfoliants and compatibilisers for PP nano-composites. In current processing of nano-composites different structures are identified such as polyethyleneoxide based nonionic surfactants [22] and amphiphilic copolymers based on long-chain acrylates [23]. Recently, more specifically poly(octadecylacrylate-co-maleic anhydride) and poly(octadecylacrylate-co-Nvinylpyrrolidone) in the form of gradient copolymers are applied with unmodified montmorillonite for processing PP nano-composites. Such obtained nano-composites show partial exfoliation, the final visual appearance is similar to the classical ammonium modified systems, however better thermal and thermo-oxidative stability is proven [23]. The most important improvement is achieved in the mechanical vales comparing to the conventional polymer system.

6. Nanocomposites use in a competitive environment of the materials

Nanocomposites materials are very attractive from the scientific and practical point of view, although some other materials are also interesting, such as plastics, fillers, blends, and different additives fulfilling the specified product profile. In such competence, the lowest cost solution comprising acceptable material structure and properties/resistances would dominate. Even more, competitive (nano)composite materials would benefit from nanocomposites developments and keep their application fields with improved features. Most of nanocomposites materials applications are intended for long-term and outdoor use. This is important aspect on the need for relevant nanocomposites stability. Namely, it is known that inorganic fillers often show a negative effect on the oxidative stability to a varying extent. The interactions of the filler and the stabilizers over adsorption/desorption mechanisms are mainly responsible for the impact. The specific surface area of the filler and pore volumes, surface functionality, hydrophilicity, thermal and photo-sensation properties of the filler and transition metal content (ex. manganese, titanium, iron) have been found to be potential factors/elements of the interaction [24].

Polypropylene/montmorillonite nanocomposites, additionally stabilized with antioxidant, degrade much faster under photo-oxidative conditions than pure polypropylene [25,26].This phenomenon is attributed to active species/sites in the clay generated by photolysis or photo-oxidation, and by consequence interaction between antioxidant, montmorillonite and maleic anhydride modified polypropylene. In natural clay present iron may additionally play an active role in the dramatic modification of material oxidation conditions [27], and nanoparticles also catalyze the decomposition process [28]. The use of so-called filler deactivators or coupling agents is potential solution for diminishing the negative influence of fillers on the (photo)oxidative stability by blocking active sites on the filler surface. Amphiphilic modifiers with reactive chemical groups in the form of polymers, olygomers or low molecular weight molecules such as bisstearylamide or dodecenylsuccinic anhydride have been proposed [29].Thus, stabilizer systems containing filler deactivators should have an affirmative effect in nano-composites for long-term stability.

7. Nano-composites materials for water treatments: State-of-the-art and perspectives

Clean drinking water is essential to human health, and also so-called technical water is a critical feedstock in a variety of key industries including electronics, pharmaceuticals and food processing industries. Taking into consideration that available supplies of fresh water are limited (due to population growth, extended deficiency, stringent health regulations, and competing demands from a variety of users/consumers) the world is facing with challenges to satisfy demands on high water quality standards and quantities (volumes). Benefits and trends in nano-scale science, chemistry and engineering impose that many of the current problems regarding green chemistry may be resolved using nano-sorbents, nano-catalysts, nanoparticles and nanostructured catalytic membranes. Nano-materials are characterized by a number of key physicochemical properties being particularly attractive for water purification treatments. Nanomaterials have much large specific surface area than bulk respect particles (mass to volume ratio), also they can be functionalized with reactive chemical groups specific in affinity to a given model compound. These materials may possess redox features and take part in shape- and structural-dependent catalyzed reactions of water purification. In aqueous solutions, they can serve as sorbents/catalysts for toxic metal ions, radionuclides, organic and inorganic solutes/anions [30]. Moreover, nano-materials can be used in selective targeting of biochemically constituents of aquatic bacteria and viruses. The nano-materials seems to be key components in future environmental friendly and cost-effective functional materials to desalinate public and polluted waters world-wide, for purification of water contaminated by pesticides, pharmaceuticals, phenol and other aromatics. The presence of heavy metals in water exhibits a variety of harmful effects on the living organisms in polluted ecosystems. The removal of heavy metals from water includes the following procedures: chemical precipitation, coagulation/flocculation, membrane processes, ion exchange, adsorption, electrochemical precipitation, etc. [31,32]. However, the application of composite materials in the controlling of pollutants in the environment and drinking water is significant [33,34], as described in further text.

The use of zeolites, natural or synthetic ones in waste water treatments is highly limited due to low adsorption capacity in the case of former and relatively small grain size in latter. Modification of natural or synthetic zeolites toward composite material which would satisfy both essential properties is a challenging task. Tailoring synthetic zeolite resulted in a composite porous host supporting microcrystalline active phase of vermiculite matrix [35]. The vermiculite-based composite showed the same hydraulic properties as natural clinoptilolite with similar grain size (2-5 mm), while the rate of adsorption and maximal adsorption capacity was improved four times. In other words, cation exchange capacity is increased when compared to natural zeolite with a comparative grain size, ion-exchange kinetics are substantially improved in comparison to natural zeolite, and hydraulic conductivity is considerably higher that synthetic powdered zeolite [35].

The development of new composite material based on use of inorganic polymeric flocculants as a combination of anionic and cationic poly-aluminium chloride (PACl) in one unique polyelectrolyte is proposed [36]. The incorporation of the anionic polyelectrolyte into PACl structure noticeably affects its initial properties (*i.e.* turbidity, Al species distribution, pH and conductivity). Interactions are taking place between Al species and polyelectrolytes molecules over hydrogen bonding (amino/amidic groups of the polyelectrolyte, and the –OH and –H groups of Al species are involved) and electrostatic forces/interactions. This resulted in new composite material. The main advantage of composite coagulants is lower residual aluminium concentration that remains in the treated sample, and more efficient treatments of waters (organic matter removal) can be realized [36]. Additional benefit is in cost effective process in the absence of specific equipment for handling the polyelectrolyte (ex. pumping system, etc.). Taking into account faster flocculation, increased efficiency and cost effectiveness, such new composite material seems to be promising one.

Porous ceramic composites can be prepared by silver nanoparticles-decoration using a silver nanoparticle colloidal solution and an aminosilane coupling agent [37]. The interaction between the nanoparticles and the ceramics comprises the coordination bonds between the $–NH_2$ group and the silver atoms on the surface of the nanoparticles. The composite can be stored for long periods without losing of nanoparticles, also being highly resistance to ultrasonic irradiation and washing. Such composite has shown high sterilization property as an antibacterial water filter [37]. This low cost composite, bearing in mind commonly available synthesis, simple preparation, the use of cheap and non-toxic reagents in the procedure, may be imposed as a potential solution for widespread use in water treatments.

Ultrafine AgO particles-decorated porous ceramic composites are prepared based on the main ingredient, cristoballite. The results on composite structure show that silver(II)oxide decorated diatomite-based porous ceramic composites possess crystal structure, and are composed of tetragonal cristoballite, monoclinic silver(II)oxide and cubic silver(I)oxide [38]. Such AgO-decorated porous ceramic composites show a strong antimicrobial activity and an algal-inhibition capacity. As the extension time is longer, the antibacterial effects are enhanced up to 99.9% [38].

Actual nanostructured composite materials based on multi-walled carbon nanotubes (MWCNT) and titania exhibited strong interphase structure between MWCNT and titania. This contact and interaction facilitated a homogeneous deposition/coverage of titania over MWCNT [39]. The photo-catalytic activity of the prepared composite materials was tested in the conversion of phenol from model watery solution under UV or visible light. The results showed higher photo-catalytic activity of the composite MWCNT and titania than over mechanical mixture proving an assumption on the existence of the interphase structure effect [39].

Nanocomposite membranes based on silica/titania nanotubes over porous alumina supports membranes were prepared [40]. An inserting of amorphous silica into nanophase titania caused the surpressed of phase transformation from anatase to rutile, and decreased the titania particle size. Good photo-catalytic activity of organic contaminants degradation, and wettability of composite membrane under UV-irradiation, helped to obtain high permeate flux across the composite membrane [40].

8. New alumosilicate based composites chemically modified by coatings/thin layers – Tested in the removal of colloidal and ionic forms of harmful heavy metals from water

Without new materials, there are no new technologies. Having in mind this fact, electrochemically active and structurally modified composites were obtained through microalloying and certain metals hydroxides layering, starting from bentonite as alumosilicate precursor. The composites have prognosed electrochemical, ion-exchanging and adsorption properties, as very sensitive structural and surface properties of materials. After the series of experiments, including composites interaction with synthetic waters, the obtained results are presented, analyzed and then systematized in the form of appropriate models of interactions.

8.1. Alumosilicate composite ceramic microalloyed by Sn for the removal of ionic and colloidal forms of Mn

Usually, manganese does not present a health hazard in the household water supply. However, it can affect the flavor and color of water because it typically causes brownish-black staining of laundry, dishes and glassware [32]. Although manganese is one of the elements that are at least toxic, concentrations of manganese much higher than the maximum allowed concentration during long-term exposure can cause health damage. A number of known procedures for the manganese removal are not suitable for an elimination of its all chemical species due to reversible release of manganese into water systems. Therefore, some of these used procedures are at the edge of techno-economical viability. In order to remove ionic and colloidal forms of manganese, a new aluminosilicate-based ceramic composite with defined electrochemical activity was synthesized [41]. Synthesis procedure of the composite material consists of two phases. Firstly, composite particles were synthesized by applying Al/Sn oxide coating on the bentonite particles in an aqueous suspension. In the second phase, aluminium powder was added to the previously obtained plastic mass and after shaping in the form of spheres 1 cm in diameter and drying, sintering was performed at 900°C. Fig. 1 a), b) and c) presents the microstructure of composite by using different magnifications.

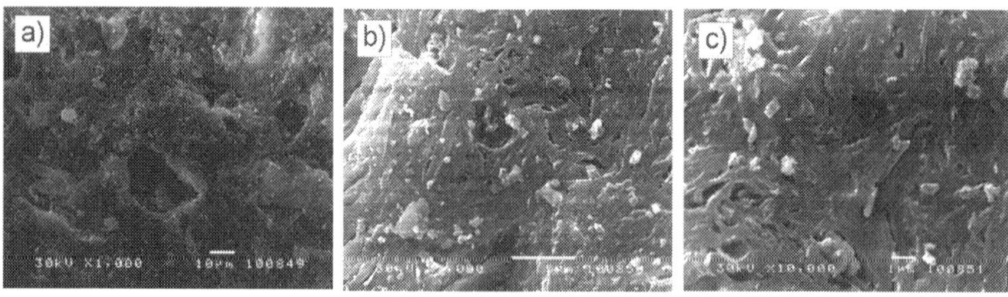

Figure 1. SEM images of the composite recorded at: a) low, b) medium and b) high magnifications.

During sintering a microalloying of composite by Sn occurred causing crystal grain surface layer amorphization and a creation of non-stoichiometric phases of Al_2O_3 with a metal excess [42,43]. In this way, microalloying causes electrochemical activity, which manifests itself in contact with the aqueous solutions of electrolytes and harmful substances in water. Therefore, the ceramics is unstable in contact with water and susceptible to corrosion because surface electrochemical processes taking place. The composite influence redox properties of water and electrochemically interacts with ionic and colloidal forms of manganese in synthetic water systems.

Alumosilicate matrix, whose particles are coated with Al/Sn oxides, was filled with a metal phase which is mostly aluminum with a small quantity of tin as a microalloying component. During the thermal treatment, liquid aluminium simultaneously reacts and penetrates the ceramics preform, resulting in metal/ceramic composite, where the all phases are interpenetrated forming a porous structure. In fact, the reduction of tin(II) occurred according to the following reaction:

$$2Al + 3SnO \rightarrow Al_2O_3 + 3Sn \tag{2}$$

The first reaction step is the reduction of Sn(II) to elemental Sn and its dispersion from the ceramics into the melt. Therefore, during the reaction, Sn is liberated into the liquid metal and diffuses towards the Al source. Moreover, oxygen partial pressure within the composite, at the Al-Al_2O_3 interface, can be estimated on the basis of thermodynamic parameters and calculated using the following equation [44]:

$$\Delta G^\circ = RT \ln PO_2 \tag{3}$$

The standard free energy of the reaction:

$$4/3Al + O_2 \rightarrow 2/3Al_2O_3 \tag{4}$$

at 900°C, given by Ellingham diagram [44] is -869 KJ/mol, and corresponding oxygen partial pressure: $PO_2 = 2.02 \cdot 10^{-39}$ Pa. Therefore, this low oxygen partial pressure during sintering provides reducing environment and the formation of nonstoichiometric oxide phases, with the metal excess, or with vacancies in oxygen sublattice. Nevertheless, Al_2O_3 belongs to the oxides of stoichiometric composition or with a negligible deviation from stoichiometry, it can occur as an amorphous and nonstoichiometric oxide with a metal excess during oxidation of aluminium. Common nonstoichiometric reactions occur at low oxygen partial pressures when one of the components (oxygen in this case) leaves the crystal [45,42]. A corresponding defect reaction is [45]:

(oxygen in this case) leaves the crystal [45,42]. A corresponding defect reaction is [45]:

$$O_O^x \rightleftharpoons \frac{1}{2}O_2(g) + V_O^{\bullet\bullet} + 2e' \tag{5}$$

As the oxygen atom escapes, an oxygen vacancy ($V_O^{\bullet\bullet}$) is created. Taking in mind that the oxygen is to be presented in neutral form, two resulting electrons would be easily excited into the conduction band.

Al–Sn alloys show a great activity compared to the thermodynamic Al^{3+}/Al potential of $-1.66V$ vs. NHE, which stands for a pure aluminium. The activation is manifested by a shifting of the pitting potential in the negative direction and significant reducing of the passive potential region [43,46]. The addition of microalloying Sn to aluminium produced a considerable shift of the open circuit potential (OCP) in the negative direction [46].

During the process of composite ceramics sintering, significant changes in the structure of alumosilicate matrix were occurred. Namely, the polycrystaline alumosilicate matrix with amorphised grain and sub-grain boundary were obtained, where a main role possesses metallic aluminum itself, then a microalloyed tin and nonstoichiometric excess of these elements in ceramics, creating macro-, meso- and micro- pores with the reduced mobility of grain boundaries and termination of grain growth [47]. Aluminum and tin in conjunction with other admixtures present in composite ceramics cause drastic changes in the structure-sensitive properties and electrochemical activity. An active composite ceramics in contact with synthetic water containing manganese reduce and deposit the manganese in the macro-, meso- and micro- pores (eq. 6). Electrochemical activity is provided by electrochemical potential of Al atoms and free electrons that participate in redox processes.

$$2Al + 3Mn^{2+} \rightarrow 2Al^{3+} + 3Mn \tag{6}$$

The deposited manganese on microcathode parts of the structure can further form separate clusters and the adsorption layer [48,49]. Reduction processes take place until the Al^{3+} ions continue to solvate themselves in water. A part of Al^{3+} ions reacts with OH^- ions giving insoluble $Al(OH)_3$.

8.1.1. Interaction of composite material with ionic and colloidal forms of Mn in synthetic water

Interaction of the composite material with water manifests itself as decreasing in the redox potential of water, as shown in Fig. 2. This confirms the fact that the composite is electrochemically active in contact with water. During the interaction with water, aluminium from the composite is electrochemically dissolved into water providing electrons which can participate in the number of redox reactions of water yielding reduced species (molecules, ions and radicals) such as H_2, $OH^\bullet$, etc. [47].

TDS value of distilled water immediately after contact with ceramics increases. It seems that increasing the TDS value is due to dissolution of Al^{3+}, Mg^{2+}, Na^+, SiO_3^{2-} from the bentonite based composite. Al^{3+} and SiO_3^{2-} ions are subjected to hydrolysis and polymerization reactions which are followed by spontaneous coagulation-flocculation processes and appearance of sludge after a prolonged period of time.

A reduction of manganese concentration in synthetic waters is shown in Fig. 3.

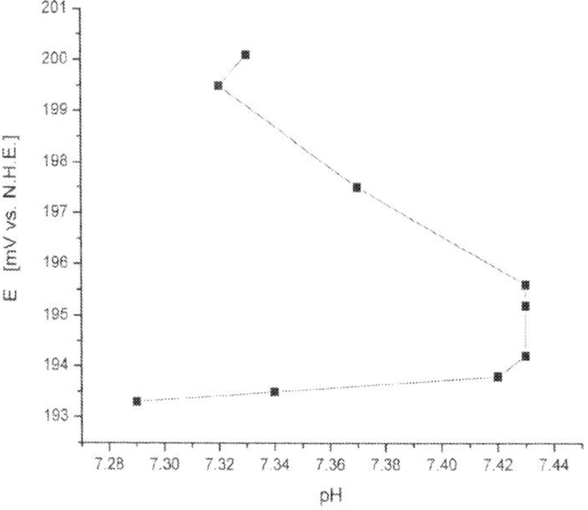

Figure 2. Redox potential of water dependence on pH during interaction of the composite with distilled water.

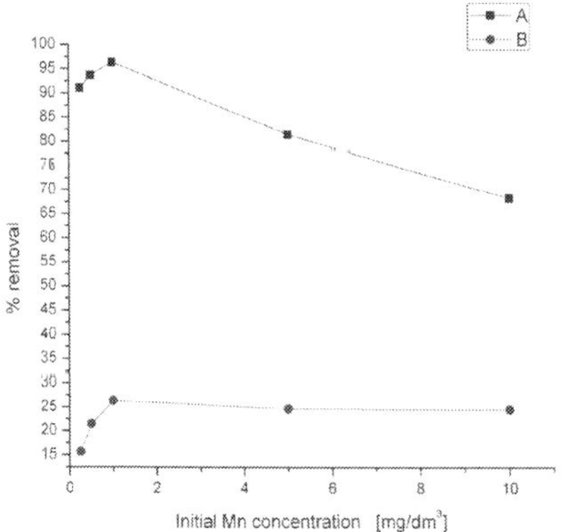

Figure 3. Percentage removal of Mn^{2+} (A) and colloidal MnO_2 (B) from synthetic waters (the composite dosage, 2 g/dm³; contacting time, 20 min; initial Mn concentrations in range 0.25 – 10 mg/dm³; initial pH 5.75 ± 0.1; temperature, 20 ± 0.5°C).

Average initial pH of the synthetic waters was 5.75. After 20 min of contact with the composite material average pH was 6.70.

During the interactions of composite with synthetic waters, the colloidal MnO_2 was removed to a lesser degree than Mn^{2+}. The authors imposed that colloidal manganese possesses the following structure of micelles:

$$\{m[MnO_2]nSO_4^{2-}\ 2(n-x)K^+\}2xK^+ \tag{7}$$

Potential-determining ions in the structure of micelles are SO_4^{2-}. They are primarily adsorbed on MnO_2 and responsible for the stability of colloids. Therefore, it is clear that the reduction of manganese is more difficult and there is an electrostatic repulsion between colloidal particles and a composite with dominantly negatively charged surface sites. Thus, the removal efficiency of colloidal manganese is significantly lower compared with the ionic form of Mn^{+2}. During the electrochemical interactions of synthetic water containing Mn^{2+} and colloidal MnO_2 with the composite material, transferring of Al^{3+} ions in a solution increases the TDS value, as shown in Table 1.

$C_0(Mn)$ mg/dm³	TDS (mg/dm³)	pH	C(Mn) mg/dm³	TDS (mg/dm³)	pH
Before Mn^{2+} synthetic water treatment			After Mn^{2+} synthetic water treatment		
0.25	3	5.75	0.0223	17	6.65
0.50	7	5.73	0.0318	21	6.71
1.0	10	5.71	0.0363	25	6.72
5.0	14	5.70	0.9271	29	6.70
10.0	28	5.76	3.9773	39	6.58
Before colloidal MnO_2 synthetic water treatment			After colloidal MnO_2 synthetic water treatment		
0.25	3	5.82	0.2108	18	6.73
0.50	6	5.75	0.3928	22	6.71
1.0	11	5.71	0.7366	25	6.72
5.0	14	5.72	3.768	29	6.75
10.0	28	5.75	7.549	39	6.67

Table 1. The results of synthetic waters analysis before and after treatment with composite material.

The initial dissolution of the Al based alloys introduces both aluminium and alloying ions into the solution, and then the reposition of microalloying tin onto active sites at surface occurs [46], so it was not detected by ICP-OES analysis.

Aluminium ions generated during electrochemical processes of manganese removal may form monomeric species such as $Al(OH)^{2+}$, $Al(OH)_2^{+}$ and $Al(OH)_4^{-}$. During the time, these monomers have tendency to polymerize in the pH range 4–7 which results in oversaturation and formation of amorphous hydroxide precipitate according to complex precipitation kinetics. Many polymeric species such as $Al_6(OH)_{15}^{+3}$, $Al_7(OH)_{17}^{+4}$, $Al_8(OH)_{20}^{+4}$, $Al_{13}O_4(OH)_{24}^{+7}$, $Al_{13}(OH)_{34}^{+5}$ have been reported [50]. Average concentration of aluminium, immediately after 20 min of composite interaction with Mn^{2+} synthetic waters, was 0.2131 mg/dm³ and included all mentioned monomeric and polymeric species which were not coagulated. After a prolonged period of time concentration of aluminum has a tendency to decrease reaching values that are below 0.1 mg/dm³, due to precipitation of $Al(OH)_3$ sludge.

The increase in the pH during the experiments can be explained in terms of the electrochemical and the chemical reactions that take place in the system composite-synthetic water. Water reduction at cathodic parts of composite (eq. 8), the electrochemical dissolution of aluminum (eq. 9) and protolytic reactions (eq. 10-14) increase the pH value [51].

$$H_2O + e^- \leftrightarrows 1/2 H_2 + OH^- \tag{8}$$

$$2Al + 6H_2O \leftrightarrows 2Al^{3+} + 3H_2 + 6\,OH^- \tag{9}$$

$$Al(OH)_4^- + H^+ \leftrightarrows Al(OH)_3 + H_2O \tag{10}$$

$$Al(OH)_3 + H^+ \leftrightarrows Al(OH)_2^+ + H_2O \tag{11}$$

$$Al(OH)_2^+ + H^+ \leftrightarrows Al(OH)^{2+} + H_2O \tag{12}$$

$$Al(OH)^{2+} + H^+ \leftrightarrows Al^{3+} + H_2O \tag{13}$$

$$Al(OH)_3(s) \leftrightarrows Al^{3+} + 3OH^- \tag{14}$$

8.2. Bentonite modified by mixed Fe, Mg (hydr)oxides coatings for the removal of ionic and colloidal forms of Pb(II)

Lead (Pb) is heavy metal which presents one of the major environmental pollutants due to its hazardous nature. It diffuses into water and the environment through effluents from lead smelters as well as from battery, paper, pulp and ammunition industries. Scientists established that lead is nonessential for plants and animals, while for humans it is a cumulative poison which can cause damage to the brain, red blood cells and kidneys [52].

In this subchapter, a cheap and effective composite material as a potentially attractive adsorbent for the treatment of Pb(II) contaminated water sources has been described. The procedure for obtaining a bentonite based composite involves the application of mixed Fe and Mg hydroxides coatings onto bentonite particles (0.375 mmol Fe and 0.125 mmol Mg per gram of bentonite) in aqueous suspension and subsequent thermal treatment of the solid phase at 498 K [53]. Bearing in mind layered structure of montmorillonite, the quite limited extent of isomorphous substitution of Mg for Fe in iron (hidr)oxides and significant differences in acid-base surface properties between these two (hydr)oxides, formation of heterogeneous coatings onto bentonite and specific structure of obtained composite have been achieved [54]. Different adsorption sites on such heterogeneous surface provide efficient removal of numerous chemical species of Pb(II) over a wide pH range.

The structural changes of montmorillonite during composite synthesis are mainly reflected in the reduction of d_{001} diffraction peak intensity in X-ray diffractograms and its shifting towards the higher values of 2θ. Moreover, it can be observed that the peak is broadened suggesting that the distance between the layers is non-uniform with disordered and partially delaminated structure. The crystallographic spacing d_{001} of montmorillonite in the native bentonite and the composite, computed by using Bragg's equation ($n\lambda = 2d \sin \theta$), is 1.54 nm and 1.28 nm, respectively. These changes in the structure took place because the d-spacing is very sensitive to the type of interlamellar cations, and the degree of their hydration [55].

The XRD patterns of the composite and starting (native) bentonite are presented in Fig. 4a and b, respectively.

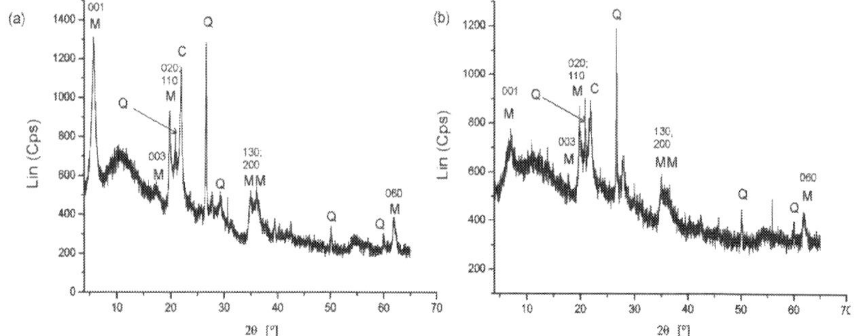

Figure 4. X-ray diffractograms of (a) composite and (b) native bentonite

SEM micrographs (Fig. 5 a, b and c) show that bentonite and composite are composed of laminar particles arranged in layered manner, forming the aggregates with diameters up to 50 μm.

Figure 5. (a) SEM of synthesized composite, (b) SEM of composite after interaction with Pb(II) solution and (c) surface morphology of the native bentonite

No significant changes in the microstructure of composite occurred during the interaction with the aqueous solution of Pb(II).

Despite a thorough washing process, a large amount of NO_3^- is retained in the composite. A vibration mode at ca. 1389 cm^{-1} in FTIR spectrum confirms the NO_3^- stretching which indicates that some positive charged sites exist on the surface of composite and that they are counterbalanced by the NO_3^- which can be exchanged by other anions [53]. In addition, the formation of poorly crystallized magnesium hydroxonitrate in pH range 9-11 [56,57], where Fe/Mg coprecipitation was performed over bentonite particles, is very likely.

8.2.1. Specific surface area determined by N_2 adsorption/desorption using BET equation

The Fig. 6. shows the comparative nitrogen adsorption-desorption isotherms of native bentonite and composite.

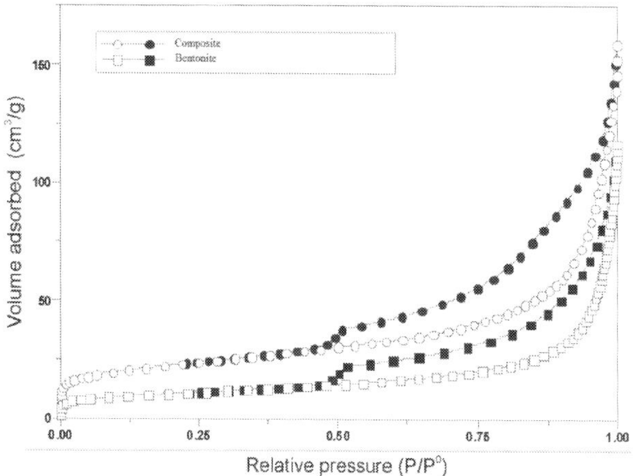

Figure 6. Nitrogen adsorption-desorption isotherms of native bentonite and composite.

The isotherms can be assigned to Type II isotherms, corresponding to non-porous or macroporous adsorbents. The hysteresis loops of Type H3 in the IUPAC classification occur at $p/p^0 > 0.5$, which is not inside the typical BET range. Furthermore, hysteresis loops of these isotherms indicate that they were given by either slit-shaped pores or, as in the present case, assemblages of platy particles of montmorillonite. Porous structure parameters are summarized in Table 2.

Sample	S_{BET} (m²/g)	Median mesopore diameter (nm)	Cumulative mesopore area (m²/g)	Cumulative mesopore volume (cm³/g)	Micropore volume (cm³/g)
Bentonite	37.865	13.629	53.329	0.1202	0.0153
Composite	80.385	11.021	82.675	0.1716	0.0316

Table 2. Specific surface area and porosity of native bentonite and composite, determined by applying BET, BJH and D-R equation to N_2 adsorption at 77 K

Compared to native bentonite, during the composite synthesis additional meso- and micropores were generated. Pore volumes (Gurvich) at p/p^0 0.999 for bentonite and composite are 0.180 cm³/g and 0.243 cm³/g, respectively. It was found that isotherms gave linear BET plots from p/p^0 0.03 to 0.21 for bentonite and from 0.03 to 0.19 for composite.

The composite has the specific surface area that is twice the size compared to the surface area of the native bentonite. This can be explained by the structural changes that occurred during the chemical and thermal modification of the native bentonite. The structural changes include delamination as well as the decrease of the distance between the layers of montmorillonite particles, because the interlayer water was lost under heating. The higher surface area of composite mainly results from the interparticle spaces generated by the three-dimensional co-aggregation of magnesium polyoxocations, iron oxide clusters and plate particles of montmorillonite. Macro- and mesopores arose from particle-to-particle interactions, while micropores were generated in the interlayer spaces of clay minerals due to irregular stacking of layers of different lateral dimensions [58].It is apparent that the changes of montmorillonite structure are responsible for the creation of new pore structure in the composite, which is then stabilized by the thermal treatment with the removal of H_2O molecules. The changes that involve partial dehydroxylation and cationic dehydration are brought about by thermal activation and they lead to various forms of cross-linking between oxides and smectite framework. As a result, composite does not swell and can be easily separated from water by filtration or centrifugation. There is a wide pore size distribution which supports disordered structure consisting of the delaminated parts with mesoporosity and the layered parts with microporosity.

The pH of the Pb(II) solution plays an important role in the adsorption process, influencing not only the surface charge of the adsorbent and the dissociation of functional groups on the active sites of the adsorbent but also the solution Pb(II) chemistry. The adsorption of Pb(II) on the composite decreased when pH decreased as shown in Fig. 7.

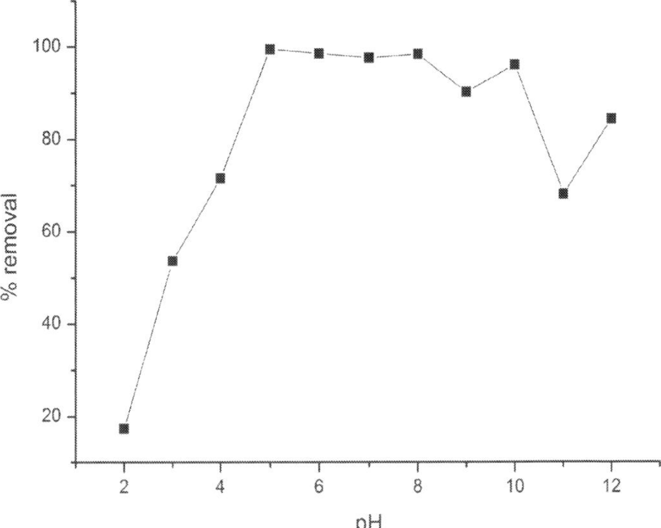

Figure 7. Effect of pH on adsorption of Pb(II) onto composite

The adsorptive decrease at pH below 5 was caused by the competition between H^+ and Pb^{2+} for the negatively charged surface sites. Maximum retention is in the pH range 5-10. The main Pb(II) species in the pH range 6.5-10 are $Pb(OH)^+$ and $Pb(OH)_2$ which can easily form colloidal micelles characterized with the following imposed structure:

$$\{m[Pb(OH)_2]nPb(OH)^+ \cdot (n-x)NO_3^-\}xNO_3^- \qquad (15)$$

The potential – determining ion is $Pb(OH)^+$ and that is the reason for the positive ZP of colloidal Pb(II) at the pH below 10 [59,60]. Therefore, colloidal micelles were easily attracted by the negatively charged composite surface. Particle size of colloidal Pb(II) at pH 7±0.1 was determined to be 268.7 ± 16.7 nm. At the pH range of 10-12 the predominant Pb(II) species are $Pb(OH)_2$ and $Pb(OH)_3^-$ which give rise to the formation of negatively charged colloidal micelles with the following structure:

$$\{m[Pb(OH)_2]nPb(OH)_3^-\cdot(n-x)Na^+\}xNa^+ \qquad (16)$$

ZP values for Pb(II) colloidal solutions at pH 11.8 were - 50.7±3.6 mV with particle size of 252.7±28.2 nm. Having in mind surface heterogeneity of the composite and high point of zero charge value of $Mg(OH)_2$ (between pH 12 and pH 13) [61], negative ions and particles can be adsorbed on the positively charged surface sites at pH 10-12. Removal efficiency of $Pb(OH)_3^-$ was higher than negatively charged colloids, probably because the ionic species were involved in the process of ion exchange and chemisorption, while colloidal micelles could be bound to the surface dominantly by electrostatic forces.

8.3. Bentonite based composite coated with immobilized thin layer of organic matter

Synthesis of bentonite based composite material, described in this section, was carried out by applying thin coatings of natural organic matter, obtained by alkaline extraction from peat, mostly comprised of humic acids [62]. Humic acids have high complexing ability with various heavy metal ions, but it is difficult to use them as the sorbent because of their high solubility in water. However, they form stabile complexes with the inorganic ingredients of bentonite (montmorillonite, quartz, oxides, etc.) and can be additionally insolubilized and immobilized by heating at 350°C. After immobilization, humic acids represent an important sorbent for heavy metals, pesticides and other harmful ingredients from water. Humic acid are insolubilized by condensation of carboxylic and phenolic hydroxyl groups. Therefore, the aim was to remove manganese from aqueous solutions by treating it with synthesized composite as well as to study and explain the mechanism of composite interaction with manganese aqueous solutions. The composite does not release significant quantity of organic matter in water because it is tightly bonded to bentonite surface [63-65]. The degree of manganese removal was more than 94% at a range of initial manganese concentrations from 0.250 to 10 mg/l.

The result of conductometric titration is given in Fig. 8. Equivalence point was located at the intercept of the first and second linear part of the titration curve. The value of the total acidic group content is calculated to be 215.18 μmol/g.

The experimental data of manganese adsorption onto composite are very well fitted by the Freundlich isotherm model (Fig. 9.) with a very high correlation coefficient value of 0.9948. The good agreement of experimental data with the Freundlich model indicates that there are several types of adsorption sites on the surface of the composite. The amount of adsorbed Mn(II) increases rapidly in the first region of adsorption isotherm and then the slope of isotherm gradually decreases in the second region. The adsorption capacity of composite is 11.86 mg/g, at an equilibrium manganese concentration of 16.28 mg/l.

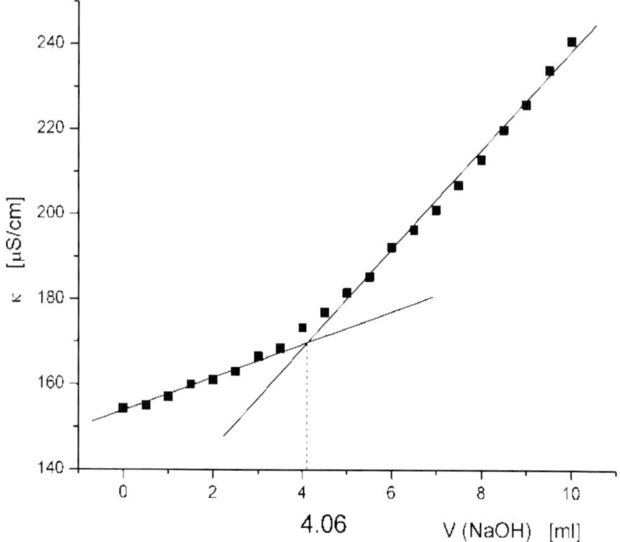

Figure 8. The conductometric titration of composite suspension (1 g in 250 ml of 1mM NaCl solution as background electrolyte) with 0.053 M NaOH.

After the treatment of model water with composite for the period of 20 min, the following results were obtained (Table 3).

Before water treatment			After water treatment			
$C_0(Mn)$ mg/l	pH	Conductivity μS/cm	pH	Conductivity μS/cm	C(Mn) mg/l	%Mn Adsorption
0	6.43	8.01	6.67	11.43	0	0
0.250	6.37	9.57	7.11	13.76	0.0030	98.8
0.490	6.32	10.67	7.15	15.31	0.0039	99.2
1.0	6.30	14.67	7.12	31.10	0.0090	99.1
2.5	6.20	20.70	6.96	37.20	0.0187	99.25
5.0	6.19	32.80	6.83	49.40	0.0646	98.71
10.0	6.16	55.30	6.70	68.90	0.5314	94.69

Table 3. The results of water analysis before and after treatment with composite

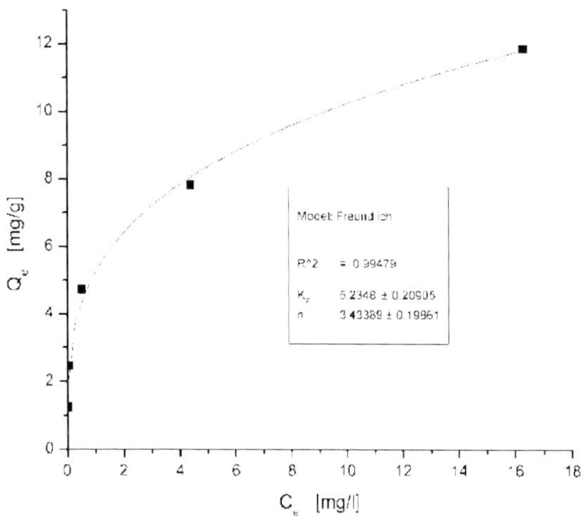

Figure 9. Freundlich adsorption isotherm for manganese adsorption onto composite.

During the thermal treatment in nitrogen atmosphere at 350 °C, the condensation of carboxyl and adjacent alcohol and phenol groups occurs. In this way the solubility of organic matter immobilized on bentonite matrix surface decreases [65]. Moreover, a part of carboxyl groups is decomposed by decarboxylation reaction, releasing CO_2 and CO. However, despite of this, a part of oxygen functional groups remains on the surface, and these groups act as sites that bind bivalent manganese forming inner-sphere complexes.

Besides organic functional groups, there are also Si-OH and Al-OH groups on the sites of crystal grain breaks, as well as permanent negative charge due to isomorphic substitution in clay minerals. They all contribute to the reduction of manganese concentration in the aqueous solution. Manganese retention by the formation of outer-sphere complexes, including ion exchange, can be showed by an Eq. (17) [66].

$$(\equiv S\text{-}O^-)_2...C^{n+}_{3-n} + Mn^{2+} \rightleftharpoons (\equiv S\text{-}O^-)_2...Mn^{2+} + (3-n)\,C^{n+} \tag{17}$$

in which C represents the cation that is exchanged.

The formation of inner-sphere complexes is represented by the Eqs. (18) and (19) and involves the release of hydrogen ions and the change of solution pH.

$$\equiv S\text{-}OH + Mn^{2+} \rightleftharpoons \equiv S\text{-}O\text{-}Mn^+ + H^+ \tag{18}$$

$$\equiv 2S\text{-}OH + Mn^{2+} \rightleftharpoons (\equiv S\text{-}O)_2\text{-}Mn + 2H^+ \tag{19}$$

According to these equations, it can be concluded that the pH value of the solutions decrease after the treatment. However, an opposite phenomenon can be experimentally observed (Table 3). The explanation for it is that hydrogen ions which are released during manganese retention participate in the protonation of surface groups:

$$\equiv\!S\text{-}OH + H^+ \rightleftharpoons \equiv\!S\text{-}OH_2^+ \tag{20}$$

$$\equiv\!S\text{-}O^- + H^+ \rightleftharpoons \equiv\!S\text{-}OH \tag{21}$$

Therefore, the pH value of the Mn^{2+} aqueous solutions after treatment with composite had a higher value than the initial pH. This indicates that more hydrogen ions are bound to the surface than released by manganese binding. Namely, the composite exhibits amphoteric character due to the surface sites that act either as proton acceptors or as proton donors.

Organic matter decreases the PZC value of bentonite and neutralizes positive electric charge that comes from interlaminated cations, thus increasing composite affinity to manganese, even at lower pH values (67). Fig. 10. presents the pH dependence of residual Mn concentration, for the initial Mn concentration of 5 mg/l. The residual concentration of Mn decreases gradually with pH increasing in the range of 3.5-7 and then increases in the range of 7-10, with the apparent minimum at pH 7.

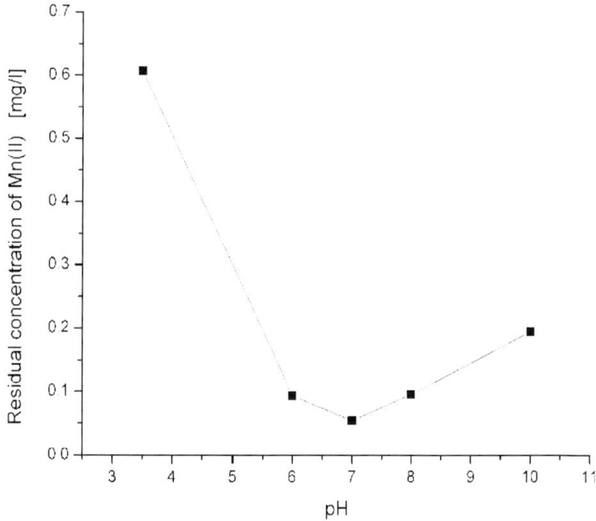

Figure 10. Residual concentration of Mn(II) as a function of model water pH.

The increase of pH value has dual effect on the removal of manganese. The increase of the pH value favours manganese removal due to increase of the number of deprotonated sites that are available for the binding of manganese. However, there is an increase in the solubility of organic matter which has been applied on the bentonite particles. The dissolved organic matter (humic acids) reacts with manganese forming complexes which bear a negative charge and have a weaker binding affinity for the composite surface than Mn^{2+}. Fig 10. indicates two opposite effects of the pH on manganese removal. The pH dependence of released organic matter (expressed as permanganate number) and turbidity (NTU) of solutions are shown in Fig. 11.

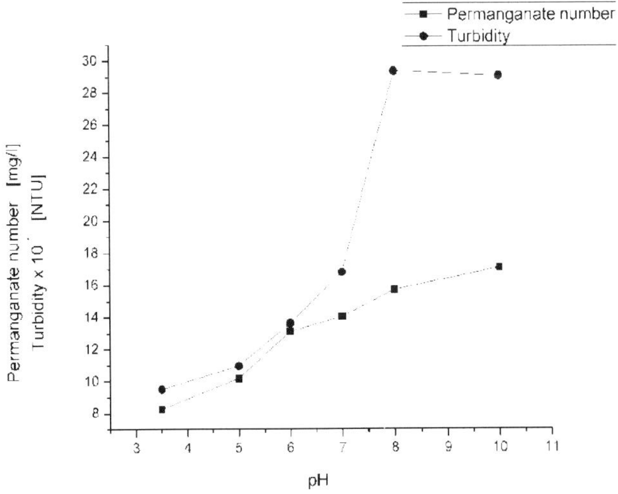

Figure 11. Premanganate number and turbidity of filtrate as function of pH (0.2 g of composite and 100 ml of 1mM Na_2SO_4 as background electrolyte).

The released organic matter contributes to the increased turbidity at higher pH values.

9. Summary

The widespread industrial areas where nanocomposites can be applied are primary and conversion industry, modern coating technologies, constructional regions, and environmental (water, air) purification. In addition to the dominant use of composites as structural elements, important application of composite materials is in the water purification technologies. In this field of application, composites usually have the role of adsorbent, electrochemically active materials, catalysts, photocatalysts etc.

Bentonite is a natural and colloidal alumosilicate with particle size less than 10 μm, which is effectively used as sorbent for heavy metals and other inorganic and organic pollutants from water. Due to its positive textural properties and high specific surface area it can be used as low-cost matrix for synthesis of adsorbents or electrochemically active composite materials for the removal of pollutants in ionic and colloidal form from water. In this respect, three new/modified bentonite based composite materials have been synthetised and characterized.

Coated or composite particles are composed of solid phase covered with thinner or thicker layer of another material. These coatings - layers covering the surface of matrix are important for several reasons. In such way, the surface and textural characteristics of the initial solid phase are modified and sintering conditions can be better controlled. An important factor in achieving the desired microstructure of ceramics is sintering procedure that includes rather complex processes. A considerable influence on sintering has been exhibited by an addition of microalloying components, which significantly determined a microstructure and resulted properties of ceramics. The presence of small amounts of impurities in the starting material can vastly influence their mechanical, optical, electrical, color, diffusive, and dielectric properties of alumosilicate matrix. In summary, the process of diffusion mass transport in ceramic crystal regions are affected by temperature, oxygen partial pressure and concentration of impurities. A procedure for the removal of manganese in ionic (Mn^{2+}) and colloidal (MnO_2) forms from synthetic waters, by reduction and adsorption processes on electrochemically active alumosilicate ceramics based composite material has been described. Synthesis procedure of the composite material consists of two phases. Firstly, composite particles were synthesized by applying Al/Sn oxide coating onto the bentonite particles in an aqueous suspension. In the second phase, aluminium powder is added to the previously obtained plastic mass and after shaping in the form of spheres 1 cm in diameter and drying, sintering was performed at 900°C. Elemental tin, resulting from the reduction of Sn^{2+}-ion, comes into contact with liquid aluminum in the pores of the matrix performing aluminum microalloying and activation. Moreover, due to a low partial pressure of oxygen, nonstoichiometric oxides with metal excess are obtained, and they play an important role in the electrochemical activity of the composite material. In accordance with this, a redox potential of water is changed in contact with composite.

Another effective composite material as a potentially attractive adsorbent for the treatment of Pb(II) contaminated water sources has been synthesized by coating of bentonite with mixed iron and magnesium (hydr)oxides. The procedure for obtaining a bentonite based composite involves the application of mixed Fe and Mg hydroxides coatings onto bentonite particles in aqueous suspension and subsequent thermal treatment of the solid phase at 225ºC. Formation of heterogeneous coatings on bentonite results in changes of bentonite acid-based properties, high specific surface area and positive adsorption characteristics. Different adsorption sites on such heterogeneous surface provide an efficient removal of numerous chemical species of Pb(II) (ionic and colloidal) over a wide pH range.

Third bentonite based composite material was obtained by applying thin coatings of natural organic matter, extracted from a peat, mostly based on humic acids. Humic acids are known due to high complexing ability to various heavy metal ions, but it is difficult to use them directly as the sorbent because of their high solubility in water. However, they form stabile complexes with the inorganic ingredients of bentonite (montmorillonite, quartz, oxides, etc.) and can be successfully insolubilized and immobilized by heating at 350°C. After immobilization, humic acids represent an important sorbent for heavy metals, pesticides and other harmful ingredients from water. Humic acid are insolubilized by condensation of carboxylic and phenolic hydroxyl groups. The composite such obtained can be effectively used as the sorbent for heavy metals.

Author details

Marjan S. Ranđelović, Aleksandra R. Zarubica and Milovan M. Purenović

University of Niš, Faculty of Science and Mathematics, Department of Chemistry, Niš, Serbia

Acknowledgement

The authors acknowledge financial support from the Ministry of Education and Science of the Republic of Serbia.

10. References

[1] Ralph B, Yuen H.C, Lee W.B. The processing of metal matrix composites - an overview. Journal of Materials Processing Technology 1997; 63 339-353

[2] Chung D. Composite Materials. Springer Science + Business Media B.V.; 2010.

[3] Usuki A, Kawasumi M, Kojima Y, Fukushima Y, Okada A, Kurauchi T. Swelling behavior of montmorillonite cation exchanged for ω-amino acids by ε-caprolactam. Journal of Materials Research 1993; 8 1179-1184.

[4] Kojima Y, Usuki A, Kawasumi M, Fukushima Y, Okada A, Kurauchi T, Mechanical properties of Nylon-6 clay hybrid. Journal of Materials Research 1993; 8 1185-1189.

[5] Giannelis E. Polymer-layered Silicate Nanocomposites. Advanced Materials 1996; 8 29-35.

[6] Pfaendner R. Nanocomposites: Industrial opportunity of challenge? Polymer Degradatio Stability 2010; 95 369-373.

[7] Nanocomposites and nanotubes conference,11-12th March, 2008; www.nanosconference.com/home.asp;. accessed 30.04.2012.

[8] Manfredi D, Pavese M, Biamino S, Antonini A, Fino P, Badini C. Microstructure and mechanical properties of co-continuous metal/ceramic composites obtained from Reactive Metal Penetration of commercial aluminium alloys into cordierite. Composites: Part A 2010; 41 639–645.

[9] Saiz E, Foppiano S, MoberlyChan W, Tomsia A.P. Synthesis and processing of ceramic–metal composites by reactive metal penetration. Composites: Part A 1999; 30 399–403.

[10] Haq I, Matijevic E. Preparation and properties of uniform coated inorganic colloidal particles. 12. Tin and its compounds on hematite. Progress in Colloid and Polymer Science 1998; 109 185–191.

[11] Bergaya F, Theng B. K. G, Lagaly G. Handbook of clay science. Elsevier; 2006.

[12] Olmos L, Martin C.L, Bouvard D. Sintering of mixtures of powders: Experiments and modelling. Powder Technology 2009; 190 134–140.

[13] Michael A, Dubois P. Polymer-layered silicate nanocomposites: preparation, properties and uses of a new class of materials. Materials Science and Engineering 2000; 28 1-63.

[14] Ray S.S, Okamoto M. Polymer/layered Silicate Nanocomposite: A Review from Preparation to Processing. Progress in Polymer Science 2003; 28 1539-1641.

[15] Hussain F, Hojjati M, Okamoto M, Gorga R.E. Review article: polymer-matrix nanocomposites, processing, manufacturing, and application: an overviewJournal of Composite Materials 2008; 40 1511-1575.

[16] Moszo J, Pukanszky B, Polymer micro and nanocomposites: Structure, interactions, properties. Journal of Industrial and Engineering Chemistry 2008; 14 535-563.

[17] Xie W, Gao Z, Pan W.P, Hunter D, Singh A, Vaia R. Thermal degradation chemistry of alkyl quaternary ammonium montmorillonite. Chemistry of Materials 2001; 13 2979-2990.

[18] Leszczynska A, Njuguna J, Pielochowski K, Banerjee J.R. Polymer/clay (montmorillonite) nanocomposites with improved thermal properties: Part II: A review

study on thermal stability of montmorillonite nanocomposites based on different polymeric matrixes. Thermochimica Acta 2007; 454(1) 1-22.

[19] Pandey J.K, Reddy K.R, Kumar A.P, Singh R.P, An overview on the degradability of polymer nanocomposites. Polymer Degradation and Stability 2005; 88 234-250.

[20] Wermter H, Pfaender R. European patent EP 1592741 assigned to Ciba Holding Inc., 2009.

[21] Leszcynska A, Njuguna J, Pielichowski K, Banerjee J.R. Thermal stability of polymer/montmorillonite nanocomposites: Part I: Factors influencing thermal stability and mechanisms of thermal stability improvement. Thermochimica Acta 2007; 453 75-96.

[22] Moad G, Dean K, Edmond L, Kukaleva N, Li G, Mayadunne RTA. Non-Ionic, Poly(ethylene oxide)-based, Surfactants as Intercalants/Dispersants/Exfoliants for Polypropylene-Clay Nanocomposites. Macromolecular Materials and Engineering 2006; 291 37-52.

[23] Moad G, Dean K, Edmond L, Kukaleva N, Li G., Mayadaunne R.T.A, Pfaendner R, Schneider A, Simon G, Wermter H. 2006. Novel Copolymers as Dispersants/Intercalants/Exfoliants for Polypropylene-Clay Nanocomposites. Macromolecular Symposia 2006; 233 170-179.

[24] Allen N.S, Edge M, Corrales T, Childs A, Liauw C.M, Catalina F, Peinado C, Minihan A, Aldcroft D. Ageing and stabilisation of filled polymers: an overview. Polymer Degradation and Stability 1998; 61 183-199.

[25] Qin H, Zhao C, Zhang C, Chen G, Yang M. Photo-oxidative degradation of polyethylene/montmorillonite nanocomposite. Polymer Degradation and Stability 2003; 81 497-500.

[26] Morlat-Therias S, Mailhot B, Gonzalez D, Gardette J.L. Photooxidation of Polypropylene/Montmorillonite nanocomposites Part II : Interactions with antioxidants. Chemistry of Materials 2005; 17 1072-1078.

[27] Morlat S, Mailhot B, Gonzalez D, Gardette J.L, Photooxidation of Polypropylene/Montmorillonite nanocomposites Part I : Influence of nanoclay and compatibilising agent.Chemistry of Materials 2004; 16(3) 377-383.

[28] Chmela S, Kleinova A, Fiedlorova A, Borsig E, Kaempfer D, Thormann R. Photo-oxidation of sPP/organoclay nanocomposites. Applied Chemistry 2005; 42(7) 821-829.

[29] Rotzinger B. Talc-filled PP: A new concept to maintain long term heat stability. Polymer Degradation and Stability 2006; 91 2884-2887.

[30] Savage N, Diallo M.S. Nano materials and water purification: opportunities and challenges. Journal of Nanoparticle Reserch 2005; 7 331-342.

[31] Da Fonseca M.G, De Oliveira M.M, Arakaki L.N.H. Removal of cadmium, zinc, manganese and chromium cations from aqueous solution by a clay mineral. Journal of Hazardous Materials 2006; B137 288–292.

[32] Dimirkou A, Doula M.K. Use of clinoptilolite and an Fe-overexchanged clinoptilolite in Zn^{2+} and Mn^{2+} removal from drinking water. Desalination 2008; 224 280–292.

[33] Bladergroen B.J, Linkov V.M. Electrosorption ceramic based membranes for water treatment. Separation and Purification Technology 2001; 25 347–354.

[34] Kanki T, Hamasaki S, Sano N, Toyoda A, Hirano K. Water purification in a fluidized bed photocatalytic reactor using TiO_2-coated ceramic particles. Chemical Engineering Journal 2005; 108 155–160.

[35] Johnson C.D, Worrall F. Novel granular materials with microcrystalline active surfaces—Waste water treatment applications of zeolite/vermiculite composites. Water Research 2007; 41 2229 – 2235.

[36] Tzoupanos N.D, Zouboulis A.I. Preparation, characterisation and application of novel composite coagulants for surface water treatment Water Research 2011; 45 3614-3626.

[37] Lv Y, Liu H, Wang Z, Liu S, Hao L, Sang Y, Liu D, Wang J, Boughton R.I. Silver nanoparticle-decorated porous ceramic composite for water treatment. Journal of Membrane Science 2009; 331 50–56.

[38] Shen W, Feng L, Feng H, Kong Z, Guo M. Ultrafine silver(II) oxide particles decorated porous ceramic composites for water treatment. Chemical Engineering Journal 2011; 175 592– 599.

[39] Wang W, Serp P, Kalck P, Gomes Silva C, Luıs Faria J. Preparation and characterization of nanostructured MWCNT-TiO$_2$ composite materials for photocatalytic water treatment applications. Materials Research Bulletin 2008; 43 958–967.

[40] Zhang H, Quan X, Chen S, Zhao H. Fabrication and characterization of silica/titania nanotubes composite membrane with photocatalytic capability. Environmental Science and Technology 2006; 40 6104-6109.

[41] Ranđelović M, Purenović M, Zarubica A, Purenović J, Mladenović I, Nikolić G. Alumosilicate ceramics based composite microalloyed by Sn: An interaction with ionic and colloidal forms of Mn in synthetic water. Desalination 2011; 279(1-3) 353-358.

[42] Jeurgens L.P.H, Sloof W.G, Tichelaar F.D, Mittemeijer E.J. Composition and chemical state of the ions of aluminium-oxide films formed by thermal oxidation of aluminium. Surface Science 2002; 506 313–332.

[43] Purenovic M.M. Influence of Some Alloying Elements and Admixtures on Electrochemical Behaviour of the System Aluminium - Oxide Layer - Electrolyte. PhD Thesis. Faculty of Technology and Metallurgy, University of Belgrade; 1978.

[44] Chan R. W, Haasen P. Physical Metallurgy. fourth edition, Elsevier Science B.V. Amsterdam, Netherlands; 1996.

[45] Rahaman M.N. Ceramic processing and sintering, second edition. Marcel Dekker, Inc. New York; 2003.

[46] Gudic S, Smoljko I, Kliskic M. Electrochemical behaviour of aluminium alloys containing indium and tin in NaCl solution. Materials Chemistry and Physics 2010; 121 561–566.

[47] Cvetković V.S, Purenović J.M, Jovićević J.N. Change of water redox potential, pH and rH in contact with magnesium enriched kaolinite–bentonite ceramics. Applid Clay Science 2008; 38 268–278.

[48] Wei Q, Ren X, Du J, Wei S, Hu S. Study of the electrodeposition conditions of metallic manganese in an electrolytic membrane reactor. Minerals Engineering 2010; 23 578–586.

[49] Cvetković V S, Purenović J M, Purenović M M, Jovićević J N. Interaction of Mg-enriched kaolinite–bentonite ceramics with arsenic aqueous solutions. Desalination 2009; 249 582–590.

[50] Can O.T, Bayramoglu M, Kobya M. Decolorization of reactive dye solutions by electrocoagulation using aluminum electrodes. Industrial and Engineering Chemistry Research 2003; 42 3391-3396.

[51] Canizares P, Martinez F, Carmona M, Lobato J, Rodrigo M. A. Continuous Electrocoagulation of Synthetic Colloid-Polluted Wastes. Industrial and Engineering Chemistry Research 2005; 44 (8171-8177.

[52] Lead in Drinking-water, Background document for development of WHO Guidelines for Drinking-water Quality, World Health Organization; 2003.

[53] Ranđelović M, Purenović M, Zarubica A, Purenović J, Matović B, Momčilović M. Synthesis of composite by application of mixed Fe, Mg (hydr)oxides coatings onto bentonite - a use for the removal of Pb(II) from water. Journal of Hazardous Materials 2012; 199-200 367-374.

[54] Cornell R. M, Schwertmann U. The Iron Oxides: Structure, Properties, Reactions, Occurences and Uses, second ed., WILEY-VCH Verlag GmbH & Co. KGaA, Weinheim; 2003.

[55] Caglar B, Afsin B, Tabak A, Eren E. Characterization of the cation-exchanged bentonites by XRPD, ATR, DTA/TG analyses and BET measurement, Chemical Engineering Journal 2009; 149 242-248.

[56] Krasnobaeva O.N, Belomestnykh I.P, Isagulyants G.V, Nosova T.A, Elizarova T.A, Teplyakova T.D, Kondakov D.F, Danilov V.P. Synthesis of Complex Hydroxo Salts of Magnesium, Nickel, Cobalt, Aluminum, and Bismuth and Oxide Catalysts on Their Base. Russian Journal of Inorganic Chemistry 2007; 52(2) 141–146.

[57] Krasnobaeva O.N, Belomestnykh I.P, Isagulyants G.V, Nosova T.A, Elizarova T.A, Kondakov D.F, Danilov V.P. Chromium, Vanadium, Molybdenum, Tungsten, Magnesium,and Aluminum Hydrotalcite Hydroxo Salts and Oxide Catalysts on Their Base, Russian Journal of Inorganic Chemistry 2009; 54(4) 495–499.

[58] Rouquerol J, Rouquerol F, Sing K.S.W. Adsorption by Powders and Porous Solids: Principles, Methodology and Applications, Academic Press, San Diego USA; 1999.

[59] Liu Q, Liu Y. Distribution of Pb(II) species in aqueous solutions, Journal of Colloid and Interface Science 2003 268 266–269.

[60] Kosmulski M. Compilation of PZC and IEP of sparingly soluble metal oxides and hydroxides from literature, Advances in Colloid and Interface Science 2009 152 14–25.

[61] Krishnan S. V, Iwasaki I. Heterocoagulation vs. surface precipitation in a quartz-$Mg(OH)_2$ system, Environmental Science and Technology 1986; 20 1224- 1229.

[62] Ranđelović M, Purenović M, Purenović J, Momčilović M. Removal of Mn^{2+} from water by bentonite coated with immobilized thin layers of natural organic matter. Journal of Water Supply: Research and Technology – AQUA 2011; 60(8) 486-493.

[63] Ayari F, Srasra E, Trabelsi-Ayadi M. Characterization of bentonitic clay and their use as adsorbent. Desalination 2005; 185 391-397.

[64] Kolokassidou C. A, Pashalidis I, Costa C.N, Efstathiou A.M, Buckau G. Thermal stability of solid and aqueous solutions of humic acid. Thermochimica Acta 2007; 454 78–83.

[65] Ghosh S, Zhen-Yu W, Kang1 S, Bhowmik P. C, Xing B. S. Sorption and fractionation of a peat derived humic acid by kaolinite, montmorillonite, and goethite. Pedosphere 2009; 19(1) 21–30.

[66] Doula M.K. Removal of Mn^{2+} ions from drinking water by using clinoptilolite and a clinoptilolite Fe oxide system. Water Research 2006; 40 3167-3176.

[67] Zhuang J, Yu G.R. Effects of surface coatings on electrochemical properties and contaminant sorption of clay minerals. Chemosphere 2002; 49 619-628.

Mechanical Coating Technique for Composite Films and Composite Photocatalyst Films

Yun Lu, Liang Hao and Hiroyuki Yoshida

Additional information is available at the end of the chapter

http://dx.doi.org/10.5772/48794

1. Introduction

1.1. Coating techniques for film materials and their applications

In the field of materials science and engineering, the investigation on film materials is becoming increasingly important. By film materials, we can develop a variety of new material properties in the fields of electrics and electronics, optics, thermotics, magnetic, and mechanics, among others (S. Yoshida et al., 2008). In recent years, without the development of film materials we could not make any great progress in the renewable energy, environment improvement, exploitation of space, and so on. The coating techniques for film materials can fall into several categories as shown in Table 1.

In these techniques, physical vapor deposition (PVD) and chemical vapor deposition (CVD) are most widely applied. PVD are atomistic deposition processes in which material is vaporized from a solid or liquid source in the form of atoms or molecules and transported in the form of a vapor through a vacuum or low pressure gaseous (or plasma) environment to the substrate, where it condenses. PVD can be used to deposit films of elements and alloys as well as compounds using reactive deposition processes (Mattox, 2010). On the other hand, CVD may be defined as the deposition of a solid on a heated surface from a chemical reaction in the vapor phase. It belongs to the class of vapor-transfer processes which is atomistic in nature, which is the deposition species are atoms or molecules or a combined of these (Pierson, 1999). Microfabrication processes widely use CVD to deposit film materials in various forms including monocrystalline, polycrystalline, amorphous and epitaxial depending on the deposition materials and the reaction conditions. As listed in Table 1, there are other coating techniques for film materials such as liquid absorption coating, thermal spraying and mechanical coating. However, their applications are narrow comparing with PVD and CVD due to their features.

Physical vapor deposition (PVD)	Vacuum deposition	Resistance heating, Flash Evaporation, Vacuum Arc, Laser heating, Highfrequency heating, Electron beam heating
	MBE (Molecular Beam Epitaxy)	
	Laser deposition	
	Sputter deposition	Ion beam sputtering, DC sputtering, Highfrequency sputtering, Magnetron sputtering, Microwave ECR plasma deposition
	Ion beam plating	Highfrequency ion plating, Activated reactive evaporation, Arc ion plating
	Ion beam deposition	
	Ionized cluster beam deposition	
Chemical vapor deposition (CVD)	Thermal CVD	Atmospheric pressure CVD, Low pressure CVD
	Plasma CVD	DC plasma CVD, Highfrequency plasma CVD, Microwave plasma CVD, ERC plasma CVD
	Photo-excited CVD	
Liquid absorption coating	Plating	Electroplating, Electroless plating
	Anodic oxide coating, Painting, Sol-gel method	
	Spin coating, Dip coating, Roll coating, Spry coating	
Thermal spraying	Flame spraying, Electrical spraying (Arc, Plasma)	
Mechanical coating	Shot coating, Powder impact plating, Aerosol deposition , Gas deposition	

Table 1. Classification of the coating techniques for film materials

1.2. Advantages and limitations of these coating techniques

Any film coating technique has its advantages and limitations. The features of these coating techniques make their application fields different. Their advantages and limitations are summarized and shown in Table 2. Thickness control and coating of large-area films can be achieved in PVD which has become the major film coating technique in the fields of electronics, electrics and optics industries due to its high production efficiency, high film purity and low production cost. However, complicated and large scale equipments are necessary. In addition, films cannot be deposit on the substrates with complex profiles. In CVD processes, films can deposit on the substrates with complex profiles and the adhesion between films and substrates is generally strong. However, the processes are performed at temperature of 600 ºC and above. Large scale equipments are also needed just as PVD processes. The advantages and limitations of other coating techniques can also be found in Table 2. We will not explain them in detail any more.

Film coating technique	Advantages	Limitations
PVD	Thickness control Large area coating	Large and complex equipment Coating on flat substrate
CVD	Coating on complex substrate Strong adhesion	Large and complex equipment Elevated temperature process
Liquid absorption coating	Simple equipment Coating on complex substrate	Complicated post treatment Elevated temperature process
Thermal spraying	Rapid coating Large specific area	Elevated temperature process Grain growth
Mechanical coating	Ambient preparation Nanoscale coating	Large and complex equipment Coating on flat substrate

Table 2. Advantages and limitations of the film coating techniques

2. A novel coating technique for composite films

2.1. Proposal of a novel coating technique

2.1.1. Contamination phenomena during Mechanical Alloying (MA)

Powder metallurgy has been widely used in the manufacturing of mechanical parts. A typical process of powder metallurgy is shown in Fig. 1. As shown in the schematic diagram, blending of powder particles is necessary before compacting. Ball milling is often used to mix powder particles and make them homogeneous. Mechanical alloying, well known as ball milling, is frequently used to improve various material properties and to prepare advanced materials that are different or impossible to be obtained by traditional techniques (Suryanarayana, 2001). Ceramic balls are often used as grinding mediums which are indispensable in MA. However, the contamination of powder from grinding mediums and the adhesion of powder particles to the grinding mediums always harass engineers. Especially, the adhesion of powder particles to the grinding mediums and the bowl is really difficult to be eliminated.

2.1.2. Concept proposal of mechanical coating technique (MCT)

From the contamination discussed above, we proposed a novel film coating technique in 2005 called mechanical coating technique (MCT) with the diagram schematic shown in Fig. 2 (Lu et al., 2005). In this technique, metal powder and ceramic grinding mediums (balls, buttons and columns) are used as the coating material and the substrates respectively. Firstly, they are charged into a bowl made of alumina. The mechanical coating is performed by a pot mill or a planetary ball mill. In the process, friction, wear and impact among metal powder particles, ceramic grinding mediums and the inner wall of the bowl occur. That results in the formation of metal films on ceramic grinding mediums. In fact, some other researchers prepared metal or alloy films on grinding mediums by this technique. Kobayashi prepared metallic films on ZrO_2 balls (Kobayashi, 1995). Romankov et al. (2006) deposited Al and Ti-Al coatings on Ti alloy substrates. Gupta et al. (2009) reported the formation of nanocrystalline Fe-Si coatings on mild steel substrates. Farahbakhsh et al. (2011) prepared Cu and Ni-Cu solid solution coatings on Ni balls. Mechanical coating technique performed by ball milling has been established and known. After the proposal of MCT in 2005, we have made some important progress in advancing it. Titanium films on ceramic balls have been fabricated and their properties have been investigated (Yoshida et al., 2009 a). By MCT and its following high-temperature oxidation, TiO_2/Ti composite photocatalyst films have been successfully prepared (H. Yoshida et al., 2008). After that, we proposed 2-step MCT to fabricate TiO_2/Ti composite photocatalyst films without high-temperature oxidation (Lu et al., 2011). In addition, 2-step MCT was also used to fabricate TiO_2/Cu composite photocatalyst films (Lu et al., 2012). In this chapter, we will give a brief introduction to MCT, 2-step MCT and the relevant processes as a novel technique to fabricate composite films and TiO_2/metal composite photocatalyst films.

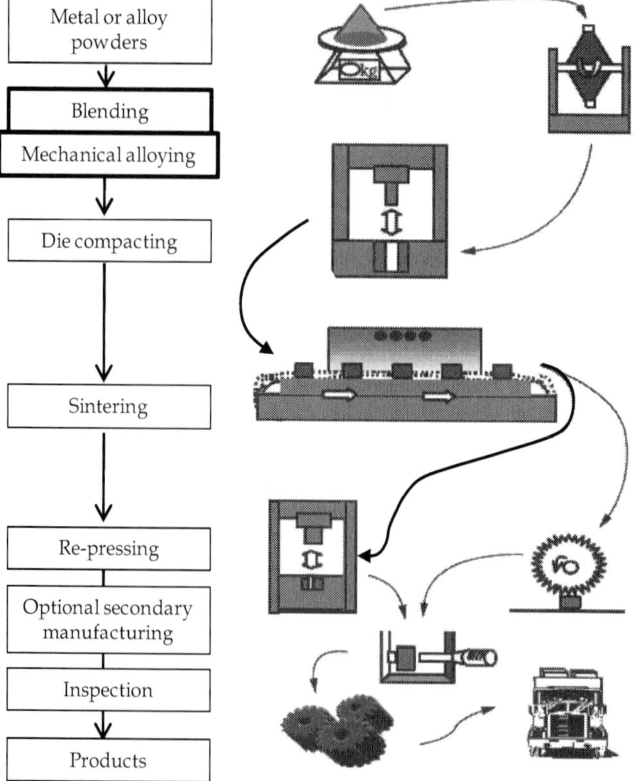

Figure 1. Simplified flowchart and schematic diagram of a typical powder metallurgy process

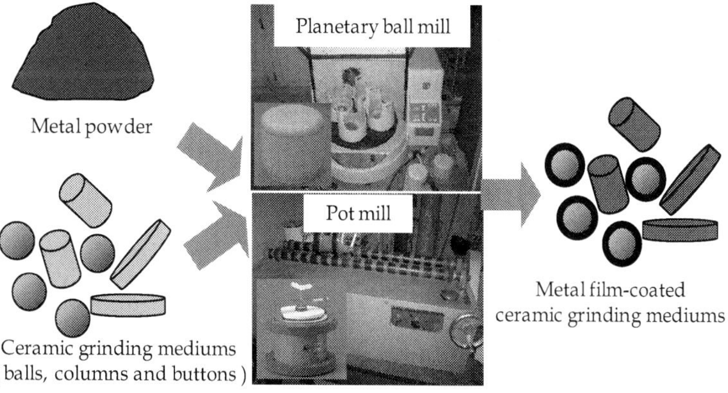

Figure 2. Schematic diagram of mechanical coating technique (MCT)

2.2. MCT and its influencing parameters

In our early work, we fabricated Ti films on alumina (Al_2O_3) grinding mediums such as balls, buttons and columns. Ti powder and Al_2O_3 grinding mediums were used as the coating materials and the substrates respectively. They were charged into a bowl made of alumina with the dimension of $\Phi75 \times 90$ mm (400 *ml*). The mechanical coating was carried out by a pot mill with a rotation speed of 80 rpm for 1000 h. Fig. 3 (a) shows the appearance comparison of the Al_2O_3 grinding mediums before and after MCT. It can be clearly seen that the Al_2O_3 grinding mediums after MCT showed metallic luster which means metal films might be formed on these grinding mediums. Also, the appearances of metal-coated Al_2O_3 balls after high-temperature oxidation are given in Fig. 3 (b).

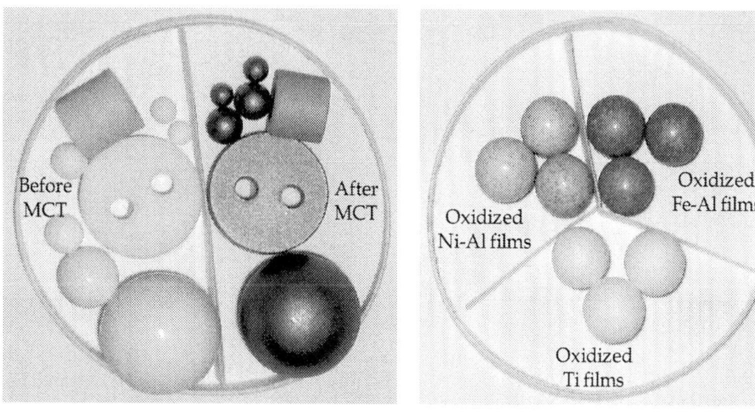

(a) Alumina objects before and after MCT

(b) Alumina balls after MCT and the following high-temperature oxidation

Figure 3. Appearances of Al_2O_3 grinding mediums after MCT and high-temperature oxidation

The influencing parameters of MCT include:

1. Impact force or impact energy: type of mill, milling speed, milling container, milling time, grinding medium, extent of filling the bowl, ball-to-powder weight ratio
2. Physic, chemical and mechanical properties of powder and grinding mediums
3. Milling atmosphere and milling temperature

2.3. Characterization of Ti films fabricated by MCT

2.3.1. Appearance, microstructure and thickness of the Ti films

Fig. 4 shows the appearances of the Al_2O_3 balls after MCT. It can be seen that the color of the Al_2O_3 balls changed from white to metallic gray as MCT time increased. It means that more Ti powder particles adhered to the surfaces of the Al_2O_3 balls. Impact force of only about 1 G can be obtained during MCT performed by pot mill. When it is carried out by planetary ball mill, impact force over 10 G or even 40 G can be realized. Therefore, the required time to form metal films can be decreased greatly in the case of planetary ball mill. The SEM images of the cross sections of the Ti-coated Al_2O_3 balls are also given in Fig. 5. With the increase of MCT time, the film thickness increased. The surface morphologies of the Ti film-coated

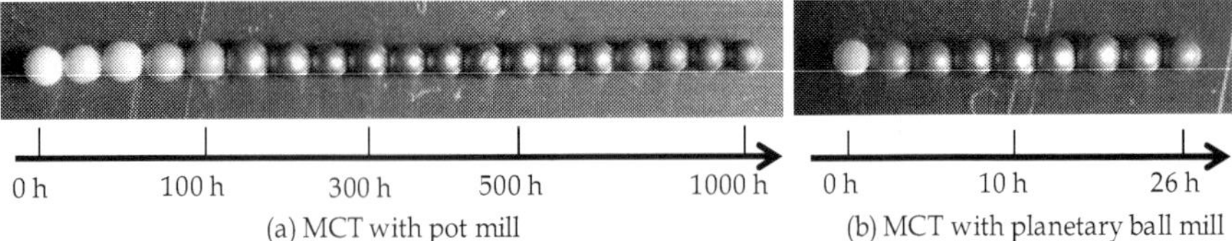

Figure 4. Appearances of the Ti-coated Al_2O_3 balls during MCT for different MCT time

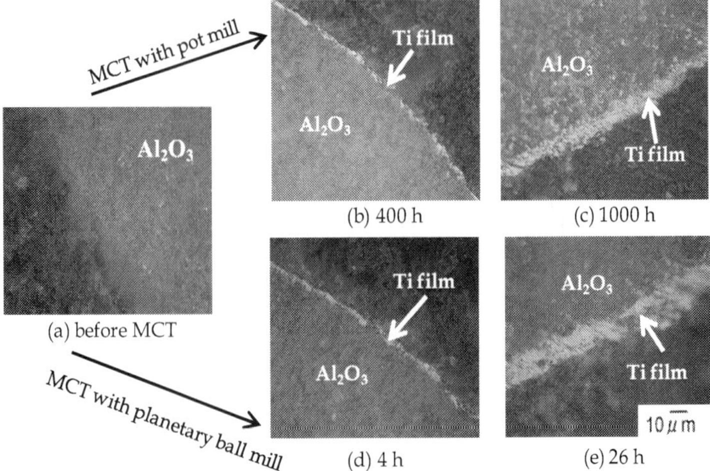

Figure 5. SEM images of the cross sections of the Ti film-coated Al_2O_3 balls during MCT

Al₂O₃ balls are shown in Fig. 6. Discrete Ti particles adhered to the surfaces of Al₂O₃ balls and they connected with each other. The irregular surface can result in high specific area. The thickness evolution of Ti films during MCT was also monitored and is illustrated in Fig. 7. No matter in the case of pot mill or planetary ball mill, the film thickness increased with the increase of MCT time. They reached 10 and 12 μm respectively in pot mill and planetary ball mill after 1000 h and 26 h. Therefore, it is possible to control the film thickness by MCT.

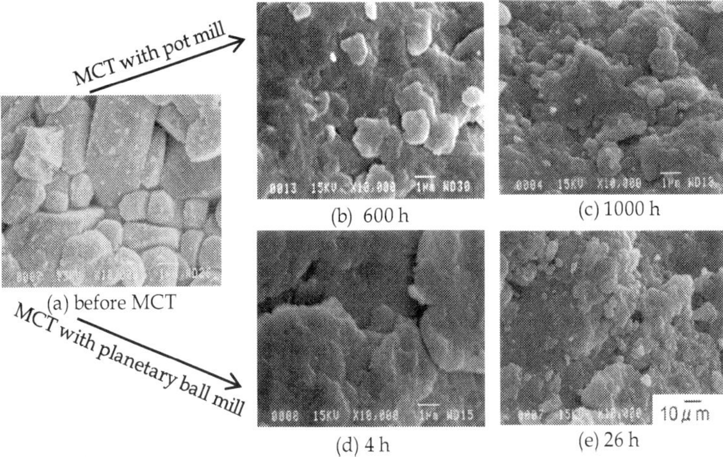

Figure 6. Surface morphologies of Ti film-coated Al₂O₃ balls during MCT

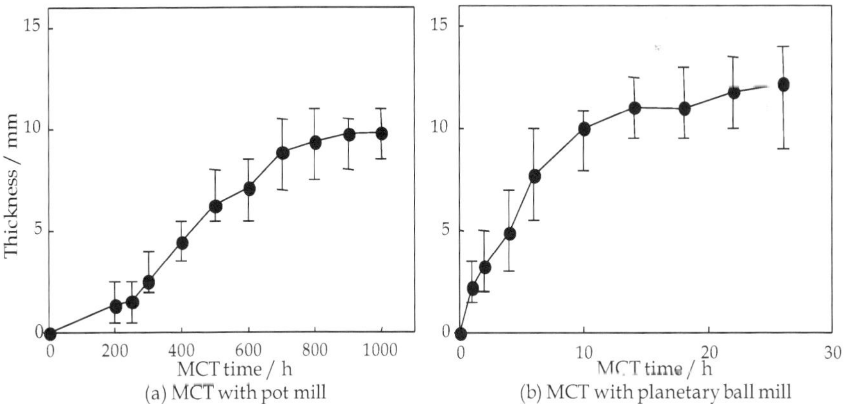

Figure 7. Thickness evolution of Ti films as a function of MCT time

2.3.2. Electrical resistance of the Ti films

Electrical resistance is one important property of film materials. Four-point probe method is frequently used to measure electrical resistance of films on a flat substrate (JIS K 7194, 1994).

However, this method is only applicable to planar films. It cannot used to measure the electrical resistance of spherical films. Therefore, we proposed a new electrical resistance determination method of spherical films such as Ti films on Al_2O_3 balls. By this method, we established the relationship between electrical resistivity and film thickness.

The determination method is shown in Fig. 8. Two plate probes contacts the Ti film-coated Al_2O_3 ball (Φ 1 mm) along the direction of tangential line. To decrease contact resistance, a pressure force of 800 gf is loaded along the normal direction of the ball. The press force is determined in pre-experiments. Electrical resistance is measured for 10 times by changing the contact points between the two plate probes and the ball. In addition, the measurement on electrical resistance is carried out for three randomly chosen Ti film-coated Al_2O_3 balls. The average value of the measurements for 30 times is used as the electrical resistance of the Ti film. Fig. 9 shows the evolution of electrical resistance of the Ti film-coated Al_2O_3 balls during MCT by pot mill and planetary ball mill. For the both cases, the electrical resistance decreased with the increase of MCT time.

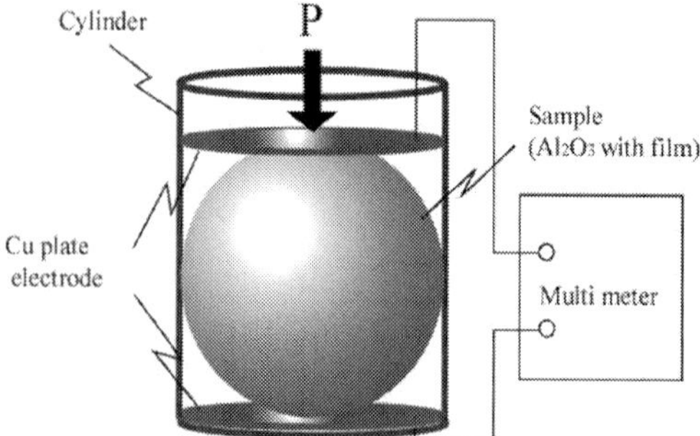

Figure 8. Measurement of electrical resistance of the Ti film-coated Al_2O_3 balls

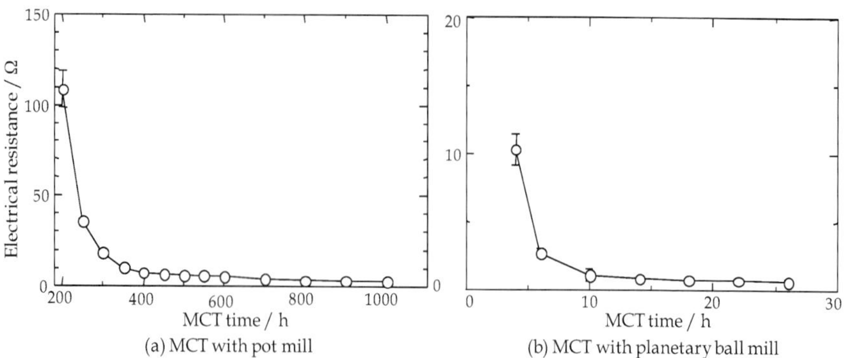

Figure 9. Electrical resistance of the Ti film-coated Al_2O_3 balls

To establish the relationship between electrical resistance and film thickness, we proposed a spherical shell model for Ti film-coated Al_2O_3 ball as shown in Fig. 10. Here r is the radius of

Al$_2$O$_3$ ball, h is the film thickness. Therefore, the electrical resistance of the spherical Ti films on Al$_2$O$_3$ ball $z = +(r+h)$ to $z = -(r+h)$ can be given by

$$R = \int_{-(r+h)}^{+(r+h)} \rho \frac{dz}{A} = 2\int_0^{+(r+h)} (\rho \frac{dz}{A_1} + \rho \frac{dz}{A_2}) \tag{1}$$

Where ρ is the electrical resistivity of the films, A_1 is the ring area of the films in the range of $0 \le z \le r$, and A_2 is the area of the circle crossed with vertical axis z in the range of $r < z \le r+h$. Electrical resistance in the range of $0 \le z \le r$ and $r < z \le r+h$ can be defined as R_1 and R_2 respectively and can be given by

$$R_1 = \int_0^r \rho \frac{dz}{A_1} = \int_0^r \frac{\rho}{\pi} \frac{dz}{(r+h)^2 - r^2} \tag{2}$$

$$R_2 = \int_r^{r+h} \rho \frac{dz}{A_2} = \int_r^{r+h} \frac{\rho}{\pi} \frac{dz}{(r+h)^2 - z^2} \tag{3}$$

During the measurement of electrical resistance shown in Fig. 8, the contact of the plate probes and Al$_2$O$_3$ ball should not be a point but a plane which has a certain area. It is proper to give the integral calculus from r to C ($r < C \le r+h$).

$$R_2 = \int_r^C \frac{\rho}{\pi} \frac{dz}{(r+h)^2 - C^2} = \frac{\rho}{2\pi(r+h)} \ln \left| \frac{(r+h)^2 - C^2}{(r+h)^2 - r^2} \right| \tag{4}$$

The electrical resistance of the spherical Ti films on an Al$_2$O$_3$ ball can be given by

$$R = 2(R_1 + R_2) = 2\left\{ \frac{\rho}{\pi} \frac{r}{(r+h)^2 - r^2} + \frac{\rho}{2\pi(r+h)} \ln \left| \frac{(r+h)^2 - C^2}{(r+h)^2 - r^2} \right| \right\} \tag{5}$$

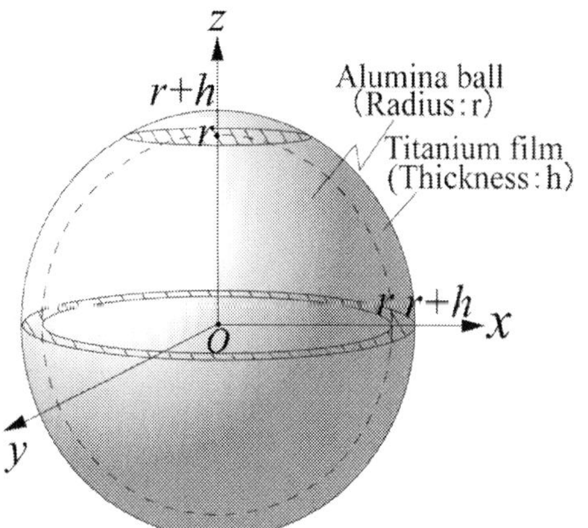

Figure 10. Spherical shell model for the electrical resistance of the Ti film-coated Al$_2$O$_3$ balls

Therefore, we can calculate electrical resistivity, ρ of the film by Eq. 5 using the measured electrical resistance of the films, R and the film thickness, h. Fig. 11 shows the relationship between the electrical resistivity of the Ti films and their thickness. It can be found that the electrical resistivity went down and then kept a constant. The evolution of the electrical resistivity should result from the density evolution of the films. In the case of planetary ball mill, the stable electrical resistivity was smaller than that in the case of pot mill. That should be due to the higher film density obtained in the circumstance of larger impact force in the case of planetary ball mill.

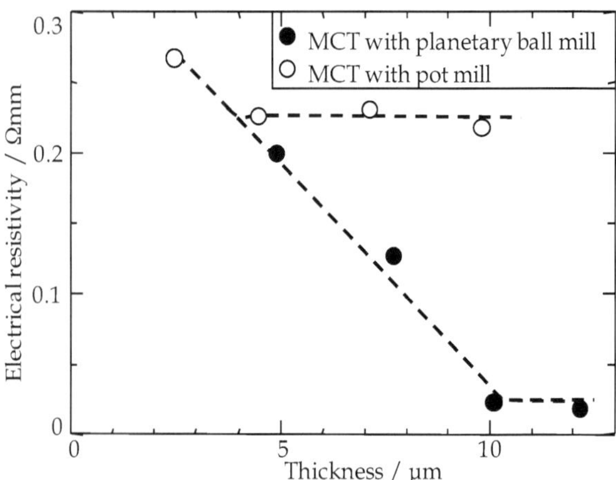

Figure 11. Relationship between electrical resistivity of Ti films and their thicknesses

2.4. TiO₂/Ti composite films fabricated by MCT and the following high-temperature oxidation

We successfully fabricated TiO₂/Ti composite films by MCT and the following high-temperature oxidation. Firstly, Ti films were prepared by MCT shown in Fig. 2. Subsequently, the Ti film-coated Al₂O₃ balls were oxidized at high temperatures. In this section, we will introduce the fabricate processes and the characterization of the TiO₂/Ti composite films.

2.4.1. Fabrication processes

Ti powder with a purity of 99.9% and an average diameter of 30 μm was used as the coating material. Al₂O₃ balls with an average diameter of 1 mm were used as the substrates. A planetary ball mill (P5/4, Fritsch) was used to perform MCT (Yoshida, 2009 b). 40 g Ti powder and 60 g Al₂O₃ balls were charged into a bowl made of alumina with a dimension of Φ75×70 mm (250 *ml*). MCT was carried out with a rotation speed of 300 rpm for 10 h. The obtained Ti film-coated Al₂O₃ balls were denoted as M10-Ti. To form TiO₂ films, the M10-Ti samples were oxidized in air at 573, 623, 673, 723, 773, 873 and 973 K for 20 h. Here the samples were denoted as M10-T-20. T means the oxidation temperature. The samples after MCT and the following high-temperature oxidation were examined by SEM (JEOL, JSM-6100) and XRD (JEOL, JDX-3530). Cu-$K\alpha$ radiation in the condition of 30 kV and 30 mA was used for XRD. Before the characterization, all the samples were cleaned in acetone by ultrasonic (frequency: 28 kHz) to remove Ti and TiO_2 that did not adhere strongly.

2.4.2. Characterization of the TiO_2/Ti composite films

Fig. 12 shows the appearances of the M10-Ti and M10-T-20 samples. The color of the M10-T-20 samples changed with increase of oxidation temperature and lost metallic luster comparing with M10-Ti. The color change indicates that the degree of oxidation was different at different oxidation temperature. The M10-573-20 samples showed brown color which was similar to that of TiO. That means the oxidation of Ti films at 573 K was insufficient. When oxidation temperature was increased to 673, 723, 773 and 873 K, the samples color changed from blue to gray. It was probably related to the growth of TiO_2 crystalline and the film thickness increase. Besides, the color of M10-973-20 samples was light yellow. It indicates that titanium was completely oxidized to titanium dioxide.

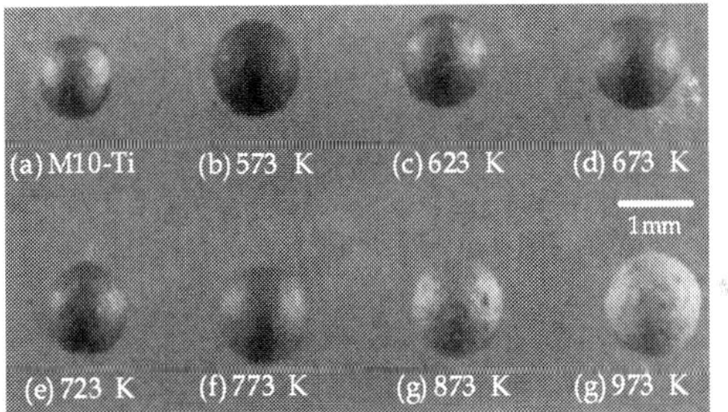

Figure 12. Appearances of M10-Ti and M10-T-20 samples

The surface SEM images of the samples are shown in Fig. 13. From Fig. 13(a), the Ti films had uneven surfaces comparing with those prepared by PVD or CVD. The surface evolution with the increase of oxidation temperature can be seen from Fig. 13(b) to (f). It seems that the surface crystals grew up with the increase of oxidation temperature. However, column nanocrystals were formed at 973 K. Fig. 14 shows the XRD patterns of the samples after MCT and the following high-temperature oxidation. The diffraction intensity of Ti peaks decreased with the increase of oxidation temperature and the Ti peak at about 41° (2θ) disappeared when oxidation temperature was increased to 973 K. Conversely, the peaks of rutile TiO_2 appeared when oxidation temperature was above 673 K and the diffraction intensity became stronger as oxidation temperature increased. From the above results, it can be concluded that the films had a composite microstructure of Ti and rutile TiO_2 when oxidation temperature was between 673 and 873 K. TiO_2/Ti composite films on Al_2O_3 balls were fabricated by MCT and the following high-temperature oxidation.

(a) M10-Ti (b) M10-573-20 (c) M10-673-20

(d) M10-773-20 (e) M10-873-20 (f) M10-973-20

Figure 13. Surface SEM images of M10-Ti and M10-*T*-20 samples

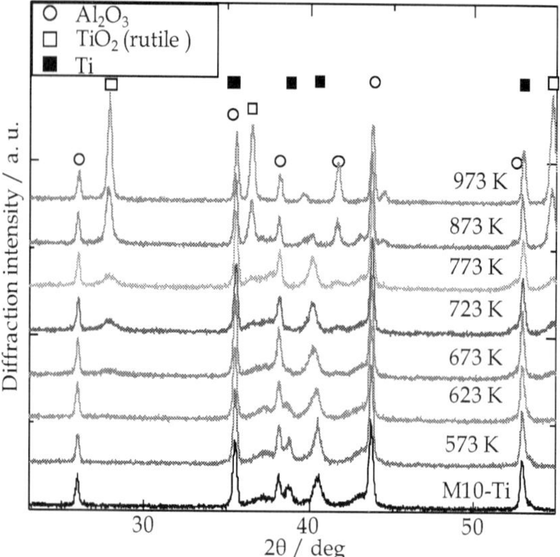

Figure 14. XRD patterns of M10-Ti and M10-T-20 samples

2.5. 2-step Mechanical Coating Technique (2-step MCT)

An advanced mechanical coating technique called 2-step mechanical coating technique (2-step MCT) was developed to fabricate TiO$_2$/Ti composite films. As anatase TiO$_2$ cannot be easily obtained by oxidation, we aim to deposit anatase TiO$_2$ on Ti films directly by MCT. In this section, we will introduce the processes of 2-step MCT and characterize the composite films. The influences of 2nd step MCT time and the introduction of ceramic impact balls on the composite films and their photocatalytic activity were also discussed.

2.5.1. Processes of 2-step MCT

The schematic diagram of 2-step MCT is shown in Fig. 15. In the first step, Ti films are prepared on the surfaces of Al_2O_3 balls as our previous work (Lu et al., 2005 & Yoshida et al., 2009 a). The source materials and their relevant parameters are listed in Table 3. 40 g Ti powder and 60 g Al_2O_3 balls are used as the coating material and the substrates respectively. A planetary ball mill (P5/4, Fritsch) is used to perform MCT. The experimental condition has been discussed in Section 2.4.1. In the second step, TiO_2/Ti composite films are fabricated. 15 g Ti film-coated Al_2O_3 balls and 13 g TiO_2 powder are used as the substrates and the coating material respectively. The relevant parameters can be found in Table 3. They are charged into a bowl made of alumina. Then the coating of TiO_2 is performed by the same planetary ball mill with a rotation speed of 300 rpm for 1, 3, 6 and 10 h. To investigate the influence of average diameter of TiO_2 powder on the photocatalytic activity of the composite films, two kinds of anatase TiO_2 powder with different average diameter are used as the coating materials. To understand the influence of impact force on the formation and the photocatalytic activity of TiO_2/Ti composite films, Al_2O_3 or WC impact balls with the diameter of 10 mm are also introduced into the second step of 2-step MCT. The relevant denotations are listed in Table 4.

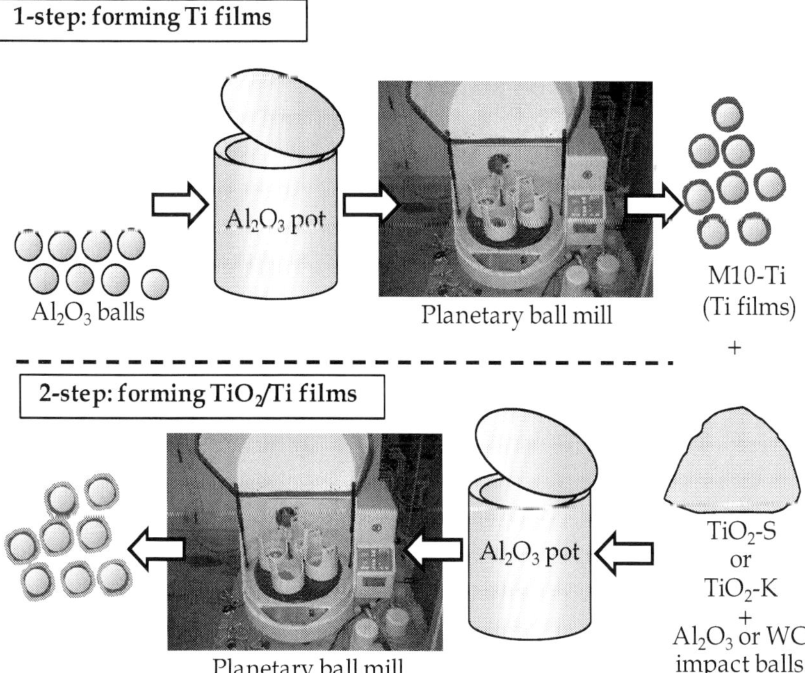

Figure 15. Schematic diagram of 2-step MCT for the fabrication of TiO_2/Ti composite films

Ti powder	Purity: 99.1% Average diameter: 30 μm
Substrate	φ1 Al_2O_3 balls Purity: 93.0%
TiO_2 powder (anatase)	Average diameter: 7 nm (TiO_2-S)
	Average diameter: 0.45 μm (TiO_2-K)

Table 3. Source materials for the fabrication of TiO_2/Ti composite films by 2-step MCT

TiO_2 powder	Impact ball	Sample denotation
		CMxS
TiO_2-S	Al_2O_3	φ10A-CMxS
	WC	φ10W-CMxS
		CMxK
TiO_2-K	Al_2O_3	φ10A-CMxK
	WC	φ10W-CMxK

Note: x is the 2nd step MCT time.

Table 4. Sample denotations for the fabrication of TiO_2/Ti composite films by 2-step MCT

2.5.2. TiO₂/Ti composite films fabricated by 2-step MCT without impact balls

Fig. 16 shows the appearances of the Al_2O_3 balls after 2-step MCT without ceramic impact balls. The color of the samples changed and lost metallic luster with the increase of the 2nd step MCT time. The CM3S samples showed different colors from place to place on the surface. However, the CM6S and CM10S samples showed uniform color respectively. It hints that uniform composite films might form at that time. Fig. 17 shows the surface SEM images of the samples fabricated by 2-step MCT without ceramic impact balls. From Fig. 17(a), the gray areas correspond to Ti. It can be seen that uniform Ti films have been formed on the surface of Al_2O_3 ball. From Fig. 17(b) and (c), the white and gray areas correspond to Ti and TiO_2 respectively. Continuous TiO_2 films were not form while TiO_2 deposited on Ti films in the form of discrete island. Meanwhile, the SEM images of the cross sections of the samples fabricated by 2-step MCT without ceramic impact balls are shown in Fig. 18. It can be clearly seen that Al_2O_3 balls were coated with Ti films and discrete islands of TiO_2 adhered to the Ti films. A composite microstructure of Ti and TiO_2 was formed. During the impact between Al_2O_3 balls or Al_2O_3 ball and the inner wall of the bowl, TiO_2 powder particles were trapped between them. Under the great impact force, TiO_2 particles were inlaid into the Ti films. It results in the formation of the TiO_2/Ti composite microstructure. Fig. 19 shows the XRD patterns of the samples after 2-step MCT without impact balls. When the 2nd step MCT time was 1 h, the peaks of anatase TiO_2 appeared which means that TiO_2 particles had adhered to Ti films. As it came to 3 h, the intensity of anatase TiO_2 peaks reached their highest values. It indicates that the loading amounts of TiO_2 in the TiO_2/Ti composite films reached the maximum values. After that, the TiO_2 peaks became lower which should be due to the exfoliation of TiO_2 that coated Ti films. From the above results, the films had a composite microstructure of Ti and TiO_2. The loading amounts of TiO_2 in the composite films changed with the increase of the 2nd step MCT time. 2-step MCT is a simple and applicable technique to fabricate TiO_2/Ti composite films.

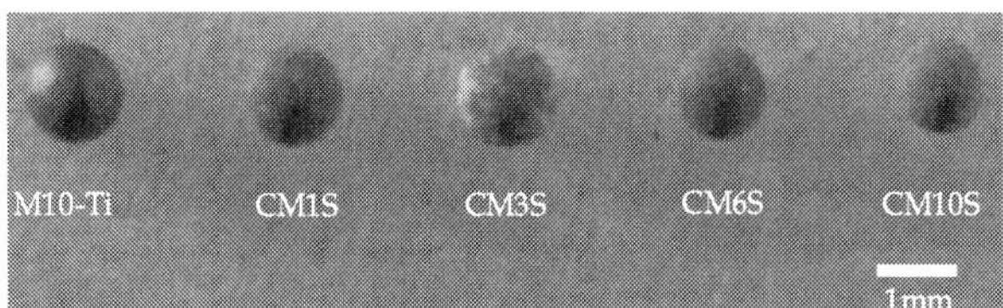

Figure 16. Appearances of the samples after 2-step MCT

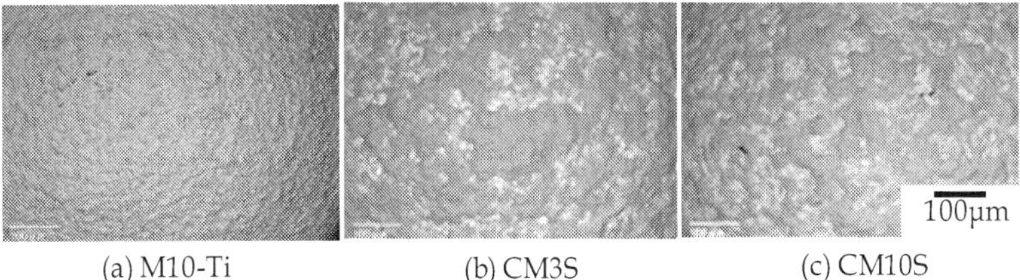

(a) M10-Ti   (b) CM3S   (c) CM10S

Figure 17. SEM images of the surfaces of the samples fabricated by 2-step MCT

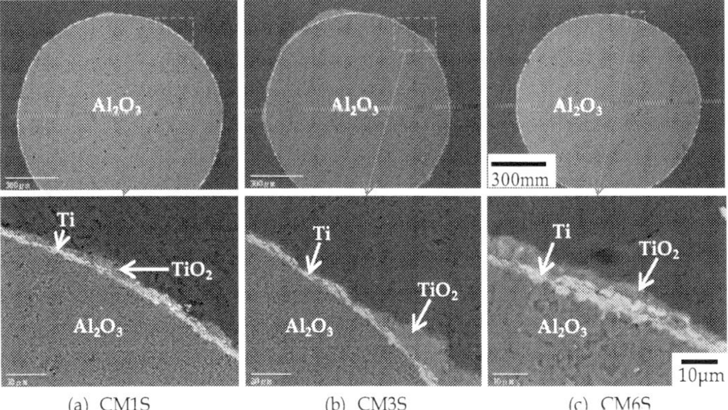

(a) CM1S   (b) CM3S   (c) CM6S

Figure 18. SEM images of the cross sections of the samples fabricated by 2-step MCT

2.5.3. TiO₂/Ti composite films fabricated by 2-step MCT with impact balls

Fig. 20 shows the appearances of the Al_2O_3 balls after 2-step MCT with ceramic impact balls. The samples lost metallic luster and their colors changed. That hints TiO_2/Ti composite films might be formed. Fig. 21 shows the SEM image of the cross section of the TiO_2/Ti composite films fabricated by 2-step MCT with WC impact balls. It can be seen that the films had a composite microstructure of Ti film and discrete islands of TiO_2. The surface condition of the composite films fabricated with TiO_2 powder with different average diameters are compared in Fig. 22. The distribution of TiO_2 powder particles with the average diameter of 0.45 μm were more uneven under the impact of WC balls compared with those with the average diameter of 7 nm. Φ10W-CM6K sample had a high hardness of 472 (dynamic hardness) on the cross section. It was close to that of alumina. It hints that the composite films were very hard.

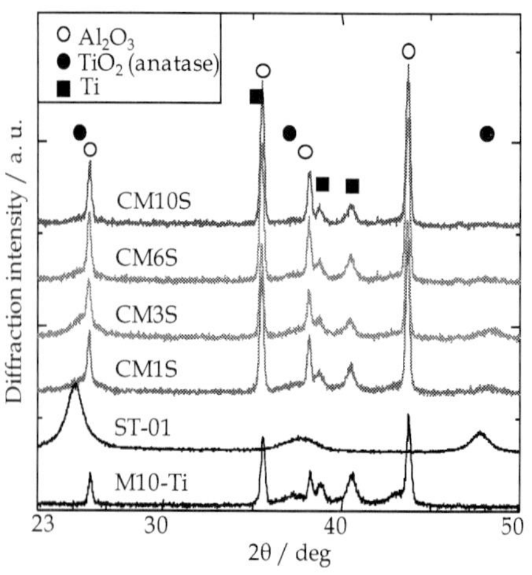

Figure 19. XRD patterns of the samples fabricated by 2-step MCT

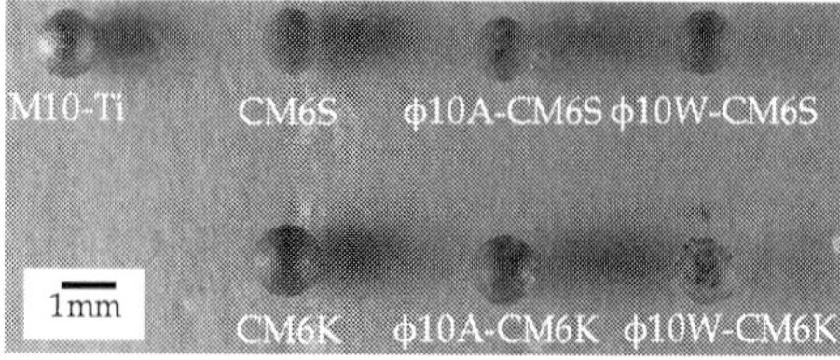

Figure 20. Appearances of the samples fabricated by 2-step MCT

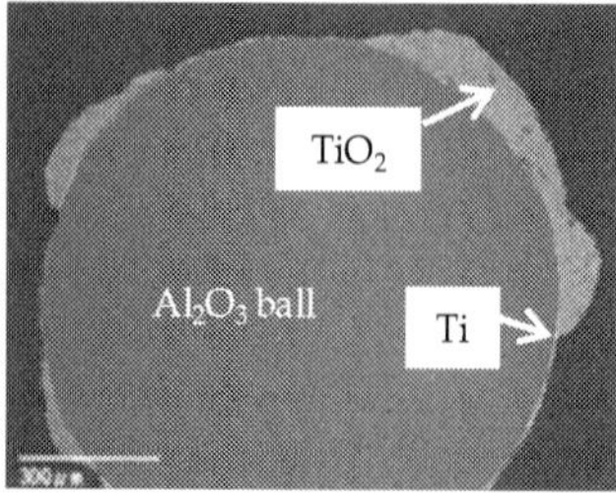

Figure 21. SEM image of the cross section of the Φ10W-CM6K sample fabricated by 2-step MCT

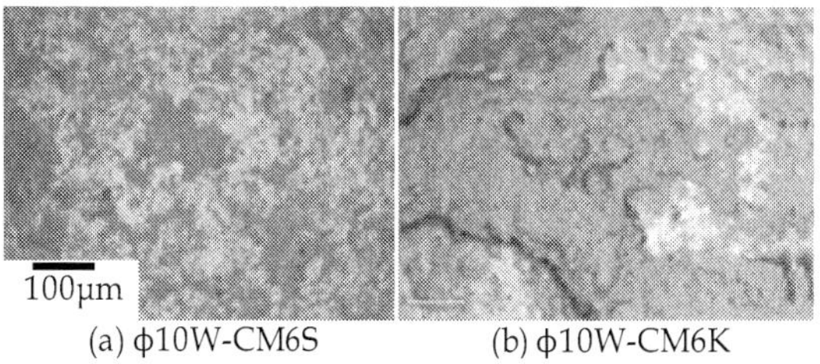

Figure 22. Surface SEM images of the samples fabricated by 2-step MCT

The XRD patterns of the samples fabricated with TiO_2 powder of different average diameters by 2-step MCT with WC impact balls are given in Fig. 23. In the case of nano-sized TiO_2 powder (Fig.23 (a)), the peaks of Ti and TiO_2 can be found. On the other hand, the diffraction peaks of TiO_2 cannot be detected for the micron-sized TiO_2 powder (Fig.23 (b)). It means the loading amounts of TiO_2 in the composite films were rather small.

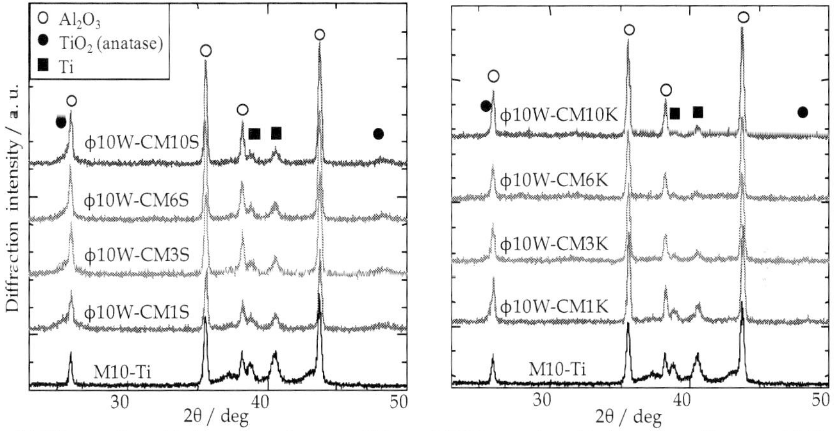

(a) TiO_2 powder with the average diameter of 7 nm   (b) TiO_2 powder with the average diameter of 0.45 μm

Figure 23. XRD patterns of the samples fabricated by 2-step MCT with WC impact balls

3. Photocatalytic activity of TiO_2/metal composite films

Although the photocatalytic activity of TiO_2 under ultraviolet and visible light irradiation has been investigated around the world, only the photocatalytic activity of TiO_2/metal composite films under ultraviolet irradiation is involved and discussed in this section. Here, we developed evaluation method of photocatalytic activity of TiO_2/metal composite films by which we evaluated the photocatalytic activity of TiO_2/Ti and TiO_2/Cu composite films fabricated by MCT.

3.1. Evaluation method of photocatalytic activity

We developed evaluation method of photocatalytic activity of the TiO$_2$/metal composite films by referring to Japan Industrial Standard (JIS R 1703-2, 2007). The evaluation procedure is as follows. Before the evaluation of photocatalytic activity, pre-adsorption of methylene blue (MB) is carried out to obtain the same initial evaluation condition for all the samples. Firstly, the cleaned samples are dispersed uniformly to form a layer of samples on the bottom of a cylinder-shaped cell with a dimension of Φ18 × 50 mm. Subsequently, 3 *ml* MB solution with a concentration of 20 *μmol/l* is poured into the cell. The cell with the samples and MB solution is kept in a totally dark place for 12 h. Then, the evaluation of the photocatalytic activity will be carried out. The samples after pre-adsorption are laid uniformly on the bottom of a same cell to form a layer of samples and 7 *ml* MB solution with a concentration of 10 *μmol/l* is poured into the cell. The schematic diagram of the evaluation of photocatalytic activity is shown in Fig. 24. A colorimeter (Sanshin Industrial Co., Ltd) of 660 nm in UV radiation wavelength, which is near the peak of absorption spectrum of MB solution (664 nm), is used to measure the absorbance of MB solution. The UV irradiation time is 24 h. Besides, both of the pre-adsorption and the evaluation of photocatalytic activity are carried out at room temperature. The gradient, k of MB solution concentration-irradiation time curve is calculated by the least-squares method with the data from 1 to 12 h and k is used as the degradation rate constants. The higher the degradation rate constants k, the higher the photocatalytic activity.

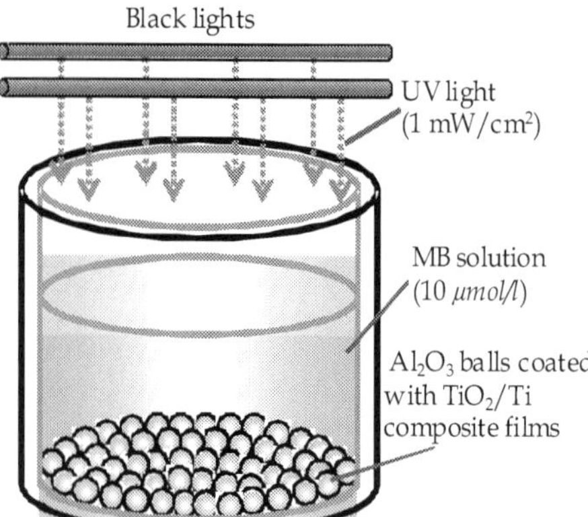

Figure 24. Photocatalytic activity evaluation of TiO$_2$/metal composite films fabricated by MCT

3.2. Photocatalytic activity of TiO$_2$/Ti composite films

Fig. 25 shows the evolution of MB solution concentration as UV irradiation time under the action of TiO$_2$/Ti composite films fabricated by MCT and its following high-temperature oxidation as described in Section 2.4. MB solution concentration slight increased in the case of M10-Ti and M10-573-20. Meanwhile, MB solution concentration decreased in varying degrees in the case of the other samples. That means MB was degraded under the action of UV light and TiO_2 samples. In other word, the samples fabricated by MCT and its following high-temperature oxidation showed photocatalytic activity.

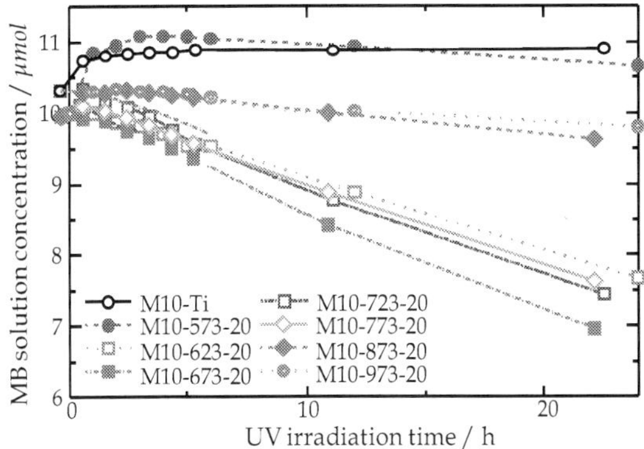

Figure 25. MB solution concentration as a function of UV irradiation time under the action of the samples fabricated by MCT and the following high-temperature oxidation

The degradation rate constants, k are also given in Fig. 26. It can be seen that the degradation rate constants, k increased with the increase of oxidation temperature and reached the peak value at 723 K above which the degradation rate constants, k decreased. Combined with the XRD patterns in Fig. 14, the evolution of the photocatalytic activity can be discussed as follows. When oxidation temperature was below 623 K, the peaks of TiO_2 were not detected which means Ti films were not oxidized or were oxidized sufficiently. Therefore, the photocatalytic activity of the films was low as shown in Fig. 26. With the increase of oxidation temperature from 623 to 773 K, more Ti films were oxidized and TiO_2/Ti composite films were formed. The improvement of photocatalytic activity should relate to the composite microstructure of TiO_2 and Ti. According to charge separation effect (Rengaraj et al., 2007), electrons in TiO_2 may transfer to metals with higher work functions. The electron transfer can decrease the recombination velocity of electron-hole pairs in TiO_2. It can improve the photocatalytic activity of TiO_2. When oxidation temperature was 723 K, the composite films obtained the optimum ratio of TiO_2 to Ti. It resulted in the highest photocatalytic activity. When oxidation temperature was above 773 K, the oxidation degree of Ti films was further increased and Ti films were completely oxidized when oxidation temperature was above a certain value. Although the amounts of TiO_2 increased, the photocatalytic activity decreased due to the weakening of charge separation effect.

The degradation rate constants, k as a function of 2nd step MCT time are shown in Fig. 27. In the case of nano-sized TiO_2 powder (Fig. 27(a)), the samples fabricated without ceramic impact balls showed the highest photocatalytic activity and the degradation rate constants, k exceeded 350 $nmol \cdot l^{-1} \cdot h^{-1}$. After the introduction of ceramic impact balls into 2-step MCT, the photocatalytic activity was decreased. On the other hand, the samples fabricated with Al_2O_3 impact balls showed the greatest photocatalytic activity in the case of micron-sized TiO_2

powder (Fig. 27(b)). Compared with the TiO$_2$/Ti composite films fabricated with micron-sized TiO$_2$ powder, the composite films fabricated with nano-sized TiO$_2$ powder showed much higher photocatalytic activity. It should relate to the higher specific area of nano-sized TiO$_2$ powder particles. From Fig. 26 and 27, anatase TiO$_2$/Ti composite films showed much higher photocatalytic activity than that of rutile TiO$_2$/Ti composite films. It is well known that anatase TiO$_2$ generally shows higher photocatalytic activity than rutile TiO$_2$.

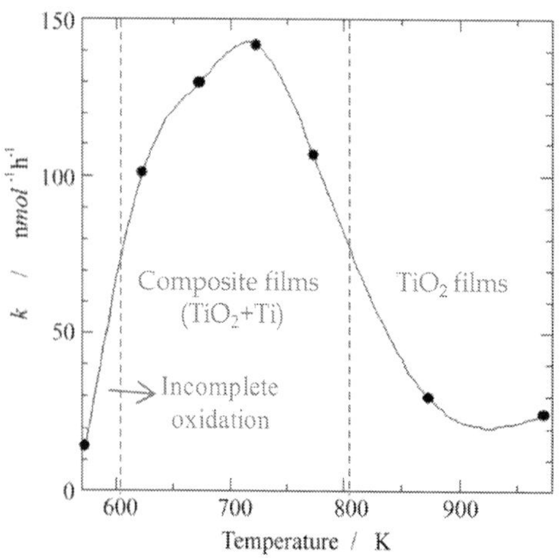

Figure 26. Degradation rate constants, k as a function of oxidation temperature

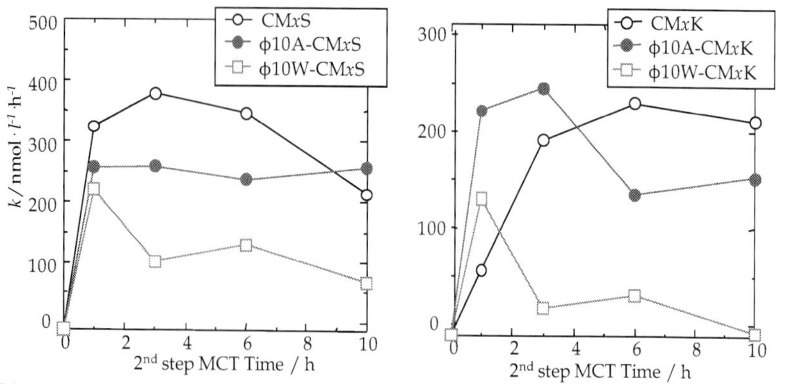

(a) TiO$_2$ powder with the average diameter of 7 nm   (b) TiO$_2$ powder with the average diameter of 0.45 μm

Figure 27. Degradation rate constants, k as a function of 2nd step MCT time

3.3. Improvement of photocatalytic activity by high-temperature oxidation

High-temperature oxidation was carried out to increase the crystallinity and the volumes of TiO$_2$ in TiO$_2$/Ti composite films fabricated by 2-step MCT and therefore improve the photocatalytic activity of the composite films (Lu et al., 2011 b). The composite films after the high-temperature oxidation were characterized and their photocatalytic activity was also evaluated. In addition, the effect on high-temperature oxidation on the microstructure and the photocatalytic activity of the composite films was also discussed.

3.3.1. Improved fabrication processes

Firstly, TiO_2/Ti composite films were prepared by 2-step MCT as described in Section 2.5.1. Ti powder with a purity of 99.1% and an average diameter of 30 μm was used as the coating material. Al_2O_3 balls with an average diameter of 1 mm were used as the substrates. After the formation of Ti films on Al_2O_3 balls, anatase TiO_2 powder with an average diameter of 0.45 μm (Kishida Chemical Co. Ltd., Japan) was used to form TiO_2/Ti composite films. To make the composite films strong enough, Al_2O_3 or WC balls with the diameter of 10 mm were introduced into the fabrication of the composite films. The schematic diagram can be seen in Fig. 15. Subsequently, high-temperature oxidation was carried out for the TiO_2/Ti composite films fabricated by 2-step MCT. The oxidation temperature was set at 673, 773 and 873 K and the oxidation time was 10 h. The denotations of the samples fabricated by 2-step MCT and the following high-temperature oxidation are listed in Table 5.

TiO_2 powder	Impact ball	Sample maker
	---	CMxK-y
TiO_2-K	Al_2O_3	φ10A-CMxK-y
	WC	φ10W-CMxK-y

Note: x means the 2nd step MCT time, y is oxidation temperature.

Table 5. Denotations of the samples fabricated by 2-step MCT and the following high-temperature oxidation

3.3.2. Characterization of the TiO_2/Ti composite films

Fig. 28 shows the appearances of the samples fabricated by 2-step MCT and the following high-temperature oxidation. The samples lost metallic luster and became white compared with the Ti film-coated Al_2O_3 balls. Also, dark and uneven areas can also be seen. The surface SEM images of the samples are shown in Fig. 29. The surface color of the samples seems to be uniform except for some point areas. It indicates that the uniform TiO_2 films were formed. However, for the TiO_2/Ti composite films fabricated by 2-step MCT there are light and dark areas corresponding to Ti and TiO_2 respectively shown in Fig. 17.

The SEM images of the cross sections of the samples are given in Fig. 30. It can be seen that a composite microstructure of TiO_2 films and Ti films formed. In other words, TiO_2/Ti composite films were fabricated. The TiO_2 films consisted of the deposited TiO_2 in the second step of 2-step MCT and the TiO_2 formed in high-temperature oxidation. From the above results, it can be concluded that the loading amounts of TiO_2 in the composite films were increased by high-temperature oxidation.

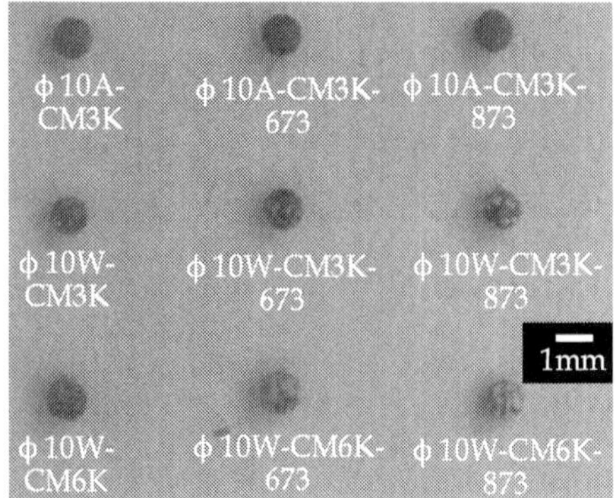

Figure 28. Appearances of the samples fabricated by 2-step MCT and the following high-temperature oxidation

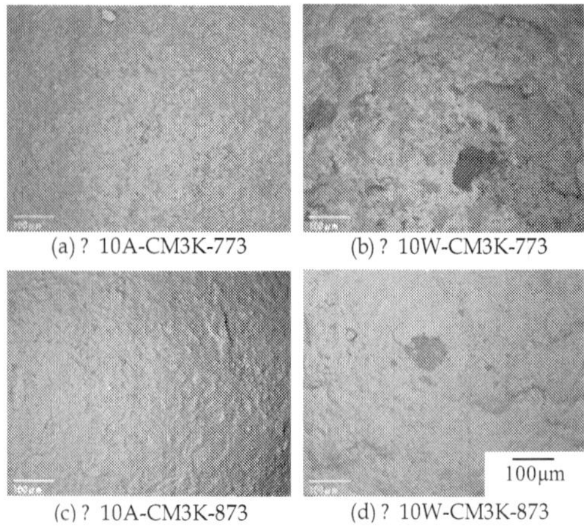

Figure 29. Surface SEM images of the samples fabricated by 2-step MCT and the following high-temperature oxidation

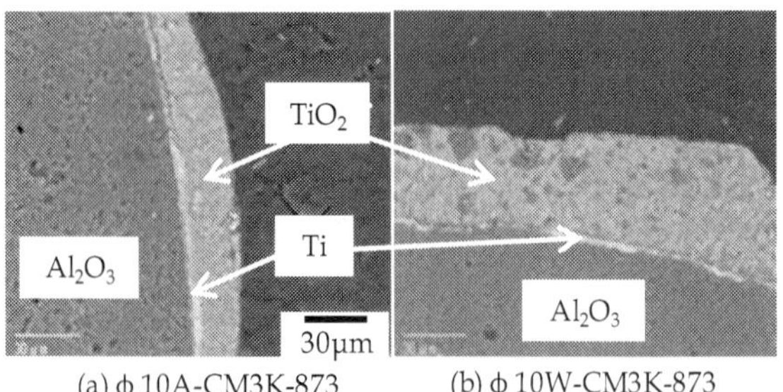

Figure 30. SEM images of the cross sections of the samples fabricated by 2-step MCT and the following high-temperature oxidation

Fig. 31 shows the XRD patterns of the samples fabricated by 2-step MCT and the following high-temperature oxidation at 673 and 873 K. For all the samples, Ti peaks and anatase TiO_2 peaks were detected while the later was rather weak due to its small loading amounts during the second step in 2-step MCT. When oxidation temperature was 873 K, the peaks of rutile TiO_2 were detected which means rutile TiO_2 was formed in the high-temperature oxidation. For the sample Φ 10W-CM6K-673 K, a weak peak of rutile TiO_2 at about 27.5º (2θ) was be found. It indicates that rutile TiO_2 formed when oxidation temperature was 673 K although the amount was rather small.

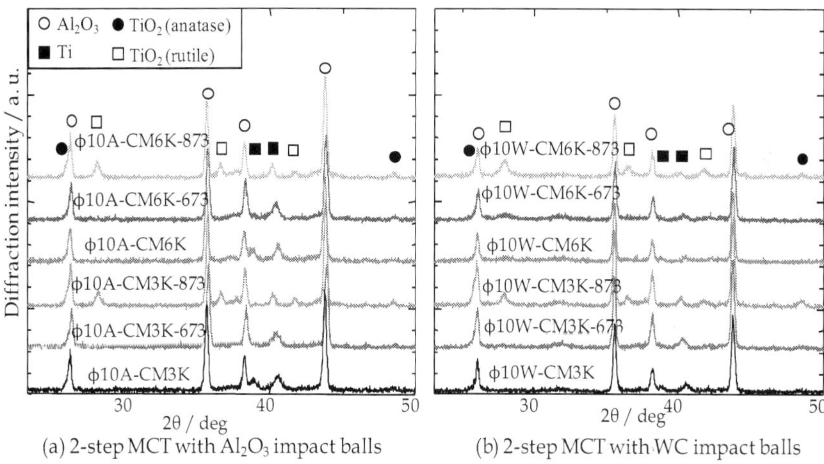

Figure 31. XRD patterns of the samples fabricated by 2-step MCT and the following high-temperature oxidation

3.3.3. Photocatalytic activity of the TiO₂/Ti composite films

The degradation rate constants, k as a function of oxidation temperature are shown in Fig. 32. For the samples fabricated by 2-step MCT with Al₂O₃ impact balls (Fig. 32(a)), they showed their highest photocatalytic activity when oxidation temperature was 673 K. When the 2nd MCT was 3 h, the composite films showed the greatest photocatalytic activity. On the other hand, for those fabricated by 2-step MCT with WC impact balls (Fig. 32(b)), the samples showed the similar evolution of photocatalytic activity except the Φ 10W-CM3K samples. However, their photocatalytic activity was much lower than those fabricated by 2-step MCT with Al₂O₃ impact balls. It may relate to the smaller loading amounts of anatase TiO_2 as WC impact balls exerted greater impact force and resulted in the exfoliation of the adhered TiO_2.

The improvement of photocatalytic activity should result from the microstructure and phase evolution of the composite films. As discussed in Fig. 31, Ti films were oxidized partly and rutile TiO$_2$ was formed when oxidation temperature was 673 K and anatase TiO$_2$ was reserved at that temperature. Therefore, a composite microstructure of Ti, anatase TiO$_2$ and rutile TiO$_2$ was formed. Charge separation effect and mixed crystal effect might work which can improve photocatalytic activity (Cao et al., 2009). With the increase of oxidation temperature to 873 K, the volume ratio of rutile TiO$_2$ in the composite films increased. It might lead that the charge separation effect and mixed crystal effect were restrained and therefore decreased the photocatalytic activity of the composite films.

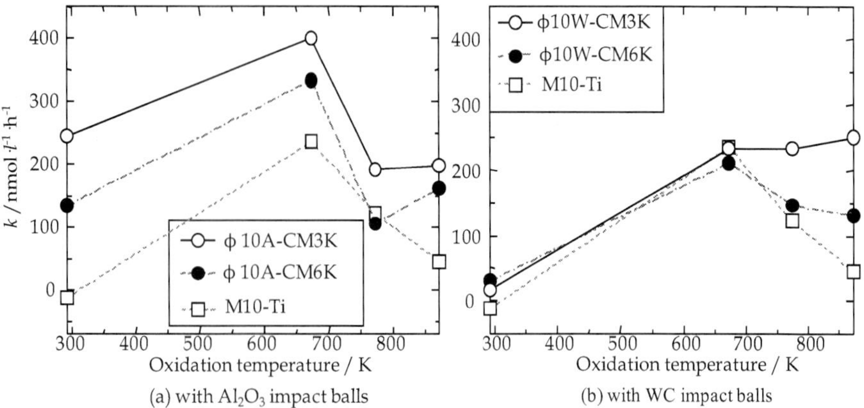

Figure 32. Degradation rate constants, k as a function of oxidation temperature

3.4. TiO$_2$/Cu composite photocatalyst films

In this section, we will describe and discuss the fabrication of TiO$_2$/Cu composite films by 2-step MCT. The composite films were characterized by XRD and SEM. The formation of Cu films and TiO$_2$/Cu composite films was also examined. The photocatalytic activity of the composite films was evaluated by measuring the degradation rate constants, k of MB under the UV irradiation.

3.4.1. Fabrication of TiO$_2$/Cu composite films

TiO$_2$/Cu composite films were fabricated by 2-step MCT as shown in Fig. 15. Firstly, Cu films were fabricated. The source materials and the experimental condition are listed in Table 6. To improve the production efficiency, Cu powders with different average diameters were used as the coating materials. Meanwhile, the loading amounts of Cu powder and the rotation speed was also changed as shown in Table 6. Secondly, TiO_2/Cu composite films were fabricated. 15 g Cu film-coated Al_2O_3 balls and 13 g anatase TiO_2 powder with the average diameter of 7 nm (ST-01, Ishihara Sangyo, Japan) were used as the substrates and the coating material respectively. To make the composite films stronger, 20 Al_2O_3 impact balls with the diameter of 10 mm were simultaneously put into the bowl. The rotation speed was set at 400 rpm and the 2nd MCT time was 1, 3, 6 and 10 h.

Experiment number	Cu powder mass [g]	Cu purity [%]	Average size of Cu [µm]	Al_2O_3 ball mass [g]	Al_2O_3 ball purity [%]	Rotation speed [rpm]
1	80	99.8	40	60	93.0	400
2	40	99.8	10	60	93.0	480

Table 6. Source materials and experimental conditions for fabrication of Cu films by MCT

3.4.2. Characterization of TiO_2/Cu composite films

Fig. 33 shows the appearances of the Cu-coated Al_2O_3 balls after MCT at 400 rpm with the relevant parameters in the experiment number 1 shown in Table 6. It can be seen that the color of the samples changed from white to red brown with the increase of 1st step MCT time. It means that more Cu powder adhered to the surfaces of Al_2O_3 balls. SEM results revealed that Cu films in this experimental condition were formed when it came to 54 h. Fig 34 shows the appearances of the samples after 2-step MCT. The color of the samples also changed with the increase of 2nd step MCT time. It indicates that the loading amounts of TiO_2 in the composite films changed. The SEM images of the cross sections of the samples are given in Fig. 35. TiO_2 adhered to Cu films (Fig. 35 (a)) and some Cu particles inlaid into TiO_2 films (Fig. 35 (b)). Therefore, it can be said that a composite microstructure of TiO_2 and Cu formed.

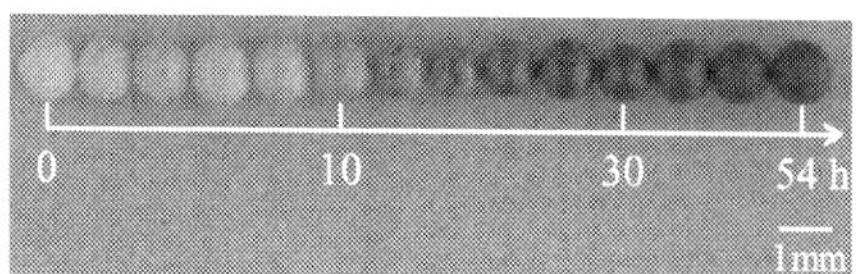

Figure 33. Appearances of the Cu-coated Al_2O_3 balls after MCT

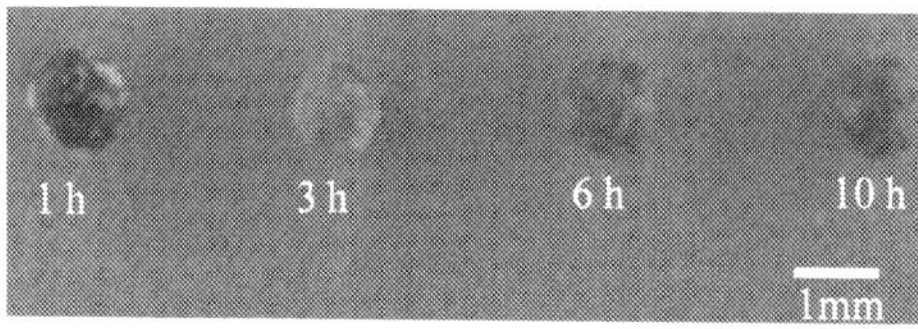

Figure 34. Appearances of the samples after 2-step MCT

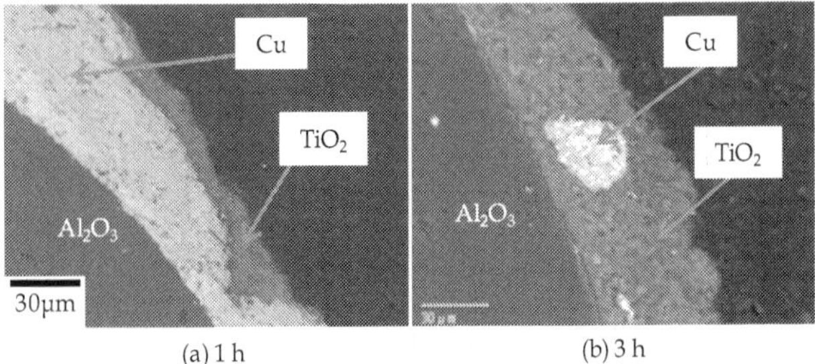

(a) 1 h   (b) 3 h

Figure 35. SEM images of the cross sections of TiO₂/Cu composite films fabricated by 2-step MCT with different the 2nd step MCT times

3.4.3. Photocatalytic activity of TiO₂/Cu composite films

MB solution concentration evolution in the evaluation of the photocatalytic activity of TiO₂/Cu composite films is shown in Fig. 36. The concentration of MB solution with the Cu film-coated Al₂O₃ balls was found to increase slightly with increase of UV irradiation time. It means Cu films did not have photocatalytic activity. On the other hand, under the action of TiO₂/Cu composite films and UV irradiation, the concentration of MB solution decreased in varying degrees. It suggests the composite films showed photocatalytic activity. For TiO₂/Cu composite films with 3 h of 2nd step MCT time, the MB solution concentration decreased to the minimum value after the same UV irradiation time for all the composite films. The degradation rate constants, k is illustrated in Fig. 37. The degradation rate constants, k increased with the increase of 2nd step MCT time and reached the peak value when it came to 3 h. After that, the degradation rate constants, k decreased with the increase of 2nd step MCT time. It means that the TiO₂/Cu composite films fabricated during MCT with 3 h of 2nd step MCT time showed the greatest photocatalytic activity for all the composite films.

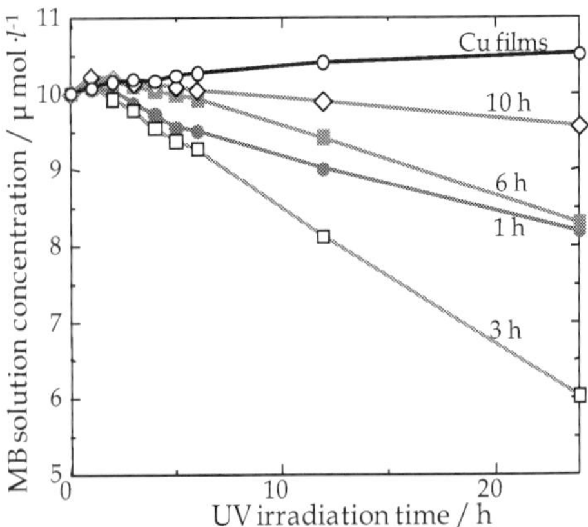

Figure 36. Evolution of MB solution concentration as a function of UV irradiation time in the evaluation of photocatalytic activity of TiO₂/Cu composite films

The photocatalytic activity of TiO_2/Cu composite films should relate to the loading amounts of TiO_2 in the composite films. With increase in 2nd step MCT time, the loading amounts of TiO_2 increased. When it came to 3 h, the loading amounts of TiO_2 might reach the peak value. After the maximum value, TiO_2 that adhered to Cu films began to peel off. The more the loading amounts of TiO_2 that deposited on Cu films, the higher the photocatalytic activity. In other words, the photocatalytic activity of TiO_2/Cu composite films should be proportional to the loading amounts of TiO_2 in the composite films. The formation of TiO_2/Cu composite microstructure is considered to be another reason why the photocatalytic activity of the composite films was improved. After formation of the interface of TiO_2/Cu, electrons in the conduction of TiO_2 will migrate to Cu films through the interface, which can decrease the recombination rate of electron-hole pairs in TiO_2. It may result in the improvement of charge separation efficiency (Rengaraj et al., 2007). Then, more electrons are trapped in Cu films for reduction reaction and more holes are held in the valence band of TiO_2 for oxidation reaction.

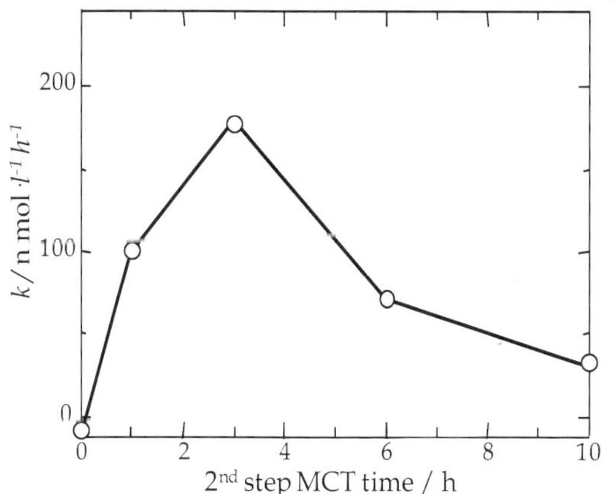

Figure 37. Degradation rate constants, k as a function of the 2nd step MCT time

4. Formation process of Cu films during MCT

Although Cu films and TiO_2/Cu composite films were fabricated by MCT, the formation process and its mechanism of Cu films are still unknown. Therefore, the formation process and its possible mechanism were examined and will be discussed in this section.

Because the formation process of Cu films happens in a closed and invisible bowl, it is difficult to determine the evolution of the films. However, it was considered to relate to the collision, friction and welding among the Cu powder particles, the inner wall of the bowl and the ceramic grinding mediums (Lü et al., 1995; Maurice and Courtney, 1990; Maurice and Courtney, 1994; Chattopadhyay et al., 2001). By now, we have tried to analyze the evolution of Cu films by observing the change of ceramic substrates. Fig. 38

shows the SEM images of Cu-coated Al₂O₃ balls after MCT. The areas of dark and light color correspond to alumina and copper respectively. More Cu particles adhered to the surfaces of Al₂O₃ balls with the increase of 1ˢᵗ step MCT time. When it came to 54 h, continuous Cu films formed and the thickness was about 10 μm. In other words, the surfaces of Al₂O₃ balls were totally coated by Cu films. The result is in good agreement with that in Fig. 33.

The coverage of Al₂O₃ ball surface with Cu is illustrated in Fig. 39. The evolution of Cu films during MCT can fall into five ranges. In the first range, the coverage hardly increased. However, Al₂O₃ balls became dark. Micron-sized Cu particles on the surfaces of Al₂O₃ balls were not found by SEM. It means that a small quantity of Cu atom clusters might transfer to the surfaces of Al₂O₃ balls. Under impact and friction force, the atom clusters adhered to the surfaces of Al₂O₃ balls and nucleated. In the second and third range, more Cu atom clusters adhered to the nuclei of Cu; these nuclei gradually grew up and could be observed by SEM (Fig. 38 (a)). Then discrete islands of Cu connected with each other (Fig. 38 (b)). The growth of Cu nuclei and the connection of discrete islands of Cu resulted in the coverage increase. After the first three ranges, the surfaces of Al₂O₃ balls were nearly coated with Cu and the coverage was close to 100%. In other words, continuous Cu films formed. Although the experiments were stopped when MCT time reached 54 h, the fourth and fifth ranges are considered to exit as the similar evolution of Fe films has been established in our published work (Hao et al., 2012). In the fourth range, the thickness of continuous Cu films may increase. As deformation of Cu particles, Cu particles become hard and adhesion between Cu particles may become difficult. Finally, exfoliation of Cu films would dominate. From the above results, the evolution of Cu films can fall into nucleation, growth of nuclei and connection, formation of continuous films and thickening, exfoliation of continuous films.

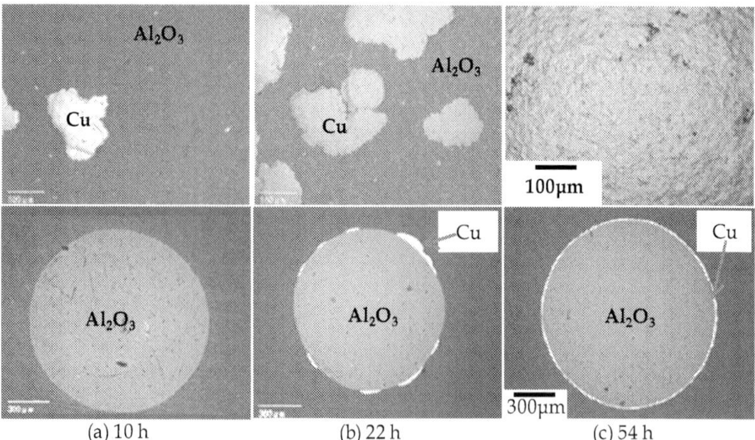

(a) 10 h　　　　(b) 22 h　　　　(c) 54 h

Figure 38. SEM images of the surfaces and the cross sections of Cu films fabricated by MCT

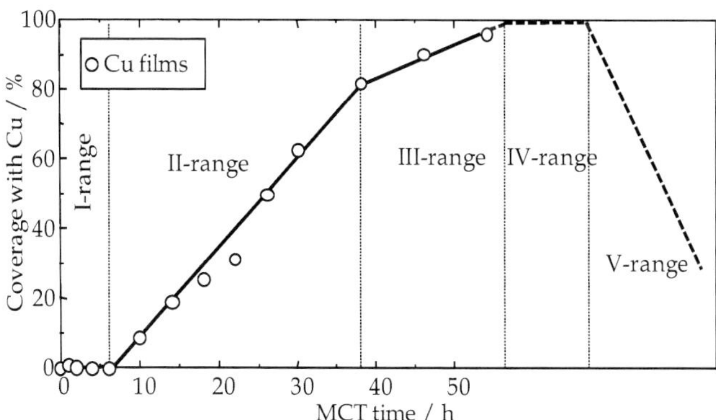

Figure 39. Coverage of Al₂O₃ ball' surface with Cu as a function of MCT time during MCT

Fig. 40 shows the SEM images of the surfaces and the cross sections of the Cu-coated Al₂O₃ balls after MCT with 480 rpm. It can be observed that continuous Cu films formed when MCT time was 20 h and the average thickness of the films was about 80 μm. Compared with the fabrication of Cu films with Cu powder of 40 μm in average particle size by MCT at 400 rpm (Fig. 38(c)), the fabrication of Cu films with Cu powder of 10 μm in average particle size by MCT at 480 rpm was quicker. In other words, the condition of Cu powder of 10 μm in average particle size and a rotation speed of 480 rpm accelerated the formation of Cu films.

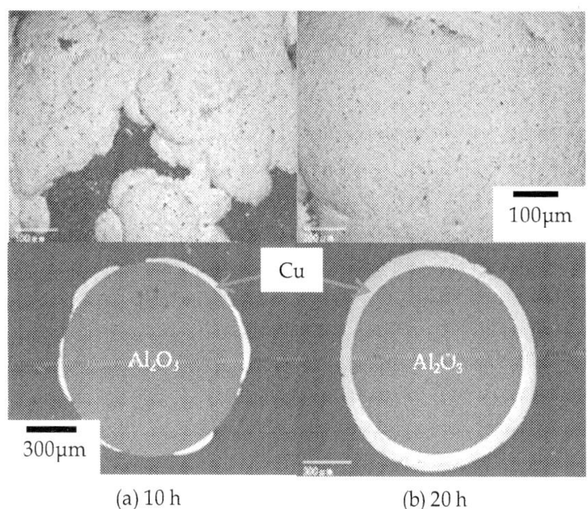

Figure 40. SEM images of the Cu-coated Al₂O₃ balls fabricated with Cu powder of 10μm in average particle size by MCT at 480 rpm

5. Prospect of MCT

Compared with the traditional film coating techniques such as PVD and CVD, our proposed mechanical coating technique (MCT) shows many advantages including inexpensive equipments, simple process, low preparation cost and large specific area, among others. In addition, it can be performed in air atmosphere at ambient temperature. It can not only fabricate metal/alloy films but also nonmetal/metal composite films such as TiO_2/Ti composite photocatalyst films. It is expected to fabricate other functional film materials in the near future.

We will continue to advance the development of MCT in the fabrication of composite films and promote their applications. Our main research subjects within next few years are listed as follows.

1. Analysis on evolution and the relevant mechanism of metal films
2. Theory construction on film formation and numerical simulation
3. Fabrication of visible light-responsive TiO_2/metal composite films
4. Improvement on photocatalytic activity of TiO_2/metal composite films
5. Application investigation of TiO_2/metal composite films in sterilization, environment purification, and so on.

Author details

Yun Lu
Graduate School & Faculty of Engineering, Chiba University, Japan

Liang Hao
Graduate School, Chiba University, Japan

Hiroyuki Yoshida
Chiba Industrial Technology Research Institute, Japan

6. References

Cao, Y. Q.; Long, H. J.; Chen, Y. M.; Cao, Y. A. (2009). Photocatalytic activity of TiO_2 films with rutile/anatase mixed crystal structures. Acta Physico-Chimica Sinica, Vol. 25, pp. 1088-1092.

Chattopadhyay, P. P.; Manna, I.; Talapatra, S.; Pabi, S. K. (2001). A mathematical analysis of milling mechanics in a planetary ball mill. Materials Chemistry and Physics, Vol. 68, pp. 85-94.

Farahbakhsh, I.; Zakeri, A.; Manikandan, P.; Hokamoto, K. (2011). Evaluation of nanostructured coating layers formed on Ni balls during mechanical alloying of Cu powder. Applied Surface Science, Vol. 257, pp. 2830-2837.

Gupta, G.; Mondal, K.; Balasubramaniam, R. (2009). In situ nanocrystalline Fe-Si coating by mechanical alloying. Journal of Alloys and Compounds, Vol. 482, pp. 118-122.

Hao, L.; Lu, Y.; Asanuma, H.; Guo, J. (2012). The influence of the processing parameters on the formation of iron thin films on alumina balls by mechanical coating technique. Journal of Materials Processing Technology, Vol. 212, pp. 1169-1176.

Japan Industrial Standard, JIS K 7194, 1994.

Japan Industrial Standard, JIS R 1703-2, 2007.

Kobayashi, K. (1995). Formation of coating film on milling balls for mechanical alloying. Materials Transactions, Vol. 36, pp. 134-137.

Lu, Y.; Hirohashi, M.; Zhang, S. (2005). Fabrication of oxide film by mechanical coating technique, *Proceedings of International Conference on Surfaces, Coatings and Nanostructured Materials*, Aveiro, Portugal.

Lu, Y.; Yoshida, H.; Nakayama, H.; Hao, L.; Hirohashi, M. (2011 a). Formation of TiO_2/Ti composite photocatalyst film by 2-step mechanical coating technique. Materials Science Forum, Vol. 675-677, pp. 1229-1232.

Lu, Y.; Yoshida, H.; Toh, K.; Hao, L.; Hirohashi, M. (2011 b). Performance improvement of TiO_2/Ti composite photocatalyst film by heat oxidation treatment. Materials Science Forum, Vol. 675-677, pp. 1233-1236.

Lu, Y.; Hao, L.; Toh, K.; Yoshida, H. (2012). Fabrication of TiO_2/Cu composite photocatalyst thin film by 2-step mechanical coating technique and its photocatalytic activity. Advanced Materials Research, Vol. 415-417, pp. 1942-1948.

Lü, L.; Lai, M. O.; Zhang, S. (1995). Modeling of the mechanical-alloying process. Journal of Materials Processing Technology, Vol. 52, pp. 539-546.

Mattox, D. M. (2010). Handbook of physical vapor deposition (PVD) processing, William Andrew Publication, ISBN 978-0-8155-2037-5, Burlington, USA

Maurice, D. R.; Courtney, T.H. (1990). Physics of mechanical alloying, a first report. Metallurgical and Materials Transactions, Vol. A21, pp. 289-303.

Maurice, D. R.; Courtney, T.H. (1994). Modeling of mechanical alloying: part I. deformation, coalescence, and fragmentation mechanisms. Metallurgical and Materials Transactions, Vol. A25, pp. 147-158.

Pierson, H. O. (1999). Handbook of chemical vapor deposition, Noyes Publications and William Andrew Publication, ISBN 0-8155-1432-8, New York, USA

Rengaraj, S.; Venkataraj, S.; Yeon, J. W.; Kim, Y.; Li, X. Z.; Pang, G. K. H. (2007). Preparation, characterization and application of Nd-TiO2 photocatalyst for the reduction of Cr (VI) under UV light illumination. Applied Catalysis, Vol. B 77, pp. 157-165.

Romankov, S.; Sha, W.; Kaloshkin, S. D.; Kaevitser, K. (2006). Formation of Ti-Al coatings by mechanical alloying method. Surface Coatings Technology, Vol. 201, pp. 3235-3245.

Suryanarayana, C. (2001). Mechanical alloying and milling. Progress of Materials Science, Vol. 46, pp. 1-184.

Yoshida, H.; Lu, Y.; Nakayama, H.; Sano, H.; Hirohashi, M. (2008). Fabrication and evaluation of composite photocatalytic film by mechanical coating technique, *Proceedings of the 6ᵗʰ International Forum on Advanced Material Science and Technology*, Hong Kong, China

Yoshida, H.; Lu, Y.; Nakayama, H.; Hirohashi, M. (2009 a). Analysis of Ti films fabricated by mechanical coating technique (In Japanese). Journal of Materials Science Society of Japan, Vol. 46, pp. 141-146.

Yoshida, H.; Lu, Y.; Nakayama, H.; Hirohashi, M. (2009 b). Fabrication of TiO_2 film by mechanical coating technique and its photocatalytic activity. Journal of Alloys and Compounds, Vol. 475, pp. 383-386.

Yoshida, S.; Taga, Y.; Kinbara, A.; et al. (2008). Handbook of thin films, Ohmsha Publication, ISBN 978-4-274-20519-4, Tokyo, Japan

Other Applications of Composites

Chapter 14

Carbon Fibre Sensor: Theory and Application

Alexander Horoschenkoff and Christian Christner

Additional information is available at the end of the chapter

http://dx.doi.org/10.5772/50504

1. Introduction

The piezoresistive[1] carbon fibre sensor (CFS) consists of a single carbon fibre roving with electrical connected endings embedded in a sensor carrier (patch) for electrical insulation. Depending on the requirements of the application different patch types (e.g. glass fibre reinforced plastic (GFRP), polyester film, neat epoxy resin) are used. In terms of the mechanical properties GFRP is a particularly suitable patch material. CFSs with a GFRP patch exhibits an improved linearity of the signal due to the supportive effect of the glass fibres to the carbon sensor fibre especially in the case of compression loading. In [6] the ex-PAN fibre T300B was identified as one suitable carbon fibre for CFSs because of the excellent linear piezoresistive behaviour, the high specific resistivity and a high breaking elongation of the fibre. Figure 1 shows a CFS with a single layer GFRP patch (UD prepreg EG/913).

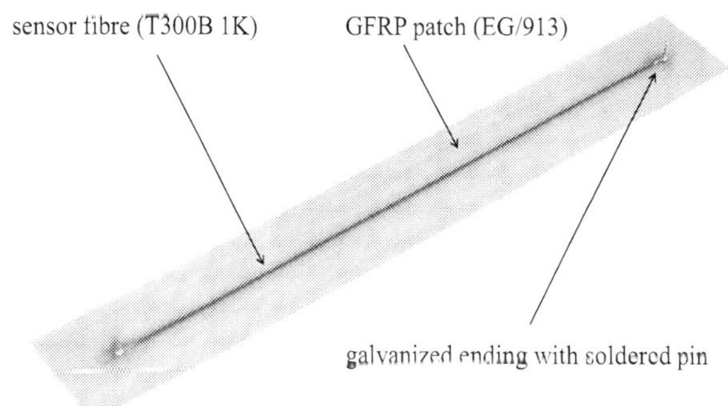

Figure 1. Carbon fibre sensor with an ex-Pan sensor fibre and a GFRP patch

---

[1] Piezoresitivity describes the change in electrical resistance of a conductor due to an applied strain.

The manufacturing process of a CFS includes three basic steps: Pre-curing of the carbon fibre, preparation of the electrical connection and embedding of the sensor fibre into the sensor carrier.

The pre-curing process is used to stabilize the carbon fibre roving and to align the filaments of the roving. For this purpose the twisted carbon fibre roving is impregnated by a resin with low viscosity and cured by using a special tooling. Good results for the impregnation of the carbon fibre roving T300B 1K were obtained by using the epoxy resin EP301 S (HBM) and a twist of 20 turns per meter. Spring elements provided a constant tension force along the roving during the curing process at 180 °C for 1.5 hours.

For preparation of electrical connections a galvanic process is applied based on a nickel electrolyte. In order to attain a homogeneous nickel coating of the filaments the resin must be removed at the fibre endings[2]. An applied current of 40 mA for 30 seconds leads to an excellent nickel coating. Depending on the application of the CFS (surface application or structural integration) the ends of the sensor fibre can be provided with soldered pins.

The embedding process of the sensor fibre depends on the used patch type and patch material. Figure 2 shows a micro section of a carbon fibre sensor in longitudinal and transverse direction.

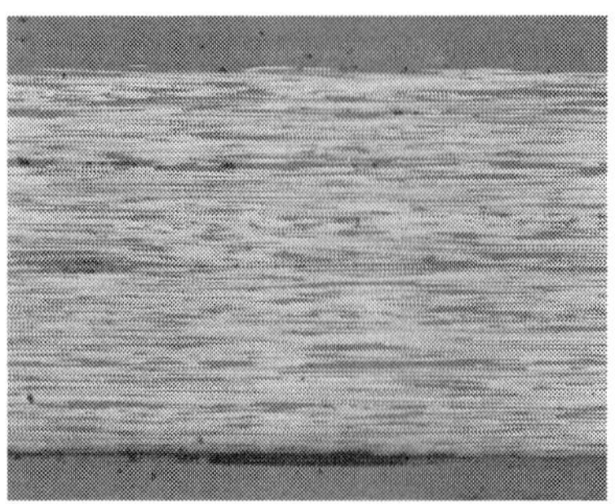

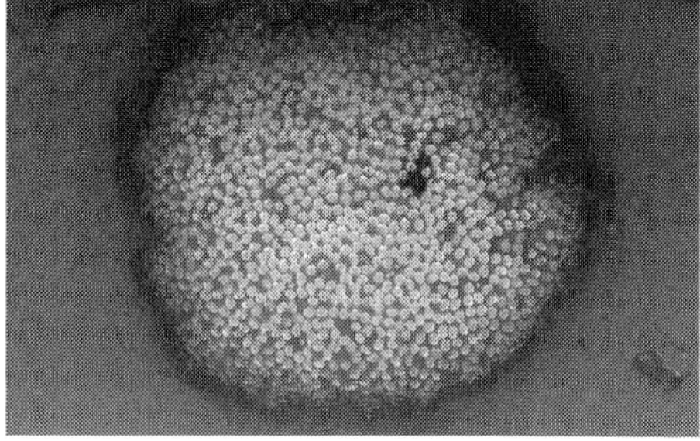

(a) Longitudinal direction   (b) Transverse direction

Figure 2. Microsection of a carbon fibre sensor in longitudinal and transverse direction
Sensorfibre: 1K roving of the carbon fibre T300B
Patch: single layer UD prepreg EG/913

2. Electromechanical properties of carbon fibres

Referring to Chung [2] and Dresselhaus [3] crystalline carbon fibres have the same crystal structure as graphite. The layered and planar structure of crystalline carbon fibres is shown in Figure 3. In each layer the sp^2 hybridized carbon atoms are arranged in a hexagonal lattice. Within a layer (x-y plane) the carbon atoms are bonded by three covalent bonds (overlapping sp^2 orbitals), and a metallic bonding is provided by the delocalization of the p_z orbitals.

---

[2] For example, the epoxy resin EP 310 S can be burned off or removed using acid.

This delocalization of the fourth valence electron leads to a good electrical conductivity of the carbon fibre. The individual layers are held together by weak van der Waals forces (z-direction). These different types of bonds within and between the layers result in the anisotropy of the mechanical, thermal and electrical material properties of carbon fibres.

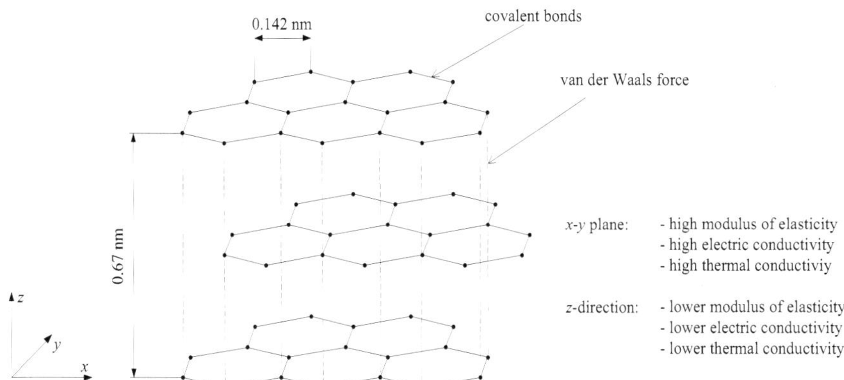

Figure 3. Graphite structure of crystalline carbon fibres

2.1. Specific resistivity

The degree of crystallinity and the micro structure of carbon fibres are controlled by the carbonisation process and determine the mechanical and electromechanical properties. Concerning ex-PAN fibres the Young's modulus ranges from 200 GPa (HT fibres) to 600 GPa (HM fibres). Ex-pitch fibres have a higher modulus up to 900 GPa. Figure 4 shows the correlation between the Young's modulus and the specific resistivity of different carbon fibre types. The specific resistivity of HT fibres (high tenacity) is in the range of 15 $\mu\Omega$m up to 18 $\mu\Omega$m. Higher orientated fibres (IM fibres and HM fibres) show a specific resistivity below 14 $\mu\Omega$m.

The specific resistance of carbon fibres is strongly temperature dependent [15]. In consequence, temperature effects have a high influence on the signals of CFSs and must be considered by an appropriate temperature compensation (see section 3).

2.2. Piezoresistivity

The electrical resistance R of a carbon fibre is given by:

$$R = \rho \frac{L}{r^2 \pi} \tag{1}$$

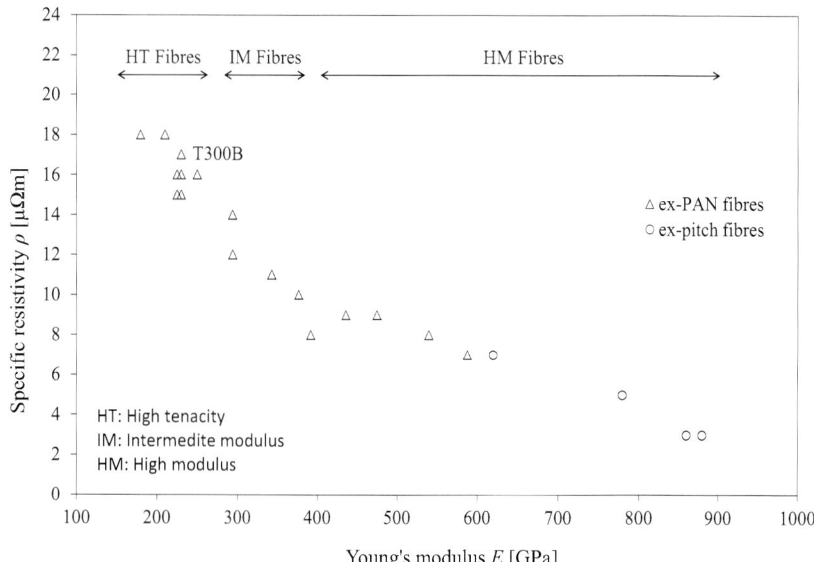

Figure 4. Correlation between the specific resistance ρ and the Young's modulus E of different ex-PAN and ex-pitch carbon fibres

where ρ is the specific resistivity, L the length and r the radius of the fibre. The total differential of $R = f(\rho, L, r)$ is then yielded by the following Equation (2).

$$\mathrm{d}R = \frac{\partial R}{\partial \rho}\mathrm{d}\rho + \frac{\partial R}{\partial L}\mathrm{d}L + \frac{\partial R}{\partial r}\mathrm{d}r = \frac{L}{r^2\pi}\mathrm{d}\rho + \rho\frac{1}{r^2\pi}\mathrm{d}L - 2\rho\frac{L}{r^3\pi}\mathrm{d}r \tag{2}$$

Using Equation (1) and Equation (2) the relative change in resistance of a carbon fibre $(\mathrm{d}R/R_0)$ can be expressed as:

$$\left(\frac{\mathrm{d}R}{R_0}\right) = \frac{\mathrm{d}\rho}{\rho} + \varepsilon(1+2\nu) = k\varepsilon \tag{3}$$

$$\text{with } \varepsilon = \frac{\mathrm{d}L}{L} \text{ and } \nu = -\frac{\mathrm{d}r/r}{\mathrm{d}L/L}$$

The term $\mathrm{d}\rho/\rho$ denotes the piezoresistive effect (material effect) and the term $\varepsilon(1+2\nu)$ represents the geometric effects. The strain sensitivity k covers both effects. It should be noted that the strain sensitivity k depends on the effective Poisson ratio ν [3]. Therefore, the strain sensitivity k must be defined for a corresponding Poisson ratio [4].

---

[3] The effective Poisson ratio ν depends on the Poisson ratio of the sensor fibre ν_f, the Poisson ratio of the sensor patch ν_p and the strain ratio $-\varepsilon_y/\varepsilon_x$ of the structure.

[4] Conventional strain gauges exhibits the same behaviour. The strain sensitivity of strain gauges is usually defined for a corresponding Poisson ratio $\nu = 0.285$ (steel).

For some applications of CFSs it can be useful to split the strain sensitivity k into the longitudinal strain sensitivity k_l and the transverse strain sensitivity k_t. The relative change in resistance is then given by:

$$\left(\frac{dR}{R}\right) = k_l\varepsilon_l + k_t\varepsilon_t \tag{4}$$

The strain sensitivity k of a T300B 1K fibre was determined as $k = 1.71$ for a corresponding Poisson ratio ν of 0.28. The sensor fibre exhibits a longitudinal strain sensitivity k_l in the range of 1.72 to 1.78 and a transverse strain sensitivity k_t in the range of 0.37 to 0.41. The piezoresistivity of the ex-PAN fibre is linear up to a strain level of approximately 6000 μm/m [1, 6]. Problems may occur at the metallized fibre endings. In order to avoid any influence of the electrical connections the strain level at the fibre endings should not exceed a level of 2500 μm/m.

This can be realized by using special tabs which exhibit a higher stiffness in the regions of the metallic connections. An example of such load relieving tabs is given in Figure 5. The lay-up of the tabs ensures a four times smaller strain level in the region of the electrical connections compared to the strain level which occurs in the testing area.

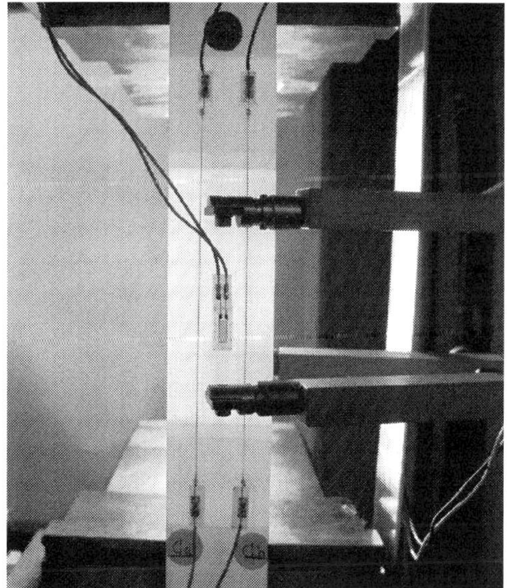

Figure 5. CFS patch with load relieving tabs
Lay-up in the testing area: $[0^\circ]$
Lay-up in the regions of the metallized fibre endings: $[0^\circ_5]$

Figure 6 shows the excellent linear piezoresistive behaviour of the ex-PAN fibre T300B 1K up to a strain level of 6000 μm/m (loading and unloading). Bending tests were performed to investigate the compression behaviour of CFSs. These investigations are not completed. First results show that the linearity of the signal depends significant on the carrier material.

Ex-pitch fibres show a nonlinear piezoresistive behaviour.

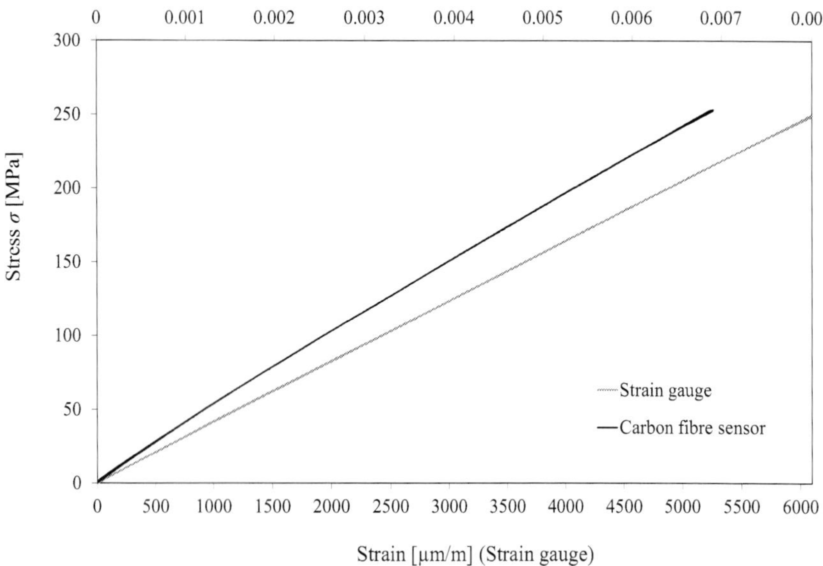

Figure 6. Characterization of the linear piezoresistivity up to a strain level of 6000 μm/m (loading and unloading) by means of a CFS patch with load relieving tabs at the endings
Sensor fibre: T300B 1K (ex-PAN fibre)
Lay-up in the testing area: $[0_8^\circ]$
Lay-up of the load relieving tabs: $[\pm45^\circ, 0_3^\circ]_{sym}$

3. The Wheatstone bridge

The change in resistance of a CFS due to an applied strain is usually small. The Wheatstone bridge is an electrical circuit which allows the determination of very small changes in electrical resistance with great accuracy. Furthermore, the Wheatstone bridge minimizes the high influence of temperature changes on the CFS signal. The measurement circuit, illustrated in Figure 7, consists of four resistances, a supply voltage and the output voltage of the bridge. The relative change in resistance can be determined by the ratio of output voltage to input voltage V_{out}/V_s. There are two configurations of the Wheatstone bridge which are of special importance for the use of CFSs. These configurations of the Wheatstone bridge are known as "half bridge" configuration and "full bridge" configuration.

In the case of a half bridge circuit (Figure 8(a)), the bridge is formed by two CFSs (R_1 and R_2) and two completion resistors (R_3 and R_4). The full bridge configuration (Figure 8(b)) is formed by four CFSs and needs no additional resistors. Both circuits will enable a compensation of

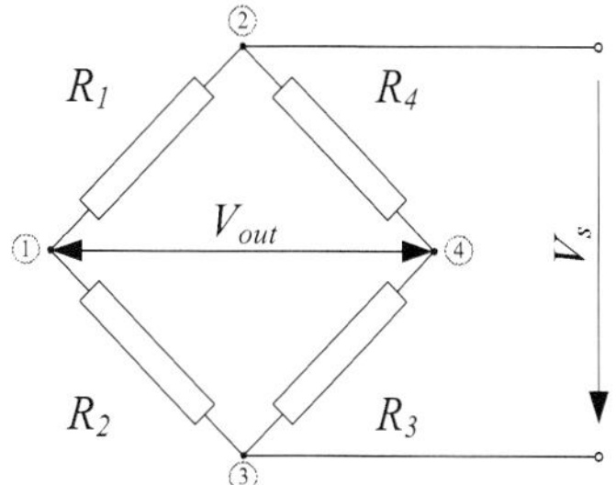

$$\frac{V_{out}}{V_s} = \frac{1}{4}\left(\frac{\Delta R_1}{R_1} - \frac{\Delta R_2}{R_2} + \frac{\Delta R_3}{R_3} - \frac{\Delta R_4}{R_4} \right)$$

Figure 7. General Wheatstone bridge circuit

temperature effects (thermal dependency of the specific resistance and thermal expansion) if the thermal conditions of the connected CFSs are identical. In the case of a half bridge circuit with one active CFS (R_1) the output voltage V_{out} is given by:

$$V_{out} = \frac{V_s}{4}\left(\frac{\Delta R_{1,mech} + \Delta R_{1,therm}}{R_1} - \frac{\Delta R_{2,therm}}{R_2} \right) \tag{5}$$

Equation (5) shows that the thermal effect is compensated by the mechanically unloaded carbon fibre sensor (R_2). Detailed information about the principle and use of the Wheatstone bridge can be found in classical textbooks on strain gauge techniques (e.g. [5, 13])

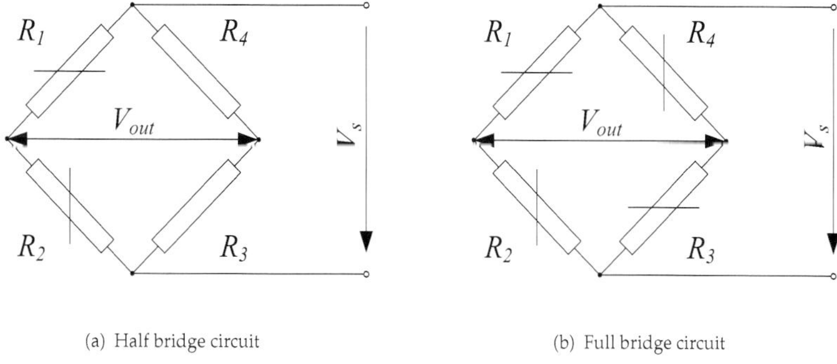

(a) Half bridge circuit   (b) Full bridge circuit

Figure 8. Half bridge and full bridge configuration of the Wheatstone bridge

4. Integral strain measurement of the carbon fibre sensor

Equation (3) describes the relative change in electrical resistance ($\Delta R/R_0$) of a carbon fibre sensor due to an applied elastic strain ε. It is important to understand that a CFS measures the strain integrally along its whole fibre length. Thus, Equation (3) can be written as:

$$\left(\frac{\Delta R}{R_0} \right) = k\frac{1}{L} \int_0^L \varepsilon(x)\, dx \tag{6}$$

Equation (6) shows that a CFS measures the displacement between the terminal points of the sensor fibre.

$$\left(\frac{\Delta R}{R_0}\right)\frac{L}{k} = \int_0^L \varepsilon(x)\,dx = [u(x=L) - u(x=0)] \tag{7}$$

This integral strain measurement of CFSs in accordance to Equation (7) can be used to create carbon fibre sensor meshes (CFS meshes). Such a CFS mesh allows the determination of the two dimensional (2D) state of strain and state of deformation of a whole structure or larger areas of a structure.

5. Carbon fibre sensor meshes

The strain and deformation analysis by means of CFS meshes are based on linear or higher order displacement approximations. The displacement functions depend on the used element type such as 3-node triangle elements, 6-node triangle elements or quadrilateral elements. The basic theory of strain analysis with CFS meshes using 3-node triangle elements with a linear displacement approximation is presented below.

A triangular CFS element defined by its vertices 1, 2, 3 and its local coordinate system is given in Figure 9.

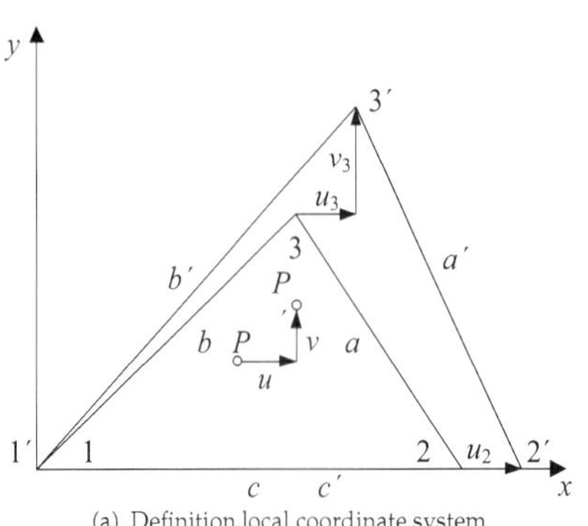

(a) Definition local coordinate system

(b) CFS element applied on aluminium

Figure 9. Triangle CFS element with 3 nodes

The displacement $u(x,y)$ and $v(x,y)$ of an inner point P can be determined by the displacements of the vertex in using a displacement function.

Assuming a linear displacement function the displacement $u(x,y)$ and $v(x,y)$ within the element can be calculated in accordance to Equation (8) and Equation (9).

$$u(x,y) = N_1 u_1 + N_2 u_2 + N_3 u_3 \tag{8}$$

$$v(x,y) = N_1 v_1 + N_2 v_2 + N_3 v_3 \tag{9}$$

Hereby u_i and v_i denote the displacements of the vertex 1, 2, 3 in the x- and y-direction while N_i represents the shape functions of the element. The shape functions N_i of a 3-node triangle can be found in classical books on the theory of finite element analysis (e.g. [16]). The engineering strains ε_x, ε_y and γ_{xy} are defined by:

$$\varepsilon_x = \frac{\delta u}{\delta x}; \qquad \varepsilon_y = \frac{\delta v}{\delta y}; \qquad \gamma_{xy} = \frac{\delta u}{\delta y} + \frac{\delta v}{\delta x} \tag{10}$$

Considering the local coordinate system ($u_1 = 0, v_1 = 0, v_2 = 0$) the strains within the triangle can be calculated by:

$$\varepsilon_x = \frac{u_2}{x_2}; \qquad \varepsilon_y = \frac{v_3}{y_3}; \qquad \gamma_{xy} = \frac{x_2 u_3 - x_3 u_2}{x_2 y_3} \tag{11}$$

In consequence of the linear displacement approximation the strains are independent of the coordinates (x and y) and thus constant within the element. The unknown displacements u_2, u_3 and v_3 of the vertex 2 and 3 can be determined by the signals of the carbon fibre sensors.

In order to verify this linear approach an experimental investigation of a CFS mesh applied on a 1000 mm x 1000 mm x 5 mm PMMA plate was performed. The simply supported plate was loaded with a single static force at the center. In addition to the experiment a finite element analysis (FEA) was performed. In [9] the results of this experiment and of the corresponding finite element simulations are presented in detail [5]. Figure 10 shows the PMMA plate and the applied CFS mesh.

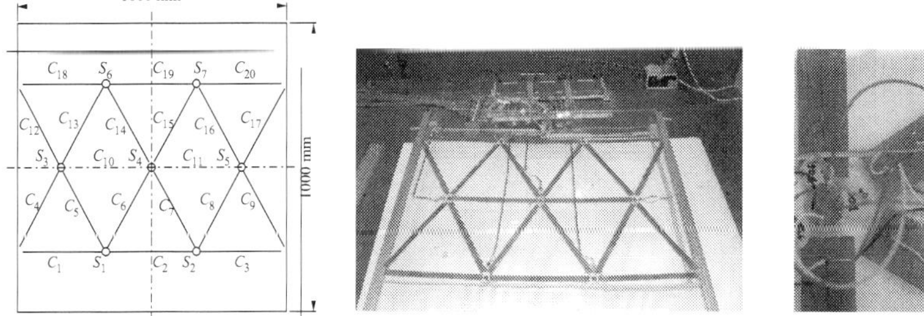

Figure 10. Carbon fibre sensor mesh applied on a 1 m x 1 m PMMA plate. Each sensor has a length of 300mm. [9]

Figure 11 shows the determined strain ε_x for each element of the mesh. There was a good correlation between the measured and the calculated strain levels. The accuracy was in the range of $\pm$ 5%.

The results of the performed investigation show that CFS meshes are a reliable instrument to determine the strain fields and principle strains of lightweight structures.

The principle strains (strain level and direction) are of particular interest in case of structures made of composite materials. For example, tailored fibre placement (TFP) is an advanced textile manufacturing process for CFRP structures in which the carbon fibre rovings are placed

---

[5] A quadratic displacement approach for a 3-node triangle is also presented and verified in [9]. However, the quadratic displacement approach of a 3-node triangle requires additional strain gauges at the nodes to determine the 12 unknown coefficients of the shape function.

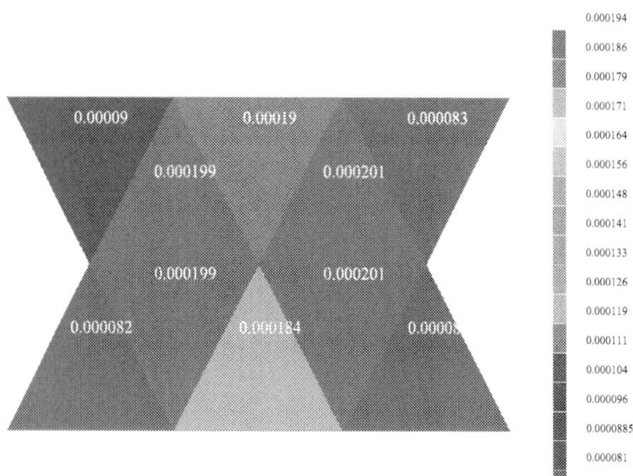

Figure 11. Strain ε_x measured by the carbon fibre sensor mesh

in accordance to the direction of principal stresses.

The finite element simulation is a powerful tool to analyse the stress fields and principle directions of lightweight structures. At the design and optimization processes the FEA is almost the only way to evaluate the structural load. However, there is a lack of techniques to review the results of the FEA. CFS meshes offer a high potential to verify the results of the finite element analysis.

6. Micro crack detection

A major failure mode of multidirectional reinforced laminates is transverse matrix cracking. Matrix cracks will reduce the effective stiffness of the laminate and will result in local stress concentrations at the crack tip. Furthermore, interlaminar crack growth and local delamination can occur. Due to its integral strain measurement method the CFS has a high potential to detect matrix crack initiation and monitor crack growth. Figure 12 shows a thin GFRP laminate (Lay-up: $[90_2^\circ, 0^\circ, 90_2^\circ]$) which has two embedded CFSs. Matrix cracks along the CFS will affect the sensor signal which will give a clear indication of the crack density. A study was performed to investigate the influence of cracks on the sensor signal [10]. The study was performed on the GFRP laminate $[0^\circ, 90_5^\circ, 0^\circ, 90_5^\circ, 0^\circ]$. Three CFSs were embedded in the mid-plane 0°-layer. At a strain level higher than 3000 μm/m matrix cracks appeared in the 90°-layers. The following techniques were applied to characterize the influence of damages on the sensor signal:

- Acoustic emission analysis in combination with pattern recognition technique
- Microscopy and micrographs
- Analytical calculations
- Finite element analysis

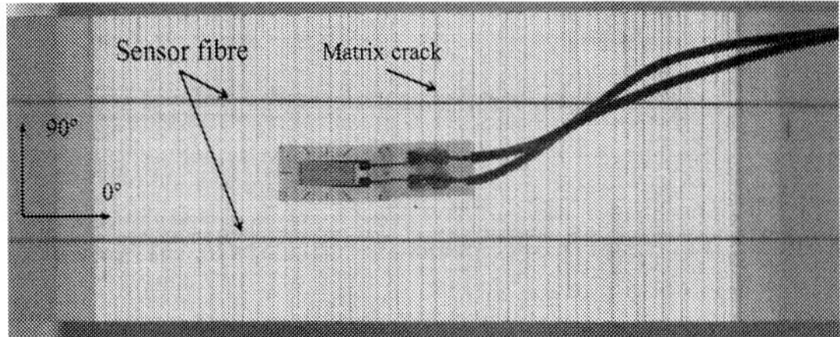

Figure 12. Multidirectional reinforced GFRP laminate with two embedded CFSs. Transverse matrix cracking will affect the sensor signal.
Lay-up: $[90_2^\circ, 0^\circ, 90_2^\circ]$

Figure 13 shows the correlation between the crack density, the CFS signals, the acoustic emission energy (AE-energy), the strain level measured by a strain gauge and the according stress level. It can be seen that the matrix crack initiation causes first AE-signals and a change of the slope of the CFS signals.

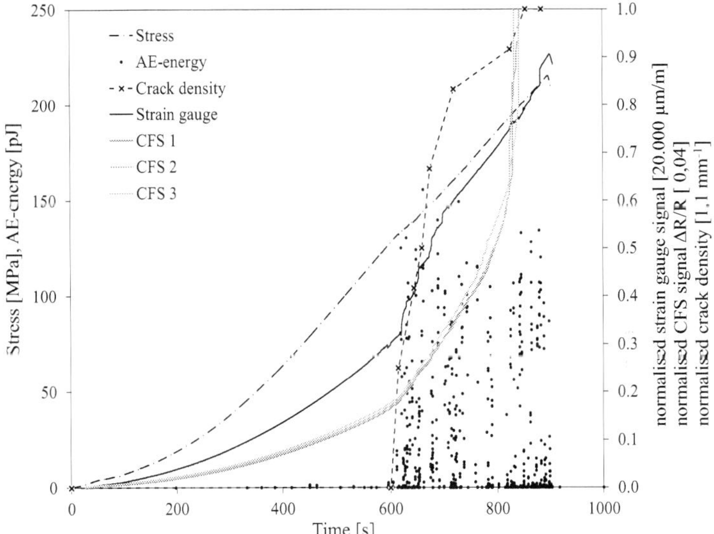

Figure 13. Acoustic emission and CFS signals measured on an GFRP laminate under uniaxial tensile load [10]
Lay-up: $[0^\circ, 90_5^\circ, 0^\circ, 90_5^\circ, 0^\circ]$

Furthermore, a good correlation between the sensor signal $(\Delta R / R_0)$ and the crack density can be observed. After having reached a crack density of 0.8 mm^{-1} the signals of the embedded carbon fibre sensors increase disproportionately. The analytical approach of Garret and Bailey [4, 11, 12] and a FEA were applied to calculate the reduction of the stiffness of the laminate. Müller showed that the global stiffness loss of the laminate due to the matrix cracking can be measured by means of the CFS. However, at high strain levels (> 70% of $\varepsilon_{ultimate}$) there is a strong influence of high local stresses at the crack tip on the CFS signal. These stress concentrations may result in filament breakage and in extremely high signal levels.

Based on this result CFSs can be used for damage monitoring and for the prediction of the lifetime of damaged structures if the damage level can be characterized by stiffness loss. Figure 14 shows the cyclic loading of a laminate to measure the stiffness loss and to determine Ladeveze's material parameters [8]. Based on this calibration the lifetime of a structure, e.g. pressure vessel, can be predicted.

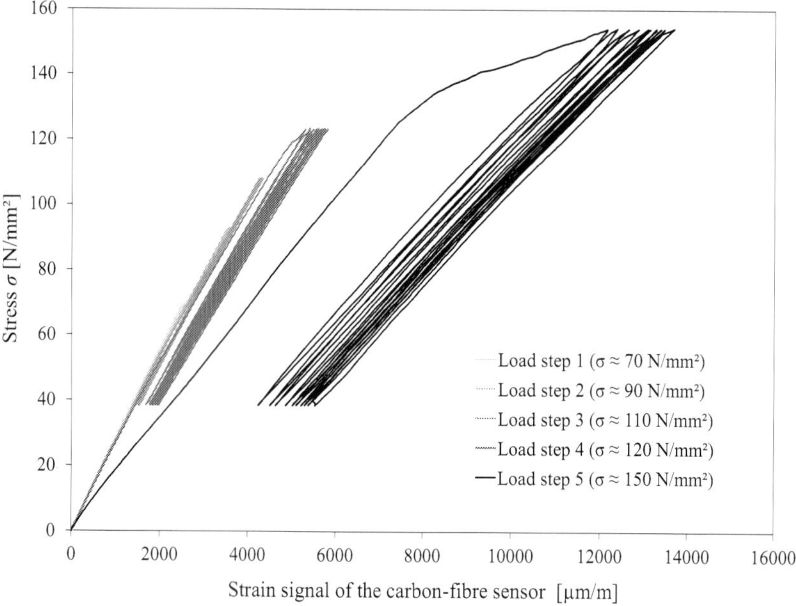

Figure 14. Damage measurement and determination of damage variables like the energy release rate
Material: GFRP, EG/913
Lay-up: $[0°, 90°_5, 0°, 90°_5, 0°]$

7. Impact and delamination detection

An impact loading can cause small damages inside the composite material which may not be found by visible inspection. One major concern is delamination damages or the disbonding of interfaces. These damages result in sublaminates having lower buckling resistance and compression strength. Although a small delamination-damage does not necessarily constitute failure, the damaged area may undergo a time-dependent growth and may attend a critical size.

The principles for achieving damage tolerant primary composite structures were established by the aircraft companies [14]. Maintenance intervals and inspection plans are determined in such a way that readily detectable damages will be repaired before damage growth can affect the fatigue strength of the structure. The influence of undetectable damages is covered by the so called barely visible impact damage (BVID) which defines the damage that establishes the strength values to be used in analysis to demonstrate compliance with the load requirements. One method to determine the influence of the BVID on the mechanical performance is the compression after impact test procedure (CAI-test procedure, i.e. Boeing BSS 7260). A 4 mm thick quasi-isotropic test specimen (150 x 100 mm) is damaged by a dropped weight impact testing machine. An impact level of about 3 J/mm will cause the BVID. This means that only a small remaining indentation is visible on the surface of the specimen, but delamination may be found inside. The strength of the damaged specimen is reduced by 10 to 20% compared to the undamaged material. This shows that in many cases the exploitation of material performance is limited, since the skin thicknesses of a composite structure are designed to absorb an impact.

For aircraft structures health monitoring systems have an extremely high potential to improve the efficiency of composite structures. Based on a monitoring system the material design values can be increased and the inspection intervals can be enlarged. Both aspects can result in a weight reduction of the structure up to 10%. A preliminary study was performed to investigate the use of CFSs as a sensor system to detect the BVID. The CAI specimen was used to integrate a rectangular CFS mesh (Figure 15 and Figure 16).

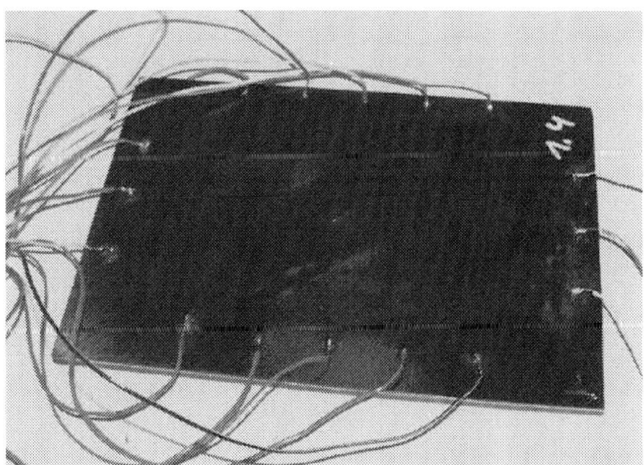

Figure 15. Compression after impact (CAI)-specimen with integrated CFS mesh to detect damages below the barely visible impact damage level (BVID)

The distance between the CFSs varied from 30 to 100 mm. Three methods were investigated to detect the impact damage:

- Online measurements of the resistivity during the impact test
- Offline measurements, comparison of the resistivity before and after the impact
- Active thermography, CFSs used as heating element

It has been shown that all three procedures are suitable to detect impact damages. A distance of 50 mm between the CFSs is necessary to detect even small damages below the

BVID. In a second step the use of CFSs will be investigated to detect debonding of skin and stringer. A health monitoring system for a complex aircraft structure will be based on several technologies like ultrasonic inspection sensors and strain sensors. The CFS technology will complement the established sensors due to its specific and simple integral measurement principle.

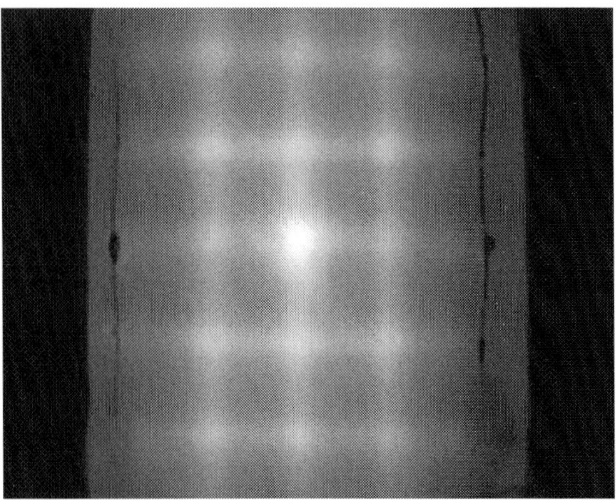

Figure 16. Compression after impact (CAI)-specimen with integrated CFS mesh. The CFSs are used as heating element for active thermography. A small delamination damage is visible (BVID)

8. Application

CFSs offer a high potential to be used as a sensor element for composite materials for stress analysis, damage detection and the monitoring of manufacturing processes. Two industrial applications have been selected to demonstrate this.

8.1. Tabletop of a CT-Scanner

Carbon fibre reinforced plastics (CFRP) are used for tabletops of computer tomography (CT) scanners, since CFRP fulfills the X-Ray transparency which is necessary to get the picture quality sufficient for medical diagnosis.

CFSs can be embedded in the tabletop of a CT scanner to measure its deflection. Based on the measured deflection the CT images can be readjusted to result in an improved medical attendance [7]. Figure 17 shows the tabletop of a CT scanner with ten u-shaped CFSs applied. For this application u-shaped CFSs are used to avoid that any metal wiring is part of the scan plane for all operation positions. The lengths of the u-shaped CFSs vary from 250 to 1250 mm. The determination of the beam deflection is based on the integral strain measurement of CFSs (Equation (6)). Considering small deformations the relation between the elastic strain of the outer fibre $\widehat{\varepsilon}$ and the beam deflection v is given by:

$$v = \iint v'' \mathrm{d}x\mathrm{d}x = \iint \frac{\widehat{\varepsilon}}{e_y} \mathrm{d}x\mathrm{d}x \tag{12}$$

Figure 17. CT tabletop with u-shaped CFSs to measure the deflection. The metallic wiring is attached to the clamping support.

where e_y denotes the distance from the neutral axis. Assuming a cantilever beam (see Figure 18) the slope v' of the beam can be determined directly by the signals of the CFSs. Assuming that the CFS starts at the clamping support ($x = 0$) the slope v' at the end of the applied CFS ($x = l_i$) becomes:

$$v'(x = l_i) = \frac{l_i}{ke_y} \left(\frac{\Delta R}{R} \right)_i \tag{13}$$

The deflection y of the beam can be calculated by numerical integration.

$$v(x = l_i) = \sum_{n=1}^{i} \left[v'(x_n) - \frac{v'(x_n) - v'(x_{n-1})}{2} \right] (l_n - l_{n-1}) \tag{14}$$

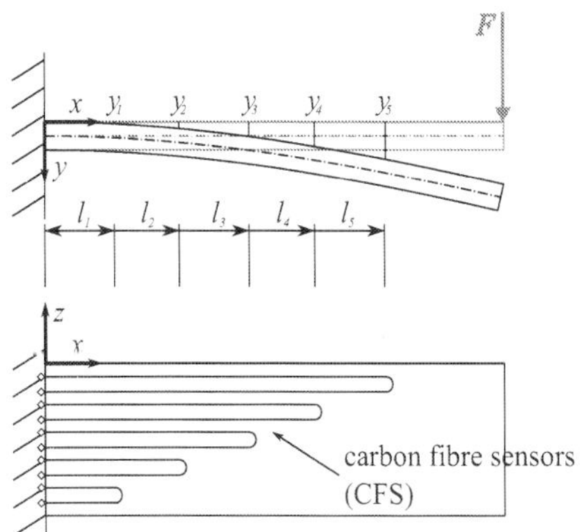

Figure 18. Side and top view of a cantilever beam with five u-shaped CFS

The index i denotes the number of CFSs applied. The quality of the approximation in accordance to Equation (14) depends on the complexity of the loading, the number of applied CFSs and the used sensor configuration.

In the case of the CT table (Figure 17) the deflection of the table can be determined with an accuracy of ± 0.3 mm for different operation positions. Figure 19 shows a typical measurement. The operating position of the CT tabletop varies stepwise from 0 mm to 2000 mm and back to 0 mm. The weight of the used dummy is 100 kg and electrical half bridges are used for temperature compensation.

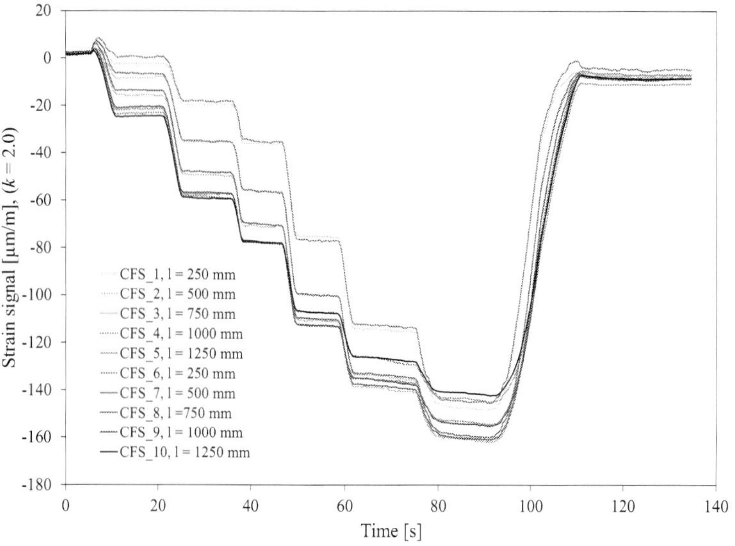

Figure 19. CFS signals of the CT table for a weight of 100 kg and different operation positions, ranging from 0 to 2000 mm

8.2. Pressure vessels

For thin walled assumptions the longitudinal stress σ_l and hoop stress σ_r in the cylindrical portion of a pressure vessel, away from the ends, are given by:

$$\sigma_l = \frac{pr}{2t} \tag{15}$$

$$\sigma_r = \frac{pr}{t} \tag{16}$$

where t is the thickness and r is the radius of the vessel. The relation σ_r/σ_l shows that the efficiency of pressure vessels can be increased if hoop wrapped vessels are used. A metal cylinder is reinforced by carbon fibres having a radial orientation (type II vessels). The strength of the vessel is increased remarkably.

There are two aspects for the use of CFSs for pressure vessels which can be easily integrated by means of the winding process:

- Determination of the pressure level of the vessel

- Monitoring of degradation processes due to fatigue or overloading of the radial fibre reinforcement

For such an application a sensor patch with four CFSs connected in full-bridge configuration is particularly suitable, since the full-bridge circuit minimizes the influences of thermal effects and shows an improved long term stability of the signal. Figure 20 shows such a hoop wrapped vessel with an embedded CFS patch.

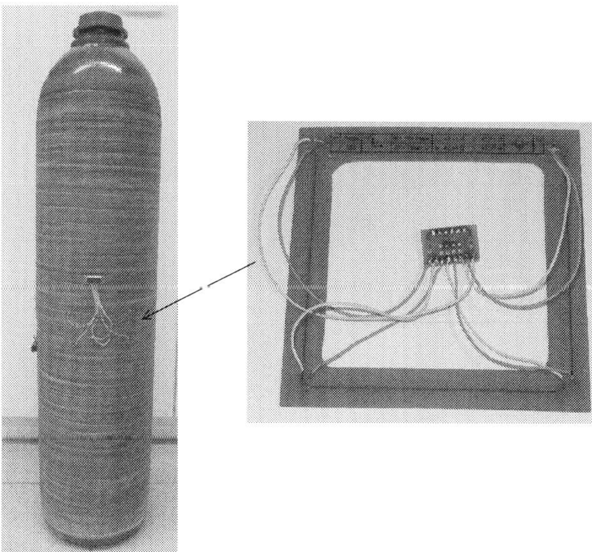

Figure 20. Hoop wrapped pressure vessel (type II vessel) with embedded CFSs

Experimental studies showed that the pressure of the vessel can be determined by using a CFS patch with a resolution of about 1 bar. A representative measurement of the performed test is shown in Figure 21. The pressure load of the vessel was increased stepwise up to a maximum pressure level of 100 bar. The subsequent pressure relief was performed in the same manner.

Micro cracks in the CFRP layers of the pressure vessel may occur as a result of mechanical or thermal overloading. The failure of the matrix causes a loss of stiffness of the CFRP layers and reduces the global stiffness of the vessel. Depending on the crack density and the crack growth the lifetime of the structure will decrease. Based on the damage master curve of the structure, an estimation of the remaining lifetime can be performed (see Figure 14).

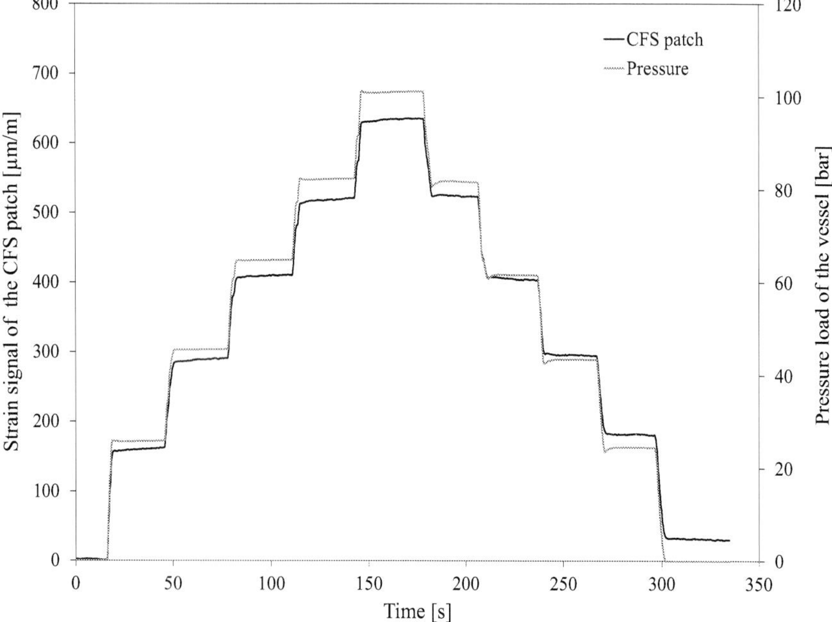

Figure 21. Strain signals of a CFS patch applied near the surface of a type II pressure vessel subjected to a pressure test

9. Conclusion

There are three main aspects which made carbon fibre sensors (CFSs) very interesting to be used for composite materials:

- Material-conformity
- Linear piezoresistivity up to high strain levels
- Integral strain measurement method

CFSs based on a T300B 1K ex-Pan fibre exhibit an excellent linear piezoresistivity up to a strain level of 6000 μm/m. The according strain sensitivity was determined as $k = 1.71$ (related to a Possion ratio of $\nu = 0.28$). The longitudinal strain sensitivity k_l of the CFS is in the range of 1.72 - 1.78. Transverse to the fibre direction CFSs exhibit a transverse strain sensitivity k_t of approximately 0.4. This significant transverse strain sensitivity must be considered in praxis. Tension load was applied for all tests, the characterization of the compression behaviour is under examination.

A disadvantage of CFSs is the high influence of temperature on the signal which will be an aspect for future research. At the moment a long term stability of ± 2 μm/m ($T = const.$) can be achieved for a half or full bridge circuit.

CFSs can be used for strain analysis, damage monitoring and the control of manufacturing processes. Concerning strain and stress analysis an approach for CFS meshes was developed based on triangular elements with linear displacement approximation. A good correlation was found between the measurement and the finite element calculation. The use of CFS meshes can be a new approach to complement finite element analysis from the experimental side.

Concerning monitoring aspects the influence of material damages on the sensor signal were studied. It has been demonstrated that based on the integral strain measurement method the CFS is an excellent sensor to detect delaminations and matrix cracks in multidirectional reinforced laminates. Therefore, the CFS offers unique features for fracture mechanics: Measurement of strain levels and detection of matrix cracks. By means of two examples the CFS technologies could be demonstrated successfully:

- Determination of the deflection of a tabletop of a CT-Scanner
- Determination of the strain level and the crack density of a pressure vessel

Strain measurement and matrix crack detection are in the focus of safe and damage tolerant composite structures making the CFS technology a complement of established sensors.

Author details

Alexander Horoschenkoff
Munich University of Applied Sciences, Germany

Christian Christner
Universität der Bundeswehr München, Germany

10. References

[1] Christner, C., Horoschenkoff, A. & Rapp, H. [2012]. Longitudinal and transverse strain sensitivity of embedded carbon-fibre sensors, *Journal of Composite Materials* Online First: DOI: 10.1177/0021998312437983.

[2] Chung, D. [1994]. *Carbon Fiber Composites*, Butterworth-Heinemann.

[3] Dresselhaus, M., Dresselhaus, G., Sugihara, K., Spain, I. & Goldberg, H. [1988]. *Graphite Fibers and Filaments*, Springer.

[4] Garrett, K. & Bailey, J. [1977]. Multiple transverse fracture in 90° cross-ply laminates of a glass fibre-reinforced polyester, *Journal of Material Science* Vol. 12: 157–168.

[5] Hoffmann, K. [1987]. *An introduction to measurements using strain gages*, Hottinger Baldwin Messtechnik.

[6] Horoschenkoff, A., Müller, T. & Kröll, A. [2009]. On the charaterization of the piezoresistivity of embedded carbon fibres, *17th International Conference on Composite Materials*, Edinburgh.

[7] Horoschenkoff, A., Müller, T., Strössner, C. & Farmbauer, K. [2011]. Use of carbon-fibre sensors to determine the deflection of composite-beams, *18th International Conference on Composite Materials*, International Conference on Composite Materials, Jeju.

[8] Ladeveze, P. & Dantec, E. L. [1992]. Damage modelling of the elementary ply for laminated composites, *Composite Science and Technology* Vol. 43: 257–267.

[9] Matzies, T., Christner, C., Müller, T., Horoschenkoff, A. & Rapp, H. [2011]. Carbon-fibre sensor meshes: Simulation and experiment, *21st International Workshop on Computational Mechanics of Materials*, Limerick.

[10] Müller, T., Horoschenkoff, A., Rapp, H., Sause, M. & Horn, S. [2010]. Einfluss von zwischenfaserbrüchen in 0/90-laminaten auf die elektrische widerstandsänderung von eingebetteten carbonfasern, *59th Deutscher Luft- und Raumfahrkongress*.

[11] Nairn, J. [1989]. The strain energy relase rate of composite microcracking: A variational approach, *Journal of Composite Materials* Vol. 23: 1106–1129.

[12] Nairn, J. & Hu, S. [1994]. Matrix microcracking, *Damage Mechanics of Composite Materials* Vol. 9: 187–243.

[13] Perry, C. & Lissner, H. [1955]. *The Strain Gauge Primer*, Mc Gram Hill.

[14] Razi, H. & Ward, S. [1996]. Principles for achieving damage tolerant primary composite aircraft structures, *11th DoD/FAA/NASA Conference on Fibrous Composites in Structural Design*.

[15] Spain, I., K., V., Goldberg, H. & Kalnin, I. [1982]. Unusual electrical resistivity behavior of carbon fibres, *Solid State Communications* Vol. 45(No.): 817–819.

[16] Zienkiewicz, O. & Taylor, R. [2000]. *Finite Element Method Volume 1: The Basis*, Elsevier.

Chapter 15

Bio-Inspired Self-Actuating Composite Materials

Maria Mingallon and Sakthivel Ramaswamy

Additional information is available at the end of the chapter

http://dx.doi.org/10.5772/47860

1. Introduction

Self-organisation is a process through which the internal organisation of the system adapts to the environment to promote a specific function without being controlled from outside. Biological systems have adapted and evolved over several billion years into efficient configurations, which are symbiotic with the environment.

Form, structure, geometry, material, and behaviour are factors, which cannot be separated from one another. For example, the veins in a leaf contribute to the overall form of the leaf, its structure and geometry. At the micro scale the fibre material organisation compliments to the responsive behaviour of the leaf. Therefore, the veins display an integral coherence within the multiple functions they perform which could be termed as 'Integrated Functionality'. Integrated Functionality occurs in nature due to multiple levels of hierarchy in the material organization.

The premise of this research is to integrate sensing and actuation functions into a fibre composite material system. Fibre composites, which are anisotropic and heterogeneous, offer the possibility for local variations in their material properties. Embedded fibre optics would be used to sense multiple parameters and Shape memory alloys integrated into composite material for actuation. The definition of the geometry, both locally and globally would complement the adaptive functions and hence the system would display 'Integrated Functionality'.

2. Less is more: Organization strategies in organic composite materials

Cellulose, collagen, chitin and silks are the only four types of fibrous tissues found in natural constructions (Figure 2). Biology is capable of building all living organisms using only these four materials. It does so, without further variation than changing the arrangement and organization of the fibres in the bonding substance to adapt to function specific requirements.

1.a -http://cybele.bu.edu/index/leaf.jpg
1.b-http://www.buffalogardens.com/historical/ Crystal_Palaces/body_crystal_palaces.html
1.c-http://www.mjausson.com/2003/img/ walk24Jun03/14gunnera_dt.jpg
1.d - http://universe-review.ca/I10-22a-stomata.jpg
(accessed on May 4th 2006)

Figure 1. Series of images featuring from top: hibiscus leaves, the structure at the bottom of a lily pad, the vein pattern of a leaf, and a micro-photo of stomata which aids photosynthesis.

Living tissues have the capability to adapt to constantly changing environmental conditions. This is achieved through iterative feedback loops, which sense, record, inform and instruct the fibre composite to alter its current configuration towards an optimized one.

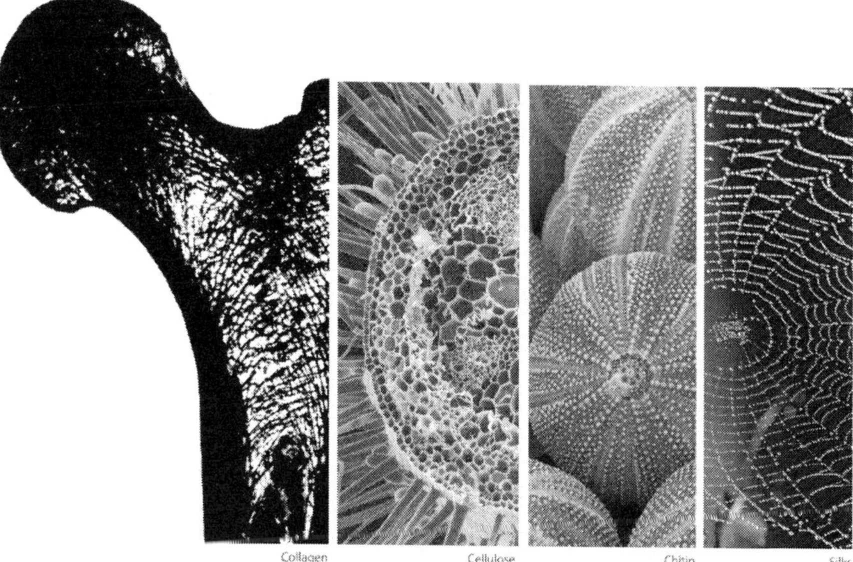

2.1 a - Emergence: Morphogenetic Design Strategies- Architectural Design, Academy Editions, London, Vol. 74 No 3 Issue May/June 2004, p. 4
2.1 b - Drew, Philip- Frei Otto – Form and Structure, Granada, London, 1976, p. 22
2.1 c -http://nanotechweb.org/articles/news/1/11/5/1/0611102 (accessed on Jun 12th 2009)
2.1 d- http://upload.wikimedia.org/wikipedia/ commons/a/ab/Spider_web_with_dew_drops04.jpg (accessed on Jun 12th 2009)

Figure 2. There are only four types of fibres in natural organisms: collagen, cellulose, chitin and silk.

In natural constructions, material is being continuously removed from places where it is not required and deposited where it can contribute to maintain the structural integrity of the structure. This concept was summarised by D'Arcy Thompson as 'growth under stresses'. Such a differentiated distribution process of fibres emerges through sensing the patterns of loading, or stresses constantly received by the natural organism.

Material self-organization and real-time optimization are both processes present in the formation and adaptation of biological tissues. They are termed as 'thigmo-morphogenesis' and are responsible of the resultant high performance and enormous capacity, found in natural fibre composites, to deal with unprecedented environmental conditions, unlike man-made composite materials commercially available till date.

Thigmo-morphogenesis refers to the changes in shape, structure and material properties that are produced in response to transient changes in environmental conditions. We are all familiar with the fact that many plants are capable of movement, sometimes slow as in the petals of flowers which open and close, tracking of the sun by the sunflowers, the convolutions of bindweed's around supporting stems, snaking of roots around obstacles; sometimes visible to the eye, as in the dropping of leaves when mimosa pudica is touched, and exceptionally very rapid, too fast to be seen, as in the closing of the leaves of the venus flytrap.

In all these examples, movement and force are generated by a unique interaction of materials, structures, energy sources and sensors. Cellulose walls of parenchyma cells non-lignified, flexible in bending but stiff in tension, constitute the material; the structures are the cells themselves and their shape with the biologically active membrane that can control the passage of fluid in and out of the cells; the energy source is the chemical potential difference between the inside and the outside of the cells; the sensors are as yet unknown. These systems are essentially working as networks of interacting mini hydraulic actuators, liquid filled bags which can become turgid or flaccid and which, owing to their shape and mutual interaction translate local deformations to global ones and are also capable of generating very high stresses. Similar mechanisms can be seen in operation when leaves emerge from buds and deploy to catch sunlight. How to package the maximum surface area of material in the bud and to expand it rapidly and efficiently is the result of smart folding strategies, turgor pressure and growth. [1]

http://www.tucsongardener.com/Year02/Fall2002/ photos/sensitivefold.JPG

Figure 3. Undisturbed delicate leaves of the sensitive plant mimosa pudica, which closes its leaves when lightly touched.

3. Alive: Hypothesis behind a smart composite material

The research presented herein proposes a bio-inspired synthetic self-actuating fibre composite, which emulates the morpho-mechanical processes found in natural fibre tissues, for its application to architectural constructions. The resulting material had to have integrated sensing capabilities and actuation competences, to be able to dynamically adapt to transient changes (Figure 4).

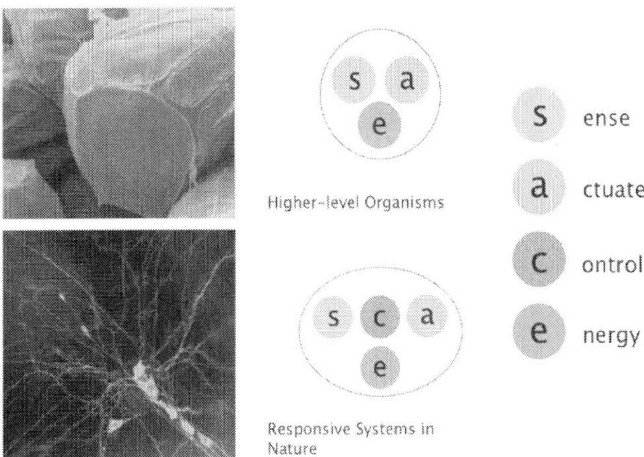

Figure 4. Diagram showing the different adaptation strategies found in natural organisms. Control being the main differentiator.

Fibre optics would be used as sensors and shape memory alloys (SMA) as actuators (Figure 5). Glass fibre mats constituted the reinforcement layer for the resin-based binding matrix, in which fibre optics and shape memory alloys were embedded. Experiments were however performed substituting fibre optics for thermocouples and strain gauges, which sensed and transmitted the energy necessary for the actuation to the shape memory alloys. Energy is supplied through a source of external heat, which forms part of the experiment setup. Further research reveals that, in hot-dry climates, sun radiation could be used as the main heat source to enable actuation, with only a small percentage of the total energy being supplied by external sources at certain periods of the day when exploiting solar heat is not possible.

In structural engineering, fibre optics is used as monitoring devices capable of measuring strain, temperature and humidity. One of their great advantages relies on their small size and fibred geometry, which converts them in perfect candidates for thin-wall structures such as fibre composite matrixes. Their receptive ability is achieved by designing the fibre optic cable to be sensitive to a specific parameter. The light pulse sent through the cable is received and further analysed by the processing unit to determine the magnitude of strain or temperature measured. Fibre optics was however substituted by strain gauges and thermocouples in the series of experiments implemented to test the performance of the self-actuating fibre composite material presented herein.

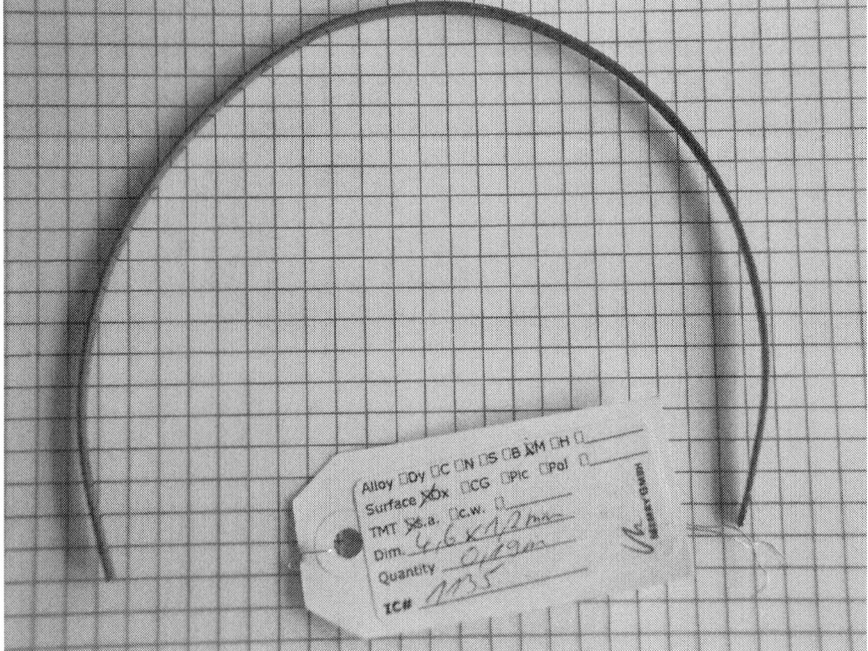

Figure 5. Nitinol based shape memory alloy. Shaped as a ribbon featuring an actuation temperature of 65°C.

Strain gauges are resistance-based sensors employed to measure variations in length of a parent component, factored by the component's original length. Strain gauges consist of a thin wire of metal foil, wrapped across a grid, which is also attached to a thin flexible backing material impregnated with glue, for adhesion to the parent component for which strain is to be monitored. This setup allows the wrapped wire to stretch or compress thus, detecting elongations and contractions felt by the parent component (Figures 6 and 7).

4. Experimental: Understanding the behaviour of SMAs

Nitinol (NiTi) is a specifically manufactured alloy of nickel and titanium, which has the ability to generate significant force upon changing shape. NiTi shape memory alloys can exist in three different crystal structures or phases called martensite, stress-induced martensite and austenite. At low temperature, the alloy exists as martensite, which is weak, soft and highly deformable. Stress-induced martensite (or super elastic NiTi) is highly elastic and is present at a temperature slightly higher than its transformation temperature. The austenite is the strongest, higher temperature phase, present in NiTi. Most of the physical properties of austenite and martensite vary during phase transformations; among these properties are the Young's modulus, the heat capacity, the latent heat and the thermal conductivity.

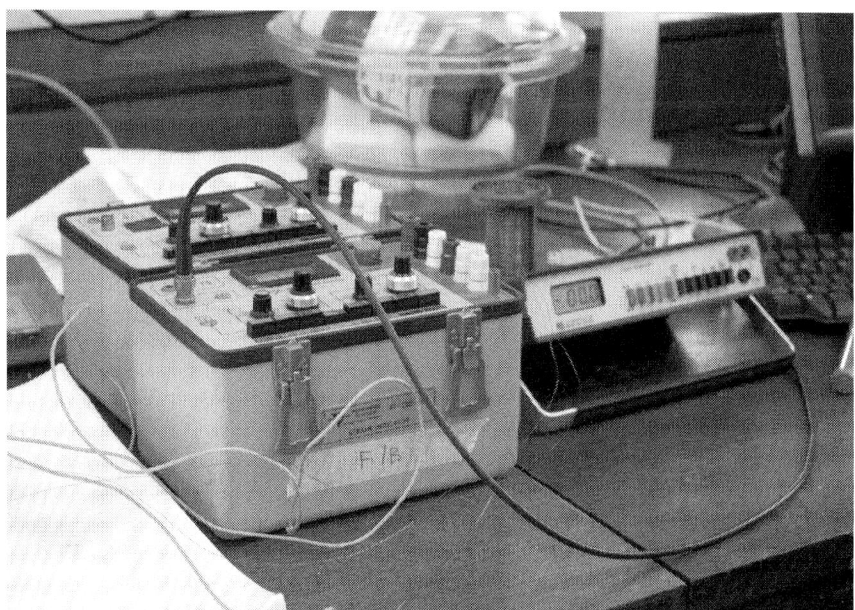

Figure 6. Picture illustrates the calibration process of the strain gauges processing unit.

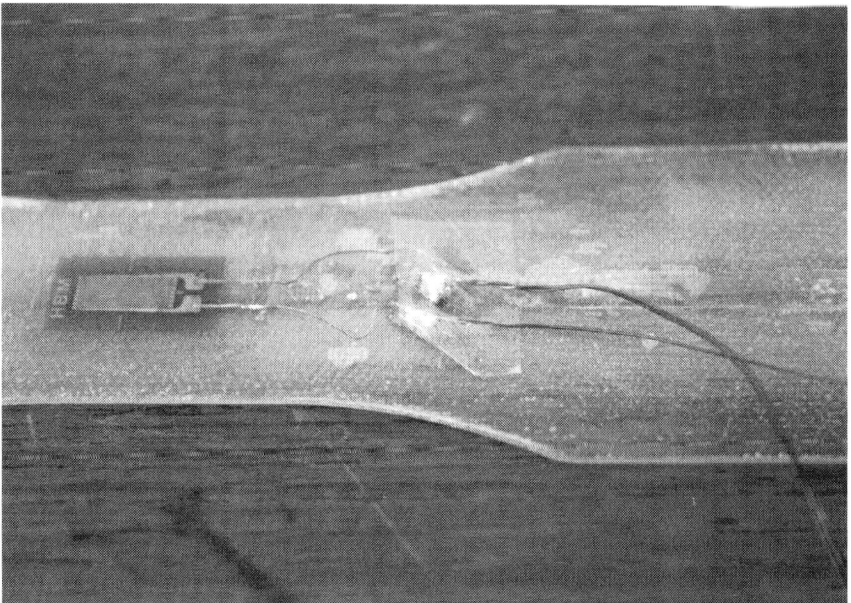

Figure 7. Picture features the strain gauge during its calibration and testing on a polyester strip.

Shape setting is essential to train the NiTi alloys to remember a specific shape. An actuator element designed for a particular purpose generally requires the setting of a custom shape. Shape setting is similar in all forms of nitinol, such as, wires, ribbons, strips, sheets, tubes or bars. It is accomplished by constraining the nitinol element on a mandrel or a fixture of the desired shape while applying an appropriate heat treatment. The heat treatment parameters and the properties of the actuator element are critical for the consequent behaviour of the NiTi, and usually need to be determined experimentally. In principle, temperatures as low as 400°C and heating times as short as 1 to 2 minutes are sufficient to set the shape, but generally one uses a temperature closer to 500°C and a heating time period of at least 5 minutes. Rapid cooling of the alloy is preferred via a water quench or rapid air-cooling. Higher heat treatment times and temperatures will increase the actuation temperature of the alloy and often results on sharper thermal responses. There is also an accompanying decrease in the ability of the actuator to resist permanent deformation. [2]

The first set of experiments focused on understanding the process of shape setting in SMAs outlined above, and their actuation behaviour against temperature. These were especially motivated by the need to quantify their lifting capacity against time for their latter integration in the fibre composite matrix. The SMAs used for these experiments were ribbons with an actuation temperature of 65°C, 1.2mm thick, 4.6mm wide and 270mm long.

The shape setting process required the SMAs to be heated at 500°C for at least 5 minutes. The SMA ribbons were trained by fixing them around metal pipes, which helped maintaining the SMAs in place during the heating process. They were subsequently removed from the oven and immediately immersed in cold water whilst maintaining the imposed bent shape throughout the cooling procedure (Figures 8 and 9).

Figure 8. Shape memory alloys and mandrels inside kiln after heating process had finalised.

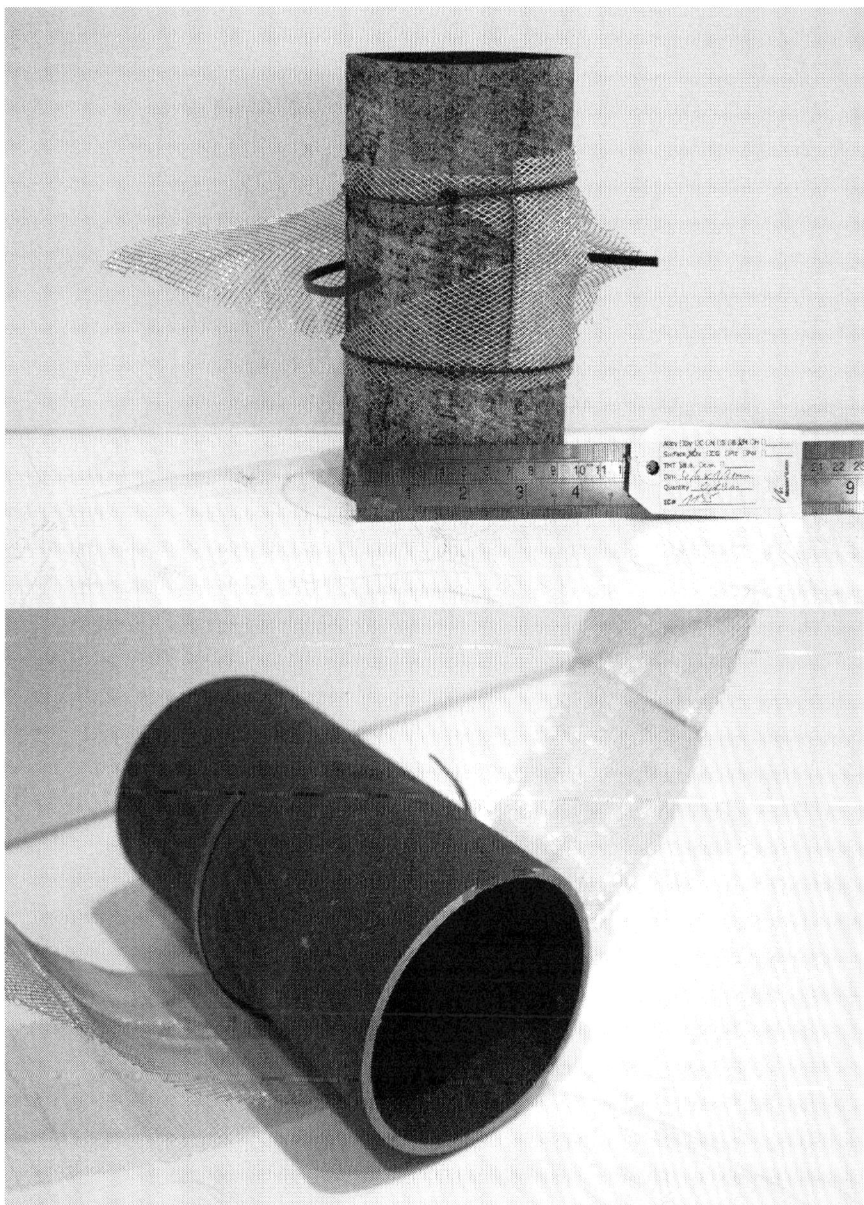

Figure 9. Shape memory alloy and mandrel following its removal from the kiln, after the cooling process.

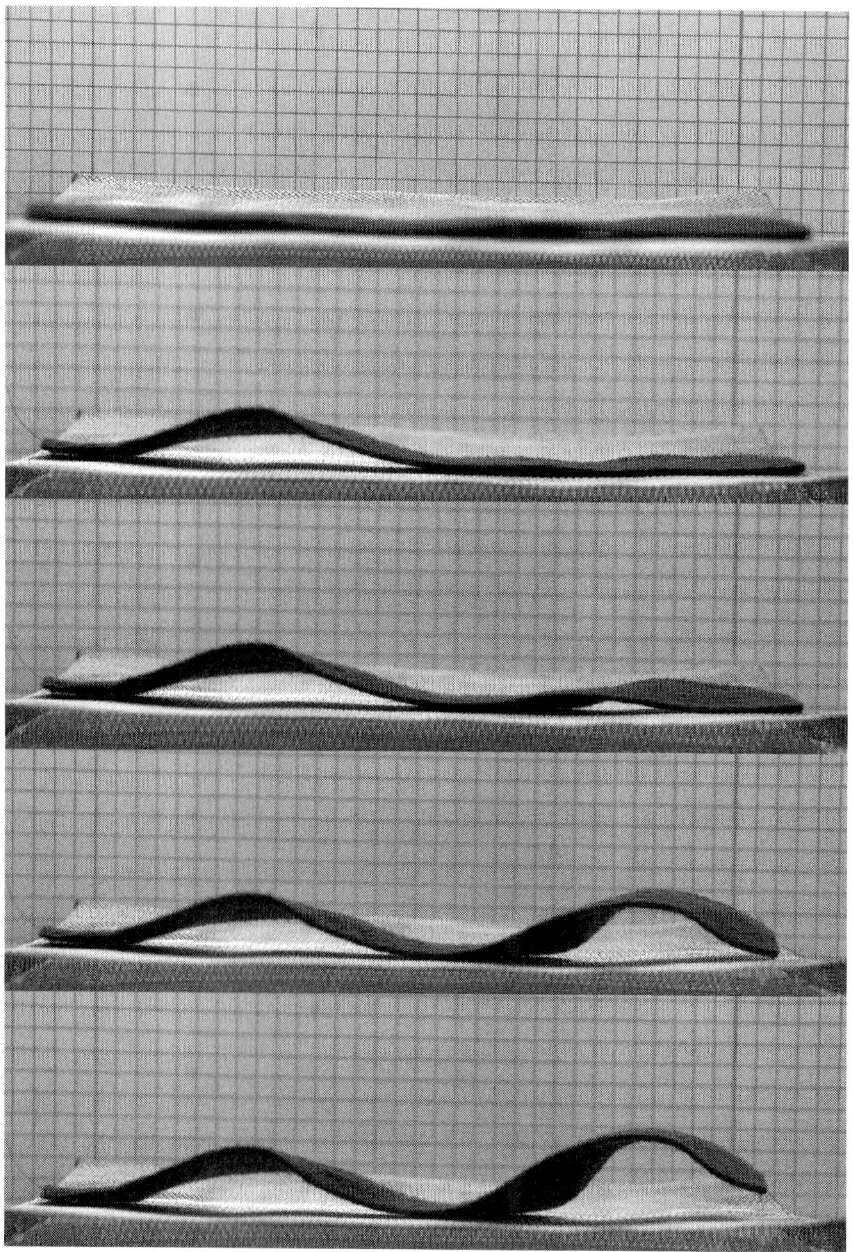

Figure 10. Sequential photograms featuring the actuation of the shape memory alloy ribbon sewed onto a piece of felt.

Subsequent experiments focused on demonstrating quite simplistically the actuation abilities of our SMAs. We first wrapped the SMAs ribbons with a piece of felt to better illustrate the resulting shape change. The outcome, featured in Figure 10, was exactly what we expected, with the SMA achieving the 'memorized' shape quite rapidly. The experiment that followed tested the behaviour of the actuator under the tension exerted by a thin fabric membrane. The fabric was anchored to a circular frame installed on a planar surface with a heat-gun pointing straight downwards onto the alloy. The ribbon curved consistently, pulling the thin membrane up steadily, as soon as the actuation temperature was reached. Figure 11 illustrates the results. While these were both very simplistic experiments, they guided our research towards quantifying the actuation force of our SMAs when embedded in a fibre composite.

Figure 11. Sequential photograms featuring the actuation of the shape memory alloy sewed to a thin fabric membrane.

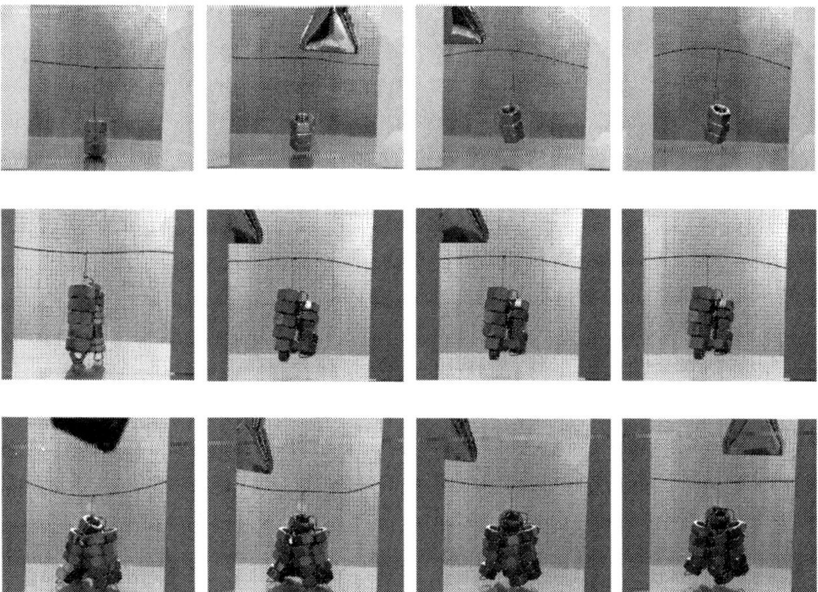

Figure 12. Sequential photograms featuring the load lifting tests undertaken.

The ribbon was this time set up as a simply supported beam under sequential loading increments, the aim being to quantify the SMAs actuation force against time (Figure 12). Whenever the load was increased, the actuation time was measured. Each load increment was quantified as 0.56 N, i.e. the weight of the nuts used. Figure 13 shows the load-time relationship when actuation occurred, lifting up the ribbon and the imposed weights. Experiments proved that the ribbon under testing could lift a load of 8.83N, which equates to 100 times its self-weight. The time taken to lift the first nut was 8 seconds, while lifting the group of nuts weighting 8.83N was 18 seconds. Assuming the 8 seconds for the first lift is the time it takes the ribbon to achieve the actuation temperature, the actual actuation time to lift 8.83N was 10 seconds. This series of experiments provided us with substantial information on the behaviour of the SMAs and particularly, how much force they could exert and how rapidly.

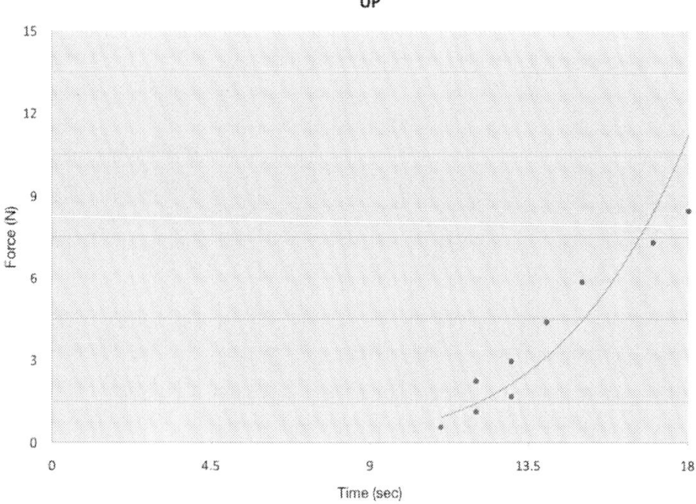

Figure 13. Graph featuring the load lifting capacity of the ribbon, against the time taken to actuate.

The next step was to calibrate the curvature change of the SMAs in relation to temperature. We had purchased ribbons with a theoretical actuation temperature of 65°C. However, we knew that the behaviour of the alloy and ultimately its actuation temperature could have been altered during the shape training process. A simple experiment using a thermocouple and a unit controlling the temperature of the alloy served to measure the curvature of the alloy at different temperature increments (Figure 14). The experiment demonstrated that the shape change did not occur instantaneously; instead, it was a steady process that started at 38°C, with the alloy reaching its maximum curvature at 58°C, as supposed to its theoretical actuation temperature of 65°C. It was clear that the training regime of the alloy had an impact on its later behaviour. The high curvature of the trained shape could have also altered the resulting actuation performance. This test also provided us with key data regarding the maximum curvature change that the alloy could achieve when setup freely. The outcome of this calibration exercise served to feed the construction of the final prototype.

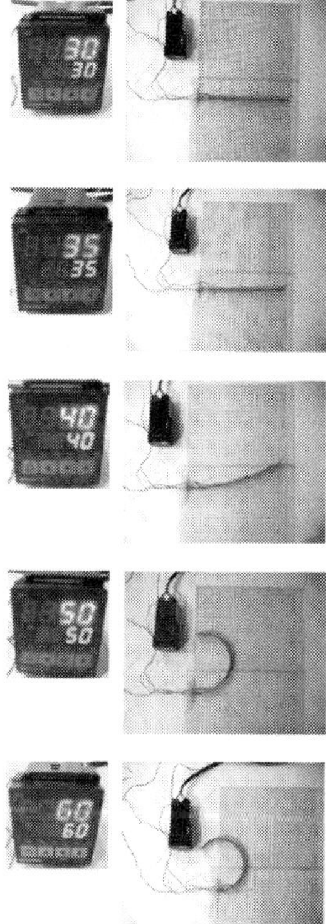

Figure 14. Series of photograms featuring the calibration of the shape memory alloy ribbon. Actuation temperature and curvature changes were measured and subsequently registered to inform the building of the final prototype.

5. Composite: Building the material system

Further experiments followed the preliminary tests described above, contributing to the refinement of the final model. These were aimed to test the elasticity of different polymers, from highly deformable to more rigid mixtures. Finding a polymer with an appropriate elasticity modulus was key to achieve the intended degree of actuation in the fibre composite material. Figure 15 features a failed setup using a silicon-based composite, which resulted extremely flexible and heavy to be anyhow actuated by four SMA ribbons. The test did serve however to guide the research towards the use of polymers with higher Young's modulus.

Our final experimental model consisted of a sandwich structure made of two layers of a glass fibre mat bonded by an epoxy-based resin. A total of four SMA alloys were embedded in between the two layers of glass fibres. Figure 16 features a diagram illustrating the set up of the model.

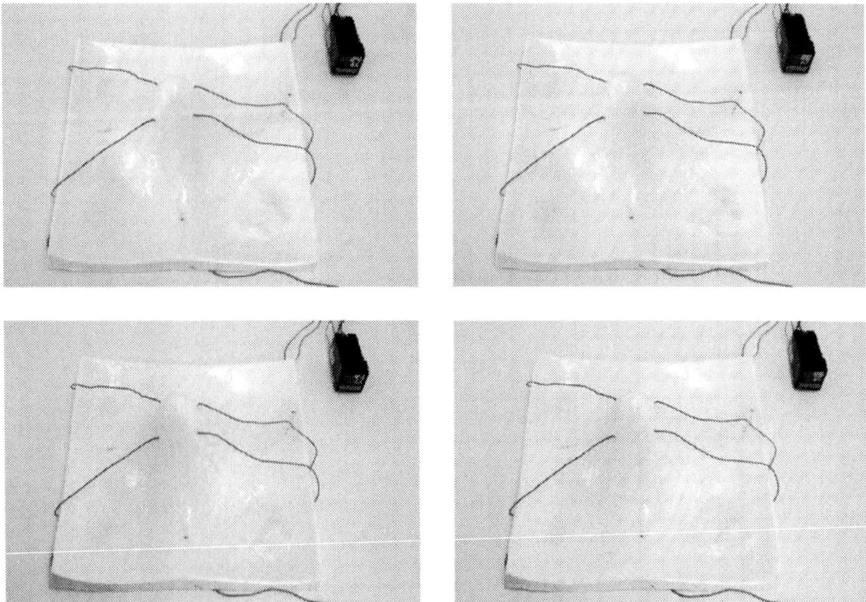

Figure 15. Sequential photograms featuring the failed experiment using a silicon-based composite.

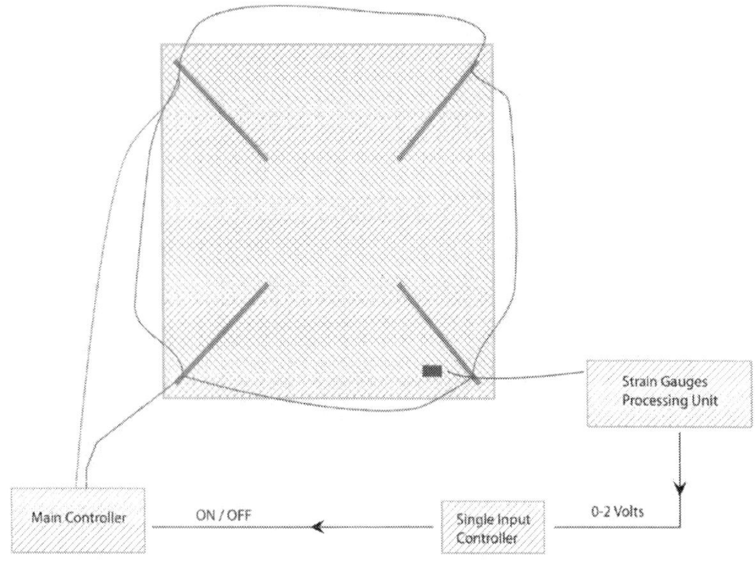

Figure 16. Diagram featuring the arrangement followed for the assembly of the final prototype.

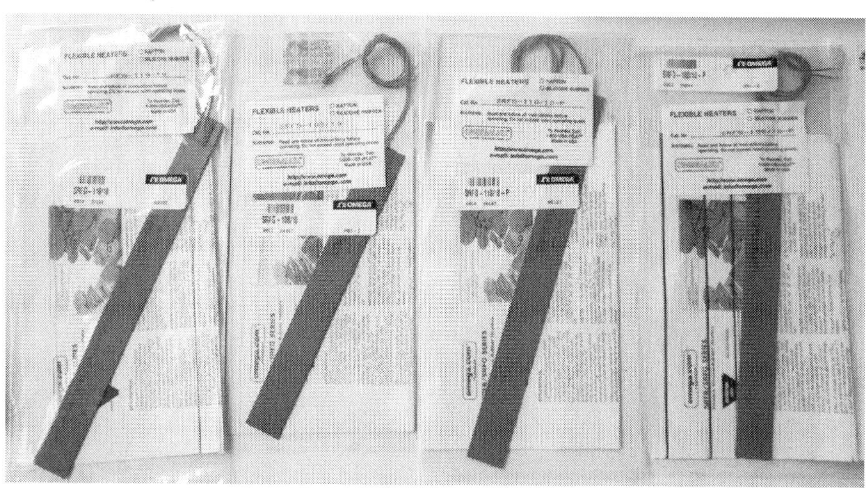

Figure 17. Silicone heating patches used in the experiments. Advantages of the use of these patches are their lightweight, thin and flexible structure, while being able to heat up to 232°C.

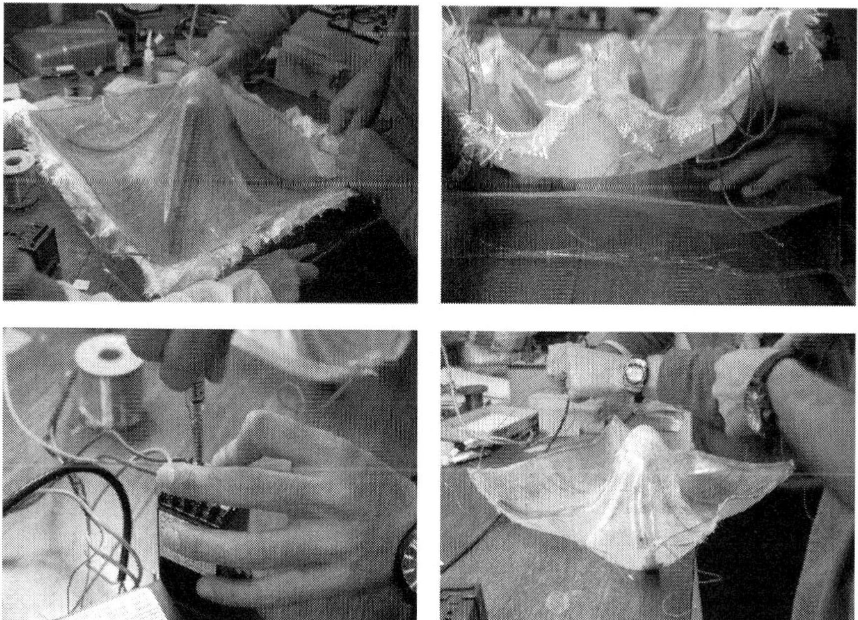

Figure 18. Pictures illustrating the construction of the final prototype.

The sensing capabilities of the model were based on an active-dummy method (also known as Wheatstone's bridge) that allowed measurement of real-time changes in strain on the parent shell structure. Two gauges were then connected to a processing unit, which translated the strain signal into a voltage output of 2 volts, when elongation was detected, and 0 volts, once it ceased. This voltage was then sent to the controller unit, which turned on and off a set of silicon heating strips upon which a shape memory alloy wire was glued (Figure 17). A set of thermocouples were used to measure the temperature of the shape memory alloys as the silicon heating strips commenced to heat them. The processing unit to which the thermocouples were connected would automatically stop the heating process once the actuation temperature of the actuators was reached (Figures 18, 19 and 20).

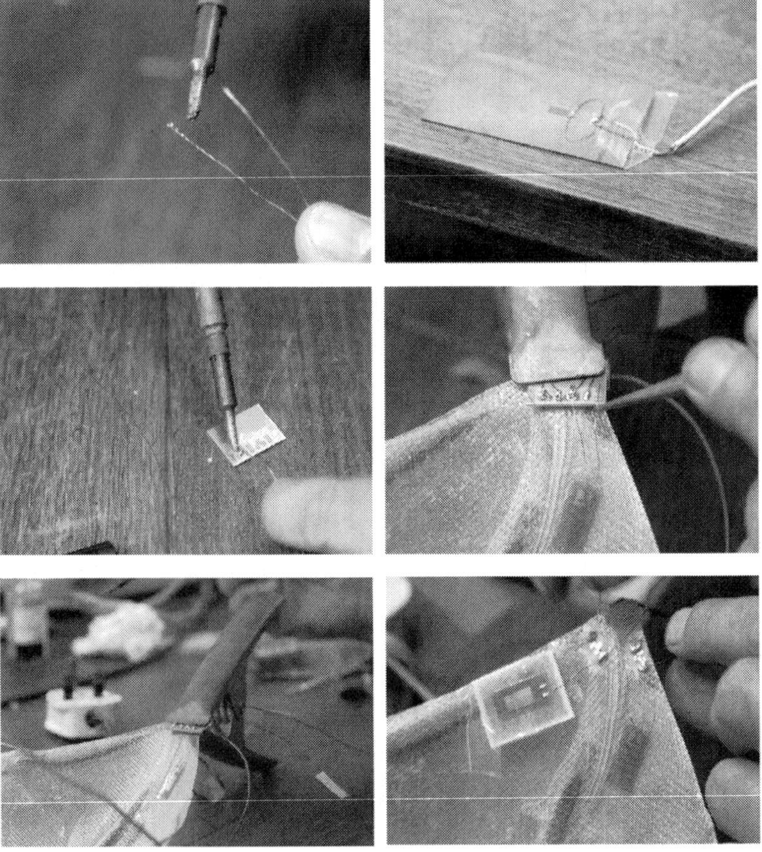

Figure 19. Pictures illustrating the soldering and preparation of the strain gauges on the final prototype.

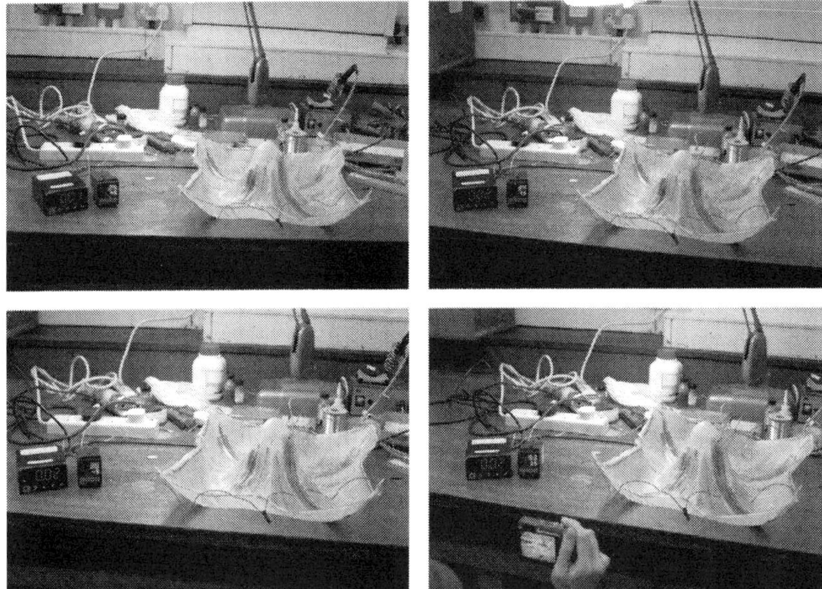

Figure 20. Series of pictures showing preliminary tests of the actuation in the laboratory.

Alternative methods to the use of silicon heating strips could have been piezoelectric fibres, which can generate electric potential in response to applied mechanical stress. The implementation of piezoelectric fibres in our setup could have potentially avoided the need for the strain gauges, the processing and control units, as well as, the heating strips.

Figure 21 features the set up of the final prototype, including the strain gauges processing unit and the different controlling devices. Figure 22 shows the actuation response of the prototype at the temperatures of 30°C, 42°C and 58°C. At 30°C the shape of the prototype has not yet been altered; it is therefore the initial shape which serves as reference of the non-actuated model. At 42°C, the strain gauges detected actuation had commenced. From previous tests, we knew that if setup freely, the alloys would start actuation at 38°C. However, the alloys were now embedded in a fibre composite, which contributed to an increase in stiffness. It was more difficult for the alloy now to start curving and as a result, to spread that deformation to the material it was embedded in. This caused a delay on the actuation of 4°C when compared to the previously tested model where the SMA did not experience external material constraints. At 58°C the prototype was about to reach its highest curvature. The shape change can be easily compared thanks to the photogram made by superimposing the actuation stages described above.

6. Architecture with smart composites

On an architectural scale, the aim was to introduce actuation not through and external heat source but by utilizing the diurnal temperature variation in the environment.

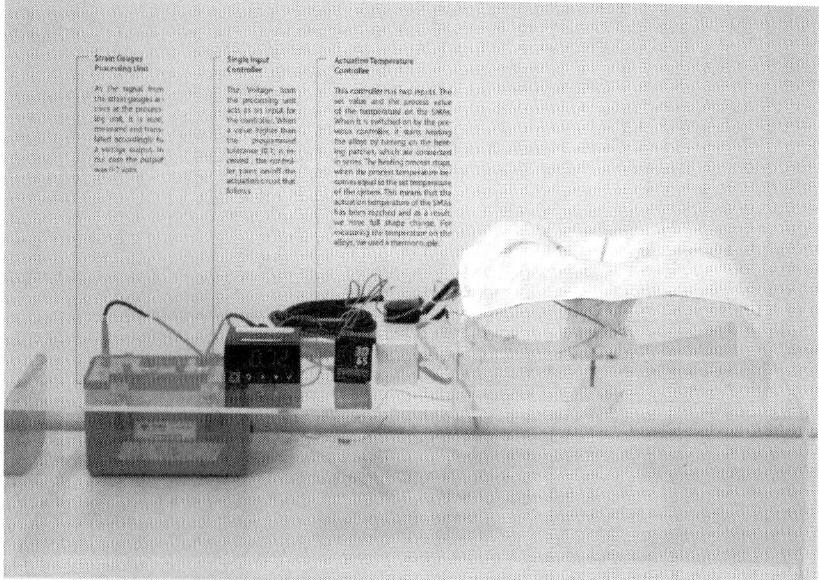

Figure 21. Final setup of the prototype including the strain gauges processing unit, the single input controller, the actuation temperature controller and the final prototype.

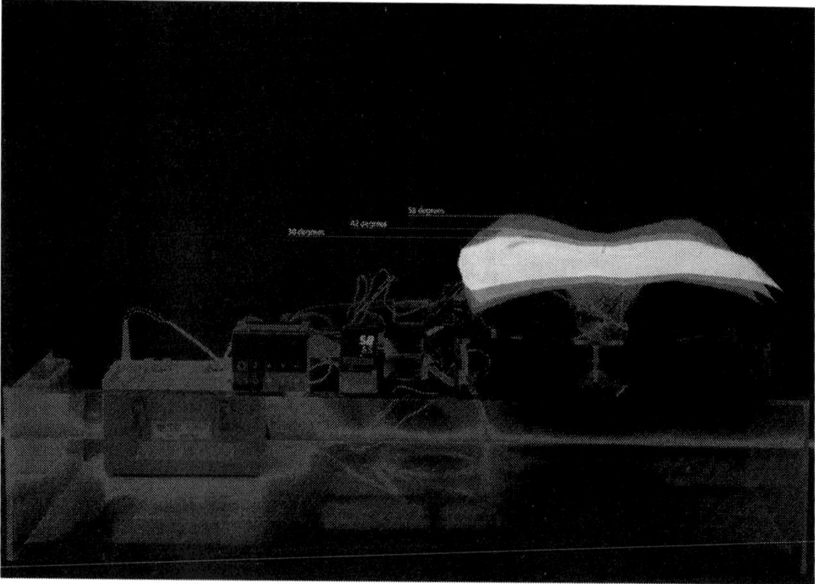

Figure 22. Superimposed photograms featuring three resulting shapes of the actuation at the indicated temperatures.

Figure 23. A view of the shell with the actuating fenestrations

Figure 24. An interior view from the shell

Hot and dry climatic zones which have a considerable diurnal variation in temperature would be the most appropriate to exploit the environmental energy for the efficient functioning of the adaptive system. While actuation temperatures of commercially available shape memory alloys range between 30°C and 95°C, the specific alloys used for the experimental setup had actuation temperatures ranging from 35°C to 65°C. In hot and dry climatic zones the variation in atmospheric temperature is measured to range from 22°C to 44°C in the summer months. Surface temperature of the alloys would therefore easily reach the required 35°C which would initiate their actuation.

Self-actuation potential of any structure invariably necessitates integrating decision-making abilities and thus intelligent behaviour into the adaptive system. The morphological definition of the structure plays a key role in adaptation and has to be coherent with the actuation logic. The material tests and experiments conducted, establish the premise for the architectural application of the self-actuating fibre composite system developed herein. The study branches further into the utilization of such a self-adaptive material system in an architectural application, with reference to context specific climatic data (Figures 23 and 24).

The image shows a large span fibre composite shell structure with actuating fenestrations. The shape memory alloys embedded in the fenestrations allow the structure to open and close based on the external environmental conditions. The opening are strategically positioned in a manner in which they do not affect the structural stability at the same time enhance the interior lighting and wind flow pattern.

Author details

Maria Mingallon
ARUP,
McGill University, School of Architecture, Canada

Sakthivel Ramaswamy
KRR Group, India

Acknowledgement

We would like to thank our tutors; Mike Weinstock for his involvement and invaluable guidance, and Michael Hensel for his help and encouragement. George Jeronimidis for his immense help in developing our research and in performing the physical experiments. His research on biomimetics is the foundation for this project. We would also like to thank Stylianos Dritsas, for his essential input in evolving the geometry. Professors Kevin Kuang and W.J. Cantwell for sharing their knowledge and research on shape memory alloys and fibre optics. We would finally like to thank our families and friends for their constant support and invaluable encouragement.

7. References

[1] Jeronimidis, George, 'Biomimetics - Differentiation / Integration / Emergence, Sensing – Actuation –Control', - Lecture at the Architectural Association, London, 2009.

[2] Smith,S. A, 'Shape Setting Nitinol', - Proceedings of the Materials and Processes for Medical Devices Conference, pp 266 – 270. edited by S. Shrivastava, ASM International, Sept., 2003.

Chapter 16

Composite Material and Optical Fibres

Antonio C. de Oliveira and Ligia S. de Oliveira

Additional information is available at the end of the chapter

http://dx.doi.org/10.5772/76664

1. Introduction

For ease of handling and polishing, optical fibres are generally mounted in some form of rigid structure. A schematic of a typical fibre termination used in astronomical instruments (typical of multiple-fibre spectrographs) is show in Fig. 1. The fibre is first placed within a flexible tube (polyimide or similar), often referred to as the strain relief tube. The fibre and tube are placed within a rigid ferrule. Adhesive is applied to the fibre and tubing to fix them in place. The ferrule can then be easily manipulated for polishing, or mounting within an instrument, without risk of damage to the fibre. The role of the strain relief tube is to prevent stresses occurring at the point where the fibre enters the ferrule given bending at this point may lead to breakages. For coupling an array of optical fibres to a microlens array, such as in IFU (Integral Field Unity), a brass plate with an array of drilled microholes may be used. Optical fibres positioned in an array of accurately drilled holes, as illustrated schematically in Fig. 2.

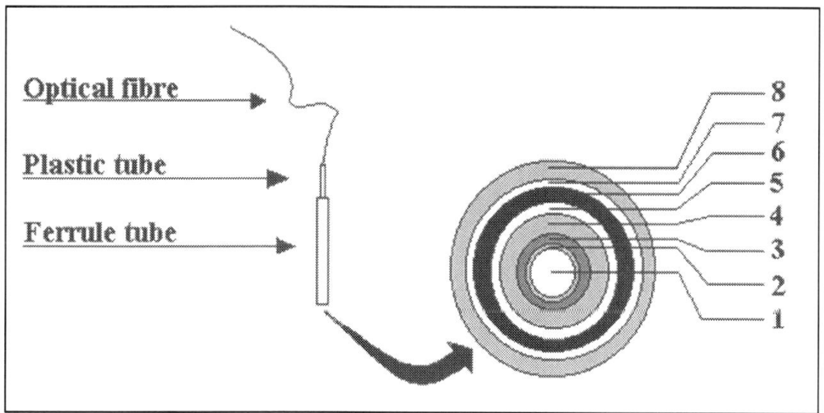

Figure 1. Schematic diagram of a single fibre mounting assembly: 1, core; 2, cladding; 3, polyamide buffer; 4, acrylate buffer; 5, epoxy; 6, plastic tube; 7, epoxy; 8, ferrule steel tube.

This system, called microholes array, contains a grid of holes spaced by the pitch of the microlens array. The holes are machined using custom made drills with two different diameters. This produces a stepped hole, with the smaller diameter hole used for fibre positioning while the larger hole is used to accommodate a ferrule. The small holes are approximately 10 μm larger than the fibre diameter to allow sufficient space for a glue to penetrate. Using a stepped hole also allows a greater depth of material to be machined than by using a small drill alone. This permits a thicker, hence more robust, piece of material to be used. A support plate is also used, positioned above the fibre-positioning array with spacers, to maintain accurate angular alignment of the ferrules with respect to the microlens optical axis. Each ferrule contains a polyimide strain relief tube to prevent mechanical stress, occurring at the point where the fibre enters the ferrule. To secure the fibres, ferrules, and polyamide tubes in place the whole input assembly is immersed in a container of EPOTEK 301-2 adhesive. This epoxy is a natural choice due to its excellent wicking properties and low shrinkage upon curing. After curing, which takes approximately three days at room temperature; any excess glue is removed prior to optical polishing.

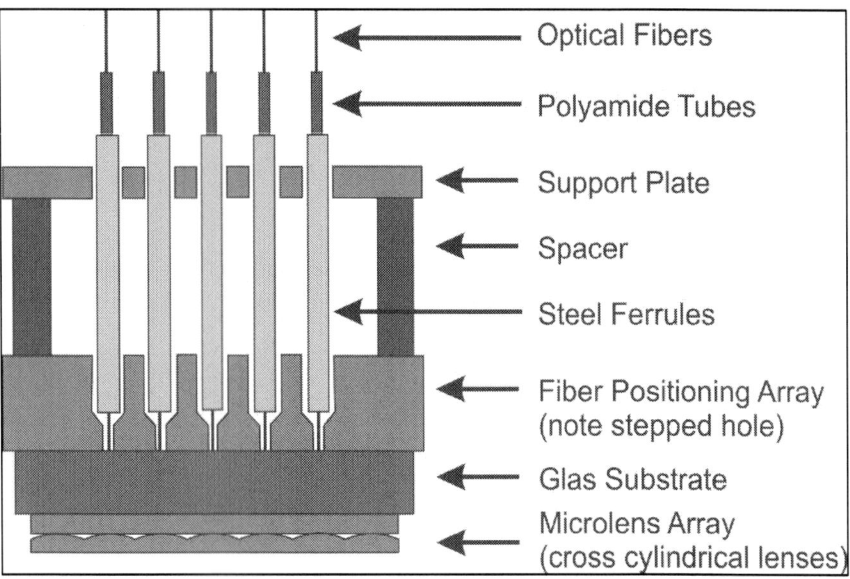

Figure 2. Schematic micro lens array and fibre positioning array used to constructed IFUs

The problem here may be an increase of FRD (Focal Ratio Degradation) caused by contraction of the metal ferrule or brass plate at low temperature causing stress on the fibres and consequent loss of throughput. Astronomic instruments like that in general work in environments with significant thermal gradients, a common characteristic of ground-based observatories. An interesting alternative to the conventional steel ferrule may be a quartz tube. Quartz material has no problems of contraction in the temperature gradients experienced in that places, -10 °C to 20 °C, but is very expensive and difficult to obtain. The ideal condition requires a material with elasticity controlled so as not to cause stress or shift the positioning of optical fibre under temperature gradients. For just such purposes, we have developed a special composite formed from a mixture of EPO-TEK 301-2 and some refractory material oxide in nano-particle form, cured and submitted to a customized thermal treatment. To avoid bubbles and points of stress, this mixture is prepared in a separate receptacle inside a vacuum chamber. The resulting material is more resistant and harder than EPO-TEK 301-2 and is found to be well suited to the fabrication of optical fibre arrays. An important secondary characteristic is the ease with which it can be polished. This feature is a result of the micro particles, which keep the polished surface very homogeneous during the final polishing procedure. The resulting composite combines the beneficial characteristics of both the epoxy and the oxide; main factor its coefficient of thermal expansion is significantly lower than simple solidified epoxy; the exact value depending on the relative concentrations. While the characteristics of this particular composite are still under study, it is clearly possible deploy this material in the construction of devices for several fibre instruments.

2. New materials to support optical fibres

Similar microholes arrays and the support plates of the Fig. 2, used to construct Eucalyptus IFU, were made with toolmakers brass (de Oliveira et al., 2002). The problem here is that differential expansion between the metal array and the glass microlens substrate may lead to the bond between them failing at low temperatures. Although the coefficient of thermal expansion of epoxy is much greater than those of steel, brass or glass, the elasticity of the epoxy accommodates the dimensional changes without breakage: however, this can introduce a small amount of stress build-up.

It is well known that mechanical deformation causes focal ratio degradation (FRD) by the formation of microbends in the fibre (Clayton 1989). FRD is a non-conservation of *étendue* such that the focal ratio is broadened by propagation in the fibre. When mounting the fibre, the appropriate epoxy and tubing should be selected, and general care must be taken to minimize mechanical stress and avoid additional FRD (de Oliveira et al., 2005). This is straightforward at room temperature, but greater care must be taken in the choice of materials for use at low temperatures. When the fibre assembly is cooled to temperatures around -10 °C, the epoxy, tubing and ferrule will all shrink differentially, and this may cause the level of FRD to increase. There is other problem when the fibre assembly is warmed to around 20 °C and cooled to around -10 °C. The UV epoxy, currently used used to cement the metal and the glass may be damaged if the system will be submitted several times at big changes of temperature. In this case the system may detach in places due the thermal gradient.

Microholes array and support plate made with solidified EPOTEK represent a first step to change the metallic base for a polymeric base. The choice of the EPOTEK 301-2 is appropriate due to its excellent wicking properties and low shrinkage upon curing. This properties are very good to use in optical fibres system, so that, if it is possible to get plates adequate to machine with this epoxy we can get total compatible in the construction of the system.

2.1. Epoxy solidified

The epoxy EPO-TEK 301-2 has low viscosity and requires a container to constrain its flow until it is solidified. Generally aluminium has been used to make these containers but it is possible to use brass, plastic or acrylic. The complete curing process takes approximately three days at room temperature and when it is dry, it is transparent to visible light. To avoid bubbles and points of stress, the epoxy is prepared in a separate container inside a vacuum chamber. The correct amount is allowed to set in the container, which is placed inside a dry environment. Once cured, some thermal treatment may be necessary. Several kinds of blocks, cylinders and plates can be made to test the polishing qualities of these test pieces. The results are very encouraging with the hardness similar to acrylic resin. The Fig. 3 and 4 shows steps to obtain samples machined to manufacture blocks of epoxy solidified.

Figure 3. EPO-TEK 301-2 solidified, during the machining procedure

Machining quality is important as burrs inside the microholes may prevent the fibres from being threaded into the holes, or cause stress, or breakage, of the fibres. The quality of such microhole arrays inspected visually using a microscope, give very encouraging results displaying a minimum of burring; scarf, remaining in the holes, may be readily removed by cleaning in an ultrasonic bath.

There are several advantages to the use of solidified epoxy as compared to brass, for example, in the fabrication and use of fibre support devices. Ease of machining and compatibility with other epoxies used to attach glass or silica, may be the most important of these advantages. Although the coefficient of thermal expansion of epoxy is much greater than that of steel or glass, Tab. 1, its elasticity accommodate thermally induced dimensional changes without breakage. This also avoids excessive stress associated with increases in FRD but, in principle, could be deleterious in compromising the critical positioning stability of optical fibres as the temperature varies.

Figure 4. Plate of EPO-TEK 301-2 solidified and machined

Material	α CTE at 20 °C	Units
Brass	19	10^{-6} / °C
Carbon Steel	10.8	10^{-6} / °C
In ox Steel	17.3	10^{-6} / °C
Quartz	0.59	10^{-6} / °C
EPO-TEK 301-2	55 at 61	10^{-6} / °C

Table 1. Coefficient of Thermal Expansion of some materials

Machining quality is important as burrs inside the microholes could prevent the fibres from entering the hole, cause stress, or breakage, of the optical fibres. The quality of the microholes array may be inspected visually using a binocular microscope and the results are often very satisfactory with minimal burring present. Swarf present in the holes is readily removed by cleaning in an ultrasonic bath. The Fig. 5 shows a sample of epoxy solidified with a microholes array.

Figure 5. Photograph of the microholes array in a sample of the EPO-TEK 301-2 solidified

After machined, this sample has a diameter of 48 mm and a thickness of 3 mm. This microholes array is matrix 30x30 holes spaced on a 1.0 mm pitch. The holes were machined using custom made drills with diameters of 0.60 mm and 0.21 mm. This produces a stepped hole, with the smaller diameter hole used for fibre positioning while the larger hole is used to accommodate a ferrule. The small holes are approximately 10 μm larger than the fibre diameter to allow sufficient space for glue penetrates. The machining error in the position of the small holes was measured to be approximately 2 μm.

2.2. Experimental stress analyses

It is possible to use a very simple experiment of Photo elasticity method to evaluate the static stress in the plates of epoxy solidified. Classical two-dimensional photo elasticity is an optical experimental technique for determining stress fields in solids bodies. For a given analysis, polarized light is passed through a transparent sample or the body in question, and stress-induced or static stress changes in the light result in an interference-like pattern, which may be analysed to determine the principal stresses at each point within the body.

The basis for photo elastic measurement is a phenomenon of double refraction (also called artificial birefringence). There are two situations that may cause this phenomenon. Certain plastics exhibit the first situation when the sample of this type is subjected to an applied load, the resulting stress /strain field causes the molecules within the transparent material to have a preferred alignment. The second situation is exhibited by certain epoxies after dried and in this case the stress may be called static stress. In both situations the light wave vibrations have two preferred directions within the material and a wave of linearly polarised light entering the field is split into two waves which are linearly polarised at right angles to each other and which propagate with different velocities. That is, two rays travel along each an original line of propagation, and their electric vectors are mutually perpendicular. In fact, each vibration is collinear with one of the principal stress directions. (See Fig. 6.) Also, since the two waves travel at different velocities, a phase difference develops between then and by using certain optical elements.

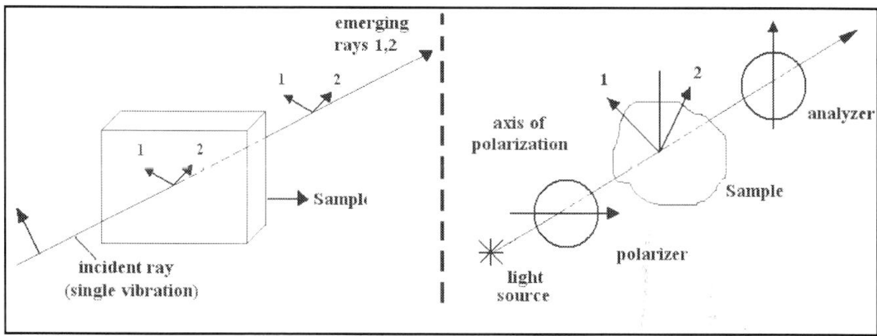

Figure 6. Left: propagation of light through photo elastic models. Right: plane polariscope used to get information of the stress inside of the sample by analyse of the artificial birefringence.

A very simple system to get qualitative images of the samples may be adapted with a transmission Polariscope, as shown in Fig. 6. The system uses a CCD camera to take the images. Stresses within solidified EPO-TEK 301-2 may be investigated with the use of the photo elasticity method whereby stress-induced birefringence is measured using polarized light. Static stress can be recorded as an interference pattern, which may be analysed to determine the principal stresses at each point within the material. Indeed, significant stress induced birefringence is detected in such samples, as is shown in Fig.7 side left.

This stress can be alleviated through thermal shock induced by warming the sample to 80 ºC for 30 min. Fig. 7 side right demonstrates the reduction in stress as the material is returned to room temperature. Of course, such experiments are only viable for transparent materials but they do give a warning that care must be taken in analysing the effects of thermally induced stress through measurement of fibre displacement in arrangements and FRD stress-induced.

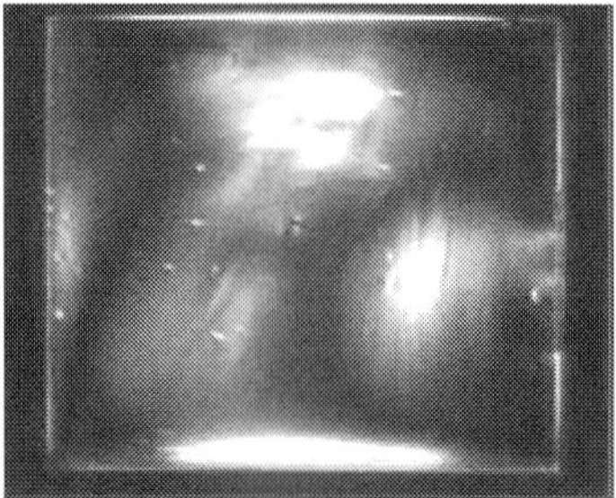 

Figure 7. Left: sample before thermal shock. Right: same sample after thermal shock

2.3. Composite

It is possible create composites using a mix of epoxy and several types of oxides in micro or nano-particle form. To avoid stress points and heterogeneous regions, the composite needs to be prepared using mixers of high speed. Ultrasonic chamber can be useful to ensure more uniformity to the mixture. Before the cure, this composite requires be subjected to a vacuum of 10^{-3} Torr to reduce bubbles inside of material. The mixture of EPO-TEK 301-2 with refractory material oxide in nano-powder, cured and submitted to a thermal treatment around 400 °C, produce a very interesting option instead simple epoxy solidified. The resulting material is more resistant and harder than EPO-TEK 301-2 and is found to be well suited to the fabrication of optical fibre arrays. Several different refractory material oxides in nano-powder may be used to produce different characteristics in this type of composite. So it is possible combine Zirconium oxide, Barium oxide, Silica oxide, Cerium oxide and others, Fig. 8, to obtain a material optimized to specific applications. The solidified mixture combines the beneficial characteristics of both the epoxy and the oxide; main factor its coefficient of thermal expansion is significantly lower than simple solidified epoxy; the exact value depending on the relative concentrations.

There are two important factors that consolidate the structure of this composite. The first is the process of cure of the liquid mixture. The second is the process of heating of the solid material obtained after the cure. The chemical reactions during the first process are limited by the time to reach the complete cure of the epoxy. Anyway, chemical analysis showed no evidence of endothermic or exothermic chemical reactions between oxides and epoxy. In fact, the materials involved in the mixture appear quite neutral. However, the heating procedure in temperatures around 400 C with slow cooling during 24 hours induces slight shrinkage on the material. Although the study still lacks depth, it is fairly simple to conclude that the structure undergoes some type of molecular rearrangement with some material loss and subsequent compaction. In fact this process carbonizes the external side of the solid material. To avoid total and destructive carbonization, both, the heating and cooling is done with the composite inside a steel container with refractory sand. After this procedure, the external part carbonized can be removed by machining process leaving the sample completely clean. The material thus obtained proves to be quite stable and resistant even

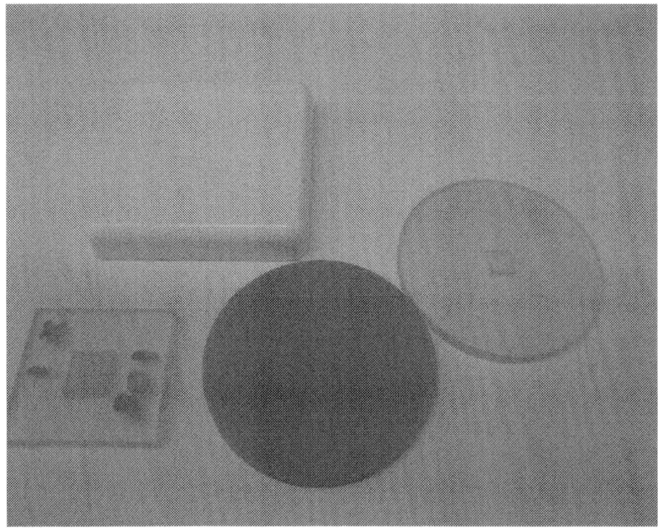

Figure 8. Samples of different composites at the centre and epoxy solidified at the borders

though it has some degree of slow oxidation on its surface. This oxidation is evident from the slight colour change after a few weeks of exposure and manipulation, but still remains a high physical stability.

This composite has two physical characteristics very interesting for the construction of optical fibres holders. The first feature is its ability to sustain their polishing, with minimum quantities of abrasives during this procedure. In other words, when the composite is subjected to a polishing of high performance, the detachment of the refractory oxide nanoparticles reinforces gently the polishing process and increasing the efficiency of this procedure. The surface roughness measured in several samples, after high performance polishing was about 0.01 microns. Furthermore, the time for obtaining a polished surface with this quality is about 10 times less than the time required to polish a surface of brass of the same size.

3. Simple composite ferrule

Mechanical deformation is a change of geometry of the optical fibre away from a straight cylinder. Large-scale bending, or macrobending is where the radius of the curvature of the bend is very large in comparison to the core diameter. On the other hand microbends are deformations of the cylindrical core shape, which are small, compared to the fibre diameter (Ransey 1988). It is well known that mechanical deformation causes FRD by the formations of microbends in the fibre (Clayton 1989). When mounting the fibre, the appropriate epoxy and tubing should be selected, and general care must be taken to minimize mechanical stress and avoid additional FRD. Currently steel ferrules tubes are used to prepare the extremities of the optical fibres for general purposes, in test lab or even as a part of some instrument. Although it is clear that inefficiencies can result in the use of metal ferrules submitted to low temperatures. Ferrules and inserts made with the composite described here, promises to be best option to handle the ends isolated of optical fibres.

3.1. Composite

While there is no direct evidence for the deterioration in Focal Ratio Degradation (FRD) of optical fibres in severe temperature gradients, the fibre ends inserted into metallic containment devices such as steel ferrules can be a source of stress, and hence increased FRD at low temperatures. In such conditions, instruments using optical fibres may suffer some increase in FRD and consequent loss of system throughput when they are working in environments with significant thermal gradients, a common characteristic of ground-based observatories. It is possible to use careful methodologies that give absolute measurements of FRD to quantify the advantages of using epoxy-based composites rather than metals as support structures for the fibre ends. This is shown to be especially important in minimizing thermally induced stresses in the fibre terminations. Furthermore, by impregnating the composites with small cerium oxide particles the composite materials supply their own fine polishing grit, Fig. 9, which aids significantly to the optical quality of the finished product. Different types of inserts are possible, Fig. 10, depending only on the precision of the machining.

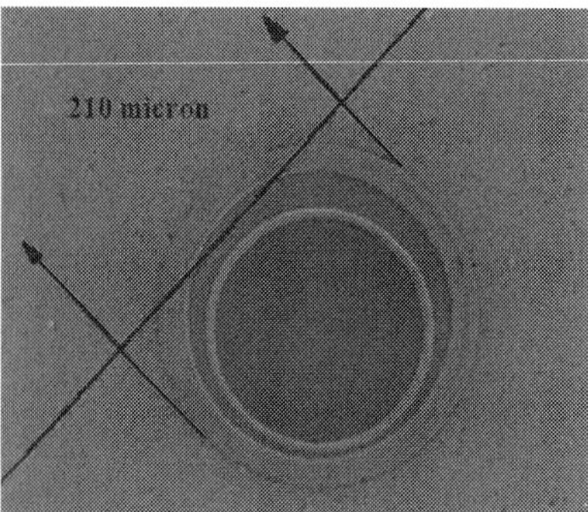

Figure 9. Microscopic photo of optical fibre inserted in a composite ferrule, after polishing procedure.

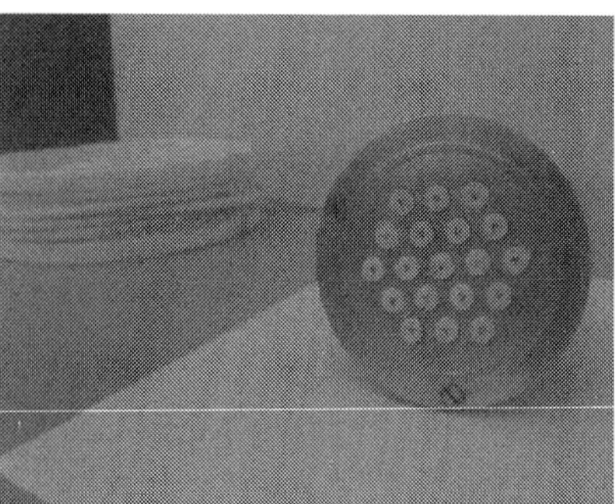

Figure 10. Inserts with optical fibres to be used in a fibres collector plate. Each insert can have several fibres.

4. Microholes arrays using plates of composite

A system like that shown in section 1, Fig. 2, presented a problem in the past: This problem was the terrific facility to detach the glass substrate of the metal brass polished. Variations of temperature at long of time cause different expansion in the metal brass and the glass. After some time, the UV epoxy normally used to glue the microlens arrays with the microholes array, cannot support more the bonding between the metal brass plates due to the successive expansions and contractions caused by temperature variations. Experiments using plates made with epoxy solidified, Fig. 11 can resolve this kind of problem in the range of temperatures between –10 ºC and 22 ºC, typical of high altitude, ground-based observatories. Although the coefficient of thermal expansion of epoxy, around 60 x 10-6 in/in/°C, is much greater than that of brass metal, steel metal or glass, its elasticity accommodates dimensional changes thermally induced. This means less pressure on the optical fibre and consequently avoids increases in FRD associated with stress but, in principle, could be deleterious in compromising the critical positioning of fibres as the temperature varies. The ideal condition requires a material with elasticity controlled so as not to cause stress or shift the positioning of optical fibre under temperature gradients.

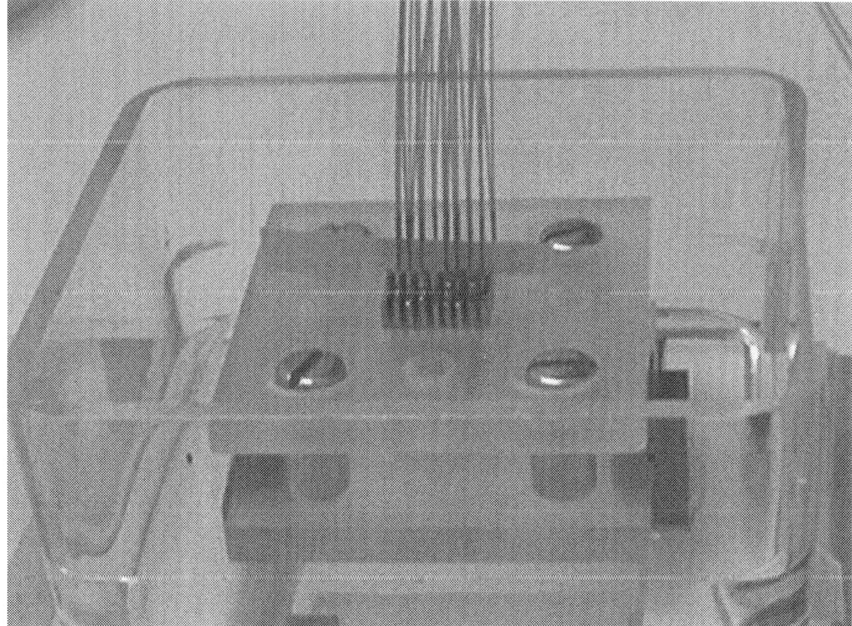

Figure 11. Microholes array device being prepared to be the input array of an IFU system

Notwithstanding the characteristics of this particular composite are still under study, this material was used successfully in the construction of devices for several fibres instruments. For example, we have used this composite to construct SIFS/IFU for the SOAR telescope in

Chile, (de Oliveira et al. 2010) and FRODOspec/IFU for the Liverpool Telescope, (Macanhan et al. 2006).

4.1. Construction of microholes arrays systems

To replace the brass metal or epoxy solidified and resolve the problems presented by both materials, we have used our composite to manufacture the parts of the microholes array device. The material composite obtained is less stressed and harder than EPOTEK 301-2 being a good choice to be used in optical fibres arrays. This material certainly has a combination of the characteristics from the epoxy and the refractory material oxides. The most important consequence of this combination is a coefficient of thermal expansion hither than metal brass but shorter than a simple epoxy solidified. The exactly number will depend of the relative concentration between the refractory material oxides and epoxy.

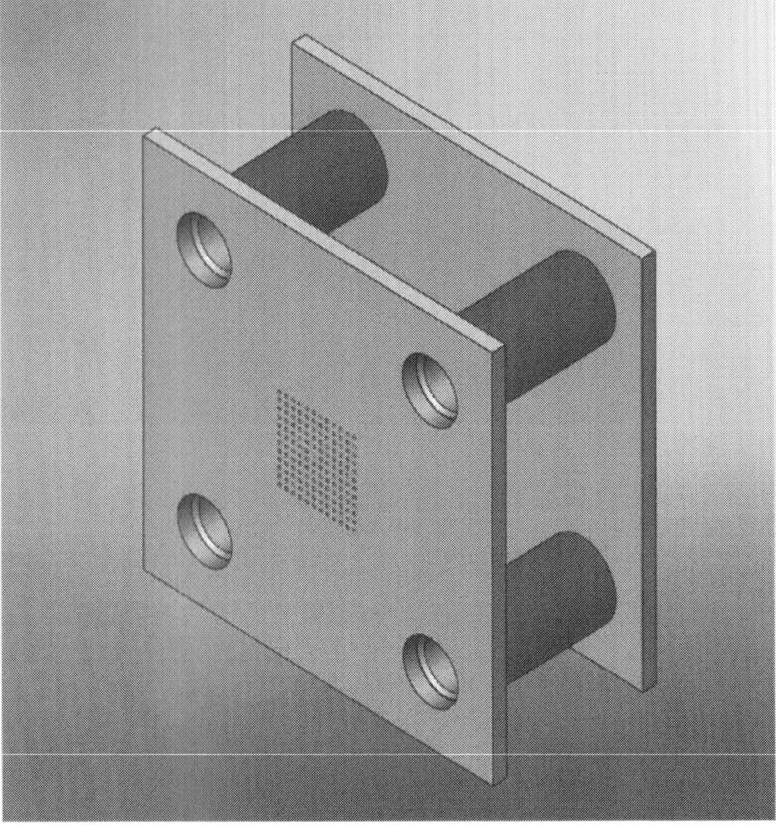

Figure 12. Schematic of the composite plates set to build the entrance device of lenslet IFU

In general, the schematic shown in Fig. 12 is the base of the entrance device of the lenslet IFU system. This device is much easier to be manufactured than the device described in section 1, Fig. 2. In fact, this new version does not require any precision in the holes confection on the composite plates. To obtain precision with the fibres position we have used a third plate called mask of precision, Fig. 13. This is a metal mask very thin obtained by a technique called electro formation. The mask obtained by this way may be configured to have holes with specifics diameters and pits, with error around 1 micron in the diameter and in the position of the holes. This technique may produce a metal nickel plate with 200 microns of thickness and the procedure is very cheap. Taking in account these facilities; the mask will define the precision of the fibres array. It is possible to obtain micro holes with the diameter exactly one or two microns larger than the diameter of the fibre used. For another hand, the diameter of the holes in the composite plates does not need to have any precision and may be much larger than the diameter of the fibre. Since that, the step holes with different diameters in the composite plates it is not more necessary, also will be not necessary to use ferrules and any kind of protection to the fibres. Eventually a device like as shown in the Fig. 13 need to be made under a microscope because the diameter of the fibre may be much small and the number of fibres involved at the assemble may be high.

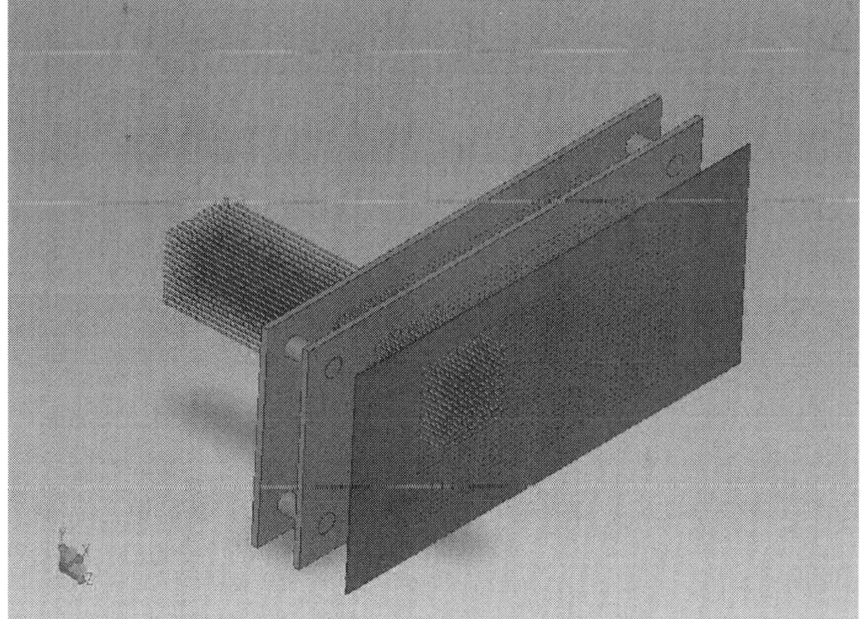

Figure 13. Entrance device during the assembling step, where the matrixes of holes in the composite plate set and in the precision mask are populated with the optical fibres terminations.

After assembled, the precision mask is glued against the composite plate and all set is immersed in EPOTEK 301-2 following the old procedure. To obtain the maximum throughput the surface of the fibres should be polished such that they are optically flat. This is the condition to attach the microlens array against the composite plate, Fig. 14 and Fig.15.

Figure 14. SIFS/IFU microlens glued

Figure 15. FRODOS IFU microlens glued

The pre-polishing process starts with the removal of excess glue with 2000 grit emery paper. Initial lapping with 6 μm diamond slurry on a copper plate and a second lapping with 1 μm diamond slurry on a tin-lead plate is used until the complete removal of the precision mask. Without the metal mask, the material of the composite plate is self-abrasive enough to produce a polishing of high performance of the optical fibres on a chemical cloth. This procedure is a basic condition to attach the microlens array against the composite plate of fibre terminations.

5. Characterization

The complete characterization of this composite may require several kinds of possible tests. However, applications with optical fibres in metrology involve analyses of displacement when the device is submitted at thermal gradients. More specifically, optical fibres arrays used in astronomic instruments need to resist low temperatures without displacement of the fibre position and without delamination problem between parts. Simples experiments show that the linear CTE assumes values between 20 and 40 x 10-6 / °C to 0 °C depending the concentration of the components. For example, a sample made with EPO-TEK 301-2, Barium oxide, Zircon oxide and Cerium oxide with proportions respectively 5:1:1:1, exhibits an α CTE around 30 x 10^{-6} 1/°C.

Analysis of the Absolut Transmission in samples shows clearly that optical fibres inserted in brass or steel ferrules suffer increases in FRD when submitted to low temperatures. On the hand, the FRD increase in optical fibres inserted in ferrule made from EPO-TEK or in composite materials is minimised when submitted at the same negative variation of temperature. In fact, the result predicts a loss of around 10 per cent for the brass ferrules and around 3 per cent for the steel. Although it is clear that inefficiencies can result in the use of metal ferrules submitted to low temperatures, the losses are not easily quantifiable. The reason for this is that the ratio of the outer diameter of the fibre and the inside diameter of the ferrule defines the amount of epoxy between the ferrule and fibre. In the final analysis, this represents more or less compression in the fibre when the metal is compressed during the reduction of temperature.

5.1. Tests on fibres in an array

It is possible to do an experiment to observe the displacement of the fibres in an array of fibres constructed in the plates described in the section 4.2. An experimental array of 10 x 8 optical fibres is chosen as a representative test since it matches the base of the input array of IFUs as described before. A displacement is likely to occur with variations in temperature since the support material may suffer from some type of mechanical distortion. The experimental assembly consists of a support to hold the input array and to control thermal dissipation. A relay lens is used to project the image of the fibre array onto a CCD as a shown in the Fig. 16. The support to hold the array under examination (Fig. 17 and 18) is made of brass and had a canal for the introduction of liquid nitrogen. A continuous flux of dry nitrogen needs to be directed towards the surface of the input array to avoid condensation. Four small temperature sensors are cemented inside holes in the tested plate, close to the optical fibres. These sensors are necessary in order to test if the temperature along the plate reach thermalized state. A digital thermometer can be used to collect information from the sensors.

In our experiment, another sensor was installed inside the brass support together with a special electrical resistor to allow for temperature control. The temperature of the input array was controlled over a range between 23 ºC and -10 ºC. Images of the illuminated optical fibres can be obtained for a set of temperatures within the allocated range. With these images it is possible to obtain information regarding the change in the position of each fibre in the array. A simple algorithm may be used to process these results.

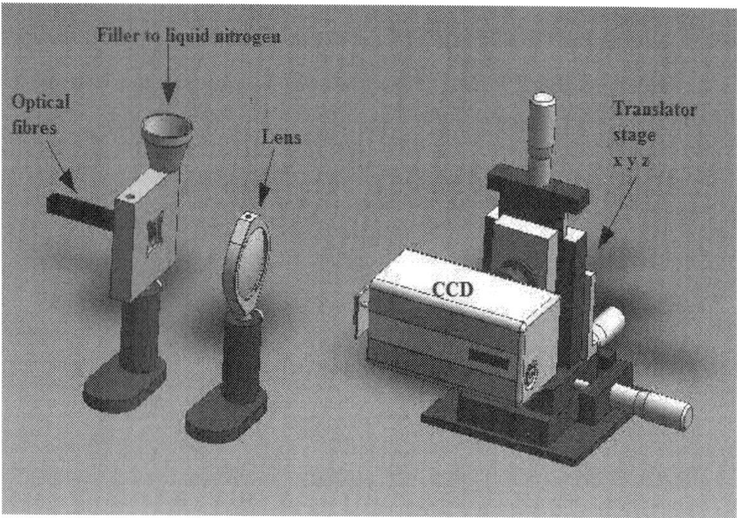

Figure 16. Diagram of the experimental set up to take images from the optical fibre array. The CCD is installed in a translation stage to put the image of the optical fibre illuminated exactly in the centre of the CCD plate. The holder support is used to keep the fibre plate array fixed and to control thermal dissipation.

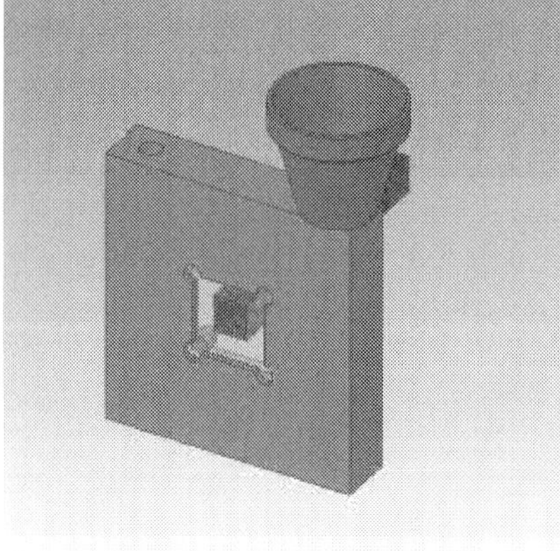

Figure 17. Schematic diagram of the holder where is fixed the fibre plate array.

5.2. Analyses of displacements of fibres in the array

An image analysis by software then determines the centroid of each bright spotlight projected by each fibre from the array on the CCD. Several images like the sample shown in the Fig. 19 are used to determine an average value in the position of each bright spot light. This associates position vectors connecting each bright spotlight with the origin. The first

Figure 18. Schematic diagram of the holder showing the canal to flow the liquid nitrogen during the procedure to decrease the temperature. The holder is made of brass.

fibre at the top/left is used as a position reference. The variations of vector's modulus during the temperature gradient are computed to produce a graph with sub pixel precision (Neal et al. 1997). In this section we present results of tests using fibre arrays submitted to negative temperature gradients. The purpose of these tests is to analyse how much the fibres in the array may be displaced from their original position as a function of the expansion of the material during the change of temperature. Figs. 20, 21 and 22 show the behaviour of arrays with 80 optical fibres when submitted to four temperature gradients. The accuracy calculated for this experiment was less than 0.2μm and the continue curve represents a fitted function.

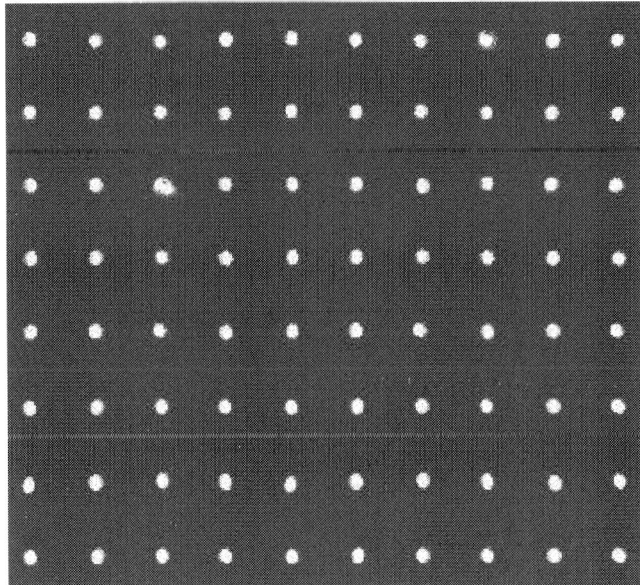

Figure 19. Image of the optical fibres matrix illuminated and projected on the CCD.

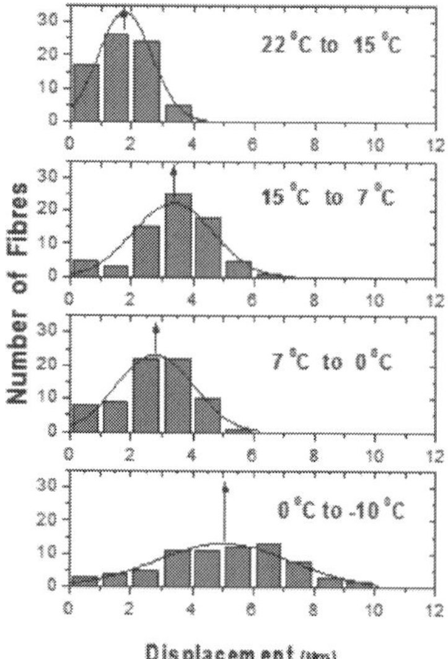

Figure 20. Distribution pattern of the optical fibres array constructed in metal brass plate submitted to four gradients of temperatures, 10 min each.

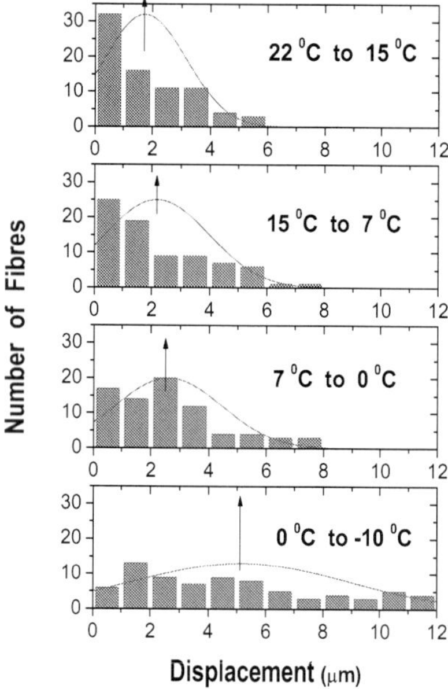

Figure 21. Distribution pattern of the optical fibres array constructed in epoxy EPO-TEK 301-2 submitted to four gradients of temperatures, 10 min each.

The bar graphs, demonstrates the distribution pattern of fibre positions as the temperature declines. It is possible to observe, the expansion of the EPO-TEK 301-2 epoxy material. The change in positions of the fibres amounts to ~12μm as the temperature approaches -10 ºC. The array made of brass almost reaches this value despite having a totally different molecular structure to the epoxy. An interesting result was obtained with the optical fibre array in composite as may be observed in the Fig. 23. The behaviour of the composite array at low temperatures, represented in the bar graphs, is less noticeable than that obtained for the brass and epoxy arrays. In fact, when submitted to -10 ºC the change of the positions at the fibres is less than 6μm. In the experimentation made with composite and brass plate samples, the temperature registered by all sensors on the plate was the same after some minutes and the error expected between the sensors would be around 0.2ºC. However, we noted variations of ~1.2°C between the sensors in the experimentation with EPO-TEK plate. This may be explained by the fact that there are regions with different degrees of stress in the solidified EPO-TEK plate. These differences imply in a possible variation of thermal conductivity along the plate. As was shown in the section 2.3, stresses within solidified EPO-TEK 301-2 can be investigated with the use of the photo elasticity method whereby stress-induced birefringence is measured using polarised light. In fact it is quite common to observe static stress in plates of EPO-TEK solidified.

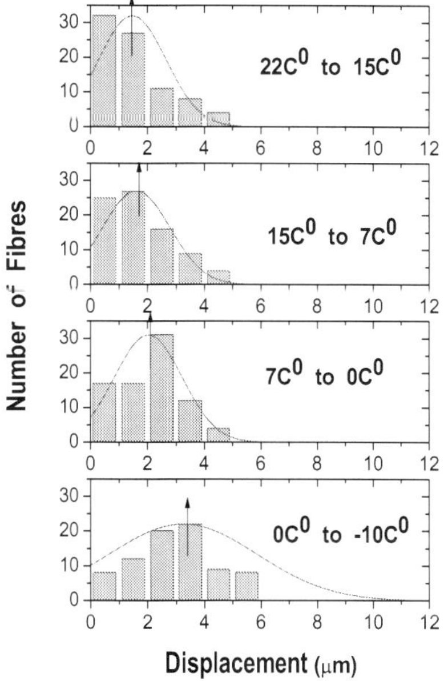

Figure 22. Distribution pattern of the fibres array in composite submitted to 4 gradients of temperatures, 10 min each.

The final conclusion for this experiment is that the composite epoxy material shows significant improvement and, in fact, has an even better performance than the brass or epoxy solidified. The chosen composite material (EPO-TEK 301-2 + zirconium oxide) retains the beneficial bonding properties of the epoxy while avoiding its thermal displacement properties.

5.3. FRD in optical fibres samples

The mode dependent loss mechanisms are the causes of focal ratio degradation (FRD) in optical fibres, and are not often addressed by manufacturers. Mode dependent losses can be divided into two basic mechanisms. The first is waveguide scattering, which causes transfer of energy into loss modes by variations of the core diameter along the length of the fibre. The second is mechanical deformation. Mechanical deformation is a change of the geometry of the fibre away from a straight cylinder. Large scale bending, or macrobendings, is where the radius of curvature of the bend is very large in comparison to the core diameter. On the other hand, microbends are deformations of the cylindrical core shape, which are small, compared to the fibre diameter (Ransey 1988). It is well known that mechanical deformation causes FRD by the formation of microbends in the fibre (Clayton 1989). FRD is a non-conservation of *étendue* (or optical entropy) such that the focal ratio is broadened by propagation in the fibre. When mounting the fibre, the appropriate epoxy and, tubing should be selected and general care must be taken to minimise mechanical stress and avoid additional FRD.

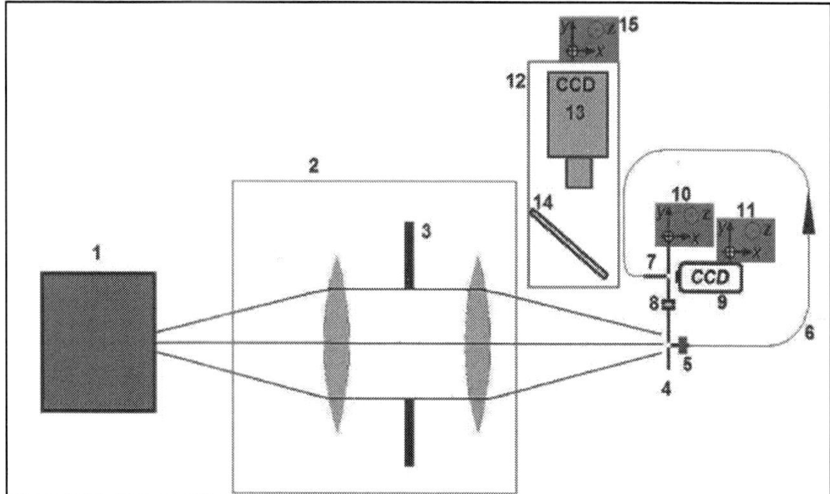

Figure 23. Diagram of the apparatus used to measure FRD – 1, light source, band pass filter and light diffuser; 2, telecentric optical system with unit magnification; 3, adjustable iris diaphragm; 4, alignment plate with a pinhole and both extremities of the tested fibre; 5, *peltier* device connected with the entrance of the optical fibre; 6, optical fibre; 7, exit of the optical fibre; 8, pinhole; 9, CCD; 10, xyz translation stage; 11, xyz translation stage; 12, microscope system; 13, CCD/lens; 14, beam splitter; 15, xyz translation stage

To measure the FRD properties of an optical fibre it is necessary to illuminate the test fibre with an input beam of known focal ratio. Then the output beam can be measured and compared with the input beam from a pinhole with the same diameter as the fibre core to determine the amount of FRD produced by the test fibre. The result is a plot of absolute transmission against output focal ratio. The experimental apparatus used to achieve this is illustrated in Fig. 23.

Illumination is provided by a 1-to-1 telecentric optical system that produces an image from an extensive uniformly illuminated source. This source is fed by a stabilized halogen lamp and has a band pass filter to provide light at 525nm, and filter's bandwidth of 100nm. An iris diaphragm placed in the collimated beam can be used to select the input focal ratio. A microscope with a CCD and beam splitter, monitored by a TV may be inserted between the pinhole/fibre plane, to be sure that the pinhole or the test fibre occupies the same position. To ensure accurate alignment of the fibre with the optical axis of the camera, the fibre is mounted in a tip-tilt translation stage. To begin the experiment, the pinhole device and the CCD are positioned to give us a reference image. In the test sequence, the pinhole is replaced with the entrance of the test fibre and the CCD is illuminated by the exit of the test fibre to give a projected image of the fibre. A distance of 9 mm between the CCD and the pinhole (or the entrance of the fibre in test) was determined as the best position to obtain images for optimal analysis. Background exposures are necessary for subtraction from the test exposures to remove the effects of hot pixels and stray light. In our experiments all fibres were tested at wavelength of 525nm, (defined using a Schott glass VG14 colour filter, **± 50 nm filter's bandwidth)**.

5.4. Reduction software

We have developed a custom software package (DEGFOC 3.0) to reduce the fibre images and to obtain throughput energy curves. This software works with PC microcomputers in a WINDOWS environment. We found this to be an effective solution for use in the optical laboratory environment allowing for ease of analysis. The DEGFOC 3.0 package gives curves of enclosed energy as is shown in the Fig. 24 with the option to save the result in ASCII format to be used in any graphic software, (eg: ORIGIN).

Fibre throughputs are automatically determined as a function of output focal ratio. The first step is an estimation of the background level to be subtracted from the test exposures to remove the effects of hot pixels and stray light. The software then finds the image centre by calculating the weighted average of all pixels. It associates a radius with each pixel and calculates the eccentricity that, in the ideal case, should be zero. Our target here is to obtain the absolute transmission of the fibre at a particular input f-ratio. After establishing the distance between the fibre test and the CCD, the software defines concentric annuli centred on the fibre image. These are then used to define the efficiency over a range of f-numbers at the exit of the fibre, where each f-number value contains the summation of all energy emergent from the fibre. Each energy value is calculated by the number of counts within each annulus divided by total number of counts from the pinhole images. The limiting focal ratio that can propagate in the tested fibre is approximately f/2.2. Therefore we have defined f/2 to be the outer limit of the external annulus within which all of the light from the test fibre will be collected. The corresponding diameters of the annulus are converted to output focal ratios, multiplying them by the appropriate constant given by the distance between the fibre output end and the detector.

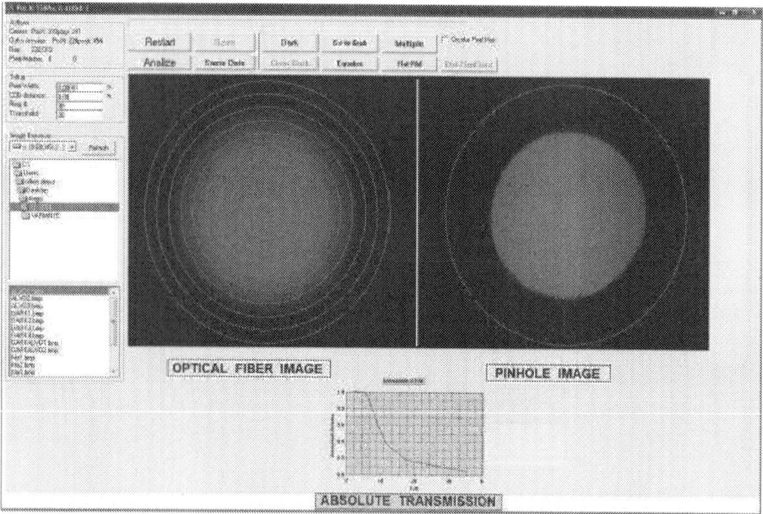

Figure 24. Print screen of the windows to the DEGFOC software.

5.5. Temperature gradient & FRD in optical fibres

In this experiment we have controlled the temperature of samples between –10 ºC and 22 ºC, typical of high altitude, ground-based observatories. To achieve this variation we have used a *Peltier* device coupled with a temperature sensor connected to an electronic controller. The end of the test fibre is placed in contact with the *Peltier* plate by a support, and to avoid problems with water condensation at low temperatures, the test ferrule is installed inside a plastic container with a glass window. A positive pressure of nitrogen gas is maintained using a flexible tube from a gas source. With these experimental arrangements it is possible to obtain images of the optical fibres with one of extremities inside a ferrule experiencing low temperatures without water condensation. This avoids the formation of ice at the end of the fibre that could attenuate the light at its termination and contaminate the results. Our aim is to measure the effect of constriction of the ferrule on the optical fibre caused by the gradient in temperature.

Plots of absolute transmission versus output focal ratio for three samples in four configurations are presented here. We have plotted graphs with the extremes curves obtained at room temperature of 23 ºC and at -10 ºC after a time interval of 30 min. chosen to stabilize the thermal effects between one measurement and next. The throughput graph obtained from one fibre with brass ferrule, in dry atmosphere, is shown in Fig. 25. These results show an increase in FRD when the brass ferrule experiences a cold temperature. The total variation observed in the hatched area is very strong and diminishes as the output focal ratio of the fibre is increased. An analysis of the results demonstrates that the loss of light at F/2.3 would be around 10 per cent. This degradation is caused, presumably, by the contraction of the brass ferrule with decreasing temperature causing compressive stress of the ferrule on the fibre. The error bars, of ± 1 per cent, together the average curves, were defined after repeating each experimentation at least six times. Some experimental uncertainty in the control of temperature causing small variations in the compression force on the ferrule/fibre and consequently cause small variations in the throughput of the sample. However is evident the presence of some dissipative process, which changes the borderline of the stress during the variations of temperature. Taking in account that the experiment is made with input focal ratio around the Numerical Aperture of the fibre (F/2.27) there is the possibility that it is changing because the stress.

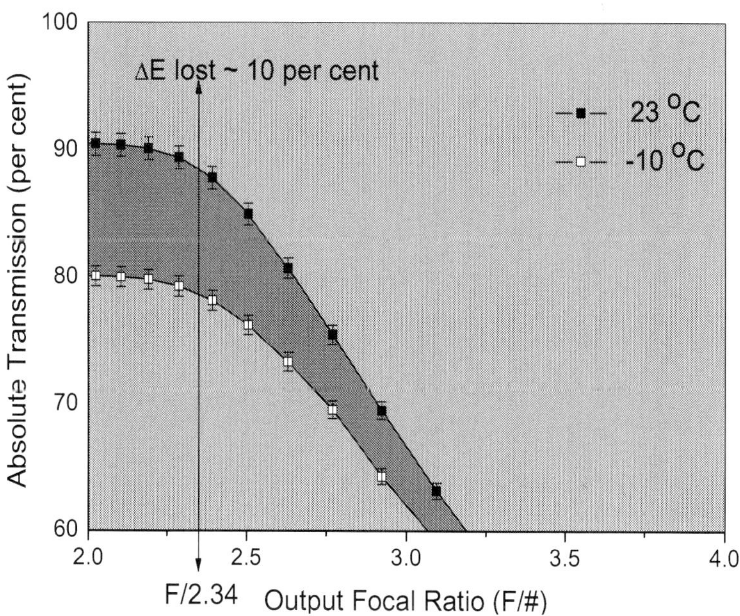

Figure 25. Performance of the optical fibre using brass ferrule. The ferrule was submitted to a negative temperature gradient of 23 °C in a dry atmosphere. The gradient was obtained, reducing the temperature, 23 °C to -10 °C, in 30 min of interval time. Two extremes curves were measured in this interval, producing the hatched area.

Such effects are critical to the design and implementation of fibre spectrographs. These results imply serious restrictions in the use of metal ferrules for optical fibres operating in ambient conditions that experience large changes of temperature typical of many observatories both during the night and throughout the year.

The throughput for samples with fibres inserted into epoxy ferrules and composite ferrules, in dry atmosphere, are shown for comparison in Figs. 26 and 27 using the same experimental procedures. Both graphs, present a very similar curves, with loss of light at F/2.34 around 2 per cent to the epoxy ferrule and 1 per cent to the composite ferrule. It seems that the loss of energy through stress-induced FRD effects is significantly less than that observed with the metal ferrule samples. The similarity of the epoxy and composite results imply that we are seeing similar effects due to the similar structure of both materials. In fact the composite material uses the same epoxy as a substrate. A natural compression happens during the cooling process, but does not produce a compressive stress of the brass ferrule on the fibre. The elastic properties of the epoxy may neutralize the mechanical stress on the fibre during the contraction process. The same error bars, of ± 1 per cent, were obtained after six repetitions of the experiment.

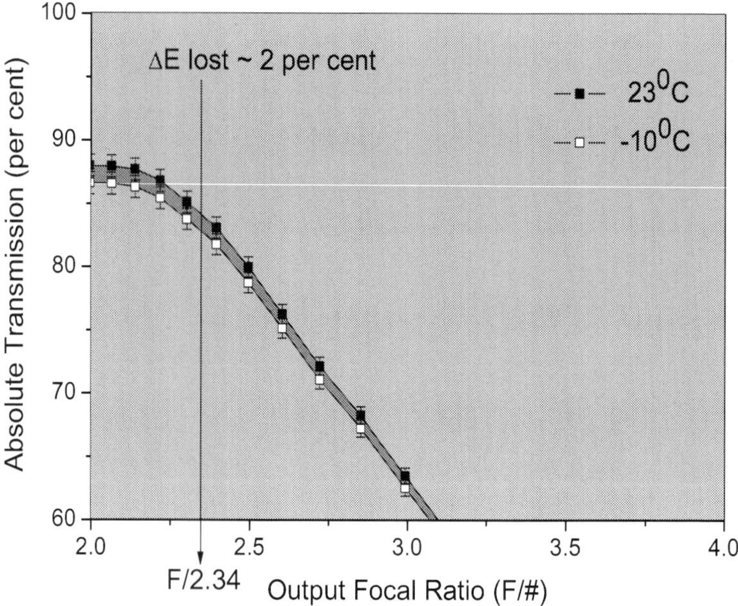

Figure 26. Performance of the optical fibre using epoxy ferrule. The ferrule was submitted to the negative temperature gradient following the same conditions of the experimentation using metal ferrule.

In general, the variation in the FRD results obtained with different samples from the same optical fibres is ±~1 per cent because the noise of the measurements. Analyse of the throughput curves obtained at room temperature from the epoxy and composite ferrules is ~ 3 per cent less on average when comparing the same curve obtained from the metal ferrules samples. The explanation for this difference may be in the aging process of the epoxy and composite ferrules. Both samples were submitted to six thermal cycles, between 50 ºC and -

20 °C after machining to avoid anomalous results during the experimentations. However, this procedure may increase the intrinsic FRD of the fibre given that the material structure of the ferrule may suffer accommodation pressing the fibre extremity. On the other hand, small differences of size in the hatched area of lost energy between similar samples could be expected. Differences like that would be explained by the difficulty to quantify the total length of the fibre immersed in epoxy inside the ferrule. The procedure of inserting fibre and epoxy into the ferrule is virtually handmade. There is no way to accurately control the amount of epoxy into opaque ferrules, because it is not possible to visualize the level of epoxy. Exception perhaps for polished quartz ferrules. Obviously, the length of fibre immersed in epoxy defines the length that would be submitted at the stress from the ferrule contraction.

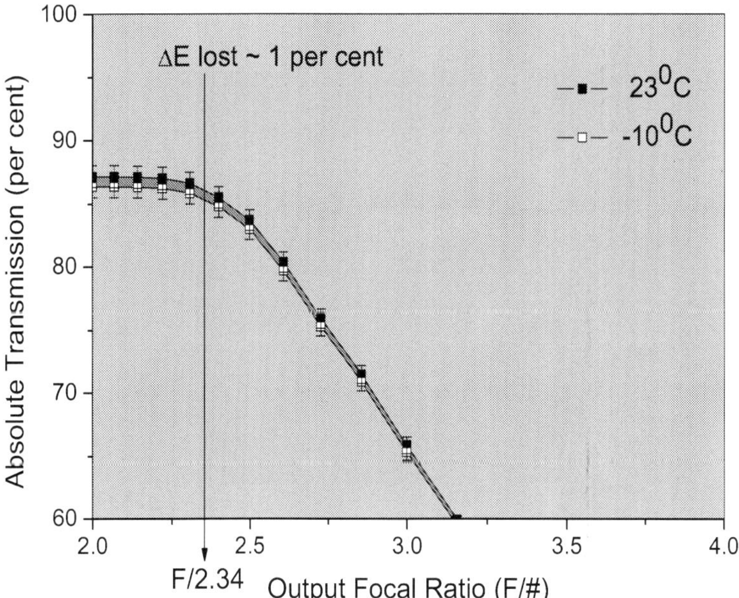

Figure 27. Performance of the optical fibre using composite ferrule. The ferrule was submitted to the same negative temperature gradient of the anterior experimentation using metal ferrule, steel ferrule and epoxy ferrule.

6. Polishing substrate

There are two ways of surface preparations in optical fibres, cleaving and polishing. In general applications directed to scientific instrumentation require optical fibres with extremities polished. This is the way where it is possible to optimize the spot light from the optical fibre. Furthermore, all fibre connectors require polishing and high performance may be reached with special machines and dedicated procedures. Currently, polishing procedures to optical fibres are based on very delicate glass paper or lapping discs soaked in abrasive liquid solutions. Often, this kind of liquid abrasive is very expensive taking in account your composition based in sophisticated chemistry keeping micro diamonds in suspension. Other options, considers abrasive silica and aluminium oxide mixed with oil solution or water solution. A very interesting application for the composite described here is its use as a high-performance abrasive disc to polish optical fibres.

6.1. Abrasive discs of composite to polish optical fibres

It is possible to fabricate composite discs, controlling the abrasive capacity through the correct choice of oxide and quantity mixed with epoxy. There are several manufacturers of oxide refractory with high purity such that it is possible to compose a complete grid of polishing discs. We can consider two major advantages in the use of polishing discs manufactured with composite: The first lies in the fact that the entire polishing process can be done using distilled water only. The second advantage is that after the polishing procedure, the disc can be restored to its original flatness and completely cleaned by machining process. The efficiency of this composite disc to polish optical fibres is based in the fact that some of the oxides of the mixture are naturally abrasives. Materials like cerium oxide or silica oxide can be prepared in liquid solutions abrasives and has been used for a long time in polishing procedures of lenses and other optical devices. Discs of composite can be made in any size and can easily be adapted in the rotation device of polishing machine. Fig. 28 shows an array of optical fibres polished using discs of composite.

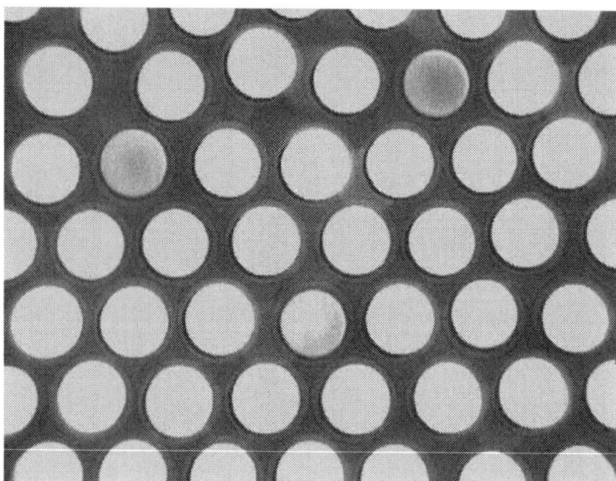

Figure 28. Microscopic photo of part of the optical fibres array, after polishing procedure, using discs of composite containing cerium oxide.

7. Conclusion

The motivation of this work was to test the performance of optical fibres inserted in ferrules made with different materials at low temperatures. The problem of finding a material best suited to securing fibres for astronomical spectrographs to cope with thermal stresses, FRD minimization and the need to achieve adequate polishing finish led us to investigate the use of composite materials. As has already been demonstrated, epoxies can be used not only as a means of holding fibres within structures (slit blocks, fibre arrays etc.) but also as a material to fabricate the structures themselves. The properties that require investigation in this context are CTE matching, machinability, bonding to glass and ease of polishing. In this context we have made several samples to evaluate FRD performance and position displacement of the inserted fibres when submitted to low temperatures.

Author details

Antonio C. de Oliveira and Ligia S. de Oliveira

Laboratório Nacional de Astrofísica / Ministério da Ciência Tecnologia e Inovação, Brazil

Acknowledgement

This work was financially supported by the FAPESP project no. 1999/03744-1 and CNPq project 62.0053/01-1- PADCT III/ Milenio. We wish to thank the staff of the Laboratório Nacional de Astrofísica/ MCT.

8. References

Clayton, C. A. (1989). The Implications of Image Scrambling and Focal Ratio Degradation in Fibre Optics on the Design of Astronomical Instrumentation, Astronomy and Astrophysics, Vol. 213, No. 1-2, (April 1989), pp. 502-515, ISSN 0004-6361

de Oliveira, A. C. et al. (2002). The Eucalyptus Spectrograph, Proceedings of SPIE Instrument Design and Performance for Optical/Infrared Ground-based Telescopes, pp. 1417-1428, Waikoloa, Hawaii, USA, August 25-28, 2002

de Oliveira, A. C. et al. (2005). Studying Focal Ratio Degradation of Optical Fibres with a Core Size of 50μm for Astronomy, Mon. Not. R. Astron. Soc., Vol. 356, No. 3, (October 2004), pp. 1079-1087, ISSN 00358711

de Oliveira, A. C. et al. (2010). The SOAR Integral Field Unit Spectrograph Optical Design and IFU Implementation, Proceedings of SPIE Modern Technologies in Space and Ground-based Telescopes and Instrumentation, pp. 77394S-1-77394S-12, San Diego, California, USA, June 27, 2010

Ransey, L. W. (1988). Focal Ratio Degradation in Optical Fibers of Astronomical Interest, In: *Fiber Optics in Astronomy*, Samuel C. Barden, pp. 26-40, Astronomical Society of the Pacific, ISBN 0-937707-20-1, San Francisco, California, USA

Macanhan, V. B. P. et al. (2006). FRODOSPEC Integral Fibre Unit, Proceedings of SAB XXXII Reunião da Sociedade Astronômica Brasileira, pp. 194-195, Atibaia, São Paulo, Brasil, August 3, 2006